Introduction to Digital Control

Biswanath Samanta

Introduction to Digital Control

An Integrated Approach

Springer

Biswanath Samanta
Department of Mechanical Engineering
Georgia Southern University
Statesboro, GA, USA

ISBN 978-3-031-66829-6 ISBN 978-3-031-66830-2 (eBook)
https://doi.org/10.1007/978-3-031-66830-2

This Springer imprint is published by the registered company Springer Nature Switzerland AG
The registered company address is: Gewerbestrasse 11, 6330 Cham, Switzerland

To my wife Rupama, son Rupayan, daughter Rupabali, my parents and grandparents

Preface

Control systems are present almost everywhere in modern societies with widespread applications ranging from consumer electronics, home appliances, heating, ventilation, and air conditioning (HVAC) systems, cars, general modes of transportation-ground, sea, and air, civilian and military defense systems to satellites and space exploration. In digital control systems, traditional hardwired analog controllers are replaced with computer algorithms and codes for greater flexibility and reliability. For successful deployment and safe, reliable, and economical operation of such digital control systems, a systematic approach is necessary for an integrated treatment of analysis, design, simulation, and implementation using relevant hardware and software platforms. This is the subject of the book.

The book evolved from course materials the author has developed while teaching undergraduate and graduate level Control Systems and related courses like Measurement Systems, Mechatronics, and Robotics at different universities over several years. Most of these courses have or used to have companion laboratory components that were also developed by the author.

This book follows the integrated approach of the author's recently published book on Mechatronics (*Introduction to Mechatronics: An Integrated Approach*, Springer, 2023) and covers a related subject of Digital Control Systems.

This book presents an integrated approach over 17 chapters covering z-transform, inverse z-transform, sampling and reconstruction, open- and closed-loop system characteristics, steady-state accuracy, stability analysis in z- and w-domains, control system specifications in time- and frequency-domains, design of controllers using time- and frequency-domain specifications, state-space methods-controllability and observability, pole placement controllers, design of observers, system identification, optimal control, implementation of controllers, and laboratory experiments for validation of analysis and design techniques on laboratory scale hardware modules. Computer programming is emphasized as an important tool integrating theoretical concepts with visualization through analysis, design, simulation, and actual laboratory experiments.

The concepts and procedures are illustrated through relevant examples in each chapter along with Matlab codes and Simulink models, as applicable. Two software platforms are covered in the book-Matlab/Simulink for analysis, design, and simulation and LabVIEW for real-time data acquisition, analysis, and digital controller implementation. Exercise problems are provided at end of each chapter to help enhance understanding of the materials covered and consolidate problem-solving skills. This also reinforces "hands-on, minds-on" learning approach emphasized throughout the book.

Topics of the book can be grouped in following modules: (A) discrete-time system fundamentals—*z*-transform, sampling and reconstruction (Chaps. 1, 2, and 3), (B) transfer function based discrete-time systems—open- and closed-loop systems, discrete-time system characteristics, stability analysis, controller design, simulation, and implementation (Chaps. 4, 5, 6, 7, 8, 9, 10, 11, and 12), and (C) state-space methods—controller and observer (estimator) design, optimal controller and estimator design, simulation and implementation, system identification, and case studies (Chaps. 13, 14, 15, 16, and 17).

In Chap. 1, basics of discrete-time systems are introduced.

In Chap. 2, an ideal sampler and *z*-transform are introduced for representation of discrete-time signals and systems along with properties of *z*-transform and methods of finding inverse *z*-transforms.

In Chap. 3, mathematical models for sampling and data reconstruction devices of zero-, first-, and fractional-order are presented.

In Chap. 4, procedures for finding discrete-time transfer functions of open-loop systems with zero-order, first-order, and fractional-order holds are covered.

In Chap. 5, discrete-time transfer functions of closed-loop systems, without and with disturbance inputs, are presented.

In Chap. 6, characteristics of closed-loop discrete-time systems including time response, closed-loop poles, mapping of *s*-plane in the *z*-plane, discrete equivalents, and steady-state accuracy are presented.

In Chap. 7, stability analysis of closed-loop discrete-time systems in *z*- and *w*-domains (discrete-time equivalent to continuous-time *s*-domain) are presented.

In Chap. 8, stability analysis techniques of discrete-time systems in *z*-domain using root locus are presented.

In Chap. 9, stability analysis techniques of discrete-time systems in *w*-domain using Bode plots and Nyquist diagrams are presented.

In Chap. 10, performance specifications of discrete-time systems including steady-state accuracy, transient response characteristics, and frequency domain specifications are covered.

In Chap. 11, methods for designing discrete-time controllers using time-domain specifications and root locus in *z*-domain are presented.

In Chap. 12, methods for designing discrete-time controllers using frequency-domain specifications in *w*-domain are presented.

In Chap. 13, methods of state-space representations of discrete-time systems along with controllability and observability assessment are covered.

In Chap. 14, methods of designing discrete-time controllers in state-space including full-state feedback pole placement controllers, full- and reduced-order prediction observers, and current observers are presented.

In Chap. 15, discrete-time system identification using least-squares, weighted least-squares, and recursive least-squares estimation is covered.

In Chap. 16, methods of designing optimal controllers in state-space along with optimal estimators (Kalman filters) are presented.

In Chap. 17, case studies covering analysis, design, and simulation of discrete-time control systems are provided as project activities.

Table of z-transforms of commonly used functions along with modified z-transforms is presented in Appendix A.

An integrated approach is adopted to cover interdisciplinary topics of digital control, emphasizing theoretical background, validation through analysis, simulation, and implementation in physical laboratory experiments.

A self-contained presentation of digital control topics without the requirement of prior knowledge of analog (continuous-time) control systems is adopted in this book. Explanations on similarities and differences between analog and digital control systems are presented for easy comparison.

End-of-chapter exercises are integrated with the topics covered and examples presented in each chapter. Answers to selected exercises are provided for ready reference to facilitate self-assessment of learning the materials of each chapter.

Solutions manual is made available through the publisher to the instructors who adopt the book as the main textbook in their courses.

The book can be used in a 3-credit hour course at senior undergraduate or beginning graduate level in Mechanical Engineering programs. To provide hands-on experience, a companion laboratory course (2 hours lab with 1 credit hour) can be offered as a co-requisite course that can be built around the laboratory experiments covered throughout the book. Alternatively, the book can be used in a 3-credit hour course with 4 contact hours (2 hours of lectures and 2 hours of labs). At Georgia Southern University, the author uses the content for a 3-credit hour course (with 4 contact hours—2 hours of lectures and 2 hours of labs) in the graduate Mechanical Engineering program (MSME).

The book materials can also be used in a 3-credit hour course at senior undergraduate and beginning graduate level in Mechanical or Mechatronics or Control Engineering programs. The author uses analog version of the contents in a cross-listed 3-credit hour (4 contact hours—2 hours of lectures and 2 hours of labs) elective course for senior undergraduate (BS) and beginning graduate (MS) students in Mechanical Engineering at Georgia Southern.

The content of the book, though primarily used in undergraduate and graduate level programs in Mechanical Engineering, can also be suitable for students from other undergraduate and graduate programs offering courses on digital control systems.

The book can also be used for self-study by practicing engineers for refreshing topics related to digital control systems.

Statesboro, GA, USA Biswanath Samanta

Acknowledgments

I had a unique opportunity to develop Control Systems and related courses like Measurement Systems, Mechatronics, and Robotics along with companion laboratory facilities from scratch and teach these courses, both undergraduate and graduate, at several universities over the years as a faculty member. The challenging but exciting and professionally satisfying journey started with the senior level elective course on Modern Control Systems (digital and state-space) at my alma mater, Indian Institute of Technology (IIT), Kharagpur, and continued to courses on Control Systems (analog and digital), both undergraduate and graduate, at Georgia Southern University, with similar and other courses in between at these and other universities I worked for. I have greatly benefited from this unique experience and am grateful to my colleagues and administration at these universities for their support.

Numerous students who took these courses in formative years of my academic career are now well established themselves in their respective professional fields. I acknowledge the positive roles of the students in development and delivery of the courses through the constructive feedback received over the years. I have also presented the development of Control Systems and related courses in international conferences, published in conference proceedings and journals, and received constructive feedback. I have incorporated all constructive feedback in organizing the content and coverage of topics in the book. I also gratefully acknowledge inspirations from my former professors and colleagues as well as many pioneering textbooks on the subject.

Solutions of examples and figures presented in the book were mostly generated using Matlab/Simulink (https://www.mathworks.com). All LabVIEW examples were generated using LabVIEW (https://www.ni.com). I thankfully acknowledge the availability of these software platforms through Georgia Southern University.

The writing of the book was possible through sacrifice, love, and support of my wife Rupama, son Rupayan, and daughter Rupabali throughout the years, especially the post-Covid years. I lovingly acknowledge the sacrifice, love, and support of my wonderful family for completion of this book project. The book manuscript was written mostly during long nights passing into early hours, in weekends, and during long breaks, with minimum to no interference from my family, except occasional tea and pizza breaks.

The final production of the book from the manuscript draft to final form was facilitated through the cooperative environment of Springer Nature. I gratefully acknowledge the cooperation and guidance from Springer Nature Publishing group in charge of this project, particularly Michael Luby (Executive Editor), Brian Halm (Project Coordinator), and Madhuvandhi Nagaraj (Production Editor).

Statesboro, GA, USA Biswanath Samanta

Contents

About the Author

Biswanath Samanta received B.Tech. (Honors) and Ph.D. in Mechanical Engineering from Indian Institute of Technology (IIT), Kharagpur. He is an Associate Professor in the Department of Mechanical Engineering at Georgia Southern University (GS), Statesboro, GA, USA. Prior to joining GS, Dr. Samanta held academic positions at Villanova University, Villanova, PA, USA; Sultan Qaboos University, Muscat, Oman (offering ABET-accredited Mechanical Engineering program); and IIT, Kharagpur, India. His expertise and research interests include broad areas of dynamic systems and control, mechatronics, robotics and intelligent systems, neuro-robotics and neuromodulation, advanced signal processing, prognostics and health management, artificial intelligence (AI) and deep learning, and applications of AI in engineering and biomedicine. Dr. Samanta has developed and taught numerous courses in these areas and supervised students at both undergraduate and graduate levels (including 3 Ph.D. and over 45 MS theses). He has over 150 refereed research articles published by professional bodies like American Society of Mechanical Engineers (ASME), Institution of Mechanical Engineers (IMechE) (UK), American Institute of Aeronautics and Astronautics (AIAA), and Institute of Electrical and Electronics Engineers (IEEE), and other publishers including Elsevier and Springer. The papers are regularly cited by independent researchers in their publications (over 5400 citations). He is a senior member of IEEE and a member of ASME, and currently the President of Robotics Technical Committee, Dynamic Systems and Control Division (DSCD), ASME. He regularly serves as a reviewer of technical papers submitted for publication in several international journals and conference proceedings, as an evaluator of book proposals submitted to different publishers, and as an examiner of Ph.D. theses submitted to different international universities. He is also regularly invited to chair sessions in international conferences. He is on the Editorial Board of international journals—Associate Editor, *ASME Letters in Dynamic Systems and Control*; Associate Editor, *Neuroimaging and Neuromodulation* (a specialty section within *Frontiers in Neuroimaging*); and Academic Editor, *Sensors*.

Chapter 1
Introduction

1.1 Basic Control Systems

Control systems are comprised of different functional components including sensors, actuators, and controllers. Control systems are present almost everywhere in modern societies with widespread applications ranging from consumer electronics, home appliances, heating, ventilation, air conditioning (HVAC) systems, cars, general modes of transportation—ground, sea, and air—and civilian and military defense systems to satellites and space exploration. In digital control systems, traditional hardwired analog controllers are replaced with computer algorithms and codes for greater flexibility and reliability along with analog-to-digital (A/D) and digital-to-analog (D/A) interfaces. The importance of a thorough study of digital control systems for better understanding, development, and applications across multiple domains cannot be overemphasized.

A schematic of a simple continuous-time closed-loop control without any disturbance inputs is shown in Fig. 1.1 where the system (plant) under consideration is represented using the transfer function $G_p(s)$, the sensor is represented using the transfer function $H(s)$, and the controller is represented as $G_c(s)$. The reference signal is represented as $R(s)$ and the closed-loop system output as $C(s)$ with the corresponding time domain representations as $r(t)$ and $c(t)$, respectively. The error between the reference signal and the sensor output is represented as $E(s)$ with its corresponding time domain representation as $e(t)$. The controller output is represented as $M(s)$ with its corresponding time domain representation as $m(t)$.

The transfer function for the plant is represented as

$$G_p(s) = \frac{C(s)}{M(s)} \tag{1.1}$$

B. Samanta, *Introduction to Digital Control*,
https://doi.org/10.1007/978-3-031-66830-2_1

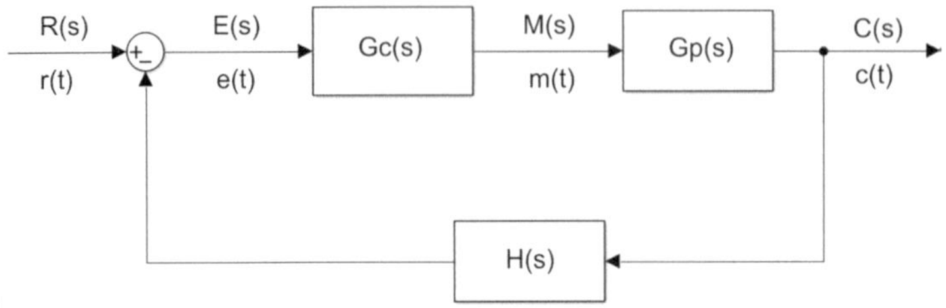

Fig. 1.1 Schematic of a continuous-time closed-loop control system

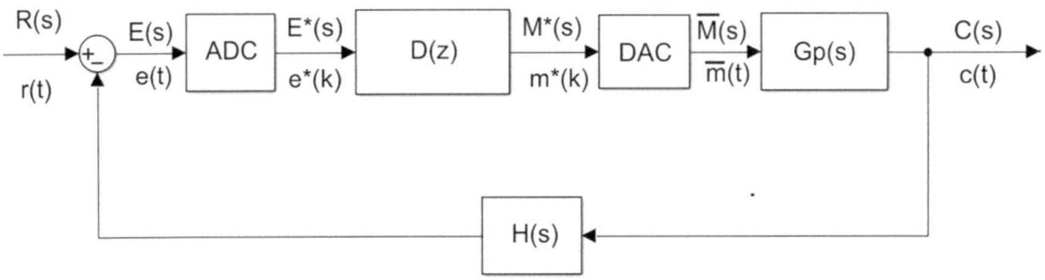

Fig. 1.2 Schematic of a discrete-time closed-loop control system with ADC and DAC interfaces

Similarly, the transfer function for the controller is represented as

$$G_c(s) = \frac{M(s)}{E(s)} \tag{1.2}$$

Though most of the physical systems are nonlinear, in most cases, the system behavior can be approximated as linear over a small range of operating points. There are many specific techniques for the analysis and design of nonlinear control systems. In this book, linear time-invariant (LTI) systems are considered.

1.2 Digital Control Systems

In modern times, traditional hardwired analog controllers are being replaced with more flexible, adaptable, and noise-tolerant digital controllers in the form of computer algorithms and codes through versatile digital devices like microcontrollers and computers, leading to the development of digital control systems. For successful deployment and safe, reliable, and economical operation of such digital control systems, a systematic approach is necessary for an integrated treatment of analysis, design, simulation, and implementation using relevant hardware and software platforms. This is what is covered in detail in this book.

A schematic of a closed-loop digital control system without any disturbance inputs is shown in Fig. 1.2 where the continuous-time controller $G_c(s)$ is replaced with the digital (discrete-time) controller transfer function $D(z)$. The analog-to-

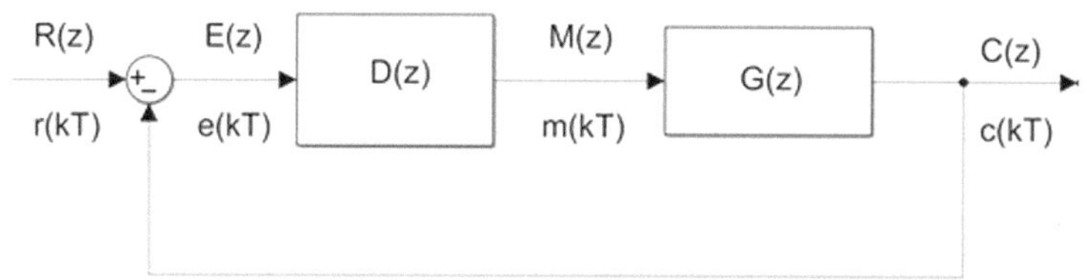

Fig. 1.3 Schematic of a discrete-time closed-loop control system with unity feedback

digital converter (ADC) and digital-to-analog converter (DAC) are required as interfaces between the continuous-time and discrete-time domains. The continuous-time domain signals and systems are represented in the s-domain using Laplace transforms. The discrete-time controller is represented in the z-domain as $D(z)$ with complex variable z defined in terms of Laplace variable s and sampling time T. The sampled version of the continuous-time signal $E(s)$ is represented as $E^*(s)$ with the corresponding continuous-time and discrete-time representations as $e(t)$ and $e^*(k)$, respectively, k being the sampling instant ($t = kT$). The output of the digital controller is represented as $M^*(s)$ with the corresponding discrete-time representation as $m^*(k)$. The DAC is used to reconstruct the discrete-time signal $m^*(k)$ to equivalent continuous time as $\overline{m}(t)$ with the corresponding representations in the s-domain as $M^*(s)$ and $\overline{M}(s)$, respectively.

The closed-loop control system of Fig. 1.2 can be represented, assuming unity feedback, in the z-domain as shown in Fig. 1.3 where the effects of sampling and data reconstruction of ADC and DAC are taken into account to convert the plant transfer function $G_p(s)$ as $G(z)$. The signals, reference, error, control, and output, are represented in the z-domain as $R(z)$, $E(z)$, $M(z)$, and $C(z)$ with the corresponding discrete-time representations as $r(kT)$, $e(kT)$, $m(kT)$, and $c(kT)$, respectively. In Fig. 1.3, a unity feedback is assumed for simplicity with the signal $c(kT)$ as the output of the sensor and the sensor transfer function $H(s)$ is included in the plant transfer $G_p(s)$ along with the sampling and data reconstruction interfaces (ADC and DAC) to get $G(z)$. In the book, discrete-time representations of signals are also shown without the sampling time T as $r(k)$, $e(k)$, $m(k)$, and $c(k)$.

Discrete-time transfer function for the plant along with ADC and DAC can be represented as

$$G(z) = \frac{C(z)}{M(z)} \tag{1.3}$$

The discrete-time transfer function of the controller can be represented as

$$D(z) = \frac{M(z)}{E(z)} \tag{1.4}$$

The rest of the book is devoted to presenting the details of the procedures for the transformation of signals and system transfer functions from continuous-time to discrete-time domain, representation and analysis of open-loop and closed-loop discrete-time systems using z-transforms, design, simulation, and implementation of discrete-time controllers. The procedures are illustrated with corresponding examples along with Matlab codes and Simulink models for analysis, design and simulation, and LabVIEW for real-time implementation on physical modules.

1.3 Organization of Book Chapters

The textbook presents an integrated approach to digital (discrete-time) control systems covering analysis, design, simulation, and real-time implementation through relevant hardware and software platforms. Topics related to discrete-time control systems include z-transform, inverse z-transform, sampling and reconstruction, open- and closed-loop system characteristics, steady-state accuracy for different system types and input functions, stability analysis in z-domain-Jury's test, bilinear transformation from z- to w-domain, stability analysis in w-domain—Routh-Hurwitz criterion, root locus techniques in z-domain, frequency domain analysis in w-domain, control system specifications in time and frequency domains, and design of controllers—PI, PD, PID, phase-lag, phase-lead, phase-lag-lead using time and frequency domain specifications, state-space methods—controllability and observability, pole placement controllers, and design of observers (estimators)—full-order prediction, reduced-order, and current observers, system identification, and optimal control—linear quadratic regulator (LQR), linear quadratic Gaussian (LQG) estimator (Kalman filter), implementation of controllers, and laboratory experiments for validation of analysis and design techniques on real laboratory scale hardware modules. Both single-input single-output (SISO) and multi-input multi-output (MIMO) systems are covered. Procedures are illustrated with examples along with Matlab codes and Simulink models. The software platform of Matlab/Simulink is used for analysis, design, and simulation, and hardware/software platforms of National Instruments (NI)/LabVIEW are used for the implementation and validation of analysis and design of digital control systems.

The rest of the book is organized into 16 chapters.

In Chap. 2, an ideal sampler and z-transform are introduced for the representation of discrete-time signals and systems along with properties of z-transform and methods of finding inverse z-transforms. Representation of difference equations from discrete-time system transfer functions along with simulation diagrams and signal flow graphs are also presented in this chapter.

Mathematical models for sampling and data reconstruction devices of zero-, first-, and fractional-order are presented in Chap. 3.

Procedures for finding discrete-time transfer functions of open-loop systems with zero-order, first-order, and fractional-order holds are presented in Chap. 4.

Discrete-time transfer functions of closed-loop systems, without and with disturbance inputs, are presented in Chap. 5. A laboratory experiment for the determination of model parameters is presented for a DC motor module in this chapter.

Characteristics of closed-loop discrete-time systems including time response, mapping of s-plane in the z-plane, and steady-state accuracy are presented in Chap. 6.

Analysis of stability of discrete-time closed-loop systems in the z-domain, conversion of discrete-time systems from the z-domain to equivalent continuous-time systems in the w-domain using bilinear transformation, and analysis of stability in the w-domain are presented in Chap. 7.

Stability analysis techniques of discrete-time systems in the z-domain using root locus are presented in Chap. 8. A laboratory experiment for stability analysis using z-domain techniques is presented for a DC motor module.

Stability analysis techniques of discrete-time systems in the w-domain using Bode plots and Nyquist diagrams are presented in Chap. 9. A laboratory experiment for stability analysis using w-domain techniques is presented for a DC motor module.

Performance specifications of discrete-time systems including steady-state accuracy, transient response characteristics, and frequency domain specifications are covered in Chap. 10.

Methods for designing discrete-time controllers using time-domain specifications and root locus in z-domain are presented in Chap. 11. Laboratory experiments for analysis, design using time-domain specifications, simulation, and real-time implementation of the controllers on a DC motor module are covered in this chapter.

In Chap. 12, methods for designing discrete-time controllers using frequency domain specifications in w-domain are presented. Laboratory experiments for analysis, design using frequency domain specifications, simulation, and real-time implementation of the controllers on a DC motor module are covered in this chapter.

Methods of state-space representations of discrete-time systems along with controllability and observability assessment are covered in Chap. 13.

In Chap. 14, methods of designing discrete-time controllers in state space including full-state feedback pole placement controllers, full- and reduced-order prediction observers, and current observers are presented. Laboratory experiments for analysis, design in state space, simulation, and real-time implementation of controllers on a DC motor module are covered in this chapter.

Discrete-time system identification using least-squares, weighted least-squares, and recursive least-squares estimation is covered in Chap. 15. Laboratory experiments for system identification are presented for a DC motor module in this chapter.

Methods of designing optimal controllers in state space along with optimal estimators (Kalman filters) taking into account relative weighting of state and control vectors and uncertainties (noise) in state and measurement vectors are presented in Chap. 16. Laboratory experiments for analysis, design in state space, simulation, and real-time implementation of optimal controllers on a DC motor module are covered in this chapter.

Case studies covering analysis, design, and simulation of discrete-time control systems are provided as project activities in Chap. 17.

A table of z-transforms of commonly used functions along with modified z-transforms is presented in Appendix A.

Answers to selected end-of-chapter exercises are presented in Appendix B.

Further Reading

1. Annaswamy AM, Johansson KH, Pappas GJ (eds) (2023) Control for societal-scale challenges: road map 2030. IEEE Control Systems Society. https://ieeecss.org/control-societal-scale-challenges-roadmap-2030
2. Antsaklis P, Gao Z (2005) Control systems. In: Christiansen D, Alexander CK, Jurgen RK (eds) Standard handbook of electronic engineering, Section 19, 5th edn. McGraw-Hill, New York, pp 19.1–19.30
3. Asadi F (2023) Signals and systems with MATLAB® and Simulink®. Springer, Cham
4. Astrom KJ, Murray RM (2021) Feedback systems: an introduction for scientists and engineers, 2nd edn. Princeton University Press, Princeton
5. Chen C-T (2006) Analog and digital control system design: transfer-function, state-space, and algebraic methods. Oxford University Press, New York
6. Dorf R, Bishop R (2016) Modern control systems, 13th edn. Pearson, Essex
7. Fadali MS, Visioli A (2020) Digital control engineering analysis and design, 3rd edn. Academic, London
8. Franklin GF, Powell JD, Emami-Naeini A (2019) Feedback control of dynamic systems, 8th edn. Pearson, Upper Saddle River
9. Franklin GF, Powell JD, Workman M (1998) Digital control of dynamic systems, 3rd edn. Addison-Wesley, Boston
10. Hall DL, Samanta B (2013) Nonlinear control of a magnetic levitation system using single multiplicative neural networks. In: ASME international mechanical engineering congress and exposition, November 18–21, 2013. https://doi.org/10.1115/IMECE2013-6406
11. Isermann R (1981) Digital control systems. Springer, Berlin
12. Jalili N, Candelino NW (2023) Dynamic systems and control engineering. Cambridge University Press, Cambridge
13. Keviczky L, Bars R, Hetthéssy J, Bányász (2019) Control engineering. Springer, Singapore
14. Koeing DM (2009) Practical control engineering: a guide for engineers, managers, and practitioners. McGraw-Hill, New York
15. Kuo BC (1995) Digital control systems, 2nd edn. Oxford University Press, Oxford
16. Landau ID, Gito G (2006) Digital control systems: design, identification and implementation. Springer, London
17. Leigh JR (2006) Applied digital control: theory, design and implementation, 2nd edn. Dover, Mineola
18. Mathworks (2023a) Control System Toolbox. https://www.mathworks.com/products/control.html
19. Mathworks (2023b) Matlab. https://www.mathworks.com/products/matlab.html
20. Mathworks (2023c) Simulink. https://www.mathworks.com/products/simulink.html
21. National Instruments (2023a) LabVIEW. https://www.ni.com/en/support/downloads/software-products/download.labview.html#487445
22. National Instruments (2023b) NI-ELVIS-II. https://www.ni.com/en-us/support/model.ni-elvis-ii.html
23. National Instruments (2023c) Labs for Quanser QNET DC Motor Add-On Board. https://education.ni.com/teach/resources/180/labs-for-quanser-qnet-dc-motor-add-on-board
24. Nise NS (2019) Control systems engineering, 8th edn. Wiley, Hoboken
25. Nof SY (ed) (2009) Springer handbook of automation. Springer, Berlin
26. Ogata K (1995) Discrete-time control systems, 2nd edn. Pearson, Hoboken
27. Phillips CL, Nagle HT, Chakrabortty A (2015) Digital control system analysis and design, 4th edn. Pearson, Essex
28. Preitl Z, Bauer P, Bokor J (2007) A simple control solution for traction motor used in hybrid vehicles. In: 4th international symposium on applied computational intelligence and informatics, May 17–18, 2007. IEEE, Timisoara. https://doi.org/10.1109/SACI.2007.375502

29. Prince A, Samanta B (2013) Neuromodulation based control of autonomous robots. In: Proceedings of international joint conference on neural networks (IJCNN), pp 1397–1403. https://doi.org/10.1109/IJCNN.2013.6706904
30. Quanser (2023) QUBE Servo 2. https://www.quanser.com/products/qube-servo-2/
31. Rabbath CA, Lechevin N (2014) Discrete-time control system design with applications. Springer, New York
32. Samanta B (2012) Intelligent control of engineering systems with situation awareness. In: ASME dynamic systems and control conference (DSCC), Ft. Lauderdale, FL, October 17–19, 2012. https://doi.org/10.1115/DSCC2012-MOVIC2012-8667
33. Samanta B, Turner J (2013) Development of a mechatronics and intelligent systems laboratory for teaching and research. Comput Educ 4(1):60–72
34. Samanta B (2023) Introduction to mechatronics – an integrated approach. Springer, Cham
35. Shertukde HM (2015) Digital control applications illustrated with MATLAB®. CRC Press, Boca Raton
36. Slotine J-JE, Li W (1991) Applied nonlinear control. Prentice Hall, Englewood Cliffs
37. Starr G (2020) Introduction to applied digital controls. Springer, Cham
38. Turner J, Samanta B (2012a) Least squares estimation of dynamic system parameters using LabVIEW. In: IEEE South East conference, Orlando, FL, March 15–18, 2012. https://doi.org/10.1109/SECon.2012.6196962
39. Turner J, Samanta B (2012b) Nonlinear control of dynamic systems using single multiplicative neuron models. In: ASME international mechanical engineering congress and exposition, Houston, TX, November 9–15, 2012. https://doi.org/10.1115/IMECE2012-87440
40. Turner J, Samanta B (2014) A controller implementation using FPGA in LabVIEW environment. Comput Educ 5(1):6–13
41. Veloni A, Miridakis NI (2018) Digital control systems-theoretical problems and simulation tools. CRC Press, Boca Raton
42. Vilanova R, Visioli A (2012) PID control in the third millennium – lessons learned and new approaches. Springer, London
43. Visioli A (2006) Practical PID control. Springer, London

Chapter 2
Discrete-Time Systems and z-Transform

2.1 Introduction

In this chapter, an ideal sampler and z-transform are introduced for representation and analysis of discrete-time signals and systems. Procedures of finding z-transform and inverse z-transform are presented along with properties of z-transform. Procedures of finding difference equations from discrete-time system transfer functions are discussed along with the representations in simulation diagrams and signal flow graphs. The concepts and processes are illustrated through examples along with Matlab codes. End-of-chapter exercises are provided to enhance understanding of the topics covered and consolidate problem-solving skills.

2.2 Discrete-Time Systems

The block diagram of a closed-loop discrete-time system is shown in Fig. 2.1 with a digital filter (controller) $D(z)$ between the sampler and data construction interfaces. The reference and the output signals are represented as $R(s)$ and $C(s)$ with the corresponding time domain representations as $r(t)$ and $c(t)$, respectively. The continuous-time domain signals and systems are represented in the s-domain using Laplace transforms. The discrete-time controller is represented in the z-domain as $D(z)$ with complex variable z defined in terms of Laplace variable s and sampling time T. The sampled version of the continuous-time signal $E(s)$ is represented as $E^*(s)$ with the corresponding continuous-time and discrete-time representations as $e(t)$ and $e^*(k)$, respectively. The output of the digital controller is represented as $M^*(s)$ with the corresponding discrete-time representation as $m^*(k)$. The data reconstructor represented as $G_h(s)$ is used to reconstruct the discrete-time signal $m^*(k)$ to equivalent continuous-time as $\overline{m}(t)$ with the corresponding representations in the s-domain as $M^*(s)$ and $\overline{M}(s)$, respectively. The system is represented in Fig. 2.2 with

B. Samanta, *Introduction to Digital Control*,
https://doi.org/10.1007/978-3-031-66830-2_2

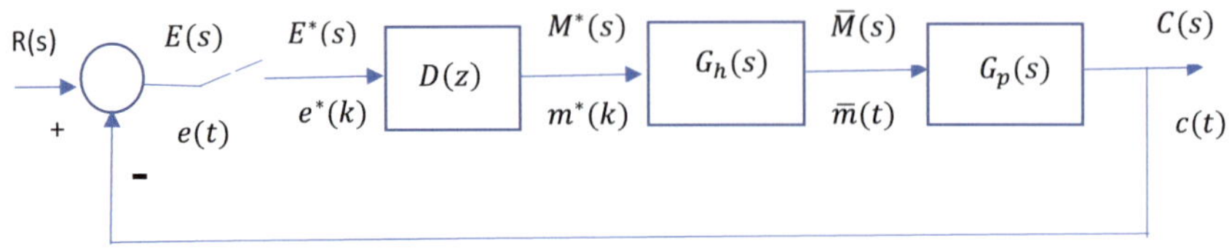

Fig. 2.1 Closed-loop discrete-time control system with a digital filter

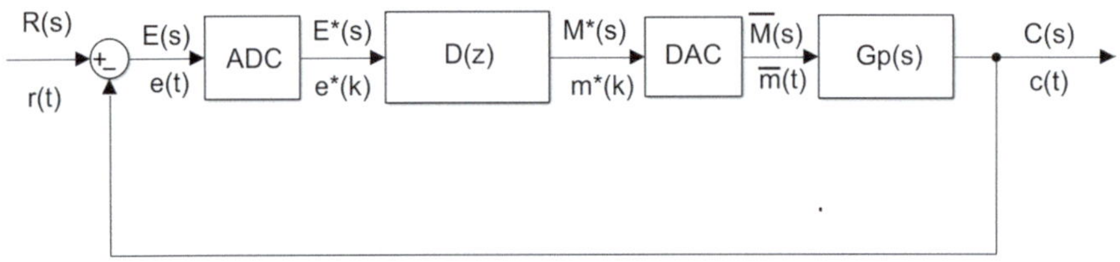

Fig. 2.2 Closed-loop discrete-time control system with ADC and DAC as interfaces

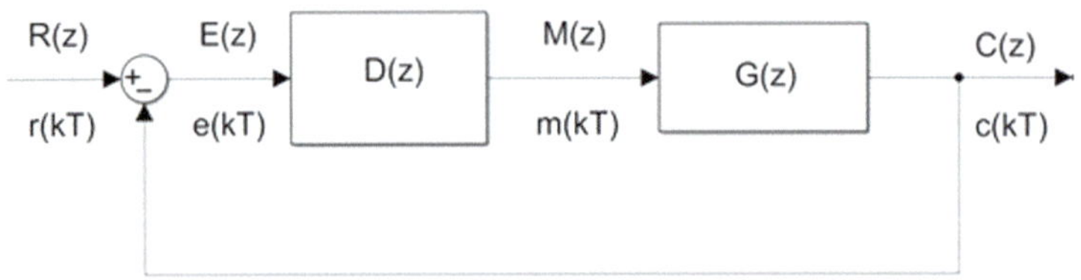

Fig. 2.3 Closed-loop discrete-time control system in the z-domain

sampler and data reconstruction interfaces represented as analog-to-digital converter (ADC) and digital-to-analog converter (DAC), respectively.

The closed-loop system of Fig. 2.2 can be represented in the z-domain as shown in Fig. 2.3, where the effects of sampling and data reconstruction of ADC and DAC are taken into account to convert the plant transfer function $G_p(s)$ as $G(z)$. In Fig. 2.3, a unity feedback is assumed with the signal $c(kT)$ as the output of the sensor, and the sensor transfer function $H(s)$ is included in the plant transfer $G_p(s)$ along with the sampling and data reconstruction interfaces (ADC and DAC) to get $G(z)$. Details of conversion from the s-domain to the z-domain are covered in the following chapters.

In the next section, an ideal sampler and z-transform are introduced.

2.3 Ideal Sampler and z-Transform

The schematic of a signal $f(t)$ being sampled using an ideal sampler with a sampling time T to get a sequence of samples $f^*(t) : f(kT)$ is shown in Fig. 2.4, where k is the sampling instant, $k = 0, \ldots, \infty$.

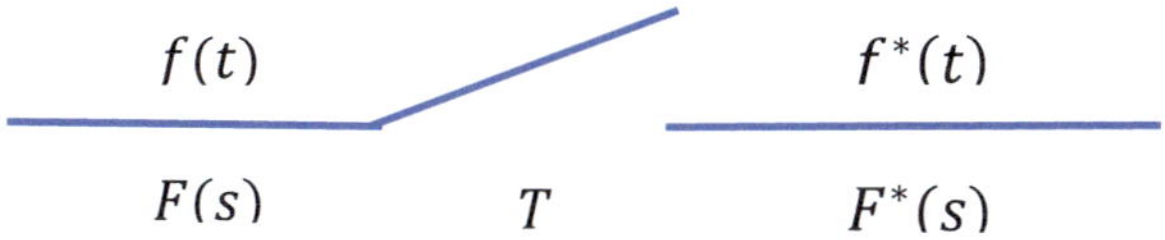

Fig. 2.4 Schematic of an ideal sampler

$$f^*(t) = \sum_{k=0}^{\infty} f(kT)\delta(t - kT) \tag{2.1}$$

Applying Laplace transform definition,

$$F(s) = \int_{-\infty}^{\infty} f(t)e^{-st}dt \tag{2.2}$$

Laplace transform can be obtained as follows:

$$L\{f^*(t)\} = L\left\{\sum_{k=0}^{\infty} f(kT)\delta(t - kT)\right\} \tag{2.3}$$

$$\begin{aligned} F^*(s) &= \sum_{k=0}^{\infty} \int_{-\infty}^{\infty} f(kT)\delta(t - kT)e^{-st}dt \\ &= \sum_{k=0}^{\infty} f(kT)e^{-skT} \\ &= \sum_{k=0}^{\infty} f(kT)\left(e^{-sT}\right)^k \end{aligned} \tag{2.4}$$

Definition of z-transform

$$F(z) = \sum_{k=0}^{\infty} f(kT)z^{-k} \tag{2.5}$$

where

$$z = e^{sT} \tag{2.6}$$

s and z are complex variables in Laplace and z domains, respectively, and T is the sampling time.

2.4 Examples of z-transform of Commonly Used Functions

In this section, some commonly used functions are considered for illustrating the process of deriving z-transform.

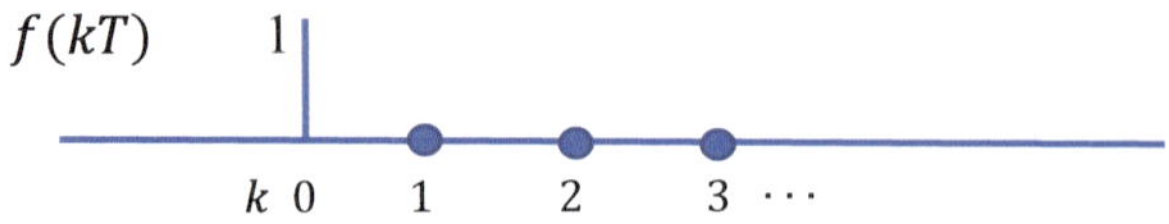

Fig. 2.5 Schematic of a discrete-time unit impulse function

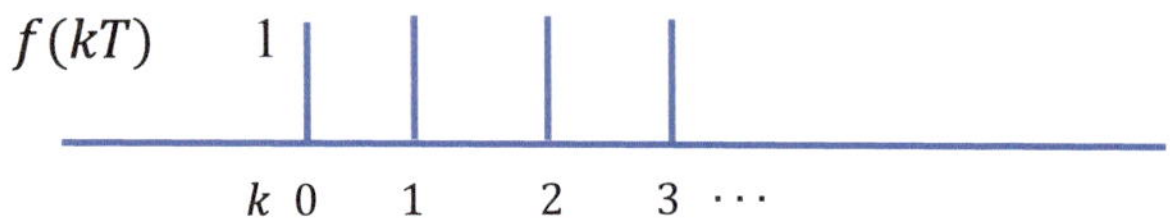

Fig. 2.6 Schematic of a discrete-time unit impulse function

2.4.1 *Unit Impulse*

The discrete-time unit impulse function can be represented mathematically using Dirac delta function in Eq. (2.7) and schematically in Fig. 2.5.

$$f(t) = \delta(t) \tag{2.7}$$

where $f(0) = 1$ and $f(t) = 0$, $t \neq 0$.

Laplace transform can be written as

$$F(s) = L(f(t)) = L(\delta(t)) = 1$$

z-transform of unit impulse can be obtained, using the definition of z-transform, Eq. (2.5), as

$$F(z) = \sum_{k=0}^{\infty} f(kT) z^{-k} = 1.z^{-0} + 0.z^{-1} + 0.z^{-2} + \ldots = 1$$

2.4.2 *Unit Step*

The discrete-time unit step function can be represented mathematically in Eq. (2.8) and schematically in Fig. 2.6.

$$f(t) = u(t) \tag{2.8}$$

where $f(t) = 1$, $t \geq 0$.

Laplace transform can be written as

$$F(s) = L(f(t)) = L(u(t)) = \frac{1}{s}$$

z-transform of unit impulse can be obtained, using the definition of z-transform, Eq. (2.5), as

$$\begin{aligned} F(z) &= \sum\nolimits_{k=0}^{\infty} f(kT) z^{-k} = 1.z^{-0} + 1.z^{-1} + 1.z^{-2} + \cdots \\ &= 1 + z^{-1} + z^{-2} + \cdots \end{aligned} \tag{2.9}$$

Using binomial expansion

$$(1+a)^n = 1 + \frac{n}{1!}a + \frac{n(n-1)}{2!}a^2 + \frac{n(n-1)(n-2)}{3!}a^3 + \ldots, \quad |a| < 1 \tag{2.10}$$

Equation (2.9) can be written using binomial expansion (2.10) with $a = -z^{-1}$, $n = -1$, as

$$F(z) = \left(1 - z^{-1}\right)^{-1} = \frac{z}{z-1}, \quad \left|z^{-1}\right| < 1 \tag{2.11}$$

2.4.3 *Unit Ramp*

The discrete-time unit ramp function can be represented mathematically in Eq. (2.12) and schematically in Fig. 2.7.

$$f(t) = kT \tag{2.12}$$

where $f(t) = t$, $t = kT$, $k \geq 0$.

Fig. 2.7 Schematic of a unit ramp function

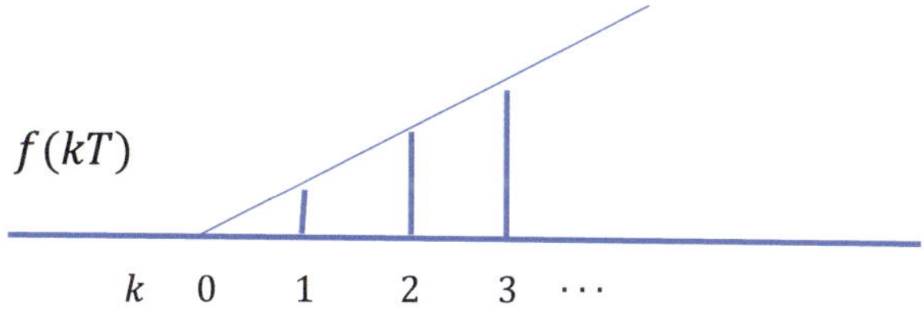

Laplace transform can be written as

$$F(s) = L(f(t)) = L(tu(t)) = \frac{1}{s^2}$$

z-transform of unit impulse can be obtained, using the definition of z-transform, Eq. (2.5), as

$$\begin{aligned} F(z) &= \sum\nolimits_{k=0}^{\infty} f(kT) z^{-k} = 0.z^{-0} + T.z^{-1} + 2T.z^{-2} + \cdots \\ &= Tz^{-1}.\left(1 + 2z^{-1} + 3z^{-2} + \cdots\right) \end{aligned} \tag{2.13}$$

Using binomial expansion of Eqs. (2.10), (2.13) can be written with $a = -z^{-1}$, $n = -2$, as

$$F(z) = Tz^{-1}.\left(1 - z^{-1}\right)^{-2} = \frac{Tz}{(z-1)^2}, \quad \left|z^{-1}\right| < 1 \tag{2.14}$$

2.4.4 Exponentially Decaying Function

The discrete-time exponentially decaying function can be represented mathematically in Eq. (2.15) and schematically in Fig. 2.8.

$$f(t) = e^{-bt}, b > 0 \tag{2.15}$$

where $f(t) = e^{-bkT} = (e^{-bT})^k = c^k$, $t = kT$, $c = e^{-bT}$ $k \geq 0$, $b > 0$, $c < 1$.

Laplace transform can be written as

$$F(s) = L(f(t)) = L\left(e^{-bt}\right) = \frac{1}{s+b} \tag{2.16}$$

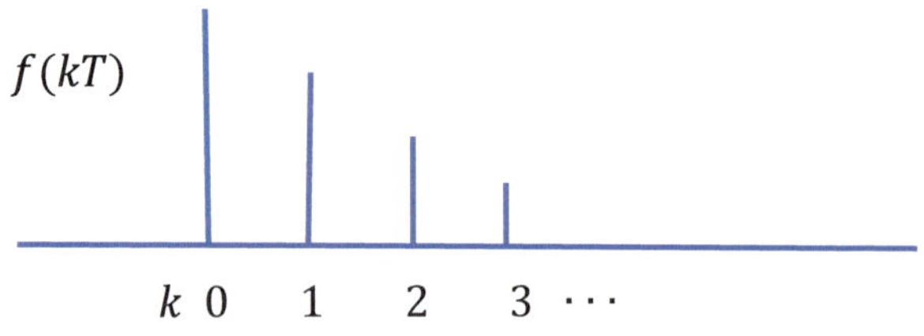

Fig. 2.8 Schematic of a discrete-time exponentially decaying function

Table 2.1 z-Transform of commonly used functions

Time function	Discrete-time sequence	Laplace transform	z-Transform
$\delta(t)$	$\delta(k)$	1	1
$u(t)$	$\{1\}$	$\frac{1}{s}$	$\frac{z}{z-1}$
t	$\{kT\}$	$\frac{1}{s^2}$	$\frac{Tz}{(z-1)^2}$
$e^{-bt}, b>0$	$\{e^{-bkT}\}, b>0$	$\frac{1}{s+b}$	$\frac{z}{z-e^{-bT}}$
$\sin(at)$	$\{\sin(akT)\}$	$\frac{a}{s^2+a^2}$	$\frac{z\ \sin(aT)}{z^2-2z\ \cos(aT)+1}$
$\cos(at)$	$\{\cos(akT)\}$	$\frac{s}{s^2+a^2}$	$\frac{z\ (z-\ \cos(aT))}{z^2-2z\ \cos(aT)+1}$

z-transform of discrete-time exponential function can be obtained, using the definition of z-transform, Eq. (2.5), as

$$\begin{aligned} F(z) &= \sum\nolimits_{k=0}^{\infty} f(kT)z^{-k} \\ &= \sum\nolimits_{k=0}^{\infty} e^{-bkT}z^{-k} \\ &= \sum\nolimits_{k=0}^{\infty} \left(e^{bT}z\right)^{-k} \end{aligned} \tag{2.17}$$

Using binomial expansion of Eqs. (2.10), (2.17) can be written with $a = -\left(e^{bT}z\right)^{-1}$, $n = -1$, and $c = e^{-bT}$ as

$$F(z) = \left(1 - \left(e^{bT}z\right)^{-1}\right)^{-1} = \frac{z}{z - e^{-bT}} = \frac{z}{z - c} \tag{2.18}$$

Some commonly used time functions along with the corresponding Laplace and z-transforms are listed in Table 2.1. A more extended list is presented in Appendix A, as Table A.1.

2.5 z-Transform Using Matlab

2.5.1 Unit Step

Matlab code finding the z-transform of a unit step consists of (1) defining k as a symbolic variable, (2) defining unit step function, (3) getting z-transform, and (4) printing in a pretty format as explained in comments. The code and the results are listed.

```
% z-transformof unit step f(k) =1
syms k % define symbolic variable k
fk=1^k % unit step
Fz=ztrans(fk);% get z transform
pretty(Fz) % print z-transform in pretty format
Results:
fk = 1; F(z)=z/(z - 1)
```

2.5.2 Unit Ramp

Matlab code finding the z-transform of a unit ramp consists of (1) defining k and T as symbolic variables, (2) defining unit ramp function, (3) getting z-transform, and (4) printing in a pretty format as explained in comments. The code and the results are listed.

```
% z-transform of unit ramp  f(t)=t   f(kT)= kT
syms k T % define symbolic variables k and T
fk = k*T % unit ramp
Fz=ztrans(fk);% get z transform
pretty(Fz) % print z-transform in pretty format

Results:
fk = T*k; F(z)= Tz/(z – 1)^2
```

2.5.3 Exponentially Decaying Function

Similarly, Matlab code for finding the z-transform of an exponentially decaying function and results are listed here.

```
% z-transform of exponentially decaying function f(k)=exp(-bkT)
syms b k T % define symbolic variables b, k and T
fk = exp(-b*k*T) % discrete-time function
Fz = ztrans(fk); % z-transform
pretty(Fz) % print z-transform in pretty format

Results:

fk = exp(-T*b*k); F(z)= z/(z - exp(-T*b))
```

Example 2.1 Use Matlab to obtain the z-transform of $f(t) = t\,e^{at}$ or $f(k) = kT\,e^{akT}$, T being the sampling time.

Solution Matlab code and results are listed here.

```
% Ch2 Example 1
% z-transform of f(t)=t*exp(at)    f(kT)= kT*exp(aKT)
syms a k T % define symbolic variables a, k and T
fk = k*T*exp(a*k*T) % discrete-time function
Fz=ztrans(fk); % z-transform
pretty(Fz) % print z-transform in pretty format

Results:

fk = T*k*exp(T*a*k); F(z)=  T*z*exp(T*a)/(z- exp(T*a))^2
```

Example 2.2 Use Matlab to obtain the z-transform of $f(t) = a^t \cos(bt)$ or $f(k) = a^{kT} \cos(bkT)$, T being the sampling time.

Solution Matlab code and results are listed here.

```
% Ch 2 Example 2
% z-transform of f(kT)=a^(kT)*cos(b*kT)
syms a b k T % define symbolic variables a, b, k and T
fk = (a^(k*T))*cos(b*k*T); % discrete-time function
Fz=ztrans(fk); % z-transform
pretty(Fz) % print z-transform in pretty format

Results:

fk = a^(T*k)*cos(T*b*k);
```

$$F(z) = \frac{z\left(\cos(Tb) - \frac{z}{a^T}\right)}{a^T\left(\frac{z^2}{a^{2T}} - \frac{2z\cos(Tb)}{a^T} + 1\right)} = \frac{za^T\cos(Tb) - z^2}{z^2 - 2a^T z\cos(Tb) + a^{2T}}$$

2.6 z-Transform from s-Domain Functions

In digital control system design applications, often the system functions are given in the Laplace domain. There are two approaches to obtain z-transform from the given s-domain function. The first method is to obtain component functions from the given function using partial fraction expansion (PFE) in the s-domain. Next, the individual component functions are converted to z-domain using table of z-transform (similar to Table 2.1 or Table A.1) or using Matlab code. The other method is to convert the given s-domain function to continuous-time domain using inverse Laplace transform. Next, the individual time-domain functions are converted to the corresponding discrete-time form, which are subsequently converted to z-domain using z-transform. Both methods are illustrated through Examples 2.3 and 2.4.

Example 2.3 Given a function in s-domain $F(s) = \frac{19s^2+35s+12}{12s^3+36s^2+24s}$ find z-transform $F(z)$ using PFE for a sampling time $T = 0.1$ s.

Solution Matlab code using partial fraction expansion (PFE) method, along with comments and results, is presented.

```
% Ch2 Example 3
% z-transform from s-function
% F(s)=(19s^2+35s+12)/(12s^3+36s^2+24s)[
clear all
num = [19 35 12]; % numerator of F(s)
den = [12 36 24 0]; % denominator of F(s)
T=0.1 % sampling time
n = length(den);
Fs =tf(num, den) %F(s)=num/den
[r,p,k] =residue(num,den) % get PFE

% convert s-domain poles (ps) to z-domain poles  pz =exp(ps*T)
for i=1:n-1
    pz(i)=exp(p(i)*T)
end
[numzz, denz]=residue(r,pz,k) % reconstruct z-domain transfer function
using residue
numz =conv(numzz,[1 0]) % multiply numerator by z
Fz=tf(numz,denz,T) % final F(z)

Results:

            19 s^2 + 35 s + 12
F(s) =  -----------------------  ;
         12 s^3 + 36 s^2 + 24 s

          1.583 z^3 - 2.897 z^2 + 1.322 z
F(z)=  ----------------------------------
       z^3 - 2.724 z^2 + 2.464 z - 0.7408
```

Example 2.4 Given a function in s-domain $F(s) = \frac{19s^2+35s+12}{12s^3+36s^2+24s}$ find z-transform $F(z)$ through inverse Laplace transform for a sampling time $T = 0.1$ s.

Solution Matlab code through inverse Laplace transform method, along with comments and results, is presented.

```
% Ch 2 Example 4
T=0.1; % sampling time
syms s k t z % define variables as symbolic
Fs = (19*s^2+35*s+12)/(12*s^3+36*s^2+24*s) % F(s) as given
ft = ilaplace(Fs) % inverse Laplace transform to get f(t)
fkT = subs(ft,t,k*T); % substitute t as k*T - discrete-time function
Fz = ztrans(fkT); % get z-transform of f(kT)
pretty(Fz)

Results:

            19 s^2 + 35 s + 12
F(s) =  -----------------------  ;
         12 s^3 + 36 s^2 + 24 s
f(t) = exp(-t)/3 + (3*exp(-2*t))/4 + 1/2

           z                 3 z                       z
F(z)= --------- + -------------------- + ---------------------
      2 (z - 1)      /          /  1 \ \     /          /   1 \ \
                   4 | z  - exp| - -  | |  3| z  - exp|  - --  | |
                     \          \  5 / /     \          \  10 / /
```

2.7 Properties of z-Transform

Properties of z-transform are presented in this section.

2.7.1 Addition and Subtraction

The z-transform of sum (or difference) of two discrete-time sequences is the sum (or difference) of the corresponding individual z-transforms, as given in Eq. (2.19)

$$Z\{f_1(kT) \pm f_2(kT)\} = F_1(z) \pm F_2(z) \tag{2.19}$$

Proof Using the definition of z-transform

$$\begin{aligned} Z\{f_1(kT) \pm f_2(kT)\} &= \sum_{k=0}^{\infty} \{f_1(kT) \pm f_2(kT)\} z^{-k} \\ &= \sum_{k=0}^{\infty} f_1(kT) z^{-k} \pm \sum_{k=0}^{\infty} f_2(kT) z^{-k} \\ &= F_1(z) \pm F_2(z) \end{aligned}$$

2.7.2 Multiplication by a Constant

The z-transform of a discrete-time sequence multiplied by a constant is equal to the product of the z-transform of the discrete-time sequence and the constant, as in Eq. (2.20)

$$Z\{af(kT)\} = aF(z) \tag{2.20}$$

Proof Using the definition of z-transform

$$\begin{aligned} Z\{af(kT)\} &= \sum_{k=0}^{\infty} \{af(kT)\} z^{-k} \\ &= a \sum_{k=0}^{\infty} \{f(kT)\} z^{-k} \\ &= aF(z) \end{aligned}$$

Linearity property of z-transform can be written using Eqs.(2.19) and (2.20) as

$$\begin{aligned} Z[a_1f_1(kT) \pm a_2f_2(kT)] &= \sum_{k=0}^{\infty} [a_1f_1(kT) \pm a_2f_2(kT)]z^{-k} \\ &= a_1 \sum_{k=0}^{\infty} f_1(kT)z^{-k} \pm a_2 \sum_{k=0}^{\infty} f_2(kT)z^{-k} \\ &= a_1F_1(z) \pm a_2F_2(z) \end{aligned} \tag{2.21}$$

2.7.3 Linear Translation

For a positive integer, n, and discrete-time sequence $\{f(k\}$ with its z-transform as $F(z)$, the relations of real translation are

$$Z[f(k-n)u(k-n)] = z^{-n}F(z) \tag{2.22}$$

and

$$Z[f(k+n)u(k+n)] = z^n\left[F(z) - \sum_{k=0}^{n-1} f(k)z^{-k}\right] \tag{2.23}$$

The discrete-time unit step function $\{u(k)\}$ is defined as

$$u(k) = \begin{cases} 0, & k < 0 \\ 1, & k \geq 0 \end{cases} \tag{2.24}$$

The time function with real translation to the right or the time-delayed function in (2.22) is expressed as

$$f(k-n)u(k-n) = f(k)u(k)|_{k \leftarrow k-n} \tag{2.25}$$

The time function with real translation to the left or the time-advanced function in (2.23) is expressed as

$$\begin{aligned} f(k+n)u(k+n) &= f(k)u(k)|_{k \leftarrow k+n} \\ &= f(k+n)u(k)\ k \geq 0 \end{aligned} \tag{2.26}$$

An example of real-time shifts of a discrete-time function $\{f(k\}$ is shown in Table 2.2.

Table 2.2 Example of real-time translations

k	$f(k)$	$f(k-3)$	$f(k+3)$
0	3.1	0	2.8
1	2.2	0	3.2
2	1.5	0	1.6
3	2.8	3.1	1.9
4	3.2	2.2	...
5	1.6	1.5	...
6	1.9	2.8	...
...	...	...	...

Proof Using the definition of z-transform Eqs. (2.5) and (2.25), the relation for the z-transform of the time-delayed function can be expressed as

$$
\begin{aligned}
z[\{f(k-n)u(k-n)\}] &= \sum_{k=0}^{\infty} f(k-n)u(k-n)z^{-k} \\
&= 0.z^{-0} + 0.z^{-1} + 0.z^{-2} + \cdots + 0.z^{-(n-1)} + f(0).z^{-n} + f(1).z^{-(n+1)} + f(2).z^{-(n+2)} + \cdots \\
&= z^{-n}\left[f(0) + f(1).z^{-1} + f(2).z^{-2} + \cdots\right] \\
&= z^{-n}F(z)
\end{aligned}
$$

Using the definition of z-transform Eqs. (2.5) and (2.26), the relation for the z-transform of the time-advanced function can be expressed as

$$
\begin{aligned}
z[f(k+n)u(k+n)] &= \sum_{k=0}^{\infty} f(k+n)u(k)z^{-k} \\
&= f(n).z^{-0} + f(n+1).z^{-1} + f(n+2).z^{-2} + \cdots \\
&= z^{n}\left[f(n).z^{-n} + f(n+1).z^{-(n+1)} + f(n+2).z^{-(n+2)} + \cdots\right] \\
&= z^{n}\Big[f(0).z^{-0} + f(1).z^{-1} + f(2).z^{-2} + \cdots + f(n-1).z^{-(n-1)} \\
&\quad + f(n).z^{-n} + f(n+1).z^{-(n+1)} + f(n+2).z^{-(n+2)} + \cdots \\
&\quad - f(0).z^{-0} - f(1).z^{-1} - f(2).z^{-2} - \cdots - f(n-1).z^{-(n-1)}\Big] \\
&= z^{n}\left[F(z) - \sum_{k=0}^{n-1} f(k)z^{-k}\right]
\end{aligned}
$$

2.7.4 Complex Translation

For a discrete-time function sequence $\{f(kT)\}$ with z-transform $F(z)$, the complex translation relation is given as

$$z\left[e^{akT}f(kT)\right] = F\left(ze^{-aT}\right) \tag{2.27}$$

Proof Using the definition of z-transform,

$$\begin{aligned} z\left[e^{akT}f(kT)\right] &= \sum_{k=0}^{\infty} e^{akT}f(kT)z^{-k} \\ &= e^{0}f(0).z^{-0} + e^{aT}f(T).z^{-1} + e^{2aT}f(2T).z^{-2} + \cdots \\ &= f(0) + f(T).\left(ze^{-aT}\right)^{-1} + f(2T).\left(ze^{-aT}\right)^{-2} + \cdots \\ &= F\left(ze^{-aT}\right) \end{aligned}$$

2.7.5 Initial Value Theorem

For a discrete-time function sequence $\{f(kT)\}$ with its z-transform as $F(z)$, the initial value of the function, $f(0)$, is obtained as

$$f(0) = \lim_{z\to\infty} F(z) \tag{2.28}$$

Proof Using the definition of z-transform,

$$\begin{aligned} F(z) &= f(0) + f(T)z^{-1} + f(2T)z^{-2} + \cdots \\ &= f(0) + \left[f(T)z^{-1} + f(2T)z^{-2} + \cdots\right] \end{aligned} \tag{2.29}$$

Taking limit $\lim_{z\to\infty}$ of both sides of Eq. (2.29), the initial value, $f(0)$, is obtained as

$$f(0) = \lim_{z\to\infty} F(z)$$

2.7.6 Final Value Theorem

For a discrete-time function sequence $\{f(k)\}$ with its z-transform as $F(z)$, the final value of the function, $\lim_{n\to\infty} f(n)$, is obtained as

$$\lim_{n\to\infty} f(n) = \lim_{z\to 1}\left[(z-1)F(z)\right] \tag{2.30}$$

provided the left-side limit exists.

Proof Using the definition of z-transform, the following relation can be expressed

$$
\begin{aligned}
Z[f(k+1)-f(k)] &= \lim_{n\to\infty}\left[\sum_{k=0}^{n} f(k+1)z^{-k} - \sum_{k=0}^{n} f(k)z^{-k}\right] \\
&= \lim_{n\to\infty}\left[\begin{array}{c} f(1)z^{-0} + f(2)z^{-1} + f(3)z^{-2} + \cdots + f(n+1)z^{-n} \\ -f(0)z^{-0} - f(1)z^{-1} - f(2)z^{-2} - \cdots - f(n)z^{-n} \end{array}\right] \\
&= \lim_{n\to\infty}\left[\begin{array}{c} -f(0) + f(1)(1-z^{-1}) + f(2)z^{-1}(1-z^{-1}) + f(3)z^{-2}(1-z^{-1}) + \cdots \\ +f(n)z^{-(n-1)}(1-z^{-1}) + f(n+1)z^{-n} \end{array}\right]
\end{aligned}
\tag{2.31}
$$

Taking limit $\lim_{z\to 1}$ of both sides of Eq. (2.31)

$$
\begin{aligned}
\lim_{z\to 1}[Z[f(k+1)-f(k)]] &= \lim_{n\to\infty}[f(n+1)-f(0)] \\
&= \lim_{n\to\infty}[f(n)] - f(0)
\end{aligned}
\tag{2.32}
$$

From the property of real translation

$$
\begin{aligned}
Z[f(k+1)-f(k)] &= z[F(z)-f(0)] - F(z) \\
&= (z-1)F(z) - zf(0)
\end{aligned}
\tag{2.33}
$$

Taking limit of $\lim_{z\to 1}$ of both sides of Eq. (2.33), the following expression can be written

$$
\lim_{z\to 1} Z[f(k+1)-f(k)] = \lim_{z\to 1}[(z-1)F(z)] - f(0) \tag{2.34}
$$

Equating two expressions of Eqs.(2.32) and (2.33), one can write

$$
\lim_{n\to\infty}[f(n)] = \lim_{z\to 1}[(z-1)F(z)]
$$

Properties of z-transform are presented in Table 2.3.

2.8 Inverse z-Transform

There are several methods for finding inverse z-transform from a given function in z-domain F(z), namely, power series method, partial fraction expansion (PFE) method, inversion-formula method, and discrete convolution. Of these, partial fraction expansion method with non-repeated simple factors is discussed in this section.

Table 2.3 Properties of z-Transform

Discrete-time function	z-Transform
$f(k)$	$\sum_{k=0}^{\infty} f(k)z^{-k}$
$a_1f_1(k) \pm a_2f_2(k)$	$a_1F_1(z) \pm a_2F_2(z)$
$f(k-n)u(k-n),\ n \geq 0$	$z^{-n}F(z)$
$f(k+n)u(k+n),\ n \geq 1$	$z^n\left[F(z) - \sum_{k=0}^{n-1} f(k)z^{-k}\right]$
$e^{akT}f(k)$	$F(ze^{-aT})$
$kf(k)$	$-z\frac{dF(z)}{dz}$
$f_1(k) \times f_1(k)$	$F_1(z)F_2(z)$
$f_1(k) = \sum_{i=0}^{k} f(i)$	$F_1(z) = \frac{z}{z-1}F(z)$
Initial value: $f(0) = \lim_{z\to\infty} F(z)$	
Final value: $\lim_{n\to\infty} f(n) = \lim_{z\to 1}\left[(z-1)F(z)\right]$	

Partial Fraction Expansion (PFE) Method

It should be recognized that each z-transform $F(z)$ has a z in the numerator. The partial fraction expansion is applied to

$$F'(z) = \frac{F(z)}{z} = \sum_{i=1}^{n} \frac{A_i}{z-p_i}, \tag{2.35}$$

where A_i are constant coefficients, $(z-p_i)$ are simple factors for $i = 1, \ldots, n$, n being the order (the highest power of z in $F(z)$).

The constant coefficients can be obtained as

$$A_i = (z-p_i)F'(z)\big|_{z=p_i} \tag{2.36}$$

Once the coefficients are obtained, the original function $F(z)$ is expressed back in terms of partial fractions with z in the numerator

$$F(z) = \sum_{i=1}^{n} \frac{A_iz}{z-p_i} \tag{2.37}$$

The inverse z-transform can be obtained by looking at each partial fraction and corresponding entry in the z-transform table, similar to Table A.1 in Appendix A,

$$f(k) = \sum_{i=0}^{n} A_i p_i^k \tag{2.38}$$

A similar approach can be adopted for repeated and complex factors, the only difference being in the process of finding the partial fractions. Once PFE is completed, inverse z-transform can be obtained by looking up each partial fraction term in the z-transform table, similar to Table A.1 in Appendix A.

Another option is to apply *iztrans()* function in Matlab to get inverse z-transform directly.

Example 2.5 Find inverse z-transform of $F(z) = \frac{0.1z}{z^2 - 1.9z + 0.9}$

Solution $F(z) = \frac{0.1z}{z^2 - 1.9z + 0.9} = \frac{0.1z}{(z-1)(z-0.9)}$

Dividing $F(z)$ by z, one can write $F'(z) = \frac{F(z)}{z} = \frac{0.1}{(z-1)(z-0.9)}$.

Partial fraction expansion of $F'(z) = \frac{A_1}{z-1} + \frac{A_2}{z-0.9}$.

Coefficients are obtained as $A_1 = (z - 1)F'(z)|_{z=1} = 1$ and $A_2 = (z - 0.9)$ $F'(z)|_{z=0.9} = -1$.

The original function $F(z)$ can be expressed in terms of partial fractions as

$$F(z) = \frac{z}{z-1} - \frac{z}{z-0.9}$$

Looking up the table of z-transform the inverse z-transform can be obtained as

$$f(k) = 1 - 0.9^k, k \geq 0$$

Example 2.6 Find inverse z-transform of $F_1(z) = \frac{0.1}{z^2 - 1.9z + 0.9}$

Solution $F_1(z) = \frac{0.1}{z^2 - 1.9z + 0.9} = z^{-1}\frac{0.1z}{(z-1)(z-0.9)} = z^{-1}F(z)$

Using PFE method similar to Example 2.5, the inverse z-transform of $F(z)$ can be obtained as

$$f(k) = 1 - 0.9^k, k \geq 0.$$

Using the real translation property, the inverse z-transform $F_1(z)$ can be obtained as

$$f_1(k) = \left(1 - 0.9^{(k-1)}\right)u(k-1), k \geq 1.$$

Example 2.7 Find inverse z-transform of $F_2(z) = \frac{0.1}{z^3 - 1.9z^2 + 0.9z}$

Solution $F_2(z) = \frac{0.1}{z^3 - 1.9z^2 + 0.9z} = z^{-2}\frac{0.1z}{(z-1)(z-0.9)} = z^{-2}F(z)$

Using PFE method similar to Example 2.5, the inverse z-transform of $F(z)$ can be obtained as

$$f(k) = 1 - 0.9^k, k \geq 0.$$

Using the real translation property, the inverse z-transform $F_2(z)$ can be obtained as

$$f_2(k) = \left(1 - 0.9^{(k-2)}\right)u(k-2), k \geq 2.$$

Example 2.8 Obtain inverse z-transform for functions given in Examples 2.5, 2.6, and 2.7.

Solution Matlab code and results are presented.

```
% Ch2 Ex2-5 inverse z-transform for F(z)=0.1z /(z-1)(z-0.9)
% f(k) = 1-0.9^k, k>=0
syms k z
% Get F'(z)=F(z)/z
numz=0.1
denz=[1 -1.9 0.9]
[r,p,k1]=residue(numz, denz) %  PFE to get coefficients r, factors p

Fpz =0.1*z/((z-1)*(z-0.9)) % given function F(z)
iztrans (Fpz,k) % inverse z-transform f(k)

Results:
r =
    1.0000
   -1.0000

p =
    1.0000
    0.9000
f(k)= 1 - (9/10)^k
```

```
% Ch 2 Ex 2-6 with delay F1(z)=(z^-1)*0.1z/(z-1)(z-0.9)
% f1{k) = f(k-1)u(k-1)=(1-0.9^(k-1))u(k-1),  k>=1
numz=0.1
denz=[1 -1.9 0.9 0]
[r,p,k1]=residue(numz, denz) %  PFE to get coefficients r, factors p

F1z =0.1/((z-1)*(z-0.9)) % given function F(z)
iztrans (F1z,k) % inverse z-transform f(k)

Results:
r = 1.0000
   -1.1111
    0.1111
p =
    1.0000
    0.9000
         0

f(k)= kroneckerDelta(k, 0)/9 - (10*(9/10)^k)/9 + 1
```

Note this gives the same values as $f_1(k) = \left(1 - 0.9^{(k-1)}\right)u(k-1), k \geq 1$.

```
% Ch2 Ex 2-7 with double delay F2(z)=(z^-2)*0.1z/(z-1)(z-0.9)
% f2{k) = f(k-2)u(k-2)=(1-0.9^(k-2))u(k-2),  k>=2
numz=0.1
denz=[1 -1.9 0.9 0 0]
[r,p,k1]=residue(numz, denz) %  PFE to get coefficients r, factors p

F2z =0.1/(z*(z-1)*(z-0.9)) % given function F(z)
iztrans (F2z,k) % inverse z-transform f(k)

Results:
r =
    1.0000
   -1.2346
    0.2346
    0.1111
p =
    1.0000
    0.9000
         0
         0
f(k)= kroneckerDelta(k - 1, 0)/9 - (100*(9/10)^k)/81 +
(19*kroneckerDelta(k, 0))/81 + 1.
```

Note this gives the same values as $f_2(k) = \left(1 - 0.9^{(k-2)}\right)u(k-2), k \geq 2$.

Example 2.9 Given z-transform $F(z) = \frac{0.1z}{z^2 - 1.9z + 0.9}$ find the initial value $f(0)$ and the final value $f(\infty)$ using corresponding theorems. Verify the values from the discrete-time function obtained from inverse z-transform.

Solution Applying initial value theorem, Eq. (2.28), the initial value of the function can be obtained as

$$f(0) = \lim_{z\to\infty} F(z) = \lim_{z\to\infty} \frac{0.1z}{z^2 - 1.9z + 0.9} = \lim_{z\to\infty} \frac{0.1z^{-1}}{1 - 1.9z^{-1} + 0.9z^{-2}} = 0$$

Applying final value theorem, Eq. (2.30), the final value of the function can be obtained as

$$\lim_{n\to\infty} f(n) = \lim_{z\to 1}[(z-1)F(z)] = \lim_{z\to 1}\left[(z-1)\frac{0.1z}{z^2-1.9z+0.9}\right]$$
$$= \lim_{z\to 1}\left[(z-1)\frac{0.1z}{(z-1)(z-0.9)}\right] = 1$$

Applying inverse z-transform the discrete-time function is obtained in Example 2.5 as

$$f(k) = 1 - 0.9^k, k \geq 0.$$

The initial value can be obtained from the discrete-time function $f(k)$ at $k = 0$ as

$$f(0) = 1 - 0.9^0 = 1 - 1 = 0.$$

The final value can be obtained from the discrete-time function $f(k)$ at $k = \infty$ as

$$f(\infty) = 1 - 0.9^\infty = 1 - 0 = 1.$$

The values are same as that obtained using the initial value and the final value theorems.

2.9 Difference Equation

A difference equation of order n can be expressed in terms of input sequence $\{e(k)\}$ and output sequence $\{m(k)\}$, as follows:

$$\begin{aligned} m(k) + a_{n-1}m(k-1) + a_{n-2}m(k-2) + \cdots + a_0m(k-n) = \\ b_ne(k) + b_{n-1}e(k-1) + b_{n-2}e(k-2) + \cdots + b_0e(k-n) \end{aligned} \tag{2.39}$$

Applying real translation, the z-transform of (2.35) can be written as

$$\begin{aligned} M(z) + a_{n-1}z^{-1}M(z) + a_{n-2}z^{-2}M(z) + \cdots + a_0z^{-n}M(z) = \\ b_nE(z) + b_{n-1}z^{-1}E(z) + b_{n-2}z^{-2}E(z) + \cdots + b_0z^{-n}E(z) \end{aligned} \tag{2.40}$$

The expression for z-transform $M(z)$ can be obtained as

$$M(z) = \frac{b_n + b_{n-1}z^{-1} + b_{n-2}z^{-2} + \ldots + b_0 z^{-n}}{1 + a_{n-1}z^{-1} + a_{n-2}z^{-2} + \ldots + a_0 z^{-n}} E(z) \tag{2.41}$$

Example 2.10 Find difference equation for output $m(k)$ in terms of input $e(k)$ and delayed values $m(k-1)$, $m(k-2)\ldots e(k-1)$, $e(k-2)$, ... for the given discrete-time transfer function $D(z) = \frac{M(z)}{E(z)} = \frac{0.5z^2 - 1.1z + 0.25}{z^2 + 0.25z + 0.5}$.

Solution Dividing both the numerator and the denominator polynomials by the highest power of z the transfer function can be written as

$$D(z) = \frac{M(z)}{E(z)} = \frac{0.5 - 1.1z^{-1} + 0.25z^{-2}}{1 + 0.25z^{-1} + 0.5z^{-2}}$$

It can be expressed as $(1 + 0.25z^{-1} + 0.5z^{-2})M(z) = (0.5 - 1.1z^{-1} + 0.25z^{-2})E(z)$.

Applying the inverse z-transform of both sides and using real translational property, the expressions for the output $m(k)$ can be obtained as

$$m(k) + 0.25m(k-1) + 0.5m(k-2) = 0.5e(k) - 1.1e(k-1) + 0.25e(k-2)$$
$$m(k) = 0.5e(k) - 1.1e(k-1) + 0.25e(k-2) - 0.25m(k-1) - 0.5m(k-2)$$

Example 2.11 Simulate the output $m(k)$ of Example 2.10 using Matlab with initial values of $m(k-1)$, $m(k-2)$, $e(k-1)$, and $e(k-2)$ as 0 for $e(k) = 0.6$.

Solution Matlab code for simulating the function $m(k)$ with initial values of $m(k-1)$, $m(k-2)$, $e(k-1)$, and $e(k-2)$ as 0 for $e(k) = 0.6$ is given. Simulated output of the difference equation is shown in Fig. 2.9.

```
%Example 2_11 difference equation
% Initialization
mk1=0; % m(k-1)=0
mk2=0; % m(k-2)=0
ek1=0; % e(k-1)=0
ek2=0; % e(k-2)=0
ek=0.6; % given constant value for e(k)
n=30;
m=zeros(n,1);
ik=zeros(n,1);
for k=1:n
    mk=0.5*ek-1.1*ek1+0.25*ek2-0.25*mk1-0.5*mk2;
    ik(k)=k;
    m(k)=mk;
    % update mk1 mk2 ek1 ek2
    mk2=mk1;
    mk1=mk;
    ek2=ek1;
    ek1=ek;
end
plot(ik,m,'*')
title('Ex 2.11 Simulation of Difference Equation')
xlabel('Sample number k')
ylabel('Output m(k)')
```

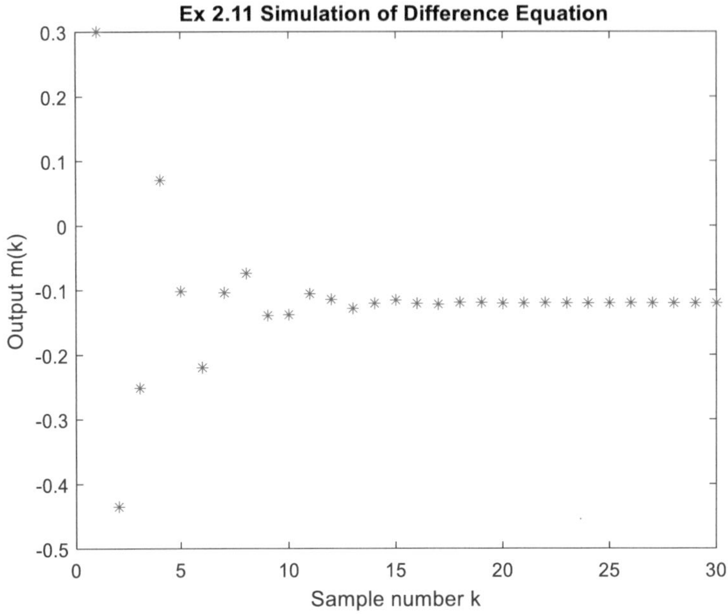

Fig. 2.9 Simulation of difference equation of Example 2.11

2.10 Simulation Diagrams and Signal Flow Diagrams

For an n-th order system, the output sequence, $\{m(k)\}$, can be expressed in terms of the input sequence, $\{e(k)\}$, from the difference equation, Eq. (2.35), as follows:

$$\begin{aligned} m(k) = {} & b_n e(k) + b_{n-1} e(k-1) + b_{n-2} e(k-2) + \cdots + b_0 e(k-n) \\ & - a_{n-1} m(k-1) - a_{n-2} m(k-2) - \cdots - a_0 m(k-n) \end{aligned} \tag{2.42}$$

The simulation diagram for Eq. (2.42) can be represented in Fig. 2.10 relating the current output $m(k)$ in terms of the current input $e(k)$ and the delayed signals $m(k-1)$, $m(k-2)$, $\ldots m(k-n)$, $e(k-1)$, $e(k-2)$, $\ldots e(k-n)$.

The signal flow diagram for the n-th order system can be represented in Fig. 2.11 replacing the delay blocks with z^{-1}

$$\begin{aligned} M(z) = {} & b_n E(z) + b_{n-1} z^{-1} E(z) + b_{n-2} z^{-2} E(z) + \cdots + b_0 z^{-n} E(z) \\ & - a_{n-1} z^{-1} M(z) - a_{n-2} z^{-2} M(z) - \cdots - a_0 z^{-n} M(z) \end{aligned} \tag{2.43}$$

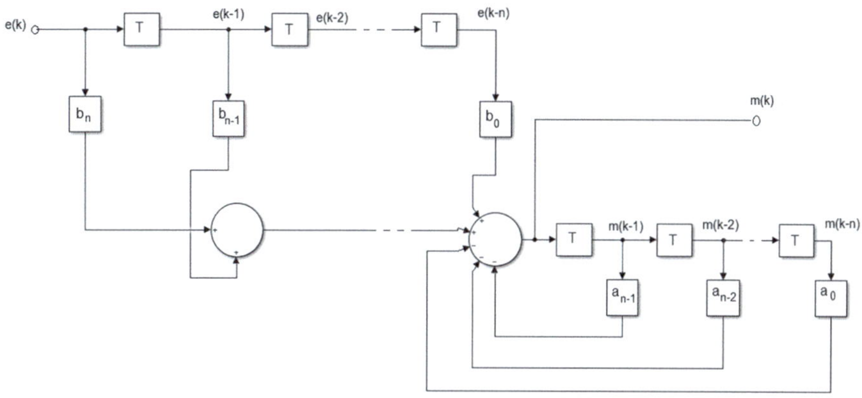

Fig. 2.10 Simulation diagram for an n-th order system

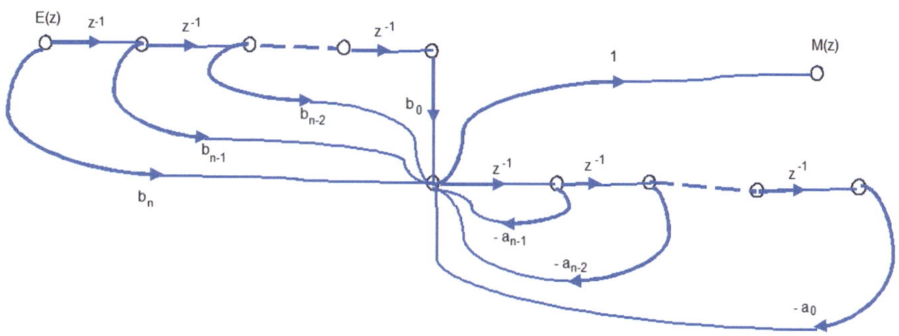

Fig. 2.11 Signal flow diagram for an n-th order system

Exercises

2.1 Use Matlab to obtain z-transform of $f(t) = te^{-at}$ or $f(k) = kT\,e^{-akT}$, T being the sampling time.

2.2 Use Matlab to obtain z-transform of $f(t) = a^t \sin(bt)$ or $f(k) = a^{kT} \sin(bkT)$, T being the sampling time.

2.3 Given a function in s-domain $F(s) = \frac{1}{s^2+3s+2}$, find z-transform $F(z)$ using PFE for a sampling time $T = 0.1$ s.

2.4 Given a function in s-domain $F(s) = \frac{4}{s^2+8s+12}$, find z-transform $F(z)$ using PFE for a sampling time $T = 0.1$ s.

2.5 Given a function in s-domain $F(s) = \frac{s^3+s+2}{s^3+3s^2+2s}$, find z-transform $F(z)$ using PFE for a sampling time $T = 0.1$ s.

2.6 Given a function in s-domain $F(s) = \frac{1}{s^2+3s+2}$, find z-transform $F(z)$ through inverse Laplace transform for a sampling time $T = 0.1$ s.

2.7 Given a function in s-domain $F(s) = \frac{4}{s^2+8s+12}$, find z-transform $F(z)$ through inverse Laplace transform for a sampling time $T = 0.1$ s.

2.8 Given a function in s-domain $F(s) = \frac{s^3+s+2}{s^3+3s^2+2s}$, find z-transform $F(z)$ through inverse Laplace transform for a sampling time $T = 0.1$ s.

2.9 Given a function in z-domain $F(z) = \frac{0.0861z}{z^2 - 1.7235z+0.7048} = \frac{z}{z-0.9048} - \frac{z}{z-0.8187}$, find inverse z-transform $f(k)$ using PFE.

2.10 Given a function in z-domain $F(z) = \frac{2z^2 - 3.4950z+1.5420}{z^3 - 2.5595z^2+2.1660z - 0.6065}$, find inverse z-transform $f(k)$ using PFE. Find $f(0)$ and $f(\infty)$. Verify these values applying initial and final value theorems to $F(z)$.

2.11 Given a function in z-domain $F(z) = \frac{z^2 - 1.5873z+0.6728}{z^3 - 2.4111z^2+1.9077z - 0.4966}$, apply initial and final value theorems to find initial and final values of the function, i.e., $f(0)$ and $f(\infty)$. Verify these values from the inverse z-transform $f(k)$.

2.12 Given a function in z-domain $F(z) = \frac{z^2 - 1.5513z+0.5686}{z^3 - 2.7235z^2+2.4643z - 0.7048}$, apply initial and final value theorems to find initial and final values of the function, i.e., $f(0)$ and $f(\infty)$. Verify these values from the inverse z-transform $f(k)$.

2.13 Given a function in z-domain $F(z) = \frac{0.3412z - 0.3165}{z^3 - 2.6465z^2+2.316z - 0.6703}$, apply initial and final value theorems to find initial and final values of the function, i.e., $f(0)$ and $f(\infty)$. Verify these values from the inverse z-transform $f(k)$.

2.14 Given a function $F(z) = \frac{2z^2 - 3.8317z+1.8630}{z^3 - 2.5751z^2+2.1816z - 0.6065}$, apply initial and final value theorems to find initial and final values of the function, i.e., $f(0)$ and $f(\infty)$. Verify these values from the inverse z-transform $f(k)$.

2.15 Find difference equation for output $m(k)$ in terms of input $e(k)$ and delayed values $m(k-2)$, $m(k-1)$...and $e(k-2)$, $e(k-1)$, ... for the given discrete-time transfer function shown in Fig. P2.1.

$$D(z) = \frac{M(z)}{E(z)} = \frac{0.60z^2 - 1.2z + 0.30}{z^2 + 0.30z + 0.60}.$$

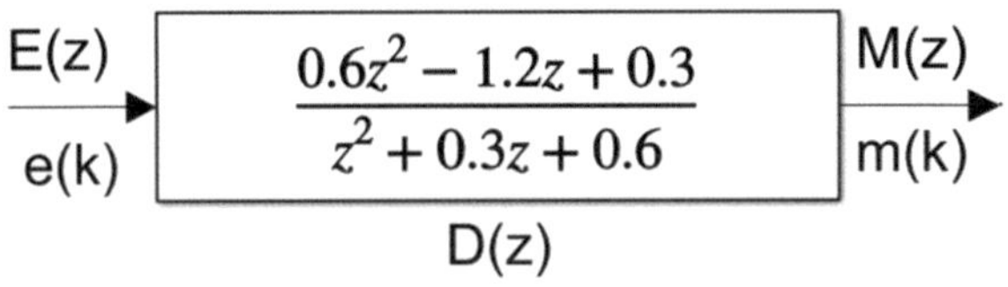

Fig. P2.1 Discrete-time transfer function for problem 2.15

2.16 Find difference equation for output $m(k)$ in terms of input $e(k)$ and delayed values $m(k-3)$, $m(k-2)$...and $e(k-3)$, $e(k-2)$, ... for the given discrete-time transfer function shown in Fig. P2.2

$$D(z) = \frac{M(z)}{E(z)} = \frac{0.50z^2 - 1.1z + 0.25}{z^3 - 0.91z^2 + 025z - 0.50}.$$

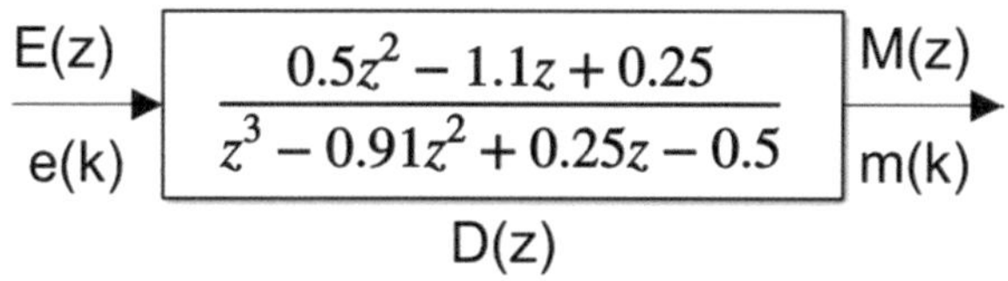

Fig. P2.2 Discrete-time transfer function for problem 2.16

2.17 Find the transfer function $D(z) = \frac{M(z)}{E(z)}$ for the digital filter shown in Fig. P2.3 that solves the difference equation: $m(k) = 1.9m(k-1) - 0.9m(k-2) + 1.5e(k) - e(k-1) + 0.3e(k-2)$.

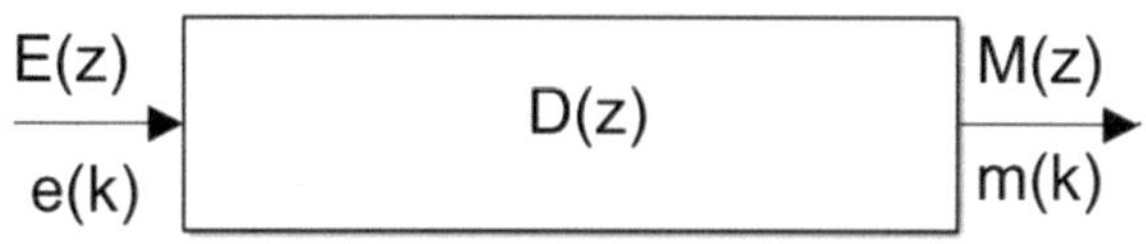

Fig. P2.3 Discrete-time transfer function for problem 2.17

2.18 Find the transfer function $D(z) = \frac{M(z)}{E(z)}$ for the digital filter shown in Fig. P2.4 that solves the difference equation: $m(k) = 1.9m(k-1) - 0.9m(k-2) + 0.5m(k-3) + 1.5e(k) - e(k-1) + 0.3e(k-2) + 0.2e(k-3)$.

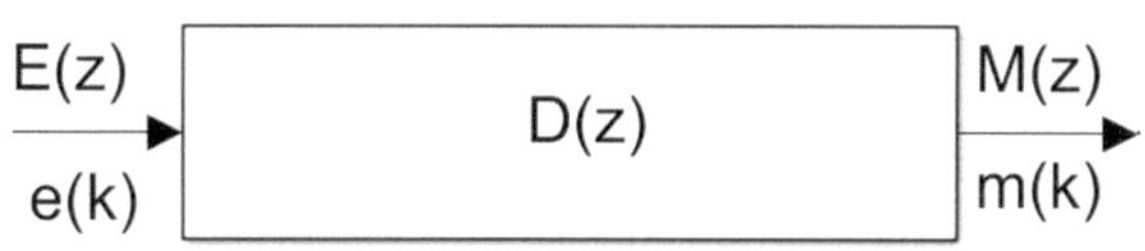

Fig. P2.4 Discrete-time transfer function for problem 2.18

2.19 Simulate the output $m(k)$ for the discrete-time transfer function for the digital filter $D(z) = \frac{M(z)}{E(z)}$ shown in Fig. P2.5.

$$D(z) = \frac{M(z)}{E(z)} = \frac{1.1z - 0.20}{z^2 + 0.25z - 0.40}$$

Use Matlab with initial values of $m(k-2)$, $m(k-1)$, $e(k-2)$, and $e(k-1)$ as 0 and $e(k) = 0.5$.

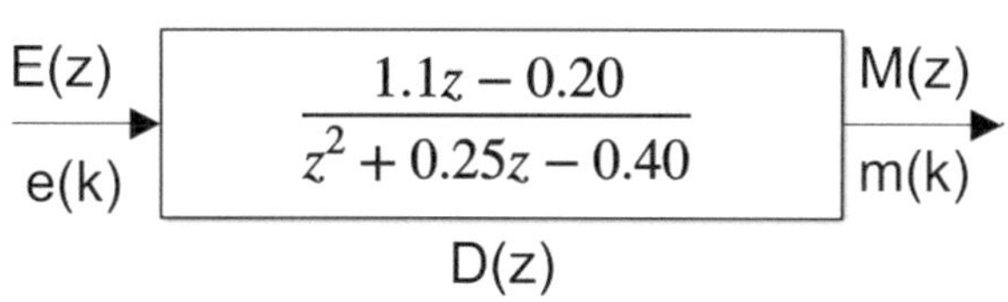

Fig. P2.5 Discrete-time transfer function for problem 2.19

2.20 Simulate the output $m(k)$ for the discrete-time transfer function for the digital filter $D(z) = \frac{M(z)}{E(z)}$ shown in Fig. P2.6.

$$D(z) = \frac{M(z)}{E(z)} = \frac{0.6z^2 - 1.1z + 0.25}{z^2 + 0.30z + 0.50}$$

Use Matlab with initial values of $m(k-2)$, $m(k-1)$, $e(k-2)$, and $e(k-1)$ as 0 and $e(k) = 0.75$.

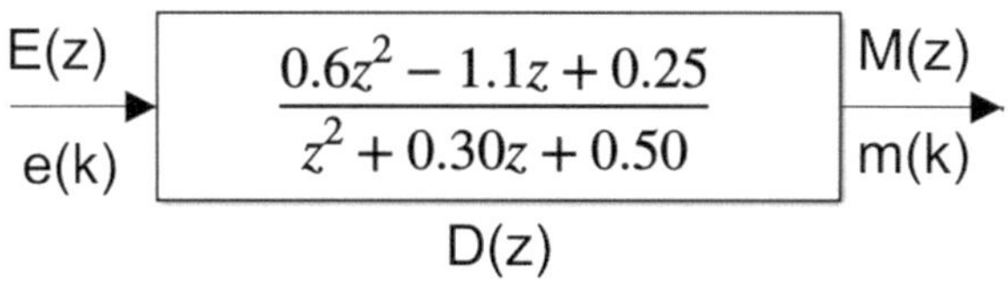

Fig. P2.6 Discrete-time transfer function for problem 2.20

Further Reading

1. Asadi F (2023) Signals and systems with MATLAB® and Simulink®. Springer, Cham
2. Astrom KJ, Murray RM (2021) Feedback systems: an introduction for scientists and engineers, 2nd edn. Princeton University Press, Princeton
3. Chen C-T (2006) Analog and digital control system design: transfer-function, state-space, and algebraic methods. Oxford University Press, New York
4. Dorf R, Bishop R (2016) Modern control systems, 13th edn. Pearson, Essex
5. Fadali MS, Visioli A (2020) Digital control engineering analysis and design, 3rd edn. Academic Press, London
6. Franklin GF, Powell JD, Workman M (1998) Digital control of dynamic systems, 3rd edn. Addison-Wesley, Boston
7. Franklin GF, Powell JD, Emami-Naeini A (2019) Feedback control of dynamic systems, 8th edn. Pearson, New Jersey
8. Isermann R (1981) Digital control systems. Springer-Verlag, Berlin
9. Jalili N, Candelino NW (2023) Dynamic systems and control engineering. Cambridge University Press, Cambridge
10. Keviczky L, Bars R, Hetthéssy J, Bányász C (2019) Control engineering. Springer, Singapore
11. Koeing DM (2009) Practical control engineering: a guide for engineers, managers, and practitioners. McGraw-Hill, New York
12. Kuo BC (1995) Digital control systems, 2nd edn. Oxford University Press, Oxford
13. Landau ID, Gito G (2006) Digital control systems: design, identification and implementation. Springer-Verlag, London
14. Leigh JR (2006) Applied digital control: theory, design and implementation, 2nd edn. Dover, Mineola
15. Mathworks (2023) Control System Toolbox. https://www.mathworks.com/products/control.html
16. Mathworks (2023) Matlab. https://www.mathworks.com/products/matlab.html
17. Nise NS (2019) Control systems engineering, 8th edn. Wiley, Hoboken
18. Nof SY (ed) (2009) Springer handbook of automation. Springer-Verlag, Berlin
19. Ogata K (1995) Discrete-time control systems, 2nd edn. Pearson, New Jersey

20. Phillips CL, Nagle HT, Chakrabortty A (2015) Digital control system analysis and design, 4th edn. Pearson, Essex
21. Rabbath CA, Lechevin N (2014) Discrete-time control system design with applications. Springer, New York
22. Samanta B (2023) Introduction to mechatronics: an integrated approach. Springer, Cham
23. Shertukde HM (2015) Digital control applications illustrated with MATLAB®. CRC Press, Boca Raton
24. Starr G (2020) Introduction to applied digital controls. Springer, Cham
25. Veloni A, Miridakis NI (2018) Digital control systems-theoretical problems and simulation tools. CRC Press, Boca Raton
26. Vilanova R, Visioli A (2012) PID control in the third millennium – lessons learned and new approaches. Springer-Verlag, London
27. Visioli A (2006) Practical PID control. Springer, London

Chapter 3
Data Sampling and Reconstruction

3.1 Introduction

In this chapter, topics of sampling and data reconstruction in relation to discrete-time control systems are presented. Shannon's sampling theorem for minimum frequency of data sampling is stated first. Next, an ideal sampler is presented along with the procedure of finding Laplace transform of sampled data. Data reconstruction from sampled data using zero-, first-, and fractional-order devices are discussed. Examples are presented to illustrate the procedures. End-of-chapter exercises are provided to enhance understanding of the topics covered and consolidate problem-solving skills.

3.2 Sampling Theorem

Shannon's sampling theorem states that the sampling frequency (f_s) should be at least twice the highest frequency ($f_{\max}$) of a sampled signal, i.e., $f_s \geq f_{\max}$. In practice, the sampling frequency is generally 5–10 times the highest frequency of the sampled signal.

3.3 Ideal Sampler

The schematic of a signal $f(t)$ being sampled using an ideal sampler with a sampling time T to get a sequence of samples $f^*(t) : f(kT)$ is shown in Fig. 3.1, where k is the sampling instant, $k = 0, \ldots, \infty$.

The sampled signal $f^*(t)$ can be expressed as a sequence of pulses at sampling instants, as represented in Eq. (3.1) and in Fig. 3.2:

B. Samanta, *Introduction to Digital Control*,
https://doi.org/10.1007/978-3-031-66830-2_3

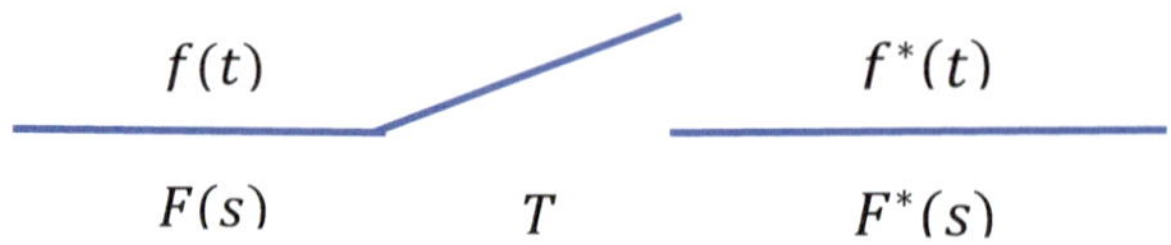

Fig. 3.1 Schematic of an ideal sampler

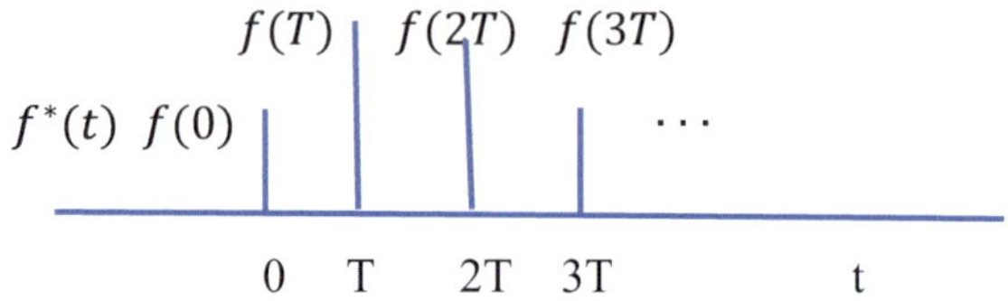

Fig. 3.2 Representation of a sampled signal through an ideal sampler

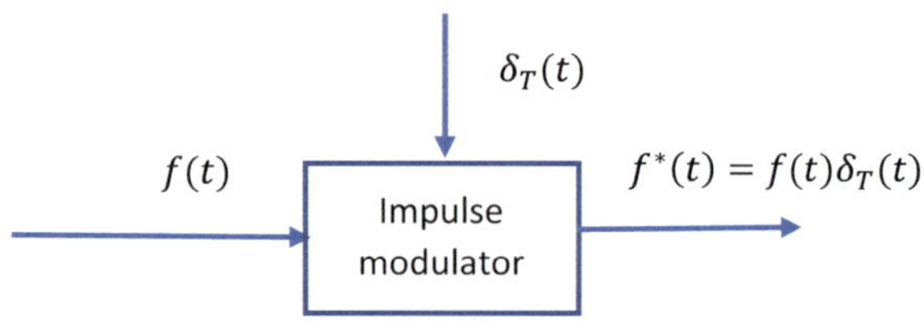

Fig. 3.3 Representation of an ideal sampler as an impulse modulator

$$f^*(t) = \sum\nolimits_{k=0}^{\infty} f(kT)\delta(t - kT) \tag{3.1}$$

The sampled signal $f^*(t)$ can be considered as the output of an impulse modulator, shown in Fig. 3.3, with δ_T defined as the carrier and the input signal $f(t)$ as the modulating signal in the modulation process.

$$\delta_T(t) = \sum\nolimits_{k=0}^{\infty} \delta(t - kT) = \delta(t) + \delta(t - T) + \delta(t - 2T) + \ldots \tag{3.2}$$

$$f^*(t) = \sum\nolimits_{k=0}^{\infty} f(kT)\delta(t - kT) = f(0)\delta(t) + f(T)\delta(t - T) + f(2T)\delta(t - 2T) + \ldots \tag{3.3}$$

Laplace transform can be obtained as

$$\{f^*(t)\} = L\left\{\sum\nolimits_{k=0}^{\infty} f(kT)\delta(t - kT)\right\} \tag{3.4}$$

$$\begin{aligned} F^*(s) &= \sum\nolimits_{k=0}^{\infty} \int_{-\infty}^{\infty} f(kT)\delta(t - kT)e^{-st}dt \\ &= \sum\nolimits_{k=0}^{\infty} f(kT)e^{-ksT} \end{aligned} \tag{3.5}$$

3.3.1 Laplace Transform of Sampled Signals

Laplace transform of a signal sampled with an ideal sampler can be evaluated using Eq. (3.5). The process is illustrated through examples in this section.

Example 3.1 Find Laplace transform of a unit step function sampled using an ideal sampler, $u^*(t) = \sum_{k=0}^{\infty} u(kT)\delta(t - kT), \;\; u(t) = 1, t \geq 0.$

Solution Using the definition of Laplace transform, one can write from Eq. (3.5):

$$\begin{aligned}
U^*(s) &= \sum_{k=0}^{\infty} \int_{-\infty}^{\infty} u(kT)\delta(t - kT)e^{-st}dt \\
&= \sum\nolimits_{k=0}^{\infty} u(kT)e^{-ksT} \\
&= u(0)e^{-0} + u(T)e^{-sT} + u(2T)e^{-2sT} + \cdots \\
&= 1 + e^{-sT} + e^{-2sT} + \cdots \\
&= \left(1 - e^{-sT}\right)^{-1} = \frac{1}{1 - e^{-sT}}, \left|e^{-sT}\right| < 1
\end{aligned}$$

Example 3.2 Find Laplace transform of an exponentially decaying function $f(t) = e^{-bt}$ sampled using an ideal sampler, $f^*(t) = \sum_{k=0}^{\infty} e^{-bkT}\delta(t - kT), \;\; b > 0, \;\; t \geq 0.$

Solution Using the definition of Laplace transform, one can write from Eq. (3.5):

$$\begin{aligned}
F^*(s) &= \sum_{k=0}^{\infty} \int_{-\infty}^{\infty} f(kT)\delta(t - kT)e^{-st}dt \\
&= \sum\nolimits_{k=0}^{\infty} e^{-bkT}e^{-skT} \\
&= e^{-0}e^{-0} + e^{-bT}e^{-sT} + e^{-b2T}e^{-2sT} + \cdots \\
&= 1 + e^{-(b+s)T} + e^{-2(b+s)T} + \cdots \\
&= \left(1 - e^{-(b+s)T}\right)^{-1} = \frac{1}{1 - e^{-(b+s)T}}, \left|e^{-(b+s)T}\right| < 1
\end{aligned}$$

Example 3.3 Find Laplace transform of a function $f(t) = 1 - e^{-bt}$ sampled using an ideal sampler, $f^*(t) = \sum_{k=0}^{\infty} \left(1 - e^{-bkT}\right)\delta(t - kT), \;\; b > 0, \;\; t \geq 0.$

Solution Using the definition of Laplace transform, one can write from Eq. (3.5):

$$\begin{aligned}
F^*(s) &= \sum_{k=0}^{\infty} \int_{-\infty}^{\infty} f(kT)\delta(t - kT)e^{-st}dt \\
&= \sum\nolimits_{k=0}^{\infty} \left(1 - e^{-bkT}\right)e^{-skT} \\
&= \sum\nolimits_{k=0}^{\infty} u(kT)e^{-ksT} - \sum\nolimits_{k=0}^{\infty} e^{-bkT}e^{-skT} \\
&= \frac{1}{1 - e^{-sT}} - \frac{1}{1 - e^{-(b+s)T}}
\end{aligned}$$

Example 3.4 Find Laplace transform of a time-delayed function $f_n(t) = (1 - e^{-b(t-nT)})u(t - nT)$ sampled using an ideal sampler, $f^*{}_n(t) = \sum_{k=0}^{\infty}\left(1 - e^{-b(kT-nT)}\right)u(t-nT)\delta(t-kT)$, $b > 0$, $t \geq 0, n > 0$, n is a positive integer.

Solution First, the Laplace transform of the sampled function without time delay is obtained, similar to Example 3.3, as

$$F^*(s) = \frac{1}{1 - e^{-sT}} - \frac{1}{1 - e^{-(b+s)T}}$$

Next, the Laplace transform of the time-delayed function with time delay as an integer multiple of sampling time is obtained as

$$F^*{}_n(s) = e^{-nTs}F^*(s) = e^{-nTs}\left(\frac{1}{1 - e^{-sT}} - \frac{1}{1 - e^{-(b+s)T}}\right)$$

3.3.2 *Laplace Transform of Sampled Signals Represented in s-Domain*

Laplace transform of a sampled signal represented in s-domain can be obtained using different methods including the method of residues. However, in this section, a method is presented where the signal is first converted to time domain from s-domain using inverse Laplace transform and then applying Eq. (3.5) to the time domain signal to find Laplace transform of the sampled signal. The procedure is illustrated through examples.

Example 3.5 Find Laplace transform of a sampled function $f(t)$ using an ideal sampler, $f^*(t) = \sum_{k=0}^{\infty} f(kT)\delta(t-kT), t \geq 0$, such that $F(s) = \frac{b}{s(s+b)}, b > 0$.

Solution The given Laplace transform of function $f(t)$ is

$$F(s) = \frac{b}{s(s+b)} = \frac{1}{s} - \frac{1}{s+b}$$

Using inverse Laplace transform the function in time domain can be expressed as

$$f(t) = 1 - e^{-bt}$$

Using Eq. (3.5), one can write Laplace transform for the sampled function $f^*(t)$ similar to Example 3.3, as

$$F^*(s) = \sum_{k=0}^{\infty} \int_{-\infty}^{\infty} f(kT)\delta(t-kT)e^{-st}dt$$
$$= \sum\nolimits_{k=0}^{\infty} \left(1 - e^{-bkT}\right)e^{-skT}$$
$$= \sum\nolimits_{k=0}^{\infty} u(kT)e^{-ksT} - \sum\nolimits_{k=0}^{\infty} e^{-bkT} e^{-skT}$$
$$= \frac{1}{1-e^{-sT}} - \frac{1}{1-e^{-(b+s)T}}$$

Example 3.6 Find Laplace transform of a sampled time-delayed function $f_n(t)$ using an ideal sampler, $f_n^*(t) = \sum_{k=0}^{\infty} f_n(kT)\delta(t-kT), t \geq 0,$ such that $F_n(s) = e^{-ns}\frac{b}{s(s+b)}, b > 0,$ $n > 0$, n is a positive integer.

Solution The given Laplace transform of function $f(t)$ without time delay can be written as

$$F(s) = \frac{b}{s(s+b)} = \frac{1}{s} - \frac{1}{s+b}$$

Using inverse Laplace transform the function in time domain without time delay can be expressed as

$$f(t) = 1 - e^{-bt}$$

Laplace transform of the sampled function without time delay can be obtained, similar to Example 3.5, as

$$F^*(s) = \frac{1}{1-e^{-sT}} - \frac{1}{1-e^{-(b+s)T}}$$

Next, the Laplace transform of the time-delayed function with time delay as an integer multiple of sampling time is obtained as

$$F^*{}_n(s) = e^{-nTs}F^*(s) = e^{-nTs}\left(\frac{1}{1-e^{-sT}} - \frac{1}{1-e^{-(b+s)T}}\right)$$

3.4 Data Reconstruction

Schematic of a closed-loop digital control system is shown in Fig. 3.4 where a sampler is used to convert the analog signal to digital (discrete-time) (analog to digital converter, ADC), and a digital controller is given by z-domain transfer

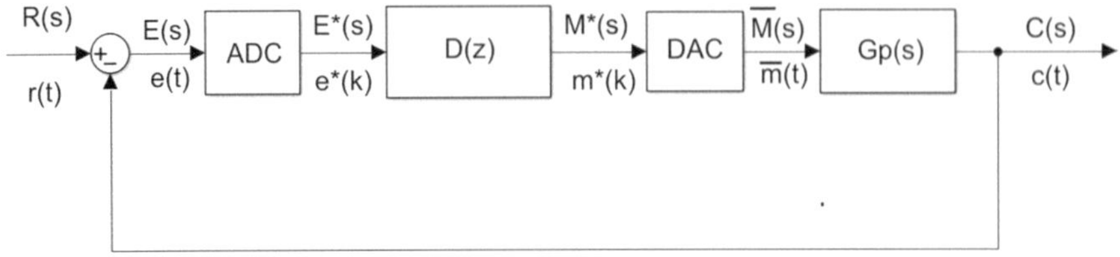

Fig. 3.4 Schematic of a closed loop discrete-time control system

function $D(z)$ and a data reconstruction (data hold) device is used to convert a digital signal to the analog domain (digital-to-analog converter, DAC). In this section, the digital controller transfer function is assumed to be unity, $D(z) = 1$ to discuss the data reconstruction from the sampled data.

One of the most common methods of data reconstruction is done through polynomial extrapolation. Taylor's series expansion can be used about $t = kT$ to express $f(t)$ as

$$f(t) = f(kT) + f'(kT)(t - kT) + \frac{f''(kT)}{2!}(t - kT)^2 + \ldots \tag{3.6}$$

where $f'(kT), f''(kT)\ldots$ represent the first, the second, ... derivatives of the function. The reconstructed function $f_k(t)$ as the output of the data hold device can be expressed as

$$f_k(t) = f(t), \quad kT \leq t \leq (k+1)T \tag{3.7}$$

The order of data hold devices is characterized by the order of the extrapolation polynomial of Eq. (3.6). For example, a zero-order hold (ZOH) device is characterized with the zero-order approximation of Eq. (3.6) consisting of the first term only. A first-order hold (FOH) device is used to approximate the function through the first two terms of Eq. (3.6), i.e., including the first derivative of the slope. A fractional-order (FrOH) device is used to approximate the function through a fractional-order (r) derivative, $0 < r < 1$. In special cases, ZOH is the fractional-order hold with $r = 0$, and FOH is a fractional-order hold with $r = 1$.

In the following subsections, transfer functions for three hold devices, namely, ZOH, FOH, and FrOH, are presented.

3.4.1 Zero-Order Hold (ZOH)

The schematic of a zero-order hold (ZOH) is shown with a unit impulse input and the corresponding step output over a sampling interval in Fig. 3.5.

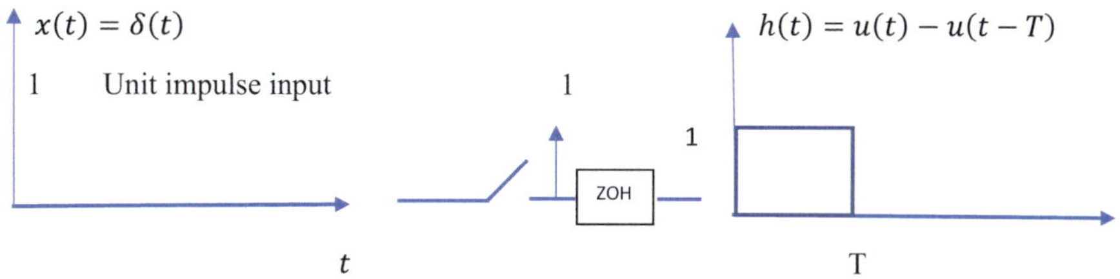

Fig. 3.5 Input and output signals of a zero-order hold

Fig. 3.6 Block diagram of a zero-order hold (ZOH)

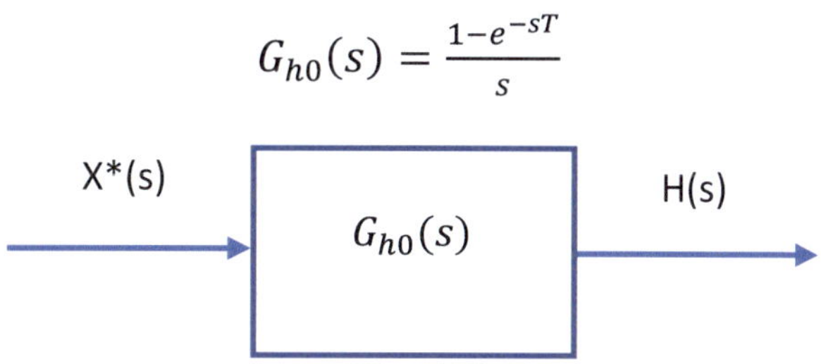

The value of the output is held constant over the sampling interval with the value being equal to the value of the function at the previous sampling instant. For a unit impulse as the input, the output is the unit pulse over the sampling interval. The input $x^*(t)$ and the output $h(t)$ of ZOH can be expressed as

$$x^*(t) = \delta(t), \quad h(t) = u(t) - u(t - T), \quad 0 \leq t < T \tag{3.8}$$

Taking Laplace transform of both input and output functions, the transfer function of ZOH, $G_{h0}(s)$, can be expressed as

$$X^*(s) = 1, H(s) = \frac{1 - e^{-sT}}{s}$$

$$G_{h0}(s) = \frac{H(s)}{X^*(s)} = \frac{1 - e^{-sT}}{s} \tag{3.9}$$

The block diagram for a ZOH along with its transfer function is shown in Fig. 3.6.

3.4.2 First-Order Hold (FOH)

The schematic of a first-order hold (FOH) is shown with a sampled unit step input and the corresponding output over a sampling interval in Fig. 3.7.

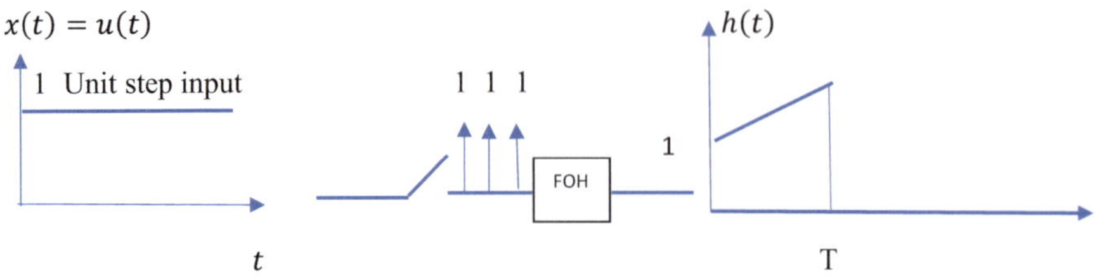

Fig. 3.7 Input and output signals of a first-order hold (FOH)

Fig. 3.8 Block diagram of a first-order hold (FOH)

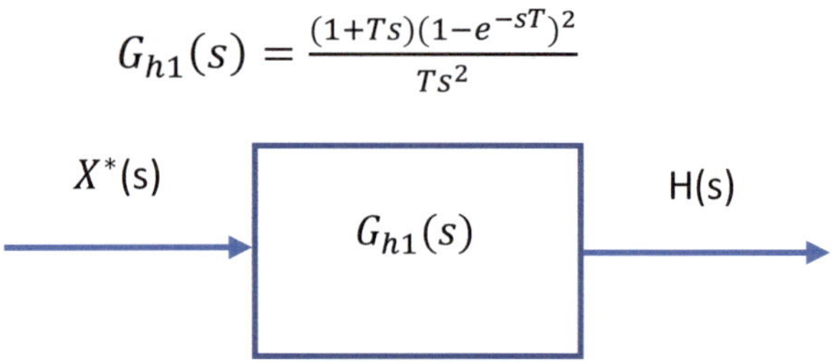

The value of the output is linearly changing over the sampling interval with the starting value being equal to the value of the function at the previous sampling instant. The input to FOH is sampled unit impulse expressed as

$$x^*(t) = \sum_{k=0}^{\infty} u(t)\delta(t - kT) \tag{3.10}$$

The output of FOH over the sampling interval can be expressed as a linearly changing function:

$$h(t) = u(t) + \frac{tu(t)}{T} - u(t - T) - \frac{(t - T)u(t - T)}{T} \tag{3.11}$$

Taking Laplace transform of both input and output functions, the transfer function of FOH, $G_{h1}(s)$, can be expressed as

$$X^*(s) = \left(1 - e^{-sT}\right)^{-1} \tag{3.12}$$

$$H(s) = \frac{1 - e^{-sT}}{s} + \frac{1 - e^{-sT}}{Ts^2} = \frac{(1 + Ts)(1 - e^{-sT})}{Ts^2} \tag{3.13}$$

$$G_{h1}(s) = \frac{H(s)}{X^*(s)} = \frac{(1 + Ts)(1 - e^{-sT})^2}{Ts^2} \tag{3.14}$$

The block diagram for a FOH along with its transfer function is shown in Fig. 3.8.

3.4.3 Fractional-Order Hold (FrOH)

A fractional-order hold is similar to a first-order hold with the difference being that a fraction (r) of the slope is used over the sampling interval, $0 < r < 1$ for the linearly changing output function. The sampled input to FrOH is the sampled unit function as given in Eq. (3.10). The output function is modified with a fraction r, $0 < r < 1$ as

$$h(t) = u(t) + \frac{rtu(t)}{T} - ru(t-T) - \frac{r(t-T)u(t-T)}{T} \tag{3.15}$$

Taking Laplace transform, the transfer function of FrOH can be expressed as

$$H(s) = \frac{1-re^{-sT}}{s} + \frac{r(1-e^{-sT})}{Ts^2} \tag{3.16}$$

$$G_{hr}(s) = \frac{H(s)}{X^*(s)} = \frac{(1-re^{-sT})(1-e^{-sT})}{s} + \frac{r(1-e^{-sT})^2}{Ts^2} \tag{3.17}$$

The block diagram for a FOH along with its transfer function is shown in Fig. 3.9.

The transfer function of FrOH is a general case for all 3 types of hold. For example, the transfer function of ZOH, Eq. (3.9), can be obtained from Eq. (3.17) with $r = 0$. Similarly, the transfer function of FOH, Eq. (3.14), can be obtained from Eq. (3.17) with $r = 1$. The shapes of output function for different r, $r = 0$, (ZOH), $0 < r < 1$ (FrOH), $r = 1$ (FOH) are shown in Fig. 3.10.

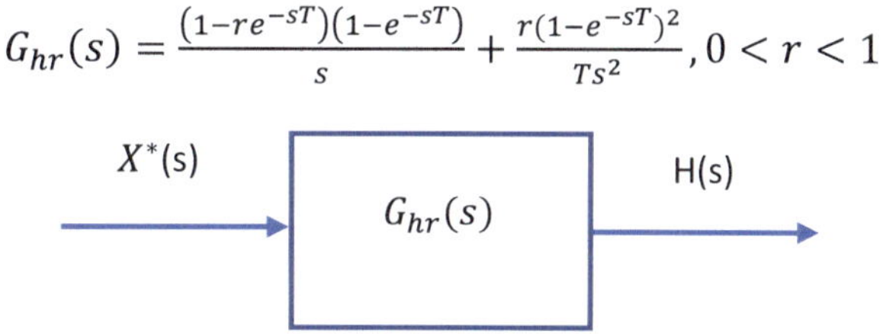

Fig. 3.9 Block diagram of a fractional-order hold (FrOH)

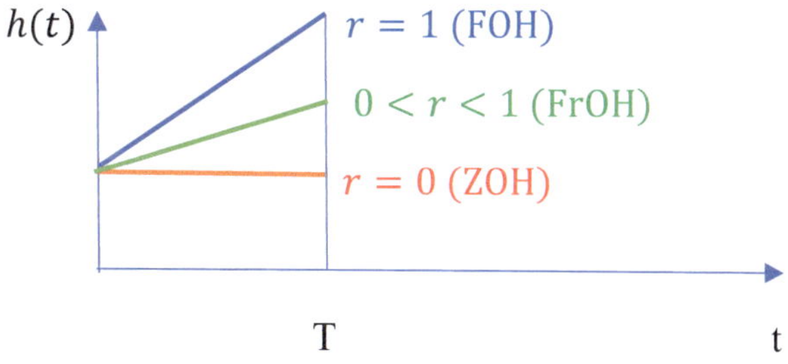

Fig. 3.10 Output functions for zero-, fractional-, and first-order holds

Exercises

3.1 Find Laplace transform of an exponentially decaying function $f(t) = e^{-3t}$ sampled using an ideal sampler, $f^*(t) = \sum_{k=0}^{\infty} e^{-3kT}\delta(t - kT), \ \ t \geq 0.$
3.2 Find Laplace transform of a function $f(t) = 1 - e^{-2t}$ sampled using an ideal sampler, $f^*(t) = \sum_{k=0}^{\infty} \left(1 - e^{-2kT}\right)\delta(t - kT), \ \ t \geq 0, T = 0.2$ s.
3.3 Find Laplace transform of a function $f_2(t) = e^{-4(t-2T)}u(t - 2T)$ sampled using an ideal sampler, $f_n^*(t) = \sum_{k=0}^{\infty} e^{-4(kT-2T)}u(t - 2T)\delta(t - kT), \ \ t \geq 0, T = 0.1$ s.
3.4 Find Laplace transform of a sampled function $f(t)$ using an ideal sampler, $f^*(t) = \sum_{k=0}^{\infty} f(kT)\delta(t - kT), t \geq 0,$ such that $F(s) = \frac{2}{s(s+2)}, T = 0.05$ s.
3.5 Find Laplace transform of a sampled function $f(t)$ using an ideal sampler, $f^*(t) = \sum_{k=0}^{\infty} f(kT)\delta(t - kT), t \geq 0,$ such that $F(s) = \frac{1}{(s+3)(s+4)}, T = 0.05$ s.
3.6 Find Laplace transform of a sampled time-delayed function $f_3(t)$ using an ideal sampler, $f_3^*(t) = \sum_{k=0}^{\infty} f_3(kT)\delta(t - kT), t \geq 0,$ such that $F_3(s) = \frac{4e^{-3s}}{s(s+4)}, \ \ T = 0.05$ s.
3.7 Find Laplace transform of a sampled function $f(t)$ using an ideal sampler $f^*(t) = \sum_{k=0}^{\infty} f(kT)\delta(t - kT), t \geq 0,$ such that $F(s) = \frac{2\left(1 - e^{-3s}\right)}{s(s+2)}, \ \ T = 0.05$ s.
3.8 Find Laplace transform of a sampled function $f(t)$ using an ideal sampler $f^*(t) = \sum_{k=0}^{\infty} f(kT)\delta(t - kT), t \geq 0,$ such that $F(s) = \frac{\left(1 - e^{-2s}\right)}{(s+1)(s+2)}, \ \ T = 0.02$ s.

Further Reading

1. Asadi F (2023) Signals and systems with MATLAB® and Simulink®. Springer, Cham
2. Astrom KJ, Murray RM (2021) Feedback systems: an introduction for scientists and engineers, 2nd edn. Princeton University Press, Princeton
3. Chen C-T (2006) Analog and digital control system design: transfer-function, state-space, and algebraic methods. Oxford University Press, New York
4. Dorf R, Bishop R (2016) Modern control systems, 13th edn. Pearson, Essex
5. Fadali MS, Visioli A (2020) Digital control engineering analysis and design, 3rd edn. Academic Press, London
6. Franklin GF, Powell JD, Emami-Naeini A (2019) Feedback control of dynamic systems, 8th edn. Pearson, Upper Saddle River
7. Franklin GF, Powell JD, Workman M (1998) Digital control of dynamic systems, 3rd edn. Addison-Wesley, Boston
8. Isermann R (1981) Digital control systems. Springer, Berlin
9. Jalili N, Candelino NW (2023) Dynamic systems and control engineering. Cambridge University Press, Cambridge
10. Keviczky L, Bars R, Hetthéssy J, Bányász C (2019) Control engineering. Springer, Singapore
11. Koeing DM (2009) Practical control engineering: a guide for engineers, managers, and practitioners. McGraw-Hill, New York
12. Kuo BC (1995) Digital control systems, 2nd edn. Oxford University Press, Oxford
13. Landau ID, Gito G (2006) Digital control systems: design, identification and implementation. Springer, London

14. Leigh JR (2006) Applied digital control: theory, design and implementation, 2nd edn. Dover, Mineola
15. Mathworks (2023) Control System Toolbox. https://www.mathworks.com/products/control.html
16. Mathworks (2023) Matlab. https://www.mathworks.com/products/matlab.html
17. Nise NS (2019) Control systems engineering, 8th edn. Wiley, Hoboken
18. Nof SY (ed) (2009) Springer handbook of automation. Springer, Berlin
19. Ogata K (1995) Discrete-time control systems, 2nd edn. Pearson, Hoboken
20. Phillips CL, Nagle HT, Chakrabortty A (2015) Digital control system analysis and design, 4th edn. Pearson, Essex
21. Rabbath CA, Lechevin N (2014) Discrete-time control system design with applications. Springer, New York
22. Samanta B (2023) Introduction to mechatronics – an integrated approach. Springer, Cham
23. Shertukde HM (2015) Digital control applications illustrated with MATLAB®. CRC Press, Boca Raton
24. Starr G (2020) Introduction to applied digital controls. Springer, Cham
25. Veloni A, Miridakis NI (2018) Digital control systems-theoretical problems and simulation tools. CRC Press, Boca Raton
26. Vilanova R, Visioli A (2012) PID control in the third millennium – lessons learned and new approaches. Springer, London
27. Visioli A (2006) Practical PID control. Springer, London

Chapter 4
Open-Loop Discrete-Time Systems

4.1 Introduction

Open-loop discrete-time systems using z-transforms are presented in this chapter. First, the relationship between Laplace transform of sampled signals and z-transform is discussed. Next, procedures of finding transfer functions of open-loop discrete-time systems using data reconstruction devices (holds) of zero-, first-, and fractional-order are presented. Examples are presented along with Matlab codes and Simulink models to illustrate the procedures. End-of-chapter exercises are provided to enhance understanding of the topics covered and consolidate problem-solving skills.

4.2 Relationship Between Sampled Signal s-Transform $F^*(s)$ and z-Transform $F(z)$

The z-transform of a discrete-time sampled function $\{f(kT)\}$ was defined in Eq. (2.5) as

$$\begin{aligned} F(z) &= Z[\{f(kT)\}] \\ &= \sum_{k=0}^{\infty} f(kT) z^{-k} \end{aligned} \tag{4.1}$$

The Laplace transform of discrete-time sampled function $f^*(kT)$ was defined in Eq. (3.5) as

$$\begin{aligned} F^*(s) &= L[f^*(kT)] \\ &= \sum_{k=0}^{\infty} f(kT) e^{-ksT} \end{aligned} \tag{4.2}$$

B. Samanta, *Introduction to Digital Control*,
https://doi.org/10.1007/978-3-031-66830-2_4

Comparing Eqs. (4.1) and (4.2), the z-transform $F(z)$ and the Laplace transform $F^*(s)$ of a discrete-time sampled function can be related as

$$F(z) = F^*(s)|_{e^{sT}=z} \tag{4.3}$$

Example 4.1 Find z-transform $F(z)$ and Laplace transform $F^*(s)$ of discrete-time sampled function $\{f\{kT)\}$, where $f(t) = e^{-2t} - e^{-3t}$.

Solution From z-transform table in Appendix Table A.1, the z-transform of the sampled function can be written as

$$\begin{aligned} F(z) &= Z\left[\left\{e^{-2kT} - e^{-3kT}\right\}\right] \\ &= \frac{z}{z - e^{-2T}} - \frac{z}{z - e^{-3T}} = \frac{z(e^{-2T} - e^{-3T})}{(z - e^{-2T})(z - e^{-3T})} \end{aligned}$$

The Laplace transform $F^*(s)$ can be written, using Eq. (4.3), as

$$F^*(s) = F(z)|_{z=e^{sT}} = \frac{e^{sT}(e^{-2T} - e^{-3T})}{(e^{sT} - e^{-2T})(e^{sT} - e^{-3T})}$$

4.3 Open-Loop Discrete-Time System Transfer Function

The schematic of an open-loop discrete-time system with an input signal $f(t)$ to the sampler, a unit digital controller with $D(z) = 1$, a hold device with a transfer function $G_h(s)$, a plant with a transfer function $G_p(s)$, and an output signal $c(t)$ is shown in Fig. 4.1. The signals and the corresponding Laplace transforms are also shown. In an actual digital control system, the ideal sampler and the data-hold device are represented as analog-to-digital converter (ADC) and digital-to-analog converter (DAC) and the digital controller is represented as a non-unity z-domain transfer function $D(z)$. Here for simplicity, $D(z)$ assumed to be unity so that the sampled data $f^*(k) = f^*(kT)$ are same as $m^*(k) = m^*(kT)$. The data-hold device converts the discrete-time data $m^*(k)$ to $\overline{m}(k)$ in continuous-time (analog) domain. The data-hold transfer function depends on the type or order of hold, i.e., zero-order hold

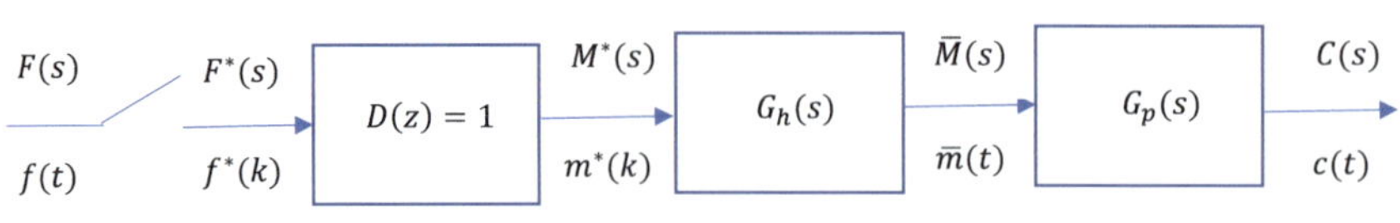

Fig. 4.1 Block diagram representation of an open-loop discrete-time system

(ZOH), first-order hold (FOH), or fractional-order hold (FrOH). The transfer functions $G_h(s)$ and $G_p(s)$ can be combined as

$$G(s) = G_h(s)\, G_p(s) \tag{4.4}$$

In the following subsections, the open-loop discrete-time systems with ZOH, FOH, and FrOH are presented.

4.3.1 Zero-Order Hold (ZOH)

In Fig. 4.1, the transfer function $G(s)$ combining a ZOH $G_{h0}(s)$ and plant $G_p(s)$ is given as

$$G(s) = \left(\frac{1 - e^{-sT}}{s}\right) G_p(s) \tag{4.5}$$

The output $C(s)$ can be expressed as

$$C(s) = G(s) F^*(s) \tag{4.6}$$

The Laplace transform of the discrete-time output signal can be expressed as

$$C^*(s) = G^*(s) F^*(s) \tag{4.7}$$

Using Eq. (4.3), the z-transform of the output signal $C(z)$ can be expressed as

$$C(z) = G(z) F(z) \tag{4.8}$$

where

$$\begin{aligned} G(z) &= Z\left[\left(\frac{1 - e^{-sT}}{s}\right) G_p(s)\right]_{e^{sT} = z} \\ &= (1 - z^{-1}) Z\left[\frac{G_p(s)}{s}\right] \end{aligned} \tag{4.9}$$

Example 4.2 Find z-transform of the output signal of an open-loop discrete-time system, similar to Fig. 4.1, with a ZOH and a plant transfer function $G_p(s) = \frac{1}{s+1}$ and a unit step function $F(s) = \frac{1}{s}$ for sampling time $T = 0.1$ s.

Solution The z-transform of the plant with ZOH can be written as, using Eq. (4.9) as

$$\begin{aligned}G(z) &= \left(1-z^{-1}\right)Z\left[\frac{G_{\mathrm{p}}(s)}{s}\right]\\ &= \left(1-z^{-1}\right)Z\left[\frac{1}{s(s+1)}\right]\\ &= \left(1-z^{-1}\right)\frac{z(1-e^{-T})}{(z-1)(z-e^{-T})} \quad \text{(using } z\text{-transform table)}\\ &= \frac{1-e^{-T}}{z-e^{-T}}\end{aligned}$$

For $T = 0.1$ s, $G(z) = \frac{1-e^{-T}}{z-e^{-T}} = \frac{0.0952}{z-0.9048}$
The z-transform of $F(s) = \frac{1}{s}$ can be written, using z-transform table, as

$$F(z) = \frac{z}{z-1}.$$

The z-transform of the output $C(z)$ can be written as

$$\begin{aligned}C(z) &= G(z)F(z)\\ &= \frac{z(1-e^{-T})}{(z-1)(z-e^{-T})}\end{aligned}$$

$$C(z) = \frac{0.0952\,z}{(z-1)(z-0.9048)} = \frac{0.0952\,z}{z^2-1.9048z+0.9048}$$

Example 4.3 Solve Example 4.2 using Matlab. Obtain open-loop response of the system. Obtain $c(kT)$ using inverse z-transform of $C(z)$.

Solution Matlab code, along with comments and results, is presented. Open-loop system response is shown in Fig. 4.2.

```
% Ch4 Example 4.3 Find G(z) and C(z) for Gp(s)=1/(s+1) with ZOH T=0.1s,
% F(s)=1/s; F(z)=z/(z-1)
% Gp(s)=1/(s+1) T=0.1
ngs = 1; dgs = [1 1]; T = 0.1;
% G(z) = Z(Gp(s) with ZOH)
[ngz,dgz]=c2dm(ngs,dgs,T,'zoh');
Gz=tf(ngz,dgz,T)
%F(z) =z/(z-1)  f(k)=1
nfz = [1 0]; dfz = [1 -1];
% G(z)*F(z)
Fz =tf(nfz, dfz, T)
%step response of G(z)
dstep(Gz,1); % step response

% inverse transform of C(z)= G(z)*F(z)
Cz = Gz*Fz
syms z k
Cz = vpa (0.0952*z/((z-1)*(z-0.9048)),4) % for using iztrans C(z) is given in z
ck = vpa(iztrans(Cz,k),4) %c(kT)

% using PFE
ngezz= 0.0952
dgfz=poly([1 0.9048])
[r,p,k1]=residue(ngezz,dgfz)
```

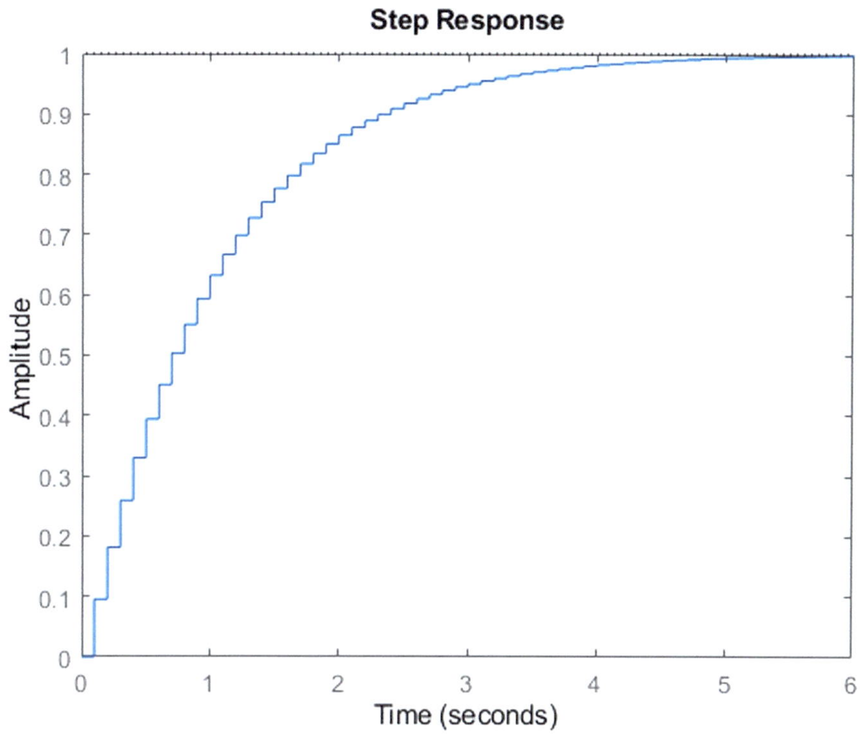

Fig. 4.2 Step response of open-loop discrete-time system with ZOH

```
Results:
 1
 Gp(s)= -----,
 s + 1

 0.0952
G(z)= ----------
 z - 0.9048

 0.0952 z
C(z) = ----------------------
 z^2 - 1.9052 z + 0.9048

c(kT)= 1-0.9048^k
```

4.3.2 First-Order Hold (FOH)

In Fig. 4.1, the transfer function $G(s)$ combining a FOH $G_{h1}(s)$ and plant $G_p(s)$ is given as

$$G(s) = \frac{(1 + Ts)(1 - e^{-sT})^2}{Ts^2} G_p(s) \tag{4.10}$$

The output $C(s)$ can be expressed as

$$C(s) = G(s)F^*(s) \tag{4.11}$$

The Laplace transform of the discrete-time output signal cane be expressed as

$$C^*(s) = G^*(s)F^*(s) \tag{4.12}$$

Using Eq. (4.3), the z-transform of the output signal $C(z)$ can be expressed as

$$C(z) = G(z)F(z) \tag{4.13}$$

where

$$\begin{aligned} G(z) &= Z\left[\frac{(1 + Ts)(1 - e^{-sT})^2}{Ts^2} G_p(s)\right]_{e^{sT}=z} \\ &= \left(1 - z^{-1}\right)^2 Z\left[\frac{(1 + Ts)G_p(s)}{Ts^2}\right] \end{aligned} \tag{4.14}$$

Example 4.4 Find the z-transform of the output signal of an open-loop discrete-time system, similar to Fig. 4.1, with a FOH and a plant transfer function $G_p(s) = \frac{1}{s+1}$ and a unit step function $F(s) = \frac{1}{s}$ in terms of sampling time T.

Solution The z-transform of the plant with FOH can be written as, using Eq. (4.14) as

$$\begin{aligned} G(z) &= \left(1 - z^{-1}\right)^2 Z\left[\frac{(1 + Ts)G_p(s)}{Ts^2}\right] \\ &= \left(1 - z^{-1}\right)^2 Z\left[\frac{1 + Ts}{Ts^2(s + 1)}\right] \end{aligned}$$

$$G(z) = \left(\frac{z - 1}{z}\right)^2 \left[Z\left[\frac{1}{Ts^2(s + 1)}\right] + Z\left[\frac{1}{s(s + 1)}\right]\right]$$

$$G(z)=\left(\frac{z-1}{z}\right)^2\left[\left[\frac{1}{T}\frac{z[(T-1+e^{-T})z+(1-e^{-T}-Te^{-T})\}}{(z-1)^2(z-e^{-T})}\right]+\left[\frac{z(1-e^{-T})}{(z-1)(z-e^{-T})}\right]\right]$$

The z-transform of $F(s)=\frac{1}{s}$ can be written, using z-transform table, as

$$F(z)=\frac{z}{z-1}.$$

The z-transform of the output $C(z)$ can be written as

$$\begin{aligned}C(z)&=G(z)F(z)\\&=\frac{z}{(z-1)}\left(1-z^{-1}\right)^2 Z\left[\frac{1+Ts}{Ts^2(s+1)}\right]\\&=\left(\frac{z-1}{z}\right)Z\left[\frac{1+Ts}{Ts^2(s+1)}\right]\\&=\left(\frac{z-1}{z}\right)\left[Z\left[\frac{1}{Ts^2(s+1)}\right]+Z\left[\frac{1}{s(s+1)}\right]\right]\end{aligned}$$

$$\begin{aligned}C(z)=&\left(\frac{z-1}{z}\right)\left[\left[\frac{1}{T}\frac{z[(T-1+e^{-T})z+(1-e^{-T}-Te^{-T})\}}{(z-1)^2(z-e^{-T})}\right]\right.\\&\left.+\left[\frac{z(1-e^{-T})}{(z-1)(z-e^{-T})}\right]\right]\ \text{(using } z\text{-transform table)}\\=&\left[\left[\frac{1}{T}\frac{[(T-1+e^{-T})z+(1-e^{-T}-Te^{-T})\}}{(z-1)(z-e^{-T})}\right]+\left[\frac{(1-e^{-T})}{(z-e^{-T})}\right]\right]\end{aligned}$$

Matlab code and results are presented. Open-loop system response is shown in Fig. 4.3.

```
Matlab code for Gp(s) with FOH
% Ch4 Example 4.3 Find G(z) and C(z) for Gp(s)=1/(s+1) with FOH T=0.1s,
% e(kT)=1
clear all; close all; clc
% Gp(s)
ngs = 1; dgs = [1 1]; % Gp(s)=1/(s+1)
Gp = tf(ngs, dgs) % Gp(s)
printsys(ngs,dgs,'s')

% G(z)
T=0.1; % sampling time T=0.1 s
[ngz,dgz]=c2dm(ngs,dgs,T,'foh') % G(z)=Z[Gh1(s)Gp(s)]
Gz=tf(ngz,dgz,T)

% F(z)= z/(z-1)
nfz=[1 0]; dfz=[1 -1]; %F(z)=z/(z-1) for e(kT)=1 unit step
Fz=tf(nfz,dfz,T)
Cz=Gz*Fz % output C(z)=G(z)F(z)

% Step response
dstep(Gz,1)
```

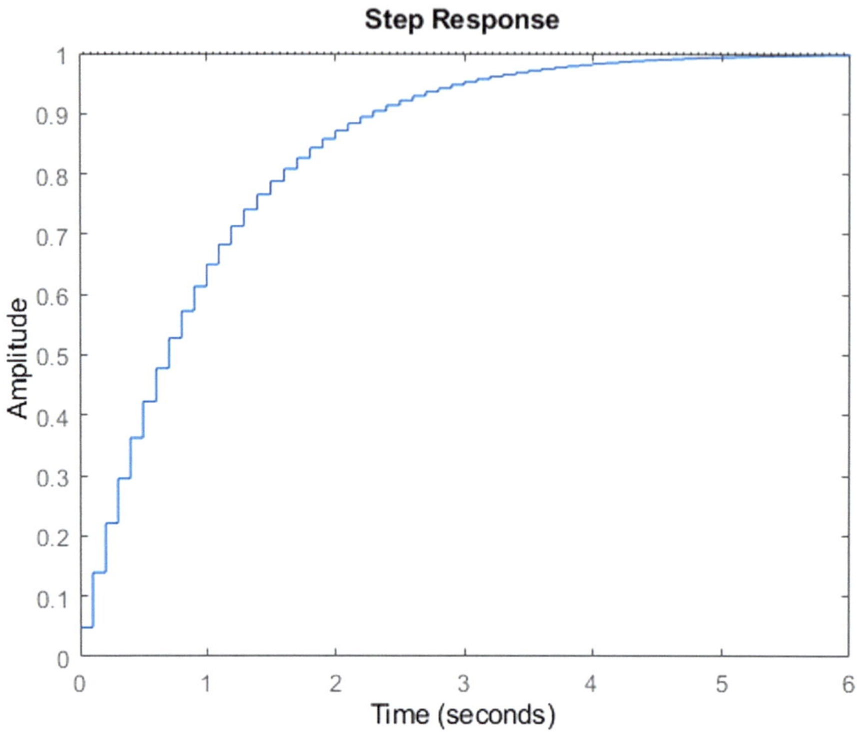

Fig. 4.3 Step response of open-loop discrete-time system with FOH

4.3.3 Fractional-Order Hold (FrOH)

In Fig. 4.1, the transfer function $G(s)$ combining a FrOH $G_{\mathrm{hr}}(s)$ and plant $G_{\mathrm{p}}(s)$ is given as

$$G(s) = \frac{(1 - re^{-sT})(1 - e^{-sT})}{s} + \frac{r(1 - e^{-sT})^2}{Ts^2} G_{\mathrm{p}}(s) \tag{4.15}$$

The output $C(s)$ can be expressed as

$$C(s) = G(s)F^*(s) \tag{4.16}$$

The Laplace transform of the discrete-time output signal cane be expressed as

$$C^*(s) = G^*(s)F^*(s) \tag{4.17}$$

Using Eq. (4.3), the z-transform of the output signal $C(z)$ can be expressed as

$$C(z) = G(z)F(z) \tag{4.18}$$

where

$$\begin{aligned} G(z) &= Z\left[\left(\frac{(1-re^{-sT})(1-e^{-sT})}{s} + \frac{r(1-e^{-sT})^2}{Ts^2}\right)G_p(s)\right]_{e^{sT}=z} \\ &= \left[(1-rz^{-1})(1-z^{-1})Z\left[\frac{G_p(s)}{s}\right] + r(1-z^{-1})^2 Z\left[\frac{G_p(s)}{Ts^2}\right]\right] \end{aligned} \tag{4.19}$$

Example 4.5 Find the z-transform of the output signal of an open-loop discrete-time system, similar to Fig. 4.1, with a FrOH and a plant transfer function $G_p(s) = \frac{1}{s+1}$ and a unit step function $F(s) = \frac{1}{s}$ in terms of sampling time T.

Solution The z-transform of the plant with ZOH can be written as, using Eq. (4.19)

$$\begin{aligned} G(z) &= \left[(1-rz^{-1})(1-z^{-1})\,Z\left[\frac{G_p(s)}{s}\right] + r(1-z^{-1})^2\,Z\left[\frac{G_p(s)}{Ts^2}\right]\right] \\ &= \left[(1-rz^{-1})(1-z^{-1})\,Z\left[\frac{1}{s(s+1)}\right] + r(1-z^{-1})^2\,Z\left[\frac{1}{Ts^2(s+1)}\right]\right] \end{aligned}$$

The z-transform of $F(s) = \frac{1}{s}$ cane be written, using z-transform table, as

$$F(z) = \frac{z}{z-1}.$$

The z-transform of the output $C(z)$ can be written as

$$\begin{aligned} C(z) &= G(z)F(z) \\ &= \frac{z}{(z-1)}\left[(1-rz^{-1})(1-z^{-1})Z\left[\frac{1}{s(s+1)}\right] + r(1-z^{-1})^2 Z\left[\frac{1}{Ts^2(s+1)}\right]\right] \\ &= \frac{z}{(z-1)}\left[(1-rz^{-1})(1-z^{-1})\left[\frac{z(1-e^{-T})}{(z-1)(z-e^{-T})}\right]\right. \\ &\quad \left. + r(1-z^{-1})^2\left[\frac{1}{T}\frac{z[(T-1+e^{-T})z+(1-e^{-T}-Te^{-T})\}}{(z-1)^2(z-e^{-T})}\right]\right] \\ &= \left[(1-rz^{-1})\left[\frac{z(1-e^{-T})}{(z-1)(z-e^{-T})}\right] + r\left[\frac{1}{T}\frac{[(T-1+e^{-T})z+(1-e^{-T}-Te^{-T})\}}{(z-1)(z-e^{-T})}\right]\right] \end{aligned}$$

4.4 Open-Loop Discrete-Time Systems of Different Configurations

Different configurations of an open-loop discrete-time system are shown in Fig. 4.4a–c

For the system in Fig. 4.4a, there is a sampler before the transfer function blocks $G_1(s)$ and $G_2(s)$. The relation can be written for the output of each block as

$$C(s) = G_2(s)M^*(s) \tag{4.20}$$

$$M(s) = G_1(s)F^*(s) \tag{4.21}$$

The starred transform can be expressed as

$$C^*(s) = G_2{}^*(s)M^*(s) \tag{4.22}$$

$$M^*(s) = G_1{}^*(s)F^*(s) \tag{4.23}$$

$$C^*(s) = G_2{}^*(s)G_1{}^*(s)F^*(s) \tag{4.24}$$

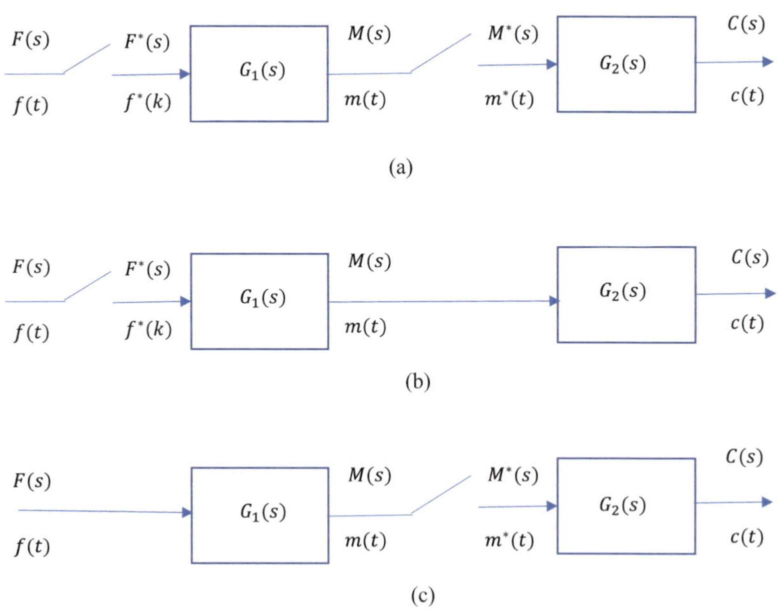

Fig. 4.4 Different configurations of an open-loop discrete-time system

z-transform can be expressed as

$$C(z) = G_1(z)G_2(z)F(z) \tag{4.25}$$

Similarly for the system in Fig. 4.4b, the output function can be expressed as

$$C(s) = G_1(s)G_2(s)F^*(s) \tag{4.26}$$

The starred transform can be expressed as

$$C^*(s) = \overline{G_1(s)G_2(s)}^*F^*(s) \tag{4.27}$$

z-transform can be expressed as

$$C(z) = \overline{G_1G_2}(z)F(z) \tag{4.28}$$

where

$$\overline{G_1G_2}(z) = Z\{G_1(s)G_2(s)] \tag{4.29}$$

Similarly for the system in Fig. 4.4c, the output function can be expressed as

$$C(s) = G_2(s)\overline{G_1(s)F^*}(s) \tag{4.30}$$

Starred transform can be expressed as

$$C^*(s) = G_2{}^*(s)\overline{G_1(s)F^*}(s)) \tag{4.31}$$

z-transform can be expressed as

$$C(z) = G_2(z)\overline{G_1F}(z) \tag{4.32}$$

Example 4.6 Find the z-transform of output function, $C(z)$ for different configurations of Fig. 4.4a–c, with $G_1(s) = \frac{1}{s+1}, G_2(s) = \frac{1}{s+2}, F(s) = \frac{1}{s}$ using ZOH and sampling time $T = 0.1$ s

Solution Matlab code

```
%Example Section 4.4
% G1(s)=1/(s+1); G2(s)=1/(s+2); F(s)=1/s
% Case Fig. 4.4(a)
ngs1=1; dgs1=[1 1]; % G1(s)
ngs2=1; dgs2=[1 2]; % G2(s)
nfs=1; dfs=[1 0];    % F(s)

T=0.1 % sampling time
[ngz1, dgz1] = c2dm(ngs1,dgs1,T,'zoh'); % G1(z)
[ngz2, dgz2] = c2dm(ngs2,dgs2,T,'zoh') ;% G2(z)
G1z=tf(ngz1, dgz1, T) % G1(z)
G2z=tf(ngz2, dgz2, T) % G2(z)
nfz=[1 0]; dfz=[1 -1]; %F(z)
Fz=tf(nfz, dfz, T) % F(z)
Cza=G1z*G2z*Fz

% Case Fig. 4.4b
% C(z)=[G1G2(z)]F(z)
[ngs12,dgs12]=series(ngs1,dgs1,ngs2,dgs2) % G1(s)*G2(s);
[ngz12,dgz12]=c2dm(ngs12,dgs12,T,'zoh'); % Z(G1(s)G2(s));
G12z=tf(ngz12,dgz12,T)
Czb=G12z*Fz

% Case Fig 4.4c
%C(z)=G2(z)*z[G1(s)F(s)]
[ng1fs,dg1fs]=series(ngs1,dgs1,nfs,dfs) % G1(s)*F(s);
[ng1fz,dg1fz]=c2dm(ng1fs,dg1fs,T,'zoh'); % Z(G1(s)F(s));
G1Fz=tf(ng1fz,dg1fz,T)
Czc=G1Fz*G2z
```

```
Results:
Cza =

 0.008625 z
 -----------------------------------
 z^3 - 2.724 z^2 + 2.464 z - 0.7408

Czb =

 0.004528 z^2 + 0.004097 z
 -----------------------------------
 z^3 - 2.724 z^2 + 2.464 z - 0.7408

Czc =

 0.0004384 z+ 0.0004241
 ----------------------------------------
 z^3 - 2.724 z^2 +2.464 z - 0.7408
```

4.5 Open-Loop Discrete-Time Systems with Digital Filters

The block diagram schematic of an open-loop discrete-time system with a digital filter and a general order hold is shown in Fig. 4.5.

The output function transform $C(s)$ can be expressed as

$$C(s) = G_{\mathrm{h}}(s)G_{\mathrm{p}}(s)M^*(s) \tag{4.33}$$

The controller output function $M^*(s)$ can be expressed in terms of input function $F^*(s)$ as

$$M^*(s) = D^*(s)F^*(s) \tag{4.34}$$

Combining Eqs. (4.33) and (4.34), the output function can be expressed as

$$C(s) = G_{\mathrm{h}}(s)G_{\mathrm{p}}(s)D^*(s)F^*(s) \tag{4.35}$$

Taking starred transform Eq. (4.35) can be written as

$$C^*(s) = \overline{G_{\mathrm{h}}(s)G_{\mathrm{p}}(s)}^*D^*(s)F^*(s) \tag{4.36}$$

z-transform can be expressed as

$$C(z) = \overline{G_{\mathrm{h}}G_{\mathrm{p}}}(z)D(z)F(z) \tag{4.37}$$

where

$$\overline{G_{\mathrm{h}}G_{\mathrm{p}}}(z) = Z[G_{\mathrm{h}}(s)G_{\mathrm{p}}(s)] \tag{4.38}$$

For ZOH Eq. (4.38) can be written as

$$\begin{aligned}\overline{G_{\mathrm{h}}G_{\mathrm{p}}}(z) &= Z\left[\frac{1-e^{-sT}}{s}G_{\mathrm{p}}(s)\right]_{e^{sT}=z}\\ &= (1-z^{-1})Z\left[\frac{G_{\mathrm{p}}(s)}{s}\right]\end{aligned} \tag{4.39}$$

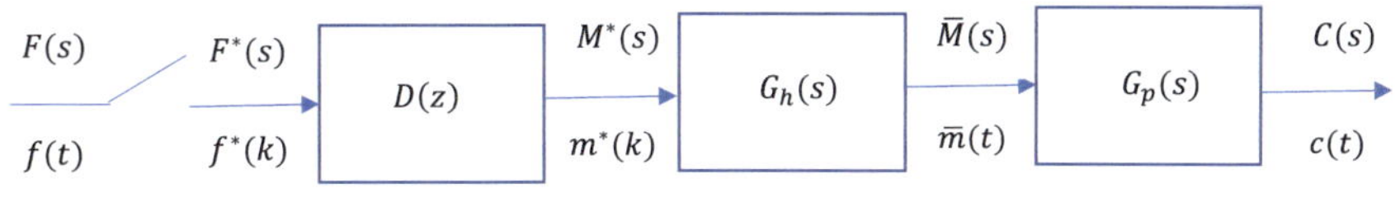

Fig. 4.5 An open-loop discrete-time system with a digital filter

Example 4.7 For an open-loop discrete-time system with a digital filter, similar to Fig. 4.5, with a ZOH and a plant transfer function $G_{\mathrm{p}}(s) = \frac{1}{s+1}$ and a unit step function $F(s) = \frac{1}{s}$, the digital filter is used to represent difference equation $m(kT) = 3f(kT) - 2f(kT - T)$. Find the transfer function $D(z)$ for the digital filter. Use sampling time $T = 0.1$ s. Find the z-transform of the output $C(z)$. Use PFE to find the inverse z-transform of the (z), $c(kT)$. Using initial and final value theorems, find the initial and the final values of the output, $c(0)$ and $c(\infty)$. Verify the values from the discrete-time function, $c(kT)$.

Solution The difference equation for the digital filter is given as

$$m(kT) = 3f(kT) - 2f(kT - T)$$

Using z-transform and applying real translation property, the transfer function of the digital filter can be expressed as

$$M(z) = 3F(z) - 2z^{-1}F(z)$$
$$D(z) = \frac{M(z)}{F(z)} = \frac{3 - 2z^{-1}}{1} = \frac{3z - 2}{z}$$

The transfer function of plant along with the ZOH can be obtained in the z-domain, similar to Example 4.2,

$$G(z) = (1 - z^{-1})Z\left[\frac{G_{\mathrm{p}}(s)}{s}\right]$$
$$= \frac{1 - e^{-T}}{z - e^{-T}}$$

For $T = 0.1$ s, $G(z) = \frac{1 - e^{-T}}{z - e^{-T}} = \frac{0.0952}{z - 0.9048}$. Similarly, the z-transform of the unit step input is obtained as

$$F(z) = \frac{z}{z - 1}.$$

The z-transform of the output $C(z)$ can be written as

$$C(z) = D(z)G(z)F(z)$$
$$= \left(\frac{3z - 2}{z}\right)\left(\frac{1 - e^{-T}}{z - e^{-T}}\right)\left(\frac{z}{z - 1}\right)$$

$$C(z) = \frac{0.2855\, z^2 - 0.1903z}{z(z - 0.9048)(z - 1)} = \frac{0.2855\, z^2 - 0.1903z}{z^3 - 1.9048z^2 + 0.9048z}.$$

Partial fraction expansion of $\frac{C(z)}{z}$ is obtained as follows:

$$\frac{C(z)}{z} = \frac{A_1}{z} + \frac{A_2}{z - 0.9048} + \frac{A_3}{z - 1}$$

$$A_1 = z\frac{0.2855\,z - 0.1903}{z(z - 0.9048)(z - 1)}\bigg|_{z=0} = -0.2103$$

$$A_2 = (z - 0.9048)\frac{0.2855\,z - 0.1903}{z(z - 0.9048)(z - 1)}\bigg|_{z=0.9048} = -0.7897$$

$$A_3 = (z - 1)\frac{0.2855\,z - 0.1903}{z(z - 0.9048)(z - 1)}\bigg|_{z=1} = 1$$

$$C(z) = A_1 + \frac{A_2 z}{z - 0.9048} + \frac{A_3 z}{z - 1}$$

Using z-transform table of Appendix, inverse z-transform can be expressed as

$$c(kT) = -0.2103\delta(kT) - 0.7897(0.9048)^k + 1, k \geq 0$$

It should be mentioned that the first term of $c(kT)$ is defined as

$$\delta(kT) = \begin{cases} 1, & \text{for } k = 0 \\ 0, & \text{otherwise} \end{cases}$$

Applying initial value theorem, the initial value $c(0)$ can be obtained as

$$c(0) = \lim_{z\to\infty} C(z) = \lim_{z\to\infty}\left(\frac{0.2855z^{-1} - 0.1903z^{-2}}{1 - 1.9048z^{-1} + 0.9048z^{-2}}\right) = 0$$

Applying final value theorem, the final value can be obtained as

$$c(\infty) = \lim_{z\to1}(z - 1)C(z) = \lim_{z\to1}\left((z - 1)\frac{0.2855z^2 - 0.1903z}{z(z - 0.9048)(z - 1)}\right) = 1$$

From the expression of the output the initial value is confirmed at $k = 0$ as

$$c(0) = -0.2103\delta(0) - 0.7897(0.9048)^0 + 1 = 0$$

Similarly, the final value of the output is confirmed for $k \to \infty$ as

$$c(\infty) = -0.2103(0) - 0.7897(0.9048)^\infty + 1 = 1$$

Example 4.8 Use Matlab to reproduce the results of Example 4.6 and find step response of the open-loop discrete-time system with the given digital filter.

Solution Matlab code, along with comments and results, is presented. Open-loop system response is shown in Fig. 4.6.

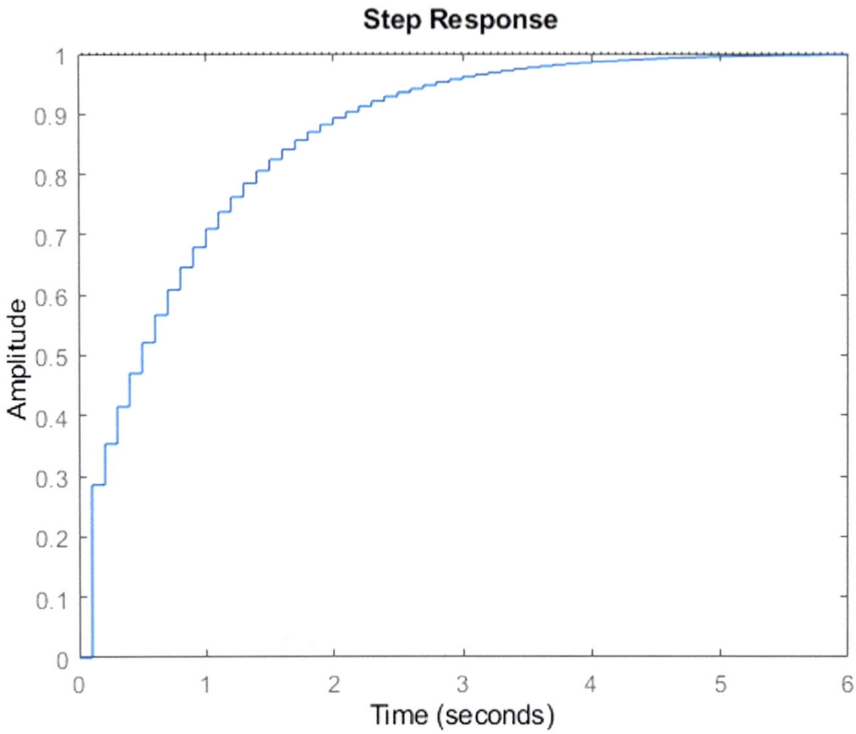

Fig. 4.6 Step response of open-loop discrete-time system with a digital filter

```
%ch4_ex4_8 open loop discrete time system with D(z)
% Given Gp(s)=1/(s+1)   T=0.1
% m(k)=3e(k)-2e(k-1) D(z)=M(z)/E(z)=(3-2/z)= (3z-2)/z
% step input e(k)=1  F(z)=z/(z-1)
% Find C(z) = G(z)D(z)F(z)
% Gp(s)=1/(s+1)
ngs = 1; dgs = [1 1]; % Gp(s)
T = 0.1; % sampling time
[ngz,dgz]=c2dm(ngs,dgs,T,'zoh'); %G(z)
Gz = tf(ngz,dgz,T) %G(z)
%Fz = z/(z-1)
nfz=[1 0]; dfz=[1 -1];
Fz =tf(nfz,dfz,T) %F(z)
%D(z) =(3*z-2)/z
ndz=[3 -2];ddz=[1 0];
Dz=tf(ndz,ddz,T) % D(z)
% C(z)=D(z)*G(z)*F(z)
Cz=Dz*Gz*Fz
%step response
DGz=Dz*Gz % D(z)G(z) : open loop system transfer function =C(z)?F(z)
dstep(DGz,1) % step response of the open-loop system
%PFE
nfpz=1
Epz=tf(nfpz,dfz,T);
Cpz=Dz*Gz*Epz % Cp(z)=C(z)/z
[numCpz,denCpz]=tfdata(Cpz,'v') % to convert to num,den form for PFE
[r,p,k]=residue(numCpz,denCpz) % PFE of Cp(z)
Results
```

```
DGz =
 0.2855 z - 0.1903
 -----------------
 z^2 - 0.9048 z

Cz =

 0.2855 z^2 - 0.1903 z
 -------------------------
 z^3 - 1.905 z^2 + 0.9048 z

r = 1.0000; -0.7897; -0.2103
p = 1.0000; 0.9048; 0
```

4.6 Modified z-Transform

Modified z-transform can be obtained considering a discrete-time sampled function $f(kT)$ with a time delay ΔT, $0 < \Delta < 1$, i.e., $f(kT - \Delta T)u(kT - \Delta T)$. The ordinary z-transform of the time-delayed function can be written as

$$Z[\{f(kT - \Delta T)u(kT - \Delta T)\}] = Z\left[F(s)e^{-\Delta Ts}\right] = \sum_{k=1}^{\infty} f(kT - \Delta T)z^{-k} \tag{4.40}$$

It should be recognized that sampling occurs as usual at $t = 0, T, 2T, \ldots$ but the samples are with values of $f(kT - \Delta T)$ at sampling instants $k = 1, 2, \ldots$

The z-transform of the delayed sampled time function can be expressed as

$$F(z, \Delta) = Z[\{f(kT - \Delta T)u(kT - \Delta T)\}] = Z\left[F(s)e^{-\Delta Ts}\right]\Big|_{z = e^{sT}} \tag{4.41}$$

The modified z-transform can be obtained from the z-transform of the delayed time function using $\Delta = 1 - m$ in Eq. (4.41) as

$$F(z, m) = F(z, \Delta)|_{\Delta = 1 - m} = Z\left[F(s)e^{-\Delta Ts}\right]\Big|_{\Delta = 1 - m} \tag{4.42}$$

Using Eqs. (4.40) and (4.42), the modified z-transform can be expressed as

$$\begin{aligned} F(z, m) &= \left[f(T - \Delta T)z^{-1} + f(2T - \Delta T)z^{-2} + f(3T - \Delta T)z^{-3} + \cdots\right]\Big|_{\Delta = 1 - m} \\ &= f(mT)z^{-1} + f([1 + m]T)z^{-2} + f([2 + m]T)z^{-3} + \cdots \end{aligned} \tag{4.43}$$

Two properties of modified z-transform are written as

$$F(z,1) = F(z,m)|_{m=1} = F(z) - f(0) \tag{4.44}$$

$$F(z,0) = F(z,m)|_{m=0} = z^{-1}F(z) \tag{4.45}$$

For $m = 1$, there is no delay but the $f(0)$ does not appear in Eq. (4.40). For $m = 0$, the sampled function is delayed by one sample period. Real translation theorem is applicable to modified z-transform, for any positive integer n as

$$Z_m\left[e^{-nTs}F(s)\right] = z^{-n}Z_m[(F(s)] = z^{-n}F(z,m) \tag{4.46}$$

Example 4.9 Find the modified z-transform of a time-delayed sampled time function $f(t) = e^{-bt}u(t)$ with time delay of $0.6T$, the sampling time $T = 0.1$ s, $\Delta = 0.6$, $b > 0$.

Solution Using Eq. (4.40), the z-transform of the time-delayed sampled function $f(kT - \Delta T)u(kT - \Delta T)$ can be expressed as

$$\begin{aligned} F(z,\Delta) &= Z[\{f(kT-\Delta T)u(kT-\Delta T)\}] = e^{-0.4bT}z^{-1} + e^{-1.4bT}z^{-2} + e^{-2.4bT}z^{-3} + \cdots \\ &= e^{-0.4bT}z^{-1}\left[1 + e^{-bT}z^{-1} + e^{-2bT}z^{-2} + \cdots\right] \\ &= e^{-0.4bT}z^{-1}\left(1 - e^{-bT}z^{-1}\right)^{-1} = e^{-0.4bT}z^{-1}\frac{1}{1-e^{-bT}z^{-1}} = \frac{e^{-0.4bT}}{z-e^{-bT}} \end{aligned}$$

The modified z-transform can be expressed, using $\Delta = 1 - m$ or $m = 1 - \Delta = 0.4$, as

$$\begin{aligned} F(z,m) &= e^{-0.4bT}z^{-1} + e^{-1.4bT}z^{-2} + e^{-2.4bT}z^{-3} + \cdots \\ &= e^{-mbT}z^{-1}\left(1 - e^{-bT}z^{-1}\right)^{-1} = \frac{e^{-mbT}}{z-e^{-bT}} \end{aligned}$$

4.7 Discrete-Time Systems with Time Delays

The block diagram of a discrete-time system with a delay t_d is shown in Fig. 4.7.

The output of the system, in s- and z-domains, can be expressed as

$$C(s) = G(s)e^{-t_d s}F^*(s) \tag{4.47}$$

$$C(z) = Z[G(s)e^{-t_d s}]F(z) \tag{4.48}$$

Assuming, in general,

$$t_d = n, T + \Delta T, 0 < \Delta < 1, n \geq 0, \tag{4.49}$$

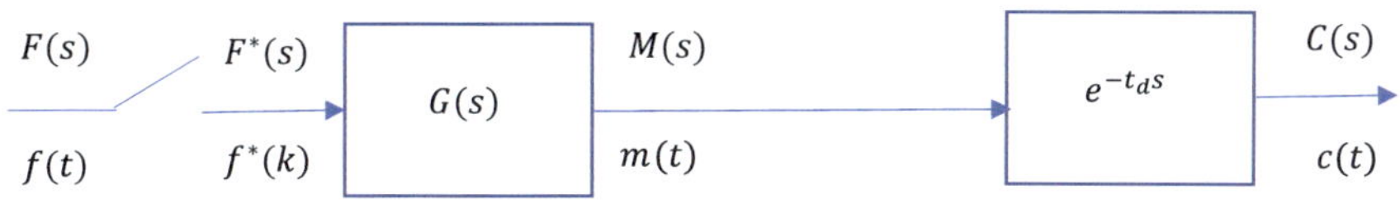

Fig. 4.7 An open-loop discrete-time system with a time delay

n is a positive integer and applying Eqs. (4.46), (4.48) can be expressed as

$$C(z) = z^{-n} Z\left[G(s)e^{-\Delta Ts}\right]F(z) = z^{-n}G(z, m)F(z) \tag{4.50}$$

Example 4.10 A discrete-time system, similar to Fig. 4.6, is with a time delay $t_d = 0.6T$, a ZOH, a plant transfer function $G_p(s) = \frac{1}{s+1}$, and a unit step function $F(s) = \frac{1}{s}$. Use sampling time $T = 0.1$ s. Find the z-transform of the output $C(z)$

Solution The open-loop discrete-time system transfer function, combining ZOH and the plant, is

$$G(s) = \left(\frac{1 - e^{-sT}}{s}\right)G_p(s) = \frac{1 - e^{-sT}}{s(s + 1)}$$

Modified z-transform can be expressed, using z-transform table from Appendix A, as

$$\begin{aligned} G(z, m) &= Z_m\left[\frac{1 - e^{-sT}}{s(s + 1)}\right] = \left(1 - z^{-1}\right)Z_m\left[\frac{1}{s(s + 1)}\right] \\ &= \left[\frac{z - 1}{z}\right]\left[\frac{z(1 - e^{-mT}) + e^{-mT} - e^{-T}}{(z - 1)(z - e^{-T})}\right] \end{aligned}$$

Applying $mT = (1 - \Delta)T = 0.4T$,

$$G(z, m) - \frac{z - 1}{z}\left[\frac{z(1 - e^{-0.4T}) + e^{-0.4T} - e^{-T}}{(z - 1)(z - e^{-T})}\right]$$

The z-transform of the output,

$$\begin{aligned} C(z) &= G(z, m)F(z) \\ &= \left[\frac{z - 1}{z}\right]\left[\frac{z(1 - e^{-0.4T}) + e^{-0.4T} - e^{-T}}{(z - 1)(z - e^{-T})}\right]\left[\frac{z}{z - 1}\right] \\ &= \frac{z(1 - e^{-0.4T}) + e^{-0.4T} - e^{-T}}{(z - 1)(z - e^{-T})} \\ &= \frac{1}{z - 1} - e^{-0.4T}\frac{1}{z - e^{-T}} = z^{-1}\left[\frac{z}{z - 1} - e^{-0.4T}\frac{z}{z - e^{-T}}\right] \end{aligned}$$

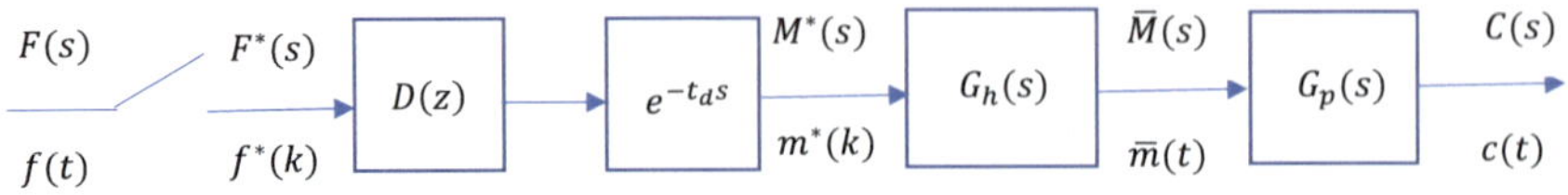

Fig. 4.8 An open-loop discrete-time system with a digital filter and a time delay

Inverse z-transform can be written as

$$c(kT) = 1 - e^{-(k-0.6)T} \; k \geq 1$$

Without delay the output is, as expected,

$$c(kT) = 1 - e^{-kT} \; k \geq 0.$$

Example 4.11 The open-loop discrete-time system with a digital filter of Example 4.6 is reconsidered with a time delay of $t_d = 0.02$ s is shown in Fig. 4.8. All other parameters are kept same as in Example 4.6. Obtain the z-transform of the output $C(z)$.

Solution From Example 4.6, the transfer function for the digital filter $D(z)$ is expressed as

$$D(z) = \frac{M(z)}{F(z)} = \frac{3 - 2z^{-1}}{1} = \frac{3z - 2}{z}$$

The modified z-transform of the ZOH and the plant combined is obtained. Similar to Example 4.8 with $mT = (1 - \Delta)T = 0.1 - 0.02 = 0.08$, as

$$\begin{aligned}
G(z, m) &= Z_m\left[\frac{1 - e^{-sT}}{s(s+1)}\right] = (1 - z^{-1})Z_m\left[\frac{1}{s(s+1)}\right] \\
&= \left[\frac{z-1}{z}\right]\left[\frac{z(1 - e^{-mT}) + e^{-mT} - e^{-T}}{(z-1)(z - e^{-T})}\right] \\
&= \left[\frac{z-1}{z}\right]\left[\frac{z(1 - e^{-0.08}) + e^{-0.08} - e^{-0.1}}{(z-1)(z - e^{-0.1})}\right]
\end{aligned}$$

The z-transform of the output can be expressed as

$$\begin{aligned}
C(z) &= G(z, m)D(z)F(z) \\
&= \left[\frac{z-1}{z}\right]\left[\frac{z(1 - e^{-0.08}) + e^{-0.08} - e^{-0.1}}{(z-1)(z - e^{-0.1})}\right]\left[\frac{3z-2}{z}\right]\left[\frac{z}{z-1}\right] \\
&= \left[\frac{z(1 - e^{-0.08}) + e^{-0.08} - e^{-0.1}}{(z-1)(z - e^{-0.1})}\right]\left[\frac{3z-2}{z}\right] \\
&= \left[\frac{0.0769z + 0.0183}{(z-1)(z-0.9048)}\right]\left[\frac{3z-2}{z}\right]
\end{aligned}$$

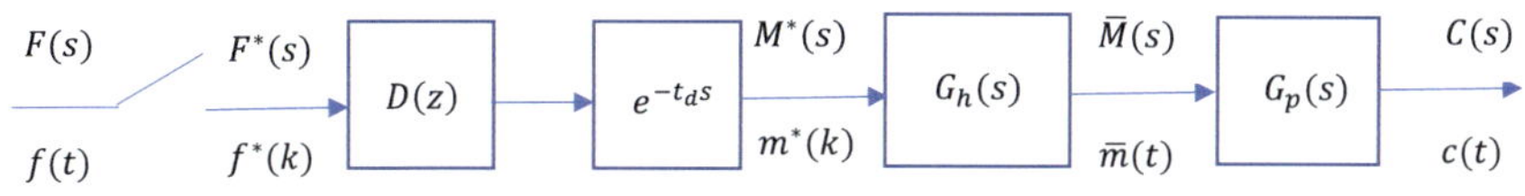

Fig. 4.9 An open-loop discrete-time system with a digital filter and a time delay

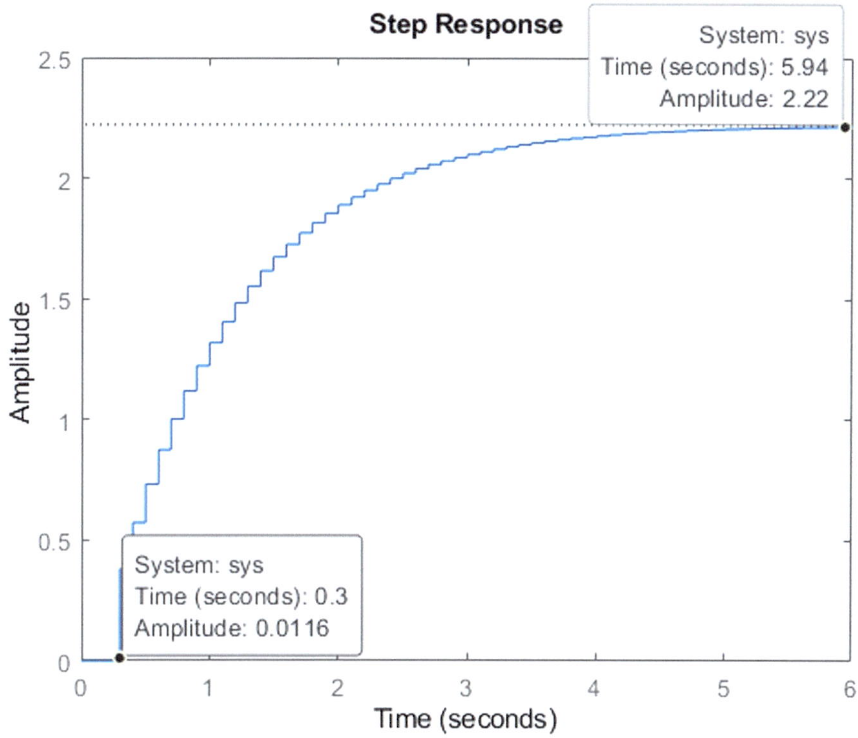

Fig. 4.10 Open-loop system step response with a digital filter and a time delay

Example 4.12 For an open-loop discrete-time system with a digital filter and a time delay $t_d = 0.2$ s, as shown in Fig. 4.9, with a ZOH and a plant transfer function $G_p(s) = \frac{1}{s+1}$ and a unit step function $F(s) = \frac{1}{s}$, the digital filter is used to represent difference equation $m(kT) = 3f(kT) - 2f(kT - T) + 0.1m(k - 1)$. Use sampling time $T = 0.1$ s. Find the z-transform of the output $C(z)$. Using initial and final value theorems find the initial and the final values of the output, $c(0)$ and $c(\infty)$. Verify the values from the open-loop system response, $c(kT)$.

Solution Matlab code, along with comments and results, is presented. Open-loop system response is shown in Fig. 4.10.

```
%ch4_ex4_12 open loop discrete time system with D(z) and delay
clear all; clc
% Given Gp(s)=1/(s+1)  T=0.1
% m(k)=3e(k)-2e(k-1)+0.1m(k-1) D(z)=M(z)/E(z)=(4-2/z)(1-0.1/z)=(4z-2)/(z-0.1)
% step input F(k)=1  F(z)=z/(z-1)
% Find C(z) = G(z)D(z)F(z)
% Gp(s)=1/(s+1)
ngs = 1; dgs = [1 1]; % Gp(s)
T = 0.1; % sampling time
[ngz,dgz]=c2dm(ngs,dgs,T,'zoh'); %G(z)
Gz = tf(ngz,dgz,T) %G(z)
dgz=[dgz 0 0] % delay of td=2T
Gz = tf(ngz,dgz,T) % G(z) with delay
%Fz = z/(z-1)
nfz=[1 0]; dfz=[1 -1];
Fz =tf(nfz,dfz,T) %F(z)
%D(z) =(4*z-2)/(z-0.1)
ndz=[4 -2];ddz=[1 -0.1];
Dz=tf(ndz,ddz,T) % D(z)

% C(z)=D(z)*G(z)*F(z)
Cz=Dz*Gz*Fz
[ncz, dcz]=tfdata(Cz,'v')
cInitial = polyval(ncz,10^10)/ polyval(dcz,10^10)
% Final Value
z1= tf([1 -1],[0 1],T)
Cz2 =Cz*z1
[numz1, denz1]=tfdata(Cz2,'v')
[z,p,k1]=tf2zp(numz1, denz1)
Cz2=minreal(Cz2,10^-4)
[numz2, denz2]=tfdata(Cz2,'v')
fFinal=ddcgain(numz2, denz2)
%step response
DGz=Dz*Gz % D(z)G(z) : open loop system transfer function =C(z)?F(z)
dstep(DGz,1) % step response of the open-loop system
```

Results:

```
Original system transfer function (without delay)
Gz =

 0.09516
 ----------
 z - 0.9048

System transfer function (with delay)

Gz =

 0.09516
 ----------------
 z^3 - 0.9048 z^2

Dz =
 4 z - 2
```

```
 -------
 z - 0.1

Cz =
 0.3807 z^2 - 0.1903 z
 ----------------------------------------
 z^5 - 2.005 z^4 + 1.095 z^3 - 0.09048 z^2
cInitial = 3.8065e-31
Cz2 =
 0.3807 z^3 - 0.571 z^2 + 0.1903 z
 ----------------------------------------
 z^5 - 2.005 z^4 + 1.095 z^3 - 0.09048 z^2
 After pole-zero cancelation
Cz2 =
 0.3807 z - 0.1903
 --------------------------
 z^3 - 1.005 z^2 + 0.09048 z

fFinal = 2.2222

DGz =
 0.3807 z - 0.1903
 -----------------------------
 z^4 - 1.005 z^3 + 0.09048 z^2
```

4.8 Simulation Using Matlab/Simulink

The open-loop system with a digital filter and a time delay of Example 4.12 can be presented in Matlab/Simulink as shown in Fig. 4.11. The plant transfer function is shown in the s-domain as $G_p(s) = \frac{1}{s+1}$. The effect of sampling and reconstruction for the discrete-time system is represented as the zero-order hold (ZOH). The open-loop system simulated response from the Simulink model is presented in Fig. 4.12. The x-axis of Fig. 4.12 denotes time in seconds and y-axis denotes the system response $c(t)$.

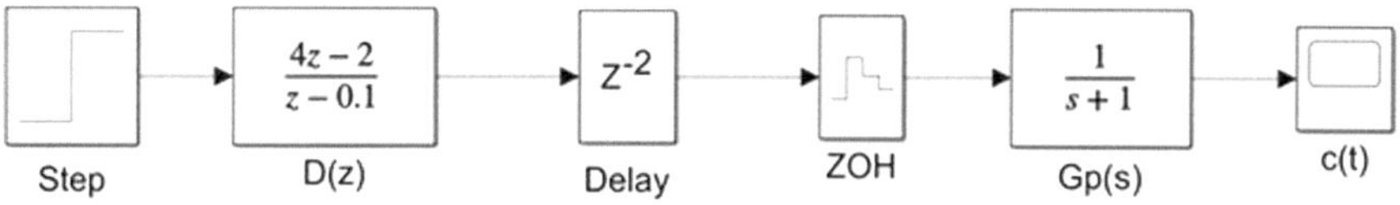

Fig. 4.11 Simulink model for the open-loop continuous-time system with a digital filter and time delay

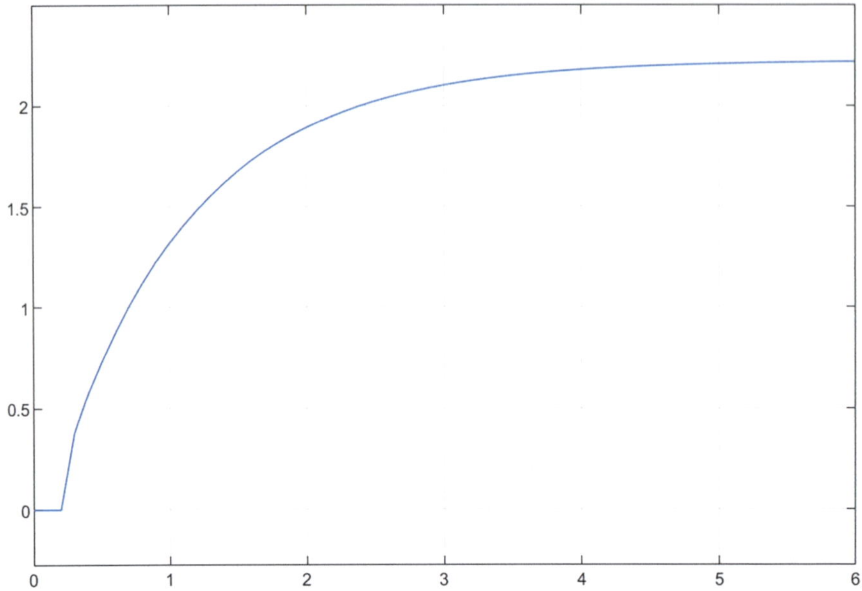

Fig. 4.12 Open-loop system response of the continuous-time system with a digital filter and time delay

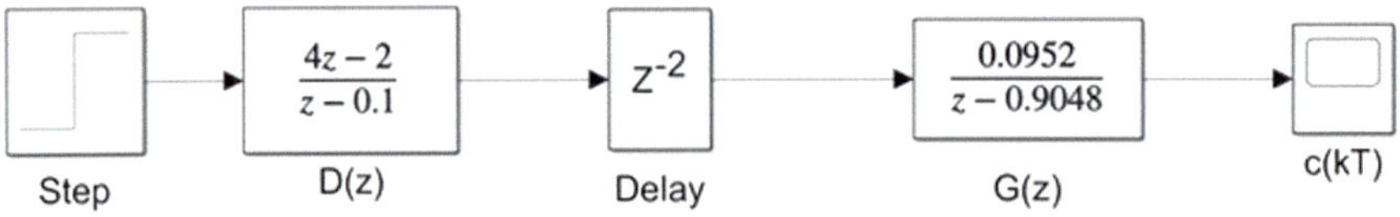

Fig. 4.13 Simulink model for the open-loop discrete-time system with a digital filter and time delay

The open-loop system can be represented in discrete-time domain replacing the continuous-time domain plant transfer function $G_{\mathrm{p}}(s) = \frac{1}{s+1}$ and ZOH with $G(z) = \frac{0.0952}{z-0.9048}$ as shown in Example 4.2. The open-loop system with digital filter $D(z)$ and the delay module can be represented in discrete-time domain in Simulink as shown in Fig. 4.13. The open-loop system step response of the discrete-time system obtained from Simulink model is presented in Fig. 4.14. The effects of sampling and time delay are evident in the simulated response of the discrete-time system in Fig. 4.14. The x-axis of Fig. 4.14 denotes time in seconds and y-axis denotes the sampled response $c(kT)$.

Example 4.13 Create a Simulink model for an open-loop continuous-time system with a transfer function $G_{\mathrm{p}}(s) = \frac{1}{s^2+3s+2}$, a ZOH, a digital filter $D(z) = \frac{3z-2}{z-0.1}$, a delay module $t_{\mathrm{d}} = 0.2$ s, a sampling time $T = 0.1$ s as shown in Fig. 4.14. Simulate the open-loop step response of the system $c(t)$.

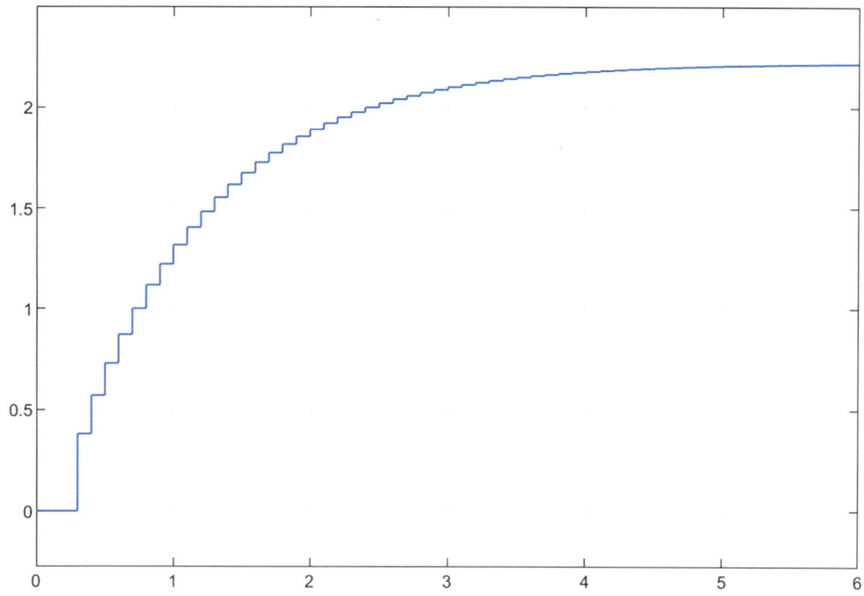

Fig. 4.14 Open-loop system response of the discrete-time system with a digital filter and time delay

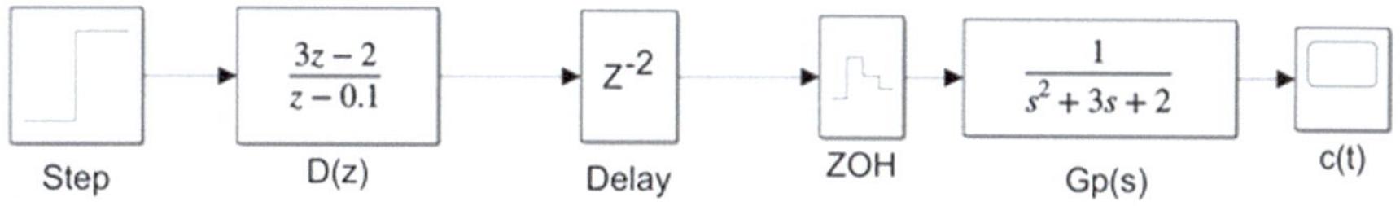

Fig. 4.15 Simulink model for the open-loop continuous-time system with a digital filter and time delay (Example 4.13)

Solution The open-loop system step response obtained from the Simulink model of Fig. 4.15 is shown in Fig. 4.16. The x-axis of Fig. 4.16 denotes time in seconds and y-axis denotes the system response $c(t)$.

The system can be represented in discrete-time domain in Simulink as shown in Fig. 4.17. The open-loop system step response of the discrete-time system obtained from Simulink model is presented in Fig. 4.18. The effects of sampling and time delay are evident in the simulated response of the discrete-time system in Fig. 4.1 compared to that in Fig. 4.15. The x-axis of Fig. 4.18 denotes time in seconds and y-axis denotes the sampled response $c(kT)$.

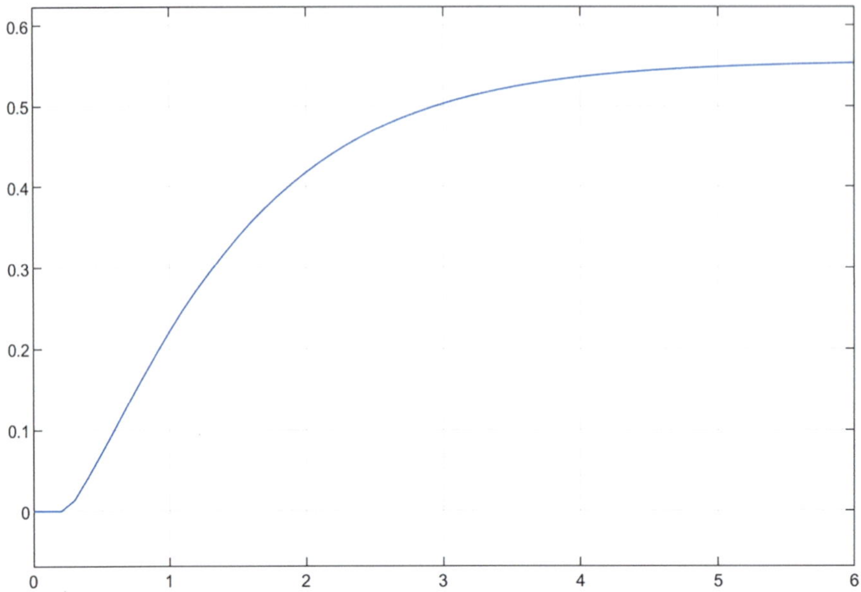

Fig. 4.16 Open-loop system response of the continuous-time system with a digital filter and time delay (Example 4.13)

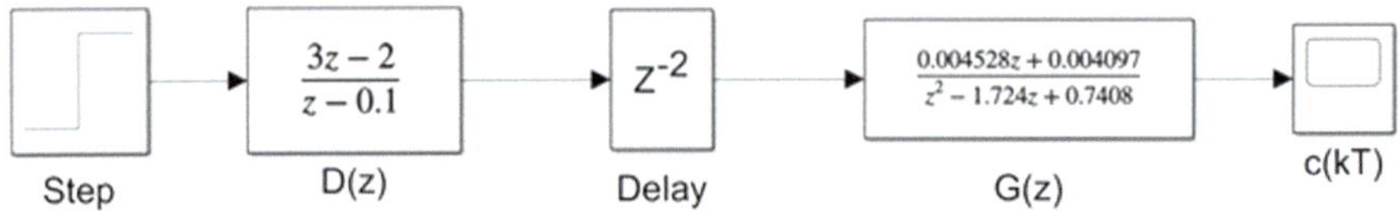

Fig. 4.17 Simulink model for the open-loop discrete-time system with a digital filter and time delay (Example 4.13)

4.9 Systems with Nonsynchronous Sampling

Schematic of a system with nonsynchronous sampling is presented in Fig. 4.19, where both samplers sample at the same rate but not synchronous. The first sampler samples at 0, T, $2T$, . . . and the second sampler samples at rT, $T + rT$, $2T + rT$, . . ., where $0 < r < 1$. The sampler and data-hold and signals are shown in Fig. 4.20a, b.

The output signal of the sampler and data-hold can be expressed, assuming ZOH, as

$$\begin{aligned}\bar{f}(t) =& f(rT)[u(t-rT)-u(t-T-rT)] + f(T+rT)\\ &\times [u(t-T-rT)-u(t-2T-rT)] + f(2T+rT)\\ &\times [u(t-2T-rT)-u(t-3T-rT)] + \ldots\end{aligned} \tag{4.51}$$

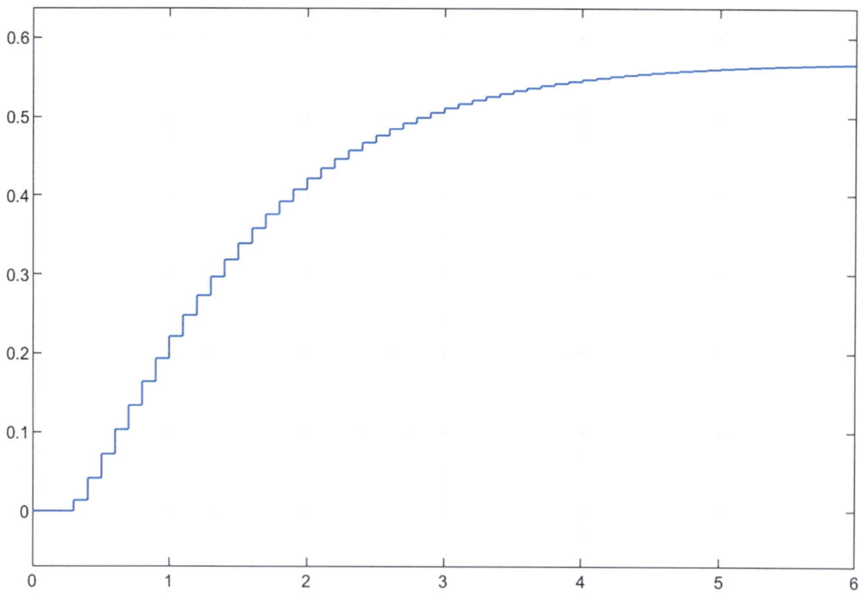

Fig. 4.18 Open-loop system response of the discrete-time system with a digital filter and time delay (Example 4.13)

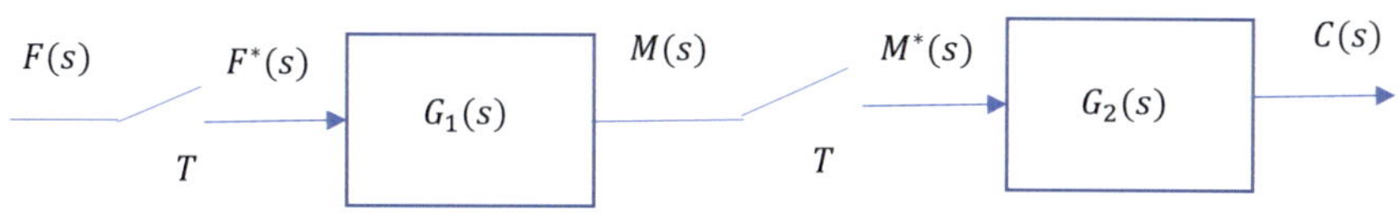

Fig. 4.19 Schematic of a system with nonsynchronous sampling

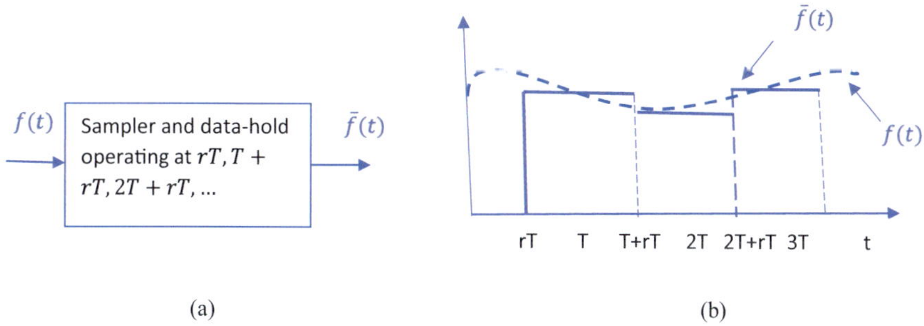

Fig. 4.20 Nonsynchronous sampling (**a**) sampler and hold, (**b**) original and sampled signals

Laplace transform of the sampled signal can be expressed as

$$\overline{F}(s) = f(rT)\left[\frac{e^{-rTs}}{s} - \frac{e^{-(T+rT)s}}{s}\right] + f(T+rT)\left[\frac{e^{-(T+rT)s}}{s} - \frac{e^{-(2T+rT)s}}{s}\right] + f(2T+rT)\left[\frac{e^{-(2T+rT)s}}{s} - \frac{e^{-(3T+rT)s}}{s}\right] + \ldots \tag{4.52}$$

After some rearrangement, Eq. (4.52) can be expressed as

$$\overline{F}(s) = \left[\frac{1}{s} - \frac{e^{-Ts}}{s}\right] e^{Ts} e^{-rTs}\left[f(rT)e^{-Ts} + f(T+rT)e^{-2Ts} + f(2T+rT)e^{-3Ts} + \ldots\right] \tag{4.53}$$

Laplace transform $\overline{F}(s)$ can be expressed using modified z-transform as

$$\overline{F}(s) = \left[\frac{1}{s} - \frac{e^{-Ts}}{s}\right] e^{Ts} e^{-rTs} F(z,m)|_{m=r, z=e^{Ts}} \tag{4.54}$$

Equation (4.54) can be rewritten using Eq. (4.42) for the modified z-transform as

$$\overline{F}(s) = \left[\frac{1}{s} - \frac{e^{-Ts}}{s}\right] e^{Ts} e^{-rTs} Z\left[F(s)e^{-\Delta Ts})\right]|_{\Delta=1-r} \tag{4.55}$$

Following Eq. (4.55), the system with nonsynchronous sampling of Fig. 4.19 can be represented as an equivalent system of Fig. 4.20 with the sampler operating at $0, T, 2T, \ldots$

The system of Fig. 4.19 with nonsynchronous sampling can be represented as an equivalent system with synchronous samplers of Fig. 4.21.

From Fig. 4.21, the intermediate signal $M(s)$, assuming ZOH after sampler (Fig. 4.22), can be expressed as

$$M(s) = e^{-(1-r)Ts}[G_1(s)]F^*(s) \tag{4.56}$$

z-transform $M(z)$ can be expressed as

$$M(z) = \left[G_1(z,m)|_{m=r, z=e^{Ts}}\right]F(z) \tag{4.57}$$

The sampled function $P^*(s)$ can be expressed as

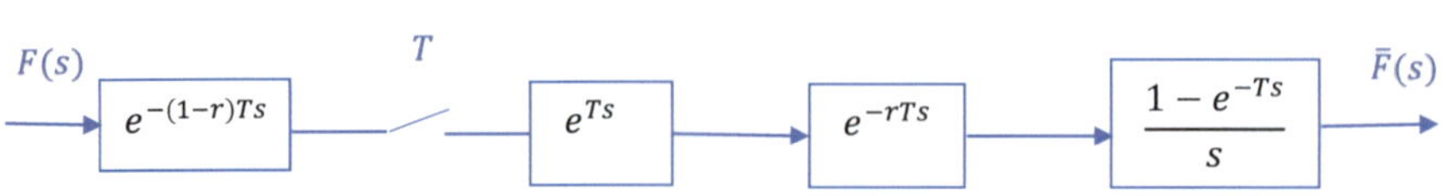

Fig. 4.21 Equivalent sampler and hold for the system with nonsynchronous sampling

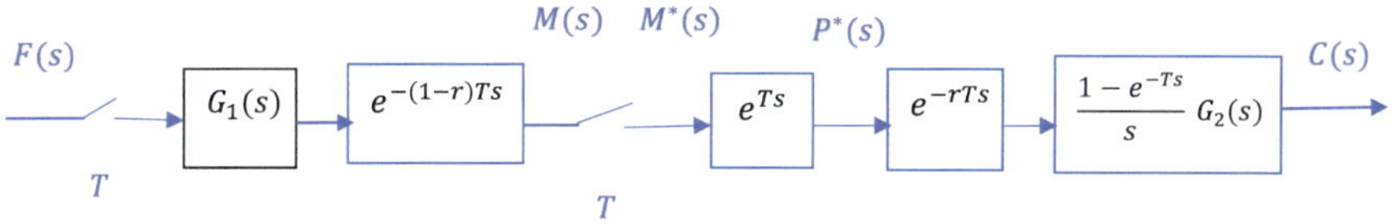

Fig. 4.22 Equivalent sampler and hold for the system with nonsynchronous sampling of Fig. 4.21

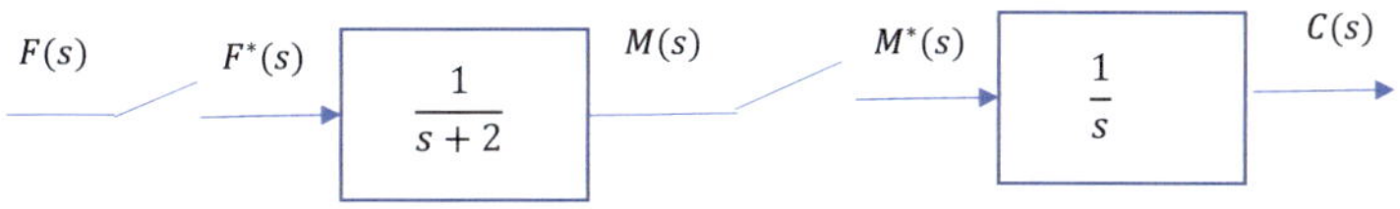

Fig. 4.23 Schematic of a system with nonsynchronous sampling

$$P^*(s) = e^{Ts} M^*(s) \tag{4.58}$$

$$P(z) = zM(z) = z\left[G_1(z,m)|_{m=r,z=e^{Ts}}\right]F(z) \tag{4.59}$$

The output function $C(s)$ can be expressed as

$$C(s) = e^{-rTs}[G_2(s)]P^*(s) \tag{4.60}$$

z-transform $C(z)$ can be obtained, assuming ZOH is included in transfer functions $G_1(s)$ and $G_2(s)$, as

$$\begin{aligned} C(z) &= \left[G_2(z,m)|_{m=1-r,z=e^{Ts}}\right]P(z) \\ &= z\left[G_1(z,m)|_{m=r,z=e^{Ts}}\right]\left[G_2(z,m)|_{m=1-r,z=e^{Ts}}\right]F(z) \end{aligned} \tag{4.61}$$

Example 4.14 Find the z-transform of the output $C(z)$ for the system of Fig. 4.23 with nonsynchronous sampling and zero-order hold (ZOH). The transfer functions are given as $G_1(s) = \frac{1}{s+2}$ and $G_2(s) = \frac{1}{s}$, and $F(s) = \frac{1}{s}$. Both samplers operate at the same sampling rate ($T = 0.1$ s), but with an offset in time ($rT = 0.02$ s).

Solution For the system $T = 0.10$ s, $rT = 0.02$ s, the modified z-transform can be obtained as

$$\begin{aligned} G_1(z,m) &= Z_m \left.\frac{1-e^{-Ts}}{s(s+2)}\right|_{m=r,z=e^{Ts}} = \left(\frac{z-1}{z}\right)\left(\frac{1}{2}\right)\left[\frac{1}{z-1} - \frac{e^{-2(0.02)}}{z-e^{-2(0.1)}}\right] \\ &= \left(\frac{z-1}{2z}\right)\left[\frac{1}{z-1} - \frac{0.9608}{z-0.8187}\right] = \left(\frac{1}{2z}\right)\left(\frac{0.0392z+0.1421}{z-0.8187}\right) \end{aligned}$$

$$\begin{aligned} G_2(z,m) &= Z_m \left.\frac{1-e^{-Ts}}{s^2}\right|_{m=1-r,z=e^{Ts}} = \left(\frac{z-1}{z}\right)\left[\frac{(1-r)\times T}{z-1} + \frac{T}{(z-1)^2}\right] \\ &= \left(\frac{z-1}{z}\right)\left[\frac{0.08}{z-1} + \frac{0.1}{(z-1)^2}\right] = \left(\frac{1}{z}\right)\left(\frac{0.08z+0.02}{z-1}\right) \end{aligned}$$

$$C(z) = zF(z)\left(G_1(z,m)\big|_{m=r,z=e^{Ts}}\right)\left(G_2(z,m)\big|_{m=1-r,z=e^{Ts}}\right)$$

$$C(z) = z\left(\frac{z}{z-1}\right)\left(\frac{1}{2z}\right)\left(\frac{0.0392z+0.1421}{z-0.8187}\right)\left(\frac{1}{z}\right)\left(\frac{0.08z+0.02}{z-1}\right)$$
$$= \left(\frac{0.0392z+0.1421}{z-0.8187}\right)\left(\frac{0.04z+0.01}{(z-1)^2}\right)$$

Exercises

4.1 Find the z-transform of the output signal of an open-loop discrete-time system as shown in Fig. P4.1 with a ZOH and a plant transfer function $G_p(s) = \frac{1}{s+2}$ and a unit step function $F(s) = \frac{1}{s}$ for sampling time $T = 0.05$ s.

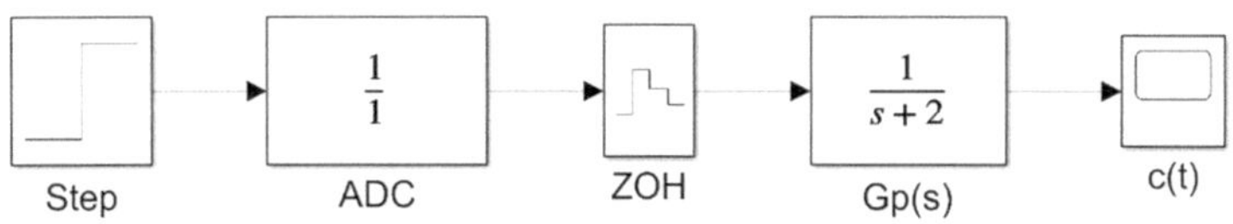

Fig. P4.1 Model of open-loop continuous-time system for Problem 4.1

4.2 Find the z-transform of the output signal of an open-loop discrete-time system as shown in Fig. P4.2, was with a ZOH and a plant transfer function $G_p(s) = \frac{1}{s^2+5s+6}$ and a unit step function $F(s) = \frac{1}{s}$ for sampling time $T = 0.05$ s.

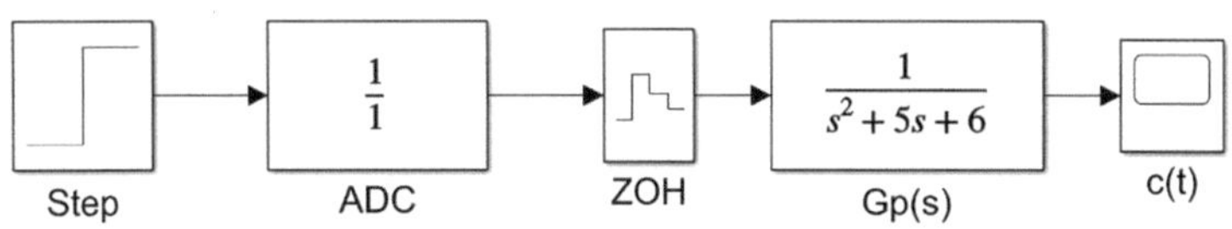

Fig. P4.2 Model of open-loop continuous-time system for Problem 4.2

4.3 Find the z-transform of the output signal of an open-loop discrete-time system as shown in Fig. P4.3 with a FOH and a plant transfer function $G_p(s) = \frac{1}{s+2}$ and a unit step function $F(s) = \frac{1}{s}$ for sampling time $T = 0.05$ s.

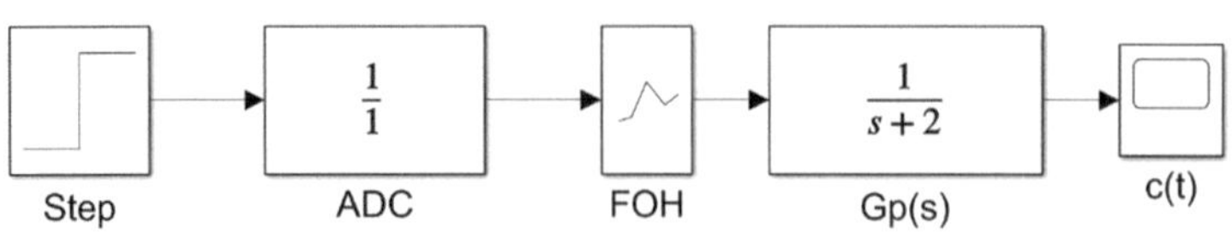

Fig. P4.3 Model of open-loop continuous-time system for Problem 4.3

4.4 Find the z-transform of the output signal of an open-loop discrete-time system as shown in Fig. P4.4 with a FOH and a plant transfer function $G_p(s) = \frac{1}{s^2+5s+6}$ and a unit step function $F(s) = \frac{1}{s}$ for sampling time $T = 0.05$ s.

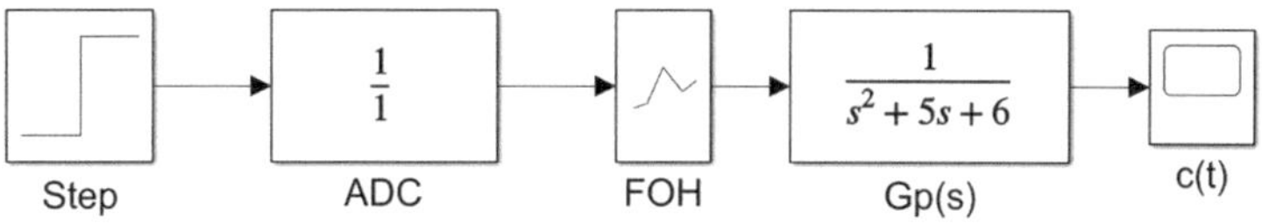

Fig. P4.4 Model of open-loop continuous-time system for Problem 4.4

4.5 Find the z-transform of output function, $C(z)$ for different configurations as shown in Fig. P4.5a–c with $G_1(s) = \frac{1}{s+3}, G_2(s) = \frac{1}{s+4}, F(s) = \frac{1}{s}$ using ZOH and sampling time $T = 0.1$ s.

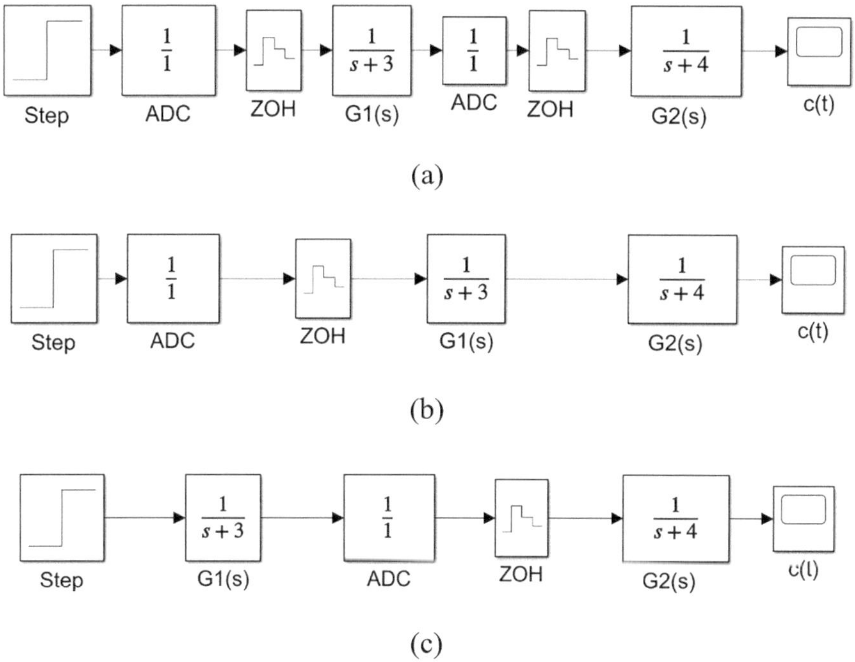

Fig. P4.5 Model of open-loop continuous-time system for Problem 4.5

4.6 Find the z-transform of output function, $C(z)$ for different configurations as shown in Fig. P4.6a–c with $G_1(s) = \frac{1}{s+5}, G_2(s) = \frac{1}{s+6}, F(s) = \frac{1}{s}$ using FOH and sampling time $T = 0.05$ s.

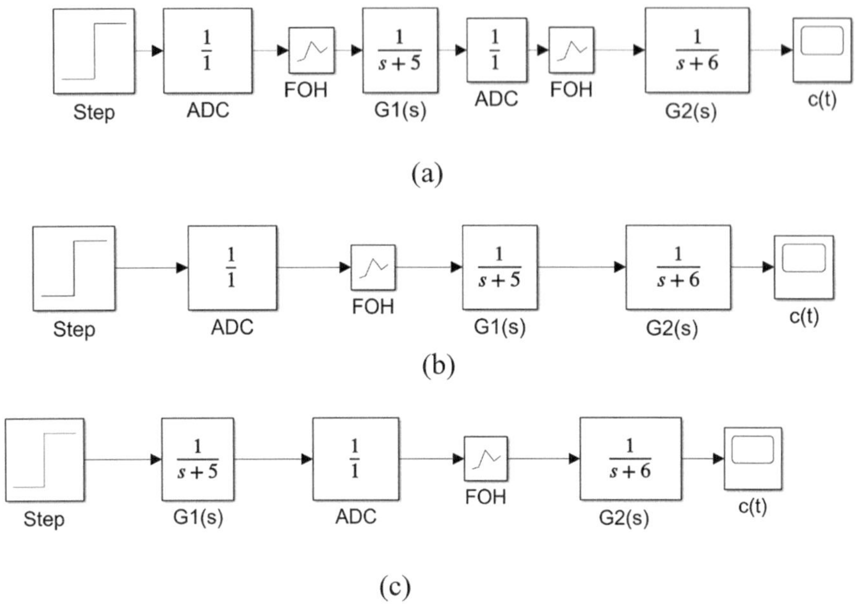

Fig. P4.6 Model of open-loop continuous-time system for Problem 4.6

4.7 For an open-loop discrete-time system with a digital filter as shown in Fig. P4.7 with a ZOH and a plant transfer function $G_p(s) = \frac{1}{s+3}$ and a unit step function $F(s) = \frac{1}{s}$, the digital filter is used to represent difference equation $m(k) = 4f(k) - 2f(k-1) + 0.3m(k-1)$. Find the transfer function $D(z)$ for the digital filter. Use sampling time $T = 0.05$ s. Find the z-transform of the output $C(z)$. Use PFE to find the inverse z-transform of the (z), $c(kT)$. Using initial and final value theorems, find the initial and the final values of the output, $c(0)$ and $c(\infty)$. Verify the values from the discrete-time function, $c(kT)$.

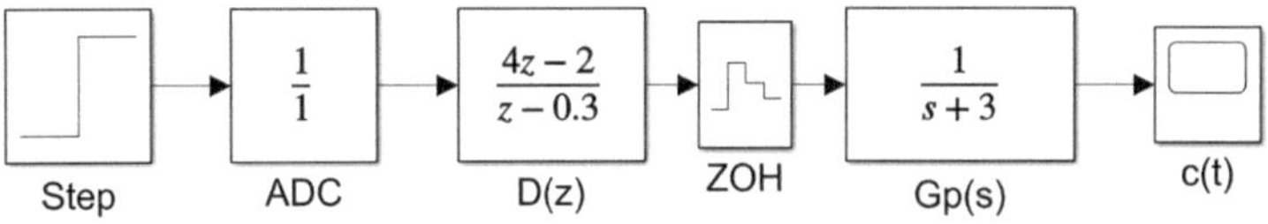

Fig. P4.7 Model of open-loop continuous-time system for Problem 4.7

4.8 For an open-loop discrete-time system with a digital filter as shown in Fig. P4.8 with a ZOH and a plant transfer function $G_p(s) = \frac{1}{s+4}$ and a unit step function $F(s) = \frac{1}{s}$, the digital filter is used to represent difference equation $m(kT) = 4f(k) - 3f(k-) + 0.4m(k-1)$. There is a delay of $t_d = 0.10$ s. Find the transfer function $D(z)$ for the digital filter. Use sampling time $T = 0.05$ s. Find the z-transform of the output $C(z)$. Obtain the open-loop system response.

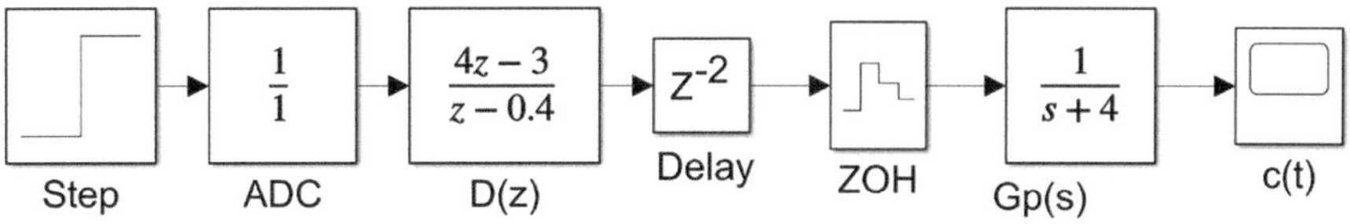

Fig. P4.8 Model of open-loop continuous-time system for Problem 4.8

4.9 Create a Simulink model for the open-loop continuous-time system for the plant transfer function $G_{\mathrm{p}}(s) = \frac{1}{s+3}$ with a ZOH, a digital filter $D(z) = \frac{4z-2}{z-0.3}$, a time delay of $t_{\mathrm{d}} = 0.15$ s, and a sampling time $T = 0.05$ s, as shown in Fig. P4.9. Simulate the open-loop system step response $c(t)$.

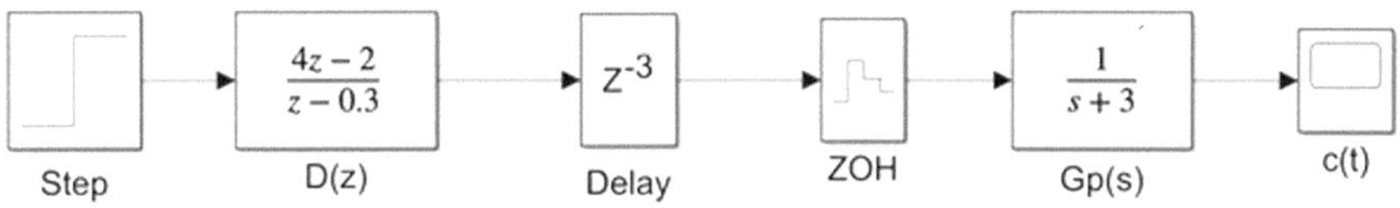

Fig. P4.9 Simulink model for the open-loop continuous-time system for Problem 4.9

4.10 Create a Simulink model for the open-loop discrete-time system for the plant transfer function $G_{\mathrm{p}}(s) = \frac{1}{s+3}$ with a ZOH, a digital filter $D(z) = \frac{4z-2}{z-0.3}$, a time delay of $t_{\mathrm{d}} = 0.15$ s, and a sampling time $T = 0.05$ s, as shown in Fig. P4.10. Simulate the open-loop discrete-time system step response $c(kT)$.

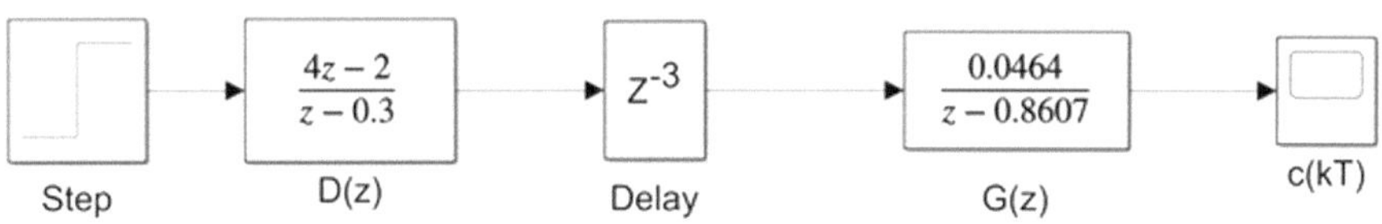

Fig. P4.10 Simulink model for the open-loop discrete-time system for Problem 4.10

4.11 Create a Simulink model for the open-loop continuous-time system for the plant transfer function $G_{\mathrm{p}}(s) = \frac{1}{s^2+5s+6}$ with a ZOH, a digital filter $D(z) = \frac{4z-2}{z-0.3}$, a time delay of $t_{\mathrm{d}} = 0.15$ s, and a sampling time $T = 0.05$ s, as shown in Fig. P4.11. Simulate the open-loop system step response $c(t)$.

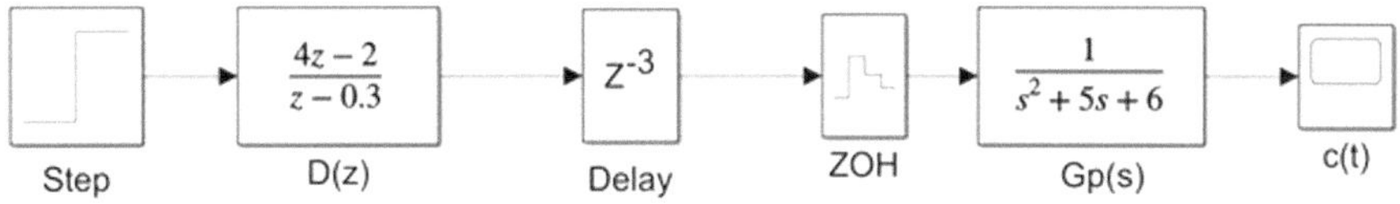

Fig. P4.11 Simulink model for the open-loop continuous-time system for Problem 4.11

4.12 Create a Simulink model for the open-loop discrete-time system for the plant transfer function $G_p(s) = \frac{1}{s^2+5s+6}$ with a ZOH, a digital filter $D(z) = \frac{4z-2}{z-0.3}$, a time delay of $t_d = 0.15$ s, and a sampling time $T = 0.05$ s, as shown in Fig. P4.12. Simulate the open-loop discrete-time system step response $c(kT)$.

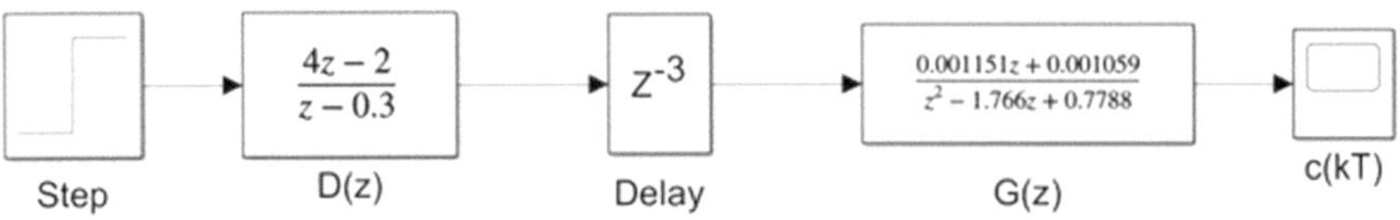

Fig. P4.12 Simulink model for the open-loop discrete-time system for Problem 4.12

4.13 Find the z-transform of the output $C(z)$ for the system of Fig. P4.13 with nonsynchronous sampling and zero-order hold (ZOH). The transfer functions are given as $G_1(s) = \frac{1}{s+3}$ and $G_2(s) = \frac{1}{s+1}$, and $F(s) = \frac{1}{s}$. Both samplers operate at the same sampling rate ($T = 0.05$ s), but with an offset in time ($rT = 0.01$ s).

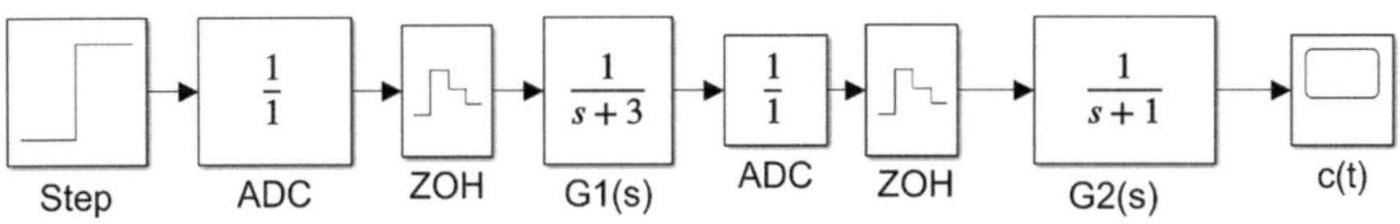

Fig. P4.13 Simulink model for the open-loop continuous-time system for Problem 4.13

Further Reading

1. Asadi F (2023) Signals and systems with MATLAB® and Simulink®. Springer, Cham
2. Astrom KJ, Murray RM (2021) Feedback systems: an introduction for scientists and engineers, 2nd edn. Princeton University Press, Princeton
3. Chen C-T (2006) Analog and digital control system design: transfer-function, state-space, and algebraic methods. Oxford University Press, New York
4. Dorf R, Bishop R (2016) Modern control systems, 13th edn. Pearson, Essex
5. Fadali MS, Visioli A (2020) Digital control engineering analysis and design, 3rd edn. Academic Press, London
6. Franklin GF, Powell JD, Workman M (1998) Digital control of dynamic systems, 3rd edn. Addison-Wesley, Boston
7. Franklin GF, Powell JD, Emami-Naeini A (2019) Feedback control of dynamic systems, 8th edn. Pearson, New Jersey
8. Isermann R (1981) Digital control systems. Springer-Verlag, Berlin
9. Jalili N, Candelino NW (2023) Dynamic systems and control engineering. Cambridge University Press, Cambridge
10. Keviczky L, Bars R, Hetthéssy J, Bányász C (2019) Control engineering. Springer, Singapore
11. Koeing DM (2009) Practical control engineering: a guide for engineers, managers, and practitioners. McGraw-Hill, New York

12. Kuo BC (1995) Digital control systems, 2nd edn. Oxford University Press, Oxford
13. Landau ID, Gito G (2006) Digital control systems: design, identification and implementation. Springer-Verlag, London
14. Leigh JR (2006) Applied digital control: theory, design and implementation, 2nd edn. Dover, Mineola
15. Mathworks (2023) Control System Toolbox. https://www.mathworks.com/products/control.html
16. Mathworks (2023) Matlab. https://www.mathworks.com/products/matlab.html
17. Nise NS (2019) Control systems engineering, 8th edn. Wiley, Hoboken
18. Nof SY (ed) (2009) Springer handbook of automation. Springer-Verlag, Berlin
19. Ogata K (1995) Discrete-time control systems, 2nd edn. Pearson, New Jersey
20. Phillips CL, Nagle HT, Chakrabortty A (2015) Digital control system analysis and design, 4th edn. Pearson, Essex
21. Rabbath CA, Lechevin N (2014) Discrete-time control system design with applications. Springer, New York
22. Samanta B (2023) Introduction to mechatronics: an integrated approach. Springer, Cham
23. Shertukde HM (2015) Digital control applications illustrated with MATLAB®. CRC Press, Boca Raton
24. Starr G (2020) Introduction to applied digital controls. Springer, Cham
25. Veloni A, Miridakis NI (2018) Digital control systems-theoretical problems and simulation tools. CRC Press, Boca Raton
26. Vilanova R, Visioli A (2012) PID control in the third millennium – lessons learned and new approaches. Springer-Verlag, London
27. Visioli A (2006) Practical PID control. Springer, London

Chapter 5
Closed-Loop Discrete-Time Systems

5.1 Introduction

In this chapter, closed-loop discrete-time systems are covered. First, the transfer function of a closed-loop discrete-time system is derived in z-domain. Next, the presence of a disturbance input to a closed-loop discrete-time system is considered, and the closed-loop system output in terms of both the reference input and the disturbance input is derived. The procedures are illustrated using examples along with the corresponding Matlab codes and Simulink models. A laboratory experiment for the determination of model parameters is presented for a DC motor module. End-of-chapter exercises are provided to enhance understanding of the topics covered and consolidate problem-solving skills.

5.2 Closed-Loop Discrete-Time System Transfer Function

The block diagram schematic of a closed-loop discrete-time system with a digital filter and a general order hold is shown in Fig. 5.1. The system is represented in Fig. 5.2 replacing the sampling and data reconstruction devices with analog-to-digital converter (ADC) and digital-to-analog converter (DAC) as interfaces.

The output function transform $C(s)$ can be expressed as

$$C(s) = G_h(s)G_p(s)M^*(s) \tag{5.1}$$

The controller output function $M^*(s)$ can be expressed in terms of input function $E(s)$ as

$$M^*(s) = D^*(s)E^*(s) \tag{5.2}$$

B. Samanta, *Introduction to Digital Control*,
https://doi.org/10.1007/978-3-031-66830-2_5

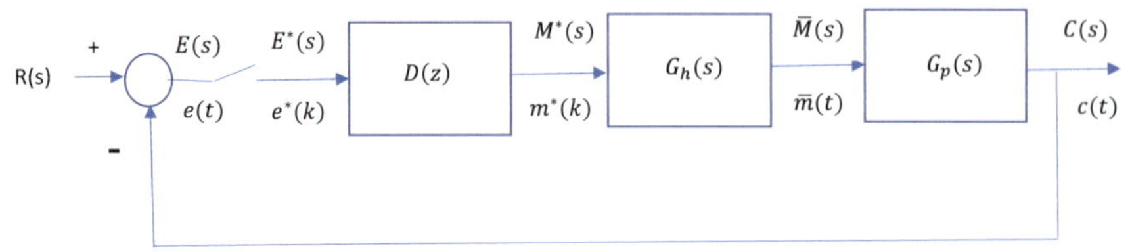

Fig. 5.1 Closed-loop discrete-time system with a general order hold and a digital filter

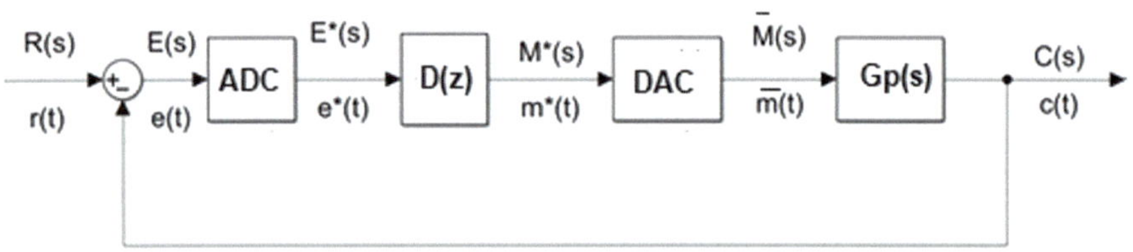

Fig. 5.2 Closed-loop discrete-time system with ADC (sampler) and DAC (reconstructor) as interfaces

Combining Eqs. (5.1) and (5.2), the output function can be expressed as

$$C(s) = G_h(s)G_p(s)D^*(s)E^*(s) \tag{5.3}$$

The error signal can be expressed as

$$E(s) = R(s) - C(s) \tag{5.4}$$

Taking starred transform Eqs. (5.4) and (5.3) can be written as

$$E^*(s) = R^*(s) - C^*(s) \tag{5.5}$$

$$C^*(s) = \overline{G_h(s)G_p}^*(s)D^*(s)E^*(s) \tag{5.6}$$

Combining Eqs. (5.5) and (5.6) one can write

$$C^*(s) = \overline{G_h(s)G_p}^*(s)D^*(s)[R^*(s) - C^*(s)] \tag{5.7}$$

z-transform can be expressed as

$$C(z) = \overline{G_hG_p}(z)D(z)[R(z) - C(z)] \tag{5.8}$$

The output of the closed-loop system can be expressed as

$$C(z) = \frac{G(z)D(z)}{1 + G(z)D(z)}R(z) \tag{5.9}$$

where

$$G(z) = \overline{G_h G_p}(z) = Z[G_h(s)G_p(s)] \tag{5.10}$$

For ZOH, Eq. (5.10) can be written as

$$\begin{aligned} G(z) = \overline{G_{h0} G_p}(z) &= Z\left[\frac{1-e^{-sT}}{s} G_p(s)\right]_{e^{sT}=z} \\ &= \left(1 - z^{-1}\right) Z\left[\frac{G_p(s)}{s}\right] \end{aligned} \tag{5.11}$$

The closed-loop system in the z-domain replacing $\overline{G_{h0}G_p}(z)$ with $G(z)$ is shown in Fig. 5.3.

In Fig. 5.4, a sensor transfer function $H(s)$ is included in the feedback path. The closed-loop discrete-time system transfer function can similarly be derived from Fig. 5.5.

The system output $C(s)$ can be expressed as in Eq. (5.3). The error can be expressed as

$$E(s) = R(s) - H(s)C(s) = R(s) - G_h(s)G_p(s)H(s)D^*(s)E^*(s) \tag{5.12}$$

Taking the starred transform of Eq. (5.12), the sampled error signal starred transform can be expressed as

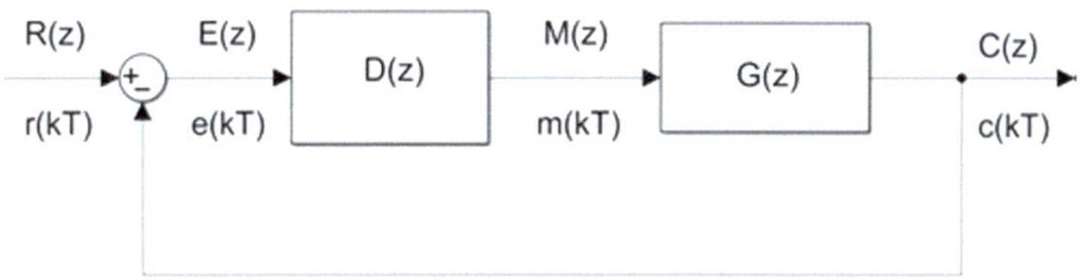

Fig. 5.3 Closed-loop discrete-time control system in z-domain

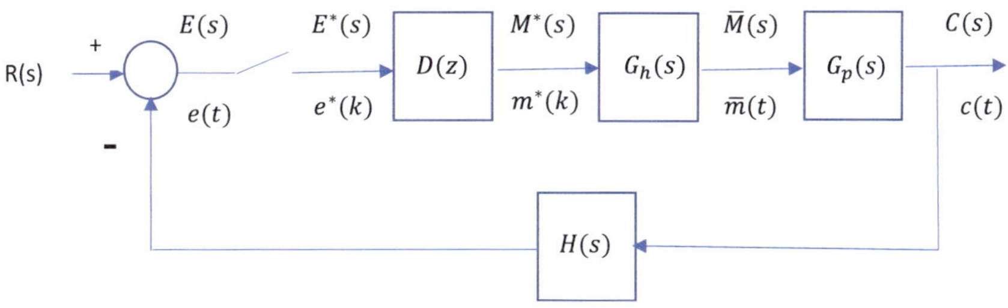

Fig. 5.4 Closed-loop discrete-time system with a digital filter and a sensor

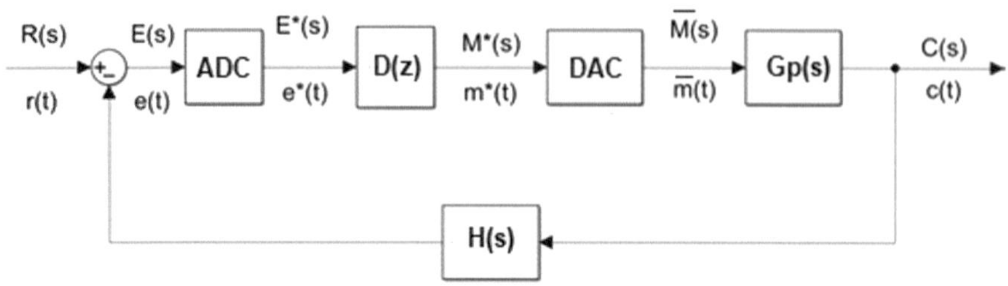

Fig. 5.5 Closed-loop discrete-time system with ADC (sampler) and DAC (reconstructor) as interfaces

$$E^*(s) = R^*(s) - \overline{G_h(s)G_p(s)H^*}(s)D^*(s)E^*(s) \tag{5.13}$$

$$E^*(s) = \frac{R^*(s)}{1 + \overline{G_h(s)G_p(s)H^*}(s)D^*(s)} \tag{5.14}$$

The system output can be expressed, similar to Eq. (5.6), as

$$C^*(s) = \overline{G_h(s)G_p{}^*}(s)D^*(s)E^*(s) \tag{5.15}$$

Combining Eqs. (5.14) and (5.15), the closed-loop system output in the presence of a non-unity sensor transfer function *H*(*s*) in the feedback loop can be expressed as

$$C^*(s) = \overline{G_h(s)G_p{}^*}(s)D^*(s)\frac{R^*(s)}{1 + \overline{G_h(s)G_p(s)H^*}(s)D^*(s)} \tag{5.16}$$

Taking *z*-transform, the closed-loop system transfer function can be expressed in the *z*-domain, as

$$C(z) = \frac{G(z)D(z)}{1 + \overline{GH}(z)D(z)}R(z) \tag{5.17}$$

where *G*(*z*) is defined in Eq. (5.10) and $\overline{GH}(z)$ is defined as

$$\overline{GH}(z) = Z\left[G_h(s)G_p(s)H(s)\right] \tag{5.18}$$

Example 5.1 For a closed-loop discrete-time system with a digital filter as shown in Fig. 5.6, with a ZOH and a plant transfer function $G_p(s) = \frac{2}{s(10s+1)}$ and a unit step function $R(s) = \frac{1}{s}$, the digital filter is given as $D(z) = \frac{z^2 - 1.2z + 0.5}{z^2 - 1.1z + 1}$. Use sampling time $T = 0.1$ s. Find the transfer function of (a) the open-loop system, (b) the closed-loop system with a feedback loop gain of $K = 1$, and (c) the closed-loop system with a feedback loop gain of $K = 0.5$. Obtain unit step response of (a) the open-loop system, (b) the closed-loop system with a feedback loop gain of 1, and (c) the closed-

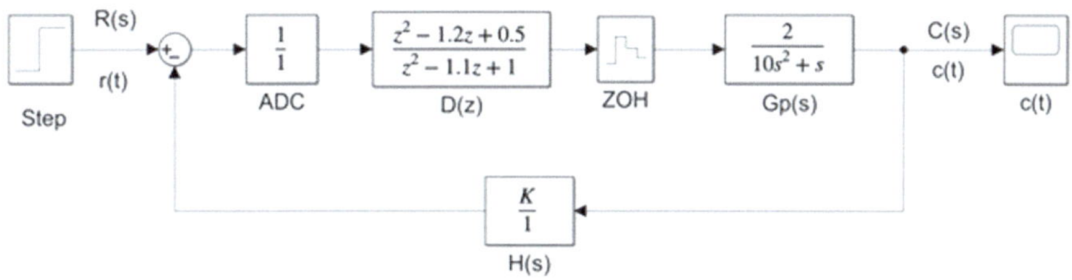

Fig. 5.6 A closed-loop discrete-time system with a digital filter

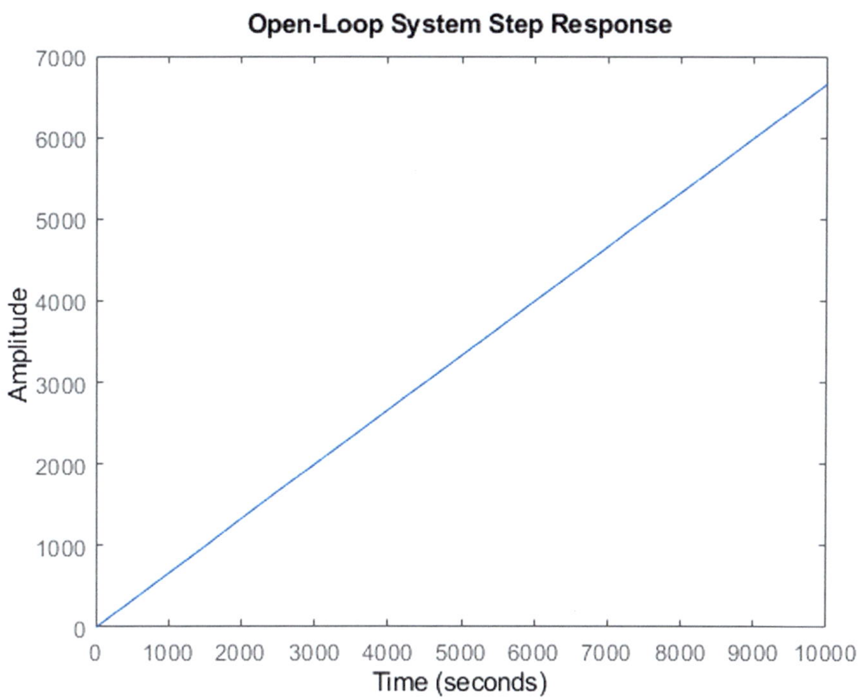

Fig. 5.7 Open-loop system step response for Example 5.1

loop system with a feedback loop gain of 0.5. Obtain steady-state values of the output for each case. Verify these values from steady-state gain (hint: use Matlab command *dcgain()*).

Matlab code and the results are presented next. The open-loop system response is shown in Fig. 5.7. The closed-loop system response obtained using Matlab code is shown in Figs. 5.8 and 5.9. The closed-loop system response obtained using the Simulink model is shown in Figs. 5.10 and 5.11.

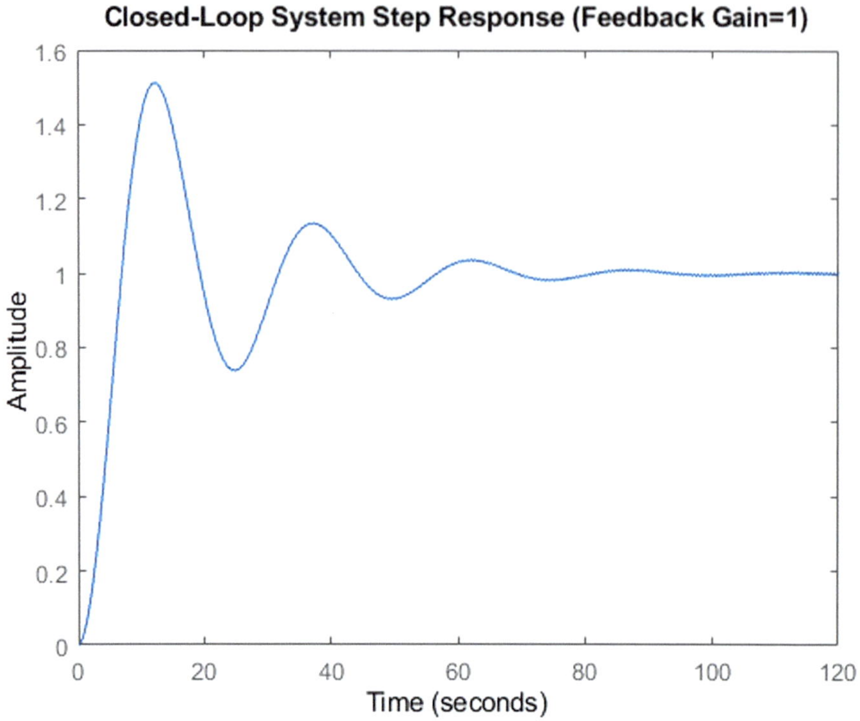

Fig. 5.8 Closed-loop system step response for Example 5.1 with feedback gain of 1

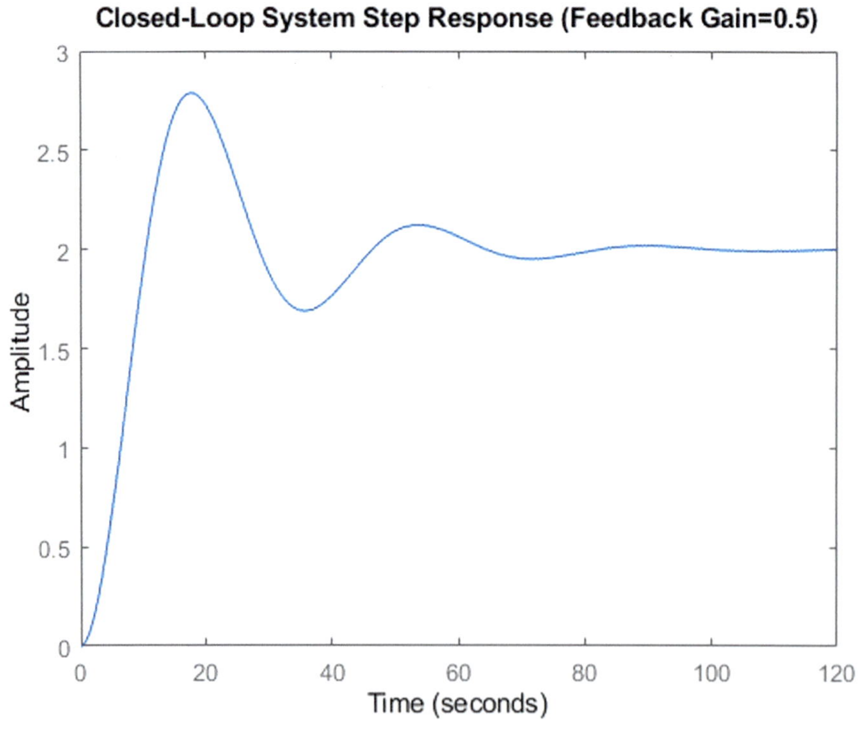

Fig. 5.9 Closed-loop system step response for Example 5.1 with feedback gain of 0.5

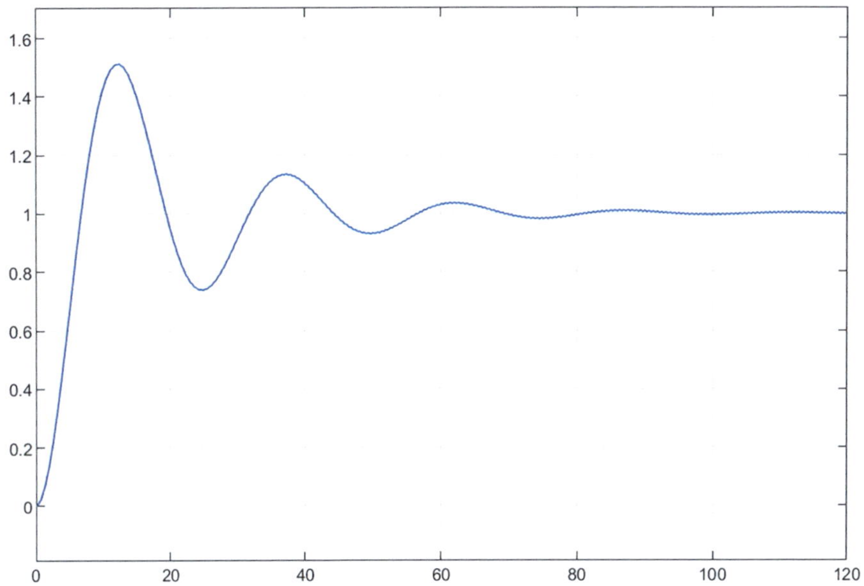

Fig. 5.10 Closed-loop system step response from Simulink for Example 5.1 with feedback gain of 1

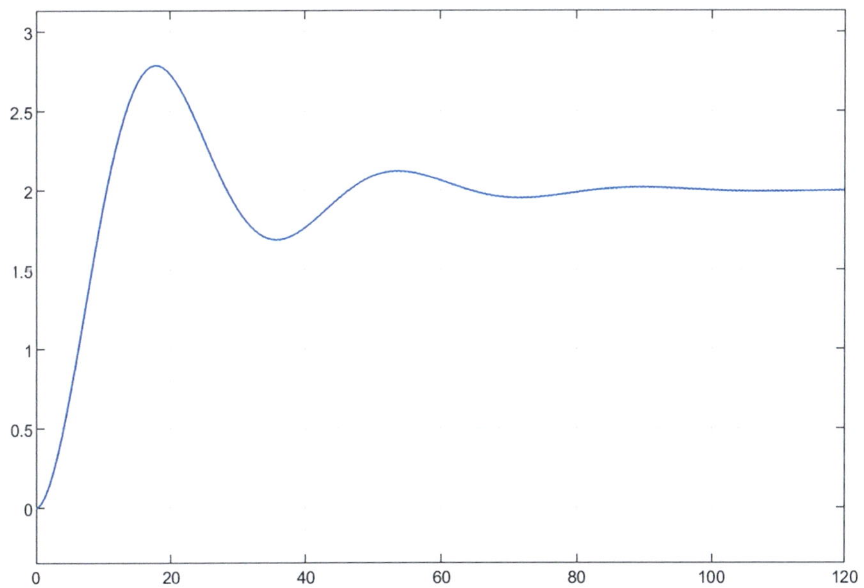

Fig. 5.11 Closed-loop system step response from Simulink for Example 5.1 with feedback gain of 0.5

```
%ch5_Ex 5.1 Closed-Loop Discrete-Time System
%  Gp(s)=2*K/(s(10s+1)) K=1
clear all; close all; clc
T = 0.1;  % sampling time
% Gp(s)=2*K/(s(10s+1))
K=1;
Gs=tf(2*K, [10 1 0]); % Gp(s)=2*K/(s(10s+1))
% G(z)
Gz=c2d(Gs,T); % G(z) for Gp(s) with ZOH

%D(z)=(z^2-1.2z+0.5)/(z^2-1.1z+1)
Dz = tf([1 -1.2 0.5], [1 -1.1 1],T) % D(z)

% open-loop system transfer function
% GD = G(z)*D(z)
GDz = Gz * Dz % Open-Loop Transfer Function
Gain_OLoop = dcgain(GDz)
[ndgz,ddgz]=tfdata(GDz,'v')
[zo,po,ko]=tf2zp(ndgz,ddgz) % zero, pole, gain form
figure(1)
dstep(GDz,1) % open-loop step response
title('Open-Loop System Step Response')

% Closed-Loop Transfer Function T(z)=D(z)G(z)/(1+D(z)G(z))

Tz = feedback(GDz,1) % closed loop TF T(z), feedback loop gain =1
Gain_CLoop1=dcgain(Tz) % steady-state gain
[ntz,dtz]=tfdata(Tz,'v')
 % zero-pole-k form
[z,p,k]=tf2zp(ntz,dtz)

% closed-loop system step response with H(s)=1
figure(2)
dstep(Tz,1) % closed loop system step response with feedback loop gain =1
title('Closed-Loop System Step Response (Feedback Gain=1)')

% Closed-loop step resonse with H(s)=0.5
Tz2 = feedback(GDz,0.5) % closed loop TF T(z), feedback loop gain = 0.5
Gain_CLoop2=dcgain(Tz2) % steady-state gain
[ntz2,dtz2]=tfdata(Tz2,'v')
 % zero-pole-k form
[z,p,k]=tf2zp(ntz2,dtz2)
figure(3)
dstep(Tz2,1) % closed loop system step response with feedback loop gain =0.5
title('Closed-Loop System Step Response (Feedback Gain=0.5)')
```

```
Open-loop system transfer function
 GDz =

  0.0009967 z^3 - 0.0002027 z^2 - 0.0006937 z + 0.0004967
  -------------------------------------------------------
        z^4 - 3.09 z^3 + 4.179 z^2 - 3.079 z + 0.99

Open-loop system zeros (zo), poles(po):
zo =
  -0.9967 + 0.0000i
   0.6000 + 0.3742i
   0.6000 - 0.3742i


po =
   0.5500 + 0.8352i
   0.5500 - 0.8352i
   1.0000 + 0.0000i
   0.9900 + 0.0000i

Steady-state gain → ∞
```

```
Closed-loop system with feedback gain of 1
 Tz =

 0.0009967 z^3 - 0.0002027 z^2 - 0.0006937 z + 0.0004967
 -------------------------------------------------------
 z^4 - 3.089 z^3 + 4.179 z^2 - 3.08 z + 0.9905

Closed-loop system zeros (z), poles(p):
z =
 -0.9967 + 0.0000i
 0.6000 + 0.3742i
 0.6000 - 0.3742i

p =
 0.5502 + 0.8358i
 0.5502 - 0.8358i
 0.9943 + 0.0251i
 0.9943 - 0.0251i

Steady-state gain = 1
```

```
Closed-loop system with feedback gain of 0.5

Tz2 =

 0.0009967 z^3 - 0.0002027 z^2 - 0.0006937 z + 0.0004967
 -------------------------------------------------------
 z^4 - 3.09 z^3 + 4.179 z^2 - 3.079 z + 0.9903
```

```
Closed-loop system zeros (z2), poles(p2):

Z2 =
 -0.9967 + 0.0000i
 0.6000 + 0.3742i
 0.6000 - 0.3742i

P2 =
 0.5501 + 0.8355i
 0.5501 - 0.8355i
 0.9947 + 0.0174i
 0.9947 - 0.0174i

Steady-state gain = 2
```

Results from the Simulink model are identical, as expected, to those obtained using the Matlab code.

5.3 Discrete-Time Systems with Disturbance Inputs

A closed-loop discrete-time system with a disturbance input is shown in Fig. 5.12 where $R(s)$ is the reference input and $R_d(s)$ is the disturbance input through the disturbance transfer function $G_d(s)$. The main control loop includes a ZOH and a digital controller $D(z)$ to control the output $C(s)$ of the plant $G_p(s)$ close to $R(s)$, even in presence of $R_d(s)$. It is necessary to obtain the output $C(s)$, its starred transform $C^*(s)$, and its z-transform $C(z)$ with $R(s)$ only, with $R_d(s)$ only, and with both $R(s)$ and $R_d(s)$.

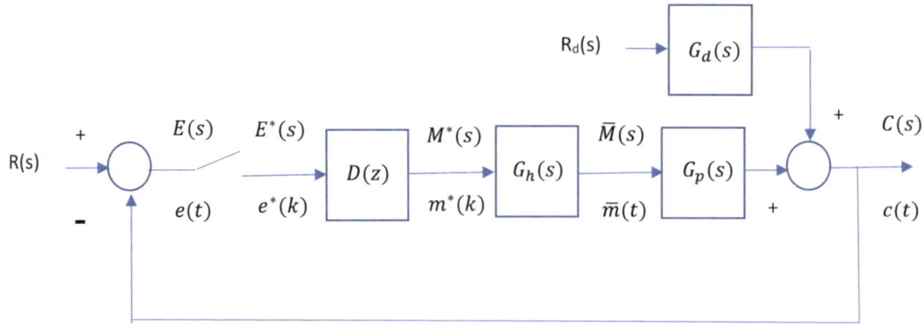

Fig. 5.12 Closed-loop discrete-time system with a digital filter and a disturbance input

First, the combination of ZOH $G_{h0}(s)$ and plan $G_p(s)$ is represented as

$$G(s) = G_{h0}(s)\ G_p(s) = \left(\frac{1 - e^{-sT}}{s}\right) G_p(s) \tag{5.19}$$

The output can be expressed as

$$C(s) = G(s)D^*(s)E^*(s) + G_d(s)R_d(s) \tag{5.20}$$

Expressing the error $E(s)$ as the difference of $R(s)$ and $C(s)$ and taking starred transform, the error can be expressed, similar to Eq. (5.5), as

$$E^*(s) = R^*(s) - C^*(s) \tag{5.21}$$

Taking the starred transform of $C(s)$ in Eq. (5.20) and using Eq. (5.21), the starred transform of the output can be expressed, after some simplification, as

$$C^*(s) = \overline{G_{h0}(s)G_p}^*(s)D^*(s)[R^*(s) - C^*(s)] + \overline{G_d R_d}^*(s) \tag{5.22}$$

$$C^*(s) = \frac{\overline{G_{h0}G_p}^*(s)D^*(s)}{1 + \overline{G_{h0}G_p}^*(s)D^*(s)} R^*(s) + \frac{\overline{G_d R_d}^*(s)}{1 + \overline{G_h(s)G_p}^*(s)D^*(s)} \tag{5.23}$$

The z-transform can be expressed as

$$C(z) = \frac{G(z)D(z)}{1 + G(z)D(z)} R(z) + \frac{\overline{G_d R_d}(z)}{1 + G(z)D(z)} \tag{5.24}$$

The closed-loop discrete-time system in the z-domain is shown in Fig. 5.13.

Example 5.2 A discrete-time closed-loop system with a reference input $R(s)$ and a disturbance input $R_d(s)$ is shown in Fig. 5.14 with the plant transfer function $G_p(s) = \frac{2}{s+1.5}$, a digital filter with transfer function $D(z) = \frac{3z-2}{z}$, a disturbance transfer function $G_d(s) = \frac{0.5}{s+1.5}$, reference input $R(s) = \frac{1}{s}$, disturbance input $R_d(s) = \frac{1}{s}$, and the feedback loop transfer function of a simple gain $H(s) = 0.5$. Use sampling time $T = 0.1$ s. Obtain the z-transform of the output $C(z)$ with $R(s)$

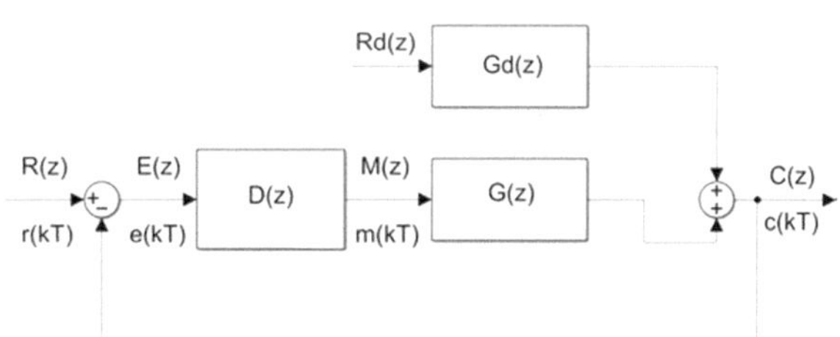

Fig. 5.13 Closed-loop discrete-time system with a digital filter and a disturbance input in z-domain

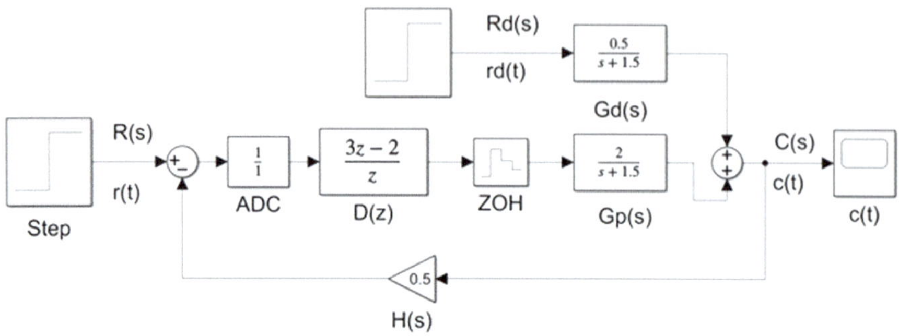

Fig. 5.14 Simulink model of a closed-loop discrete-time system with a disturbance input

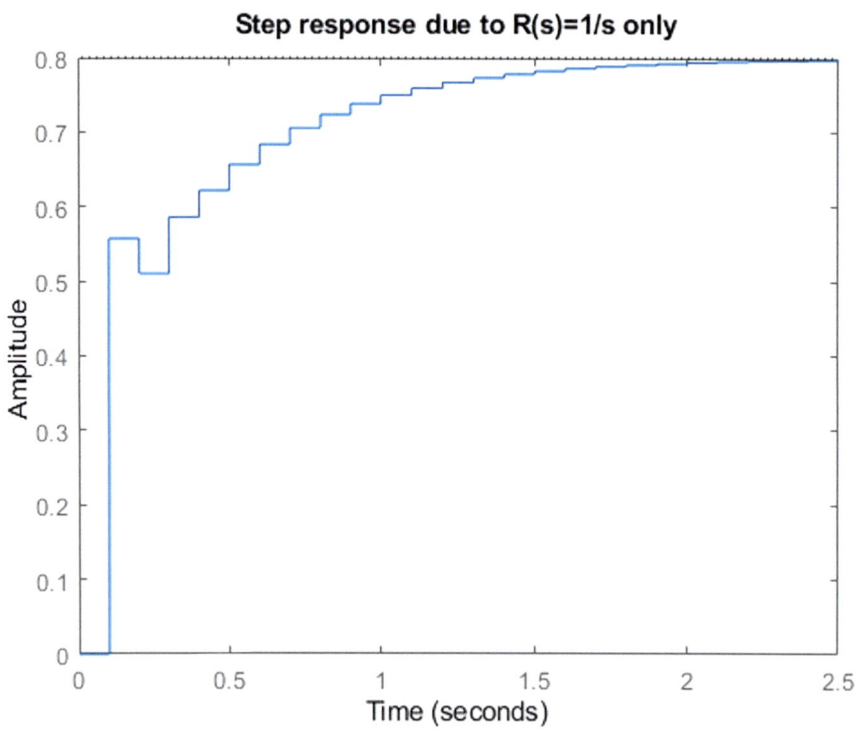

Fig. 5.15 Closed-loop system step response for Example 5.2 with reference input only

only, with $R_d(s)$ only, and with both $R(s)$ and $R_d(s)$. Obtain the closed-loop system response with $R(s)$ only.

Solution Matlab code and results are presented next. The closed-loop system step response with reference input only is shown in Fig. 5.15.

```
%ch5_ex5_2 with disturbance input
clear all; close all; clc
T=0.1;
Gp = tf(2, [1 1.5]); % Gp(s)
Gd = tf(0.5, [1 1.5]) % Gd(s)
Rds = tf(1, [1 0]); % Rd(s)=1/s
GRds= Gd*Rds
[ngdrs, dgdrs]=tfdata(GRds,'v')
[r,p,k]=residue(ngdrs,dgdrs); % PFE
n=length(dgdrs);
% convert s-domain poles (ps) to z-domain  pz =exp(ps*T)
for i=1:n-1
    pz(i)=exp(p(i)*T)
end
[numzz, denz]=residue(r,pz,k) % reconstruct z-domain transfer function using residue
numz =conv(numzz,[1 0]) % multiply by z
GdRdz=tf(numz,denz,T) %  GdRd(z)

nhs=0.5; dhs=1; % H(s)=0.4Feedback path gain
Gz=c2d(Gp, T) % G(z)=Z(Gho(s)Gp(s))

% D(z)
Dz = tf([3 -2], [1 0], T) % D(z)
%D(z)G(z)
DGz = Dz*Gz

% Tr(z) for R(z) only
Trz = feedback(DGz, nhs) % Tr(z)=C(z)/R(z)
% step response
dstep(Trz, 1)
title('Step response due to R(s)=1/s only')
Rz = tf([1 0], [1 -1], T) % R(z)=z/(z-1) unit step
% Output C(z) with R(s) only
Crz =Trz*Rz % C(z) due to R(z) only

% Output with Disturbance only
Cdz = GdRdz/(1+DGz*nhs)

% Output C(z) with both R(s) and Rd(s)
Cz = Crz + Cdz % C(z) with both R(s) and Rd(s)

Results:
Cr(z) =
         0.5572 z^2 - 0.3714 z
  ----------------------------------
  z^3 - 1.582 z^2 + 0.3964 z + 0.1857

Cd(z) =
            0.04643 z^3 - 0.03996 z^2
  ---------------------------------------------
  z^4 - 2.443 z^3 + 1.758 z^2 - 0.1555 z - 0.1599

C(z) =
    0.6036 z^6 - 1.846 z^5 + 1.969 z^4 - 0.7469 z^3 - 0.03874 z^2 + 0.05938 z
--------------------------------------------------------------------------------
z^7 - 4.025 z^6 + 6.019 z^5 - 3.72 z^4 + 0.3294 z^3 + 0.5178 z^2 - 0.09224 z -0.02969
```

Example 5.3 Use Simulink to simulate the closed-loop system response of the system shown in Fig. 5.14 for (a) $R(s) = \frac{1}{s}$ only, (b) $R_d(s) = \frac{1}{s}$, and (c) both $R(s) = \frac{1}{s}$ and $R_d(s) = \frac{1}{s}$ applied simultaneously. Use sampling time $T = 0.1$ s.

The Simulink model of Fig. 5.14 was used to obtain a simulated closed-loop system response as shown in Fig. 5.16 with $R(s)$ only, Fig. 5.17 with $R_d(s)$ only, and Fig. 5.18 with both $R(s)$ and $R_d(s)$ simultaneously.

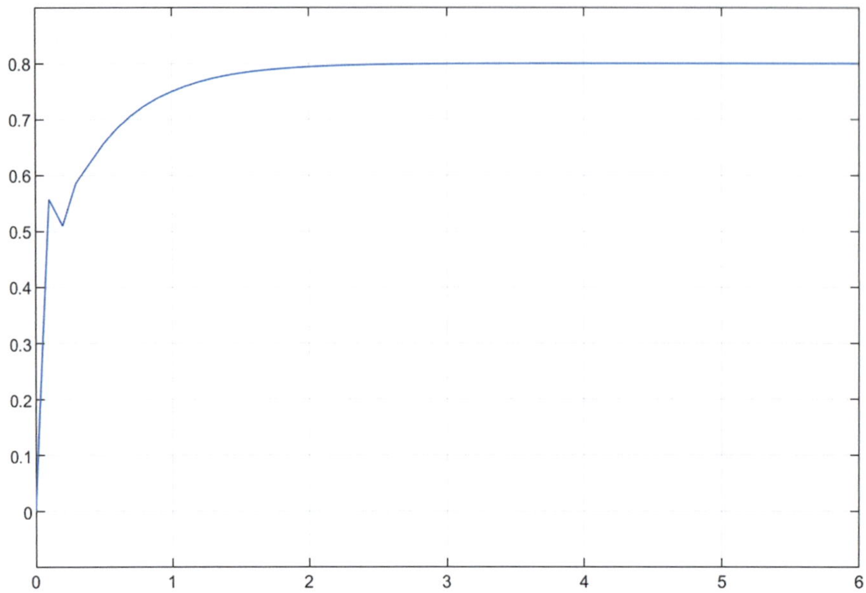

Fig. 5.16 Closed-loop system response with reference input $R(s) = \frac{1}{s}$ only

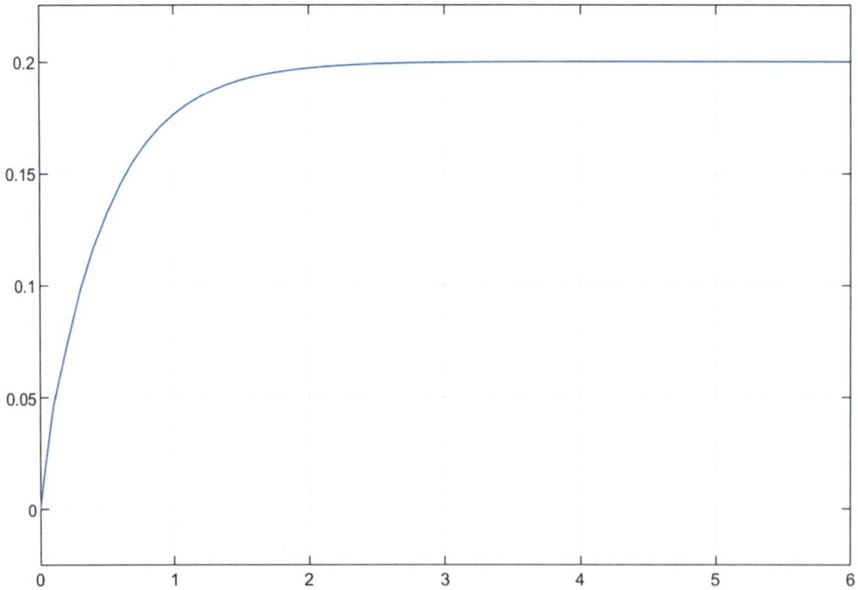

Fig. 5.17 Closed-loop system response with disturbance input $R_d(s) = \frac{1}{s}$ only

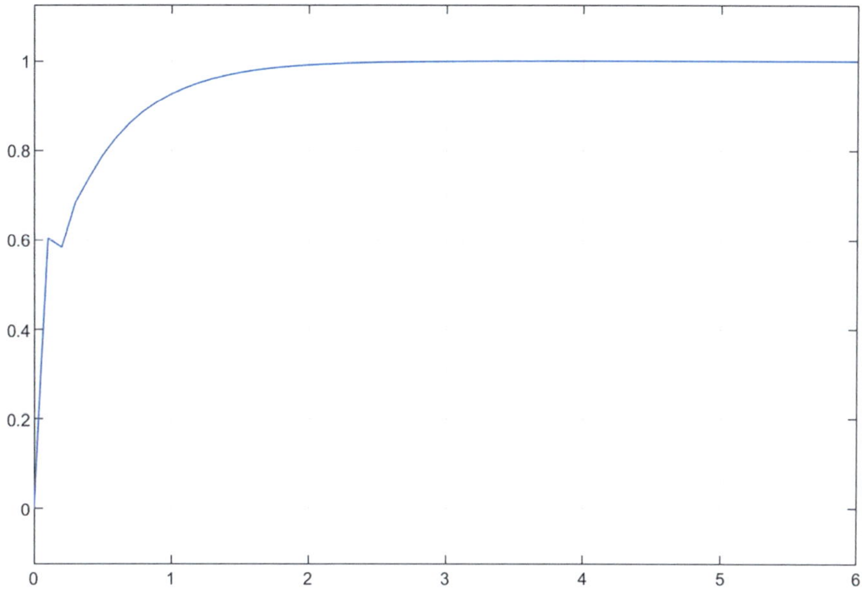

Fig. 5.18 Closed-loop system response with both reference and disturbance inputs $R(s) = R_d(s) = \frac{1}{s}$

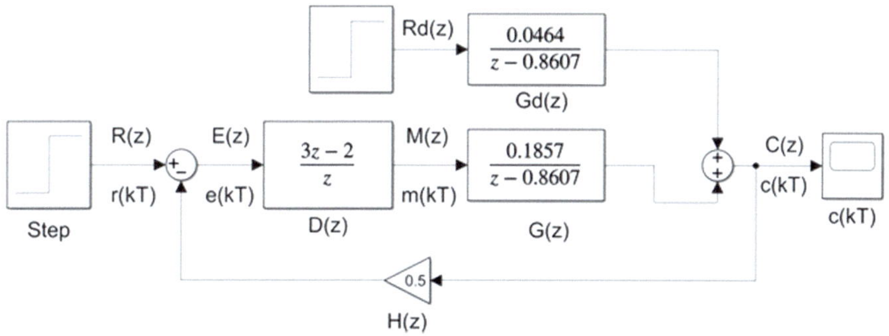

Fig. 5.19 Simulink model of a discrete-time closed-loop system with a disturbance input

Example 5.4 Use Simulink to simulate the closed-loop system response of the system shown in Fig. 5.19 with the system represented in z-domain for (a) $R(s) = \frac{1}{s}$ only, (b) $R_d(s) = \frac{1}{s}$ only, and (c) both $R(s) = \frac{1}{s}$ and $R_d(s) = \frac{1}{s}$ applied simultaneously. Use sampling time $T = 0.1$ s.

The Simulink model of Fig. 5.19 was used to simulate closed-loop system response in Fig. 5.20 with reference input only, Fig. 5.21 with disturbance input only, and Fig. 5.22 with both reference and disturbance inputs simultaneously.

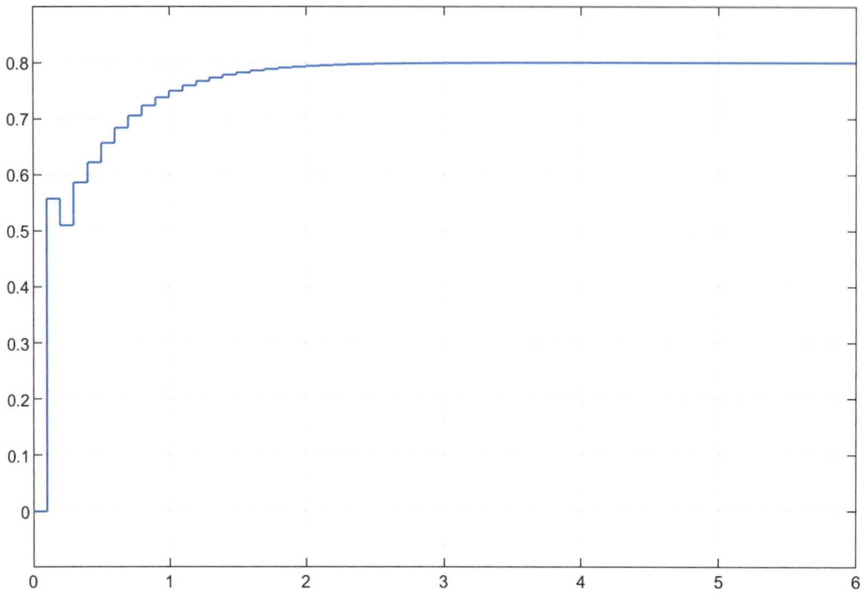

Fig. 5.20 Closed-loop discrete-time system response with step reference input $R(z) = \frac{z}{z-1}$ only

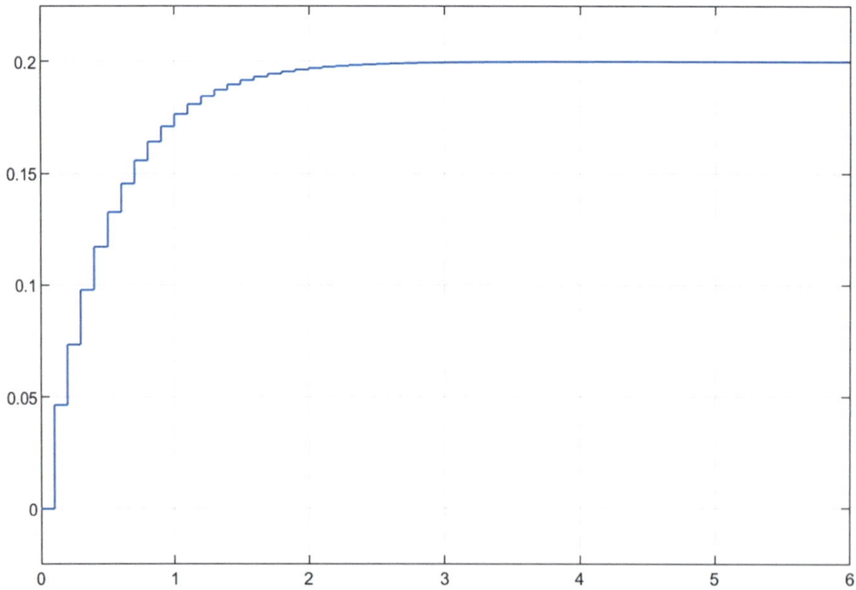

Fig. 5.21 Closed-loop discrete-time system response with step disturbance input $R_d(z) = \frac{z}{z-1}$ only

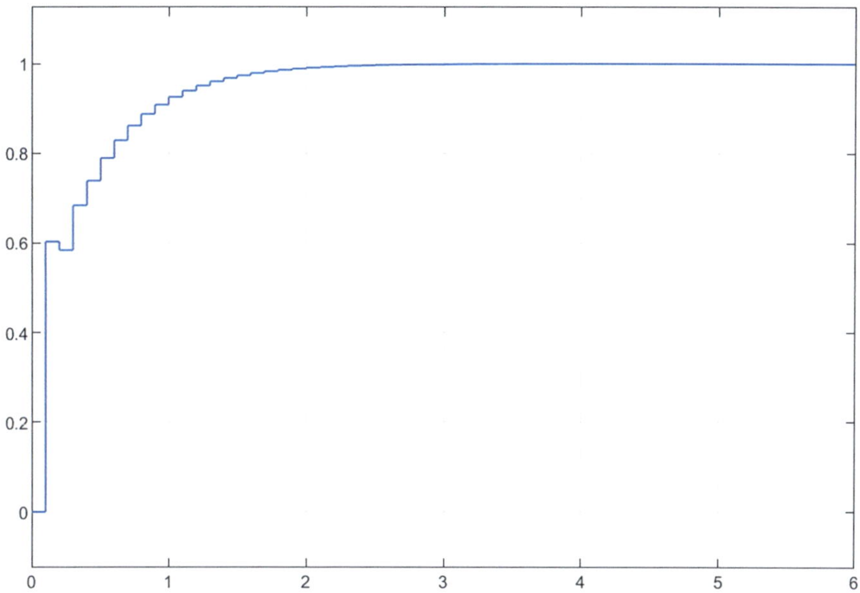

Fig. 5.22 Closed-loop discrete-time system response with step reference and disturbance inputs $R(z) = R_d(z) = \frac{z}{z-1}$

5.4 Estimation of Model Parameters and Validation for Discrete-Time Systems

Objectives The objectives are to determine the system parameters for modeling the dynamics of a DC motor from its time response characteristics of speed and to validate the system model parameters by comparing the actual and the simulated responses.

Introduction The speed versus armature voltage of a small DC motor can be modeled as a first-order system with a transfer function of $G_s(s) = \frac{K}{\tau s+1}$, where K is the steady state gain and τ is the time constant. Using time response characteristics of a first-order system for a step input, the parameters K and τ can be experimentally determined for the DC motor. The experiment is conducted on a small DC motor module mounted on a NI ELVIS II board in a LabVIEW environment. ELVIS II board has all the necessary interfaces between the DC motor module and the PC running the LabVIEW. A newer version of the DC motor module and NI ELVIS board can also be used for this experiment.

I. *Set Up LabVIEW Virtual Instrument*

(a) Ensure power supplies are connected to the NI ELVIS II board and the DC motor module. Ensure the USB cable is connected to both the ELVIS II board and the PC.
(b) Turn on both power switches of the NI ELVIS II board.
(c) The PC should automatically begin installing the drivers for the ELVIS II board.
(d) When the drivers have been installed, open *01-QNET_DCMCT_Modeling.vi* in LabVIEW.
(e) Set sampling rate. For the DC motor control trainer (DCMCT) module, the sampling rate can be set to 100 Hz as the time constant of the DC motor module (τ) is around 0.10 s. Select a device such that the front panel responds to any change in the motor position.

II. *Determine Static Gain (K) and Time Constant (τ)*

(a) Select your input parameters. Voltage should be high enough to get a distinct response curve from the motor, while frequency should be small enough to allow the motor to achieve a steady state. For the DCMCT module, typical values for the input square wave voltage can be set as follows: amplitude of 1.0 V, frequency of 0.40 Hz, and offset of 3.0 V.
(b) Run the VI to obtain simulated and measured speed responses.
(c) Click the stop button when a complete cycle of the response curve is shown on the graph.
(d) Click the Measurement Graphs tab. Use measurement cursors to get the initial time (t_0), initial motor speed (ω_0), steady-state motor speed (ω_{ss}) from the square wave response, and the input voltage step (ΔV). Complete Table 5.1 to obtain static gain (K) and time constant (τ).

Table 5.1 Experimental data for determination of static gain and time constant

Initial time t_0 (s)	Initial motor speed, ω_0 (rad/s)	Steady-state motor speed, ω_{ss} (rad/s)	Motor speed at $t = \tau$, $\omega_\tau = \omega_0 + 0.632(\omega_{ss} - \omega_0)$, (rad/s)	Time at ω_τ, t_τ (s)	Time constant, $\tau = t_\tau - t_0$ (s)	Input voltage step, $\Delta V = V_{max} - V_{min}$ (V)	Static gain, $K = \frac{\omega_{ss} - \omega_0}{\Delta V}$, $\left(\frac{rad}{s}/V\right)$

III. *Validate Measured Parameters*

(a) Click the Scopes tab to return to the initial screen.
(b) Insert your measured parameters into the Model Parameters box, ensuring the units of each match up. The simulated and measured responses should match up closely. Make sure to set the number of significant figures to at least 5 for the display format of each graph on the front panel VI (right-click on the graph → display format → decimal → 6). Export the time response graph of motor rotation $\theta(t)$ to Excel using right-click on the graph and selecting the Export to Excel option (right-click → Export to Excel). Compare it to the simulated response.

IV. *Deliverables*

A. Include Table 5.1 in the lab report.
B. *Develop the Open-Loop Transfer Functions of the DC Motor*

(a) K and τ are directly used to obtain the open-loop transfer function $G_s(s)$ for speed of the DC motor as

$$G_s(s) = \frac{K}{\tau s + 1} \tag{5.25}$$

(b) Obtain the open-loop transfer function $G_p(s)$ for the position of the DC motor as

$$G_p(s) = \frac{K}{s(\tau s + 1)} \tag{5.26}$$

C. *Develop z-Domain Transfer Functions of the DC Motor*

(a) Select a suitable sampling time (T_s) for the system. Generally, T_s should be in the order of $\frac{\tau}{5}$ to $\frac{\tau}{10}$. For the DC motor control trainer module, sampling time can be set to 0.01 s ($T_s = 0.01$ s) as the time constant is around 0.10 s. This sampling rate corresponds to the sampling rate of 100 Hz ($f_s = 1/T_s = 100$ Hz).
(b) Using Matlab, convert the s-domain transfer functions $G_s(s)$ and $G_p(s)$ to $G_s(z)$ and $G_p(z)$, respectively.

D. *Simulate Open-loop System Responses in Matlab/Simulink*

(a) Obtain the open-loop step response for the discrete-time system model of the DC motor speed in Matlab.
(b) Create a Simulink model for simulating the step response of the discrete-time system model of the DC motor speed.

(c) Compare the system response from simulation results with that obtained from the experiment.
(d) Repeat the simulation of steps (a) and (b) for motor position.
(e) Include system response graphs from simulation and experimental validation in the lab report.

Lab Report
Present all deliverables in the lab report and discuss your results.

Exercises

5.1 For a closed-loop discrete-time system with a digital filter as shown in Fig. P5.1, with a ZOH and a plant transfer function $G_p(s) = \frac{1}{s(5s+1)}$ and a unit step function $R(s) = \frac{1}{s}$, the digital filter is given as $D(z) = \frac{z^2 - 1.5z + 0.9}{z^2 - 1.3z + 1}$. Use sampling time $T = 0.1$ s. Find the transfer function of (a) the open-loop system, (b) the closed-loop system with a feedback loop gain of $K = 1$, and (c) the closed-loop system with a feedback loop gain of $K = 0.5$. Obtain unit step response of (a) the open-loop system, (b) the closed-loop system with a feedback loop gain of 1, and (c) the closed-loop system with a feedback loop gain of 0.5. Obtain steady-state values of the output for each case. Verify these values from steady-state gain (hint use Matlab command *dcgain()*).

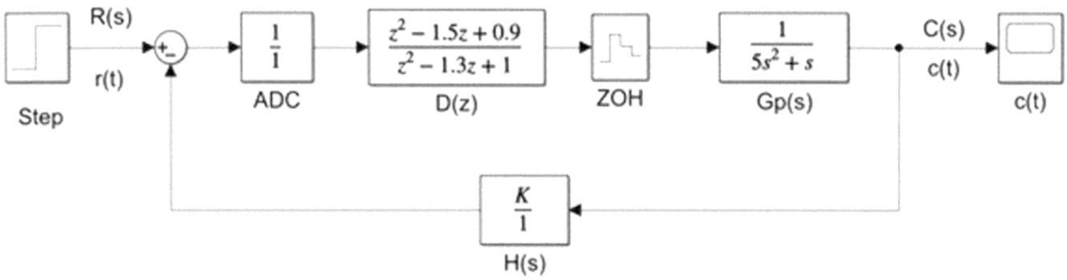

Fig. P5.1 A closed-loop discrete-time system with a digital filter

5.2 A discrete-time closed-loop system with a reference input $R(s)$ and a disturbance input $R_d(s)$ is shown in Fig. P5.2 with the plant transfer function $G_p(s) = \frac{3}{s+2.5}$, a digital filter with transfer function $D(z) = \frac{4z-2}{z-0.3}$, a disturbance transfer function $G_d(s) = \frac{0.5}{s+2.5}$, reference input $R(s) = \frac{1}{s}$, disturbance input $R_d(s) = \frac{1}{s}$, and the feedback loop transfer function of a simple gain $H(s) = 0.3$. Use sampling time $T = 0.1$ s. Obtain the z-transform of the output $C(z)$ with $R(s)$ only, with $R_d(s)$ only, and with both $R(s)$ and $R_d(s)$. Obtain the closed-loop system response with $R(s)$ only.

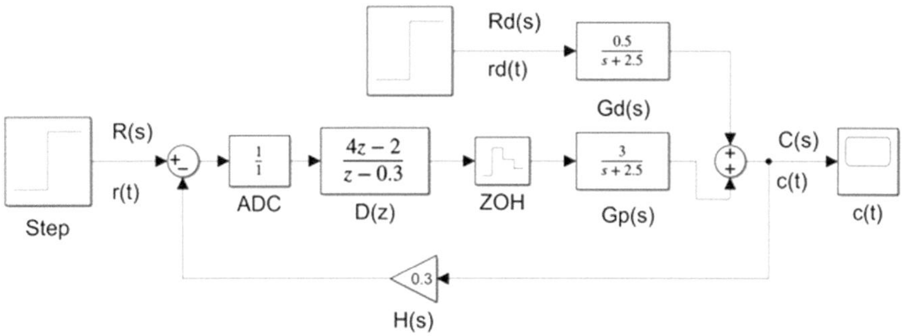

Fig. P5.2 A closed-loop discrete-time system with a digital filter and a disturbance input

5.3 Use Simulink to simulate the closed-loop system response of the system shown in Fig. P5.3 for (a) $R(s) = \frac{1}{s}$ only, (b) $R_d(s) = \frac{1}{s}$ only, and (c) both $R(s) = \frac{1}{s}$ and $R_d(s) = \frac{1}{s}$ applied simultaneously. Use sampling time $T = 0.1$ s.

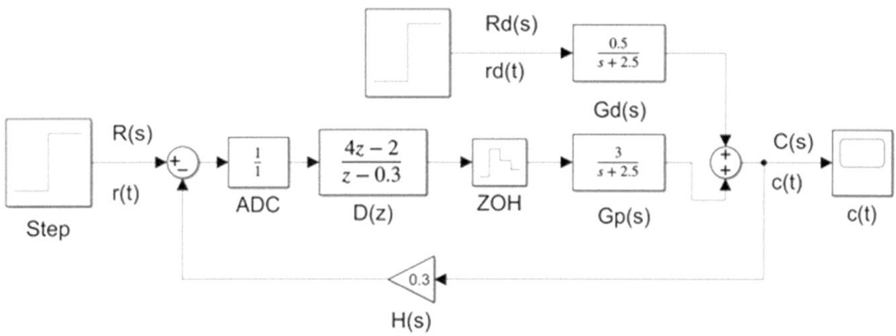

Fig. P5.3 Simulink model of a closed-loop discrete-time system for Problem 5.3

5.4 Use Simulink to simulate the closed-loop system response of the system shown in Fig. P5.4 with the system represented in z-domain for (a) $R(z) = \frac{z}{z-1}$ only, (b) $R_d(z) = \frac{z}{z-1}$ only, and (c) both $R(z) = \frac{z}{z-1}$ and $R_d(z) = \frac{z}{z-1}$ applied simultaneously. Use sampling time $T = 0.1$ s.

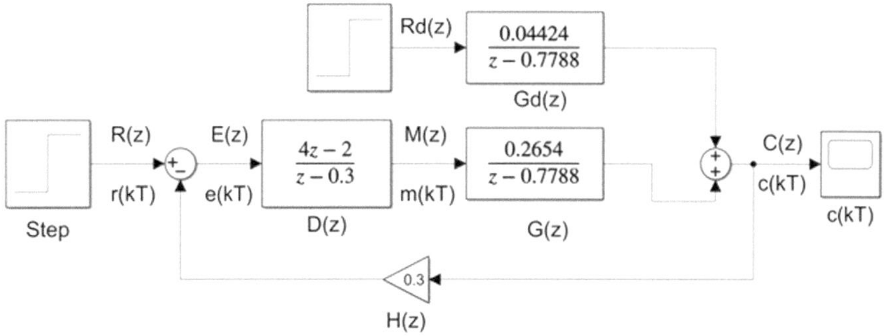

Fig. P5.4 Simulink model of a closed-loop discrete-time system for Problem 5.4

Further Reading

1. Asadi F (2023) Signals and systems with MATLAB® and Simulink®. Springer, Cham
2. Astrom KJ, Murray RM (2021) Feedback systems: an introduction for scientists and engineers, 2nd edn. Princeton University Press, Princeton
3. Chen C-T (2006) Analog and digital control system design: transfer-function, state-space, and algebraic methods. Oxford University Press, New York
4. Dorf R, Bishop R (2016) Modern control systems, 13th edn. Pearson, Essex
5. Fadali MS, Visioli A (2020) Digital control engineering analysis and design, 3rd edn. Academic Press, London
6. Franklin GF, Powell JD, Emami-Naeini A (2019) Feedback control of dynamic systems, 8th edn. Pearson, Upper Saddle River
7. Franklin GF, Powell JD, Workman M (1998) Digital control of dynamic systems, 3rd edn. Addison-Wesley, Boston
8. Isermann R (1981) Digital control systems. Springer, Berlin
9. Jalili N, Candelino NW (2023) Dynamic systems and control engineering. Cambridge University Press, Cambridge
10. Keviczky L, Bars R, Hetthéssy J, Bányász C (2019) Control engineering. Springer, Singapore
11. Koeing DM (2009) Practical control engineering: a guide for engineers, managers, and practitioners. McGraw-Hill, New York
12. Kuo BC (1995) Digital control systems, 2nd edn. Oxford University Press, Oxford
13. Landau ID, Gito G (2006) Digital control systems: design, identification and implementation. Springer, London
14. Leigh JR (2006) Applied digital control: theory, design and implementation, 2nd edn. Dover, Mineola
15. Mathworks (2024) Control System Toolbox. https://www.mathworks.com/products/control.html
16. Mathworks (2024) Matlab. https://www.mathworks.com/products/matlab.html
17. National Instruments (2024) LabVIEW. https://www.ni.com/en/support/downloads/software-products/download.labview.html#487445
18. National Instruments (2024) NI-ELVIS-II. https://www.ni.com/en-us/support/model.ni-elvis-ii.html
19. National Instruments (2024) Labs for Quanser QNET DC Motor Add-On Board. https://education.ni.com/teach/resources/180/labs-for-quanser-qnet-dc-motor-add-on-board
20. Nise NS (2019) Control systems engineering, 8th edn. Wiley, Hoboken
21. Nof SY (ed) (2009) Springer handbook of automation. Springer, Berlin
22. Ogata K (1995) Discrete-time control systems, 2nd edn. Pearson, Hoboken
23. Phillips CL, Nagle HT, Chakrabortty A (2015) Digital control system analysis and design, 4th edn. Pearson, Essex
24. Quanser (2024) QUBE Servo 2. https://www.quanser.com/products/qube-servo-2/
25. Rabbath CA, Lechevin N (2014) Discrete-time control system design with applications. Springer, New York
26. Samanta B (2023) Introduction to mechatronics – an integrated approach. Springer, Cham
27. Shertukde HM (2015) Digital control applications illustrated with MATLAB®. CRC Press, Boca Raton
28. Starr G (2020) Introduction to applied digital controls. Springer, Cham
29. Veloni A, Miridakis NI (2018) Digital control systems-theoretical problems and simulation tools. CRC Press, Boca Raton
30. Vilanova R, Visioli A (2012) PID control in the third millennium – lessons learned and new approaches. Springer, London
31. Visioli A (2006) Practical PID control. Springer, London

Chapter 6
Discrete-Time System Characteristics

6.1 Introduction

Characteristics of discrete-time closed-loop systems are presented in this chapter. Time response characteristics of discrete-time closed-loop systems, mapping of the s-plane in the z-plane for characterizing closed-loop system time response from the closed-loop poles, discrete equivalents, and steady-state accuracy for different types of systems and inputs are covered. Examples are presented along with Matlab codes and Simulink models to illustrate the analysis and simulation of closed-loop system characteristics. End-of-chapter exercises are provided to enhance understanding of the topics covered and consolidate problem-solving skills.

6.2 Discrete-Time System Time Response

A closed-loop discrete-time system with a digital filter $D(z)$ is shown in Fig. 6.1 where a general-order hold device is represented with a transfer function $G_h(s)$ and a plant with a transfer function $G_p(s)$. The sampling and data hold devices are replaced with analog-to-digital converter (ADC) and digital-to-analog converter (DAC). The signals and the corresponding Laplace transforms are also shown. The z-domain version of the closed-loop discrete-time system is shown in Fig. 6.2. The effects of sampling and hold of Fig. 6.1 are incorporated in the z-domain representation of the closed-loop discrete-time system in Fig. 6.2 through the z-transform of the general-order hold device transfer function $G_h(s)$ and the plant transfer function $G_p(s)$ as

$$G(z) = \overline{G_h G_p}(z) = Z\left[G_h(s)G_p(s)\right] \tag{6.1}$$

For a zero-order-hold (ZOH), Eq. (6.1) can be written as

B. Samanta, *Introduction to Digital Control*,
https://doi.org/10.1007/978-3-031-66830-2_6

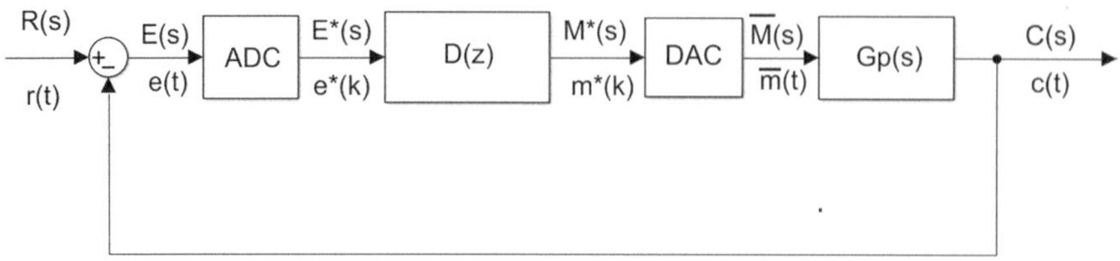

Fig. 6.1 Closed-loop discrete-time control system with a digital filter

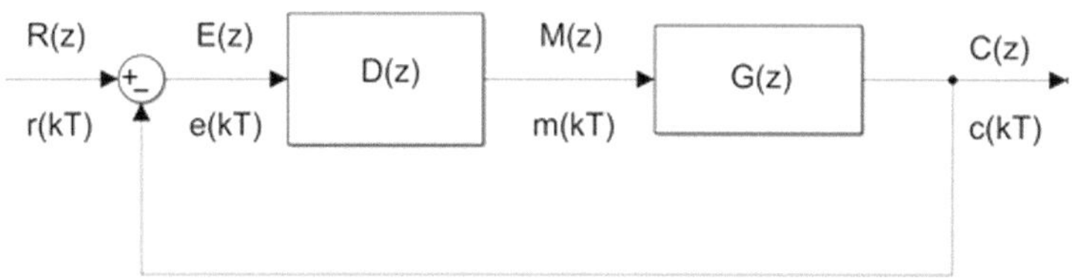

Fig. 6.2 Closed-loop discrete-time control system in z-domain

$$\begin{aligned} G(z) &= \overline{G_{h0}G_p}(z) = Z\left[\frac{1-e^{-sT}}{s}\,G_p(s)\right]_{e^{sT}=z} \\ &= (1-z^{-1})Z\left[\frac{G_p(s)}{s}\right] \end{aligned} \tag{6.2}$$

The closed-loop discrete-time system transfer function in z-domain can be derived from Fig. 6.2. The output in z-domain, $C(z)$, can be expressed as

$$C(z) = G(z)M(z) \tag{6.3}$$

The controller output $M(Z)$ can be expressed as

$$M(z) = D(z)E(z) \tag{6.4}$$

The error between the reference input and the system output can be expressed as

$$E(z) = R(z) - C(z) \tag{6.5}$$

Substituting $M(z)$ and $E(z)$ from Eqs. (6.4) and (6.5), the system output can be written as

$$C(z) = G(z)D(z)[R(z) - C(z)] \tag{6.6}$$

The closed-loop system output, $C(z)$, can be expressed as

$$C(z) = \frac{G(z)D(z)}{1 + G(z)D(z)} R(z) \tag{6.7}$$

If there is a transfer function $H(s)$ in the feedback path of Fig. 6.1, similar to Fig. 5.2, the closed-loop discrete-time system transfer function can similarly be expressed as

$$C(z) = \frac{G(z)D(z)}{1 + \overline{GH}(z)D(z)} R(z) \tag{6.8}$$

where $\overline{GH}(z)$ is defined, similar to Eq. (5.18), as

$$\overline{GH}(z) = Z\left[G_h(s)G_p(s)H(s)\right] \tag{6.9}$$

Example 6.1 Obtain step response of a closed-loop discrete-time system with the plant transfer function $G_p(s) = \frac{3}{s+2.5}$ and a zero-order hold (ZOH) for a sampling time $T = 0.1$ s. Obtain the output $c(kT)$ using inverse z-transform.

Solution The open-loop transfer function of the discrete-time system with ZOH is obtained as

$$\begin{aligned} G(z) &= \overline{G_{h0}G_p}(z) = Z\left[\frac{1 - e^{-sT}}{s} G_p(s)\right]_{e^{sT} = z} = \left(1 - z^{-1}\right) Z\left[\frac{G_p(s)}{s}\right] \\ &= \left(1 - z^{-1}\right) Z\left[\frac{3}{s(s + 2.5)}\right] = \frac{z - 1}{z}\left(\frac{3}{2.5}\right)\left[\frac{z}{z - 1} - \frac{z}{z - e^{-2.5T}}\right] \\ &= \frac{z - 1}{z}\left(\frac{3}{2.5}\right)\left[\frac{z(z - e^{-2.5T} - z + 1)}{(z - 1)(z - e^{-2.5T})}\right] = \frac{0.2654}{z - 0.7788} \end{aligned}$$

Closed-loop transfer with unity feedback is obtained as

$$T(z) = \frac{G(z)}{1 + G(z)} = \frac{\frac{0.2654}{z - 0.7788}}{1 + \frac{0.2654}{z - 0.7788}} = \frac{0.2654}{z - 0.5134}$$

z-transform of a unit step input is obtained as

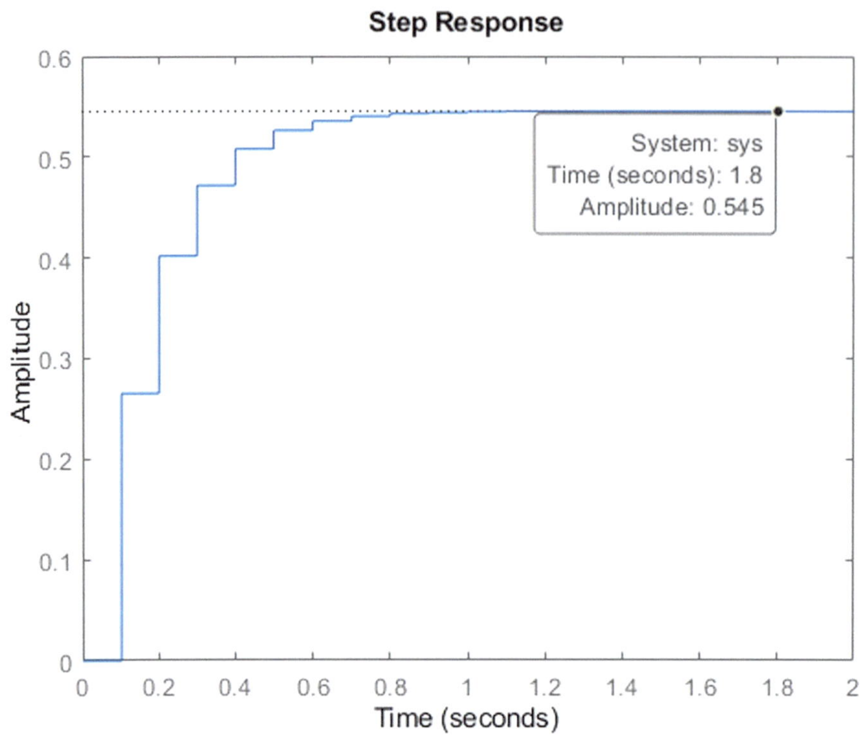

Fig. 6.3 Closed-loop discrete-time system step response for Example 6.1

$$R(z) = \frac{z}{z-1}$$

The z-transform of the output of the closed-loop discrete-time system is obtained as

$$C(z) = T(z)R(z) = \left(\frac{0.2654}{z-0.5134}\right)\left(\frac{z}{z-1}\right) = \frac{0.2654z}{z^2 - 1.5134z + 0.5134}$$

Inverse z-transform of $C(z)$ is obtained, using PFE, as

$$c(kT) = 0.5455 * \left(1 - (0.5134)^k\right)$$

Matlab code and results are presented next. The step response of the closed-loop discrete-time system is shown in Fig. 6.3.

```
% Ch6 Ex 6_1 Time response of a closed-loop discrete-time system
% Gp(s)=3/(s+2.5) First order system
T=0.1 ; % Sampling time
Gp = tf(3, [1 2.5]) %Gp(s)

% R(s)=1/s R(z)= z/(z-1)
Rz = tf([1 0], [1 -1], T) % R(z)

% G(z) = Z(Gho(s)Gp(s))
Gz =c2d(Gp,T) % G(z)

% Closed loop Transfer function
Tz = feedback(Gz, 1) % Closed loop TF with unity feedback
css = dcgain(Tz) % steady-state value of c(kT) with unit step input

% C(z) =T(z)*R(z)
Cz = Tz * Rz

% zpk for of C(z)
Cz = zpk(Cz) % zero-pole-gain form of C(z)

% Closed-loop step response
dstep(Tz,1)

% Inverse z-transform of C(z) using iztrans
syms z k
Cz =0.26544*z/(z^2-1.5134*z +0.5134);
ck = vpa(iztrans(Cz, k),4)

Results:

G(z) =
    0.2654
  ----------
  z - 0.7788
T(z) =
     0.2654
  ----------
  z - 0.5134

css =     0.5455
C(z) =
     0.26544 z
  ----------------
  (z-1) (z-0.5134)

Inverse z-transform
C(k) = 0.5455 (1 - 0.5134^k), k≥ 0
```

Example 6.2 Obtain step response of a closed-loop discrete-time system with the plant transfer function $G_p(s) = \frac{3}{s(s+2)}$ and a zero-order hold (ZOH) for a sampling time $T = 0.1$ s. Compare the discrete-time closed-loop step response with the corresponding continuous-time closed-loop system response.

Solution The open-loop transfer function of the discrete-time system with ZOH is obtained as

$$
\begin{aligned}
G(z) &= \overline{G_{h0}G_p}(z) = Z\left[\frac{1-e^{-sT}}{s}G_p(s)\right]_{e^{sT}=z}\\
&= (1-z^{-1})Z\left[\frac{G_p(s)}{s}\right] = (1-z^{-1})Z\left[\frac{3}{s^2(s+2)}\right]\\
&= \frac{z-1}{z}\left(\frac{3}{2}\right)\left[\frac{z\left[(2T-1+e^{-2T})z+\left(1-e^{-2T}-2Te^{-2T}\right)\right]}{2(z-1)^2(z-e^{-2T})}\right]\\
&= \left(\frac{3}{4}\right)\left[\frac{\left[(2T-1+e^{-2T})z+\left(1-e^{-2T}-2Te^{-2T}\right)\right]}{(z-1)(z-0.8187)}\right]\\
&= \frac{0.01405z+0.01314}{z^2-1.8187z+0.8187}
\end{aligned}
$$

Closed-loop transfer with unity feedback is obtained as

$$
T(z) = \frac{G(z)}{1+G(z)} = \frac{\frac{0.01405z+0.01314}{z^2-1.8187z+0.8187}}{1+\frac{0.01405z+0.01314}{z^2-1.8187z+0.8187}} = \frac{0.01405z+0.01314}{z^2-1.8047z+0.8319}
$$

z-transform of the unit step input is obtained as

$$
R(z) = \frac{z}{z-1}
$$

The z-transform of the output of the closed-loop discrete-time system is obtained as

$$
C(z) = T(z)R(z) = \frac{0.01405z+0.01314}{z^2-1.8047z+0.8319}\left(\frac{z}{z-1}\right)
$$

Closed-loop continuous-time system transfer function is obtained as

$$
T(s) = \frac{G_p(s)}{1+G_p(s)} = \frac{\frac{3}{s(s+2)}}{1+\frac{3}{s(s+2)}} = \frac{3}{s^2+2s+3}
$$

Matlab code and results are presented next. The step response of the closed-loop discrete-time system is shown in Fig. 6.4, and the corresponding continuous-time closed-loop system response is shown in Fig. 6.5. The effects of sampling can be observed comparing Figs. 6.4 and 6.5.

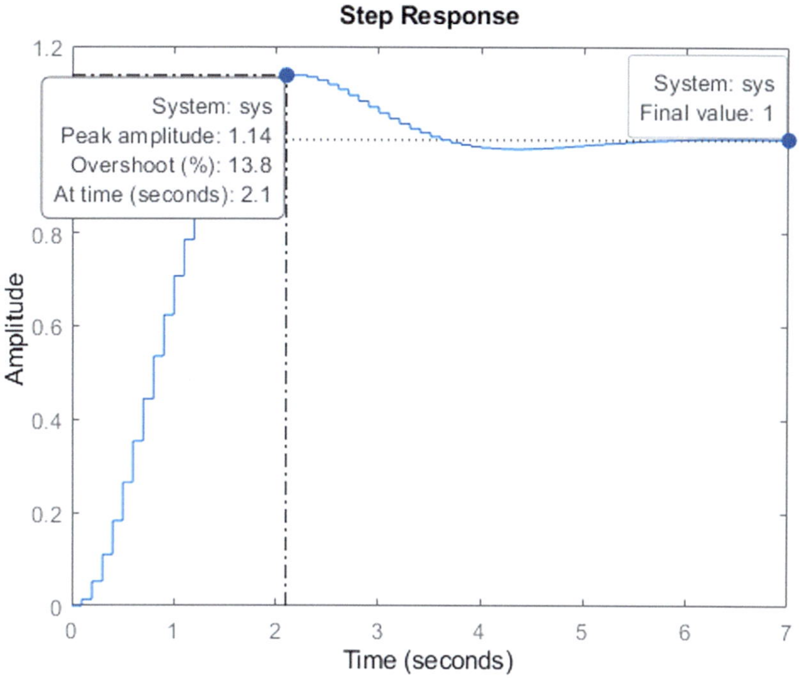

Fig. 6.4 Closed-loop discrete-time system step response for Example 6.2

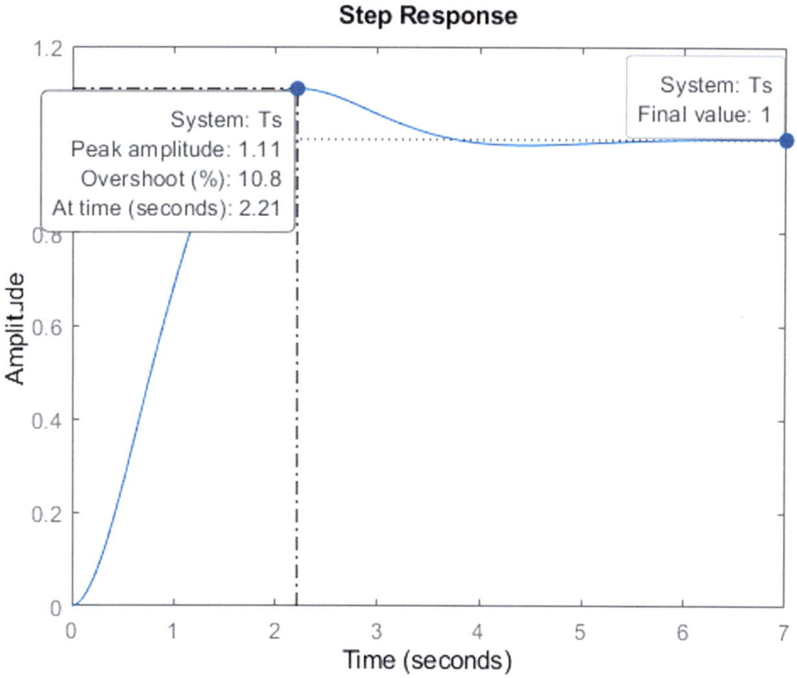

Fig. 6.5 Closed-loop continuous-time system step response for Example 6.2

```
% Ch6 Ex 6_2 Time response of a 2nd order closed-loop discrete-time system
% Gp(s)=3/s(s+2) First order system
clear all; close all; clc
T=0.1 ; % Sampling time
Gp = tf(3, [1 2 0]) %Gp(s)

% R(s)=1/s R(z)= z/(z-1)
Rz = tf([1 0], [1 -1], T) % R(z)

% G(z) = Z(Gho(s)Gp(s))
Gz =c2d(Gp,T) % G(z)

% Closed loop Transfer function
Tz = feedback(Gz, 1) % Closed loop TF with unity feedback
css = dcgain(Tz) % steady-state value of c(kT) with unit step input

% C(z) =T(z)*R(z)
Cz = Tz * Rz

% zpk for of C(z)
Cz = zpk(Cz) % zero-pole-gain form of C(z)

% Closed-loop step response
figure(1)
dstep(Tz,1)

% comparison with continuous-time system
Ts = feedback(Gp, 1) % closed-loop system transfer function
figure(2)
stepplot(Ts) % closed-loop system step response
```

```
Results:

G(z) =
   0.01405 z + 0.01314
  ----------------------
  z^2 - 1.819 z + 0.8187

T(z) =
    0.01405 z + 0.01314
  ----------------------
  z^2 - 1.805 z + 0.8319

css =    1.0000



C(z) =
      0.014048 z (z+0.9355)
  -----------------------------
  (z-1) (z^2 - 1.805z + 0.8319)

T(s) =
         3
  -------------
  s^2 + 2 s + 3
```

6.3 Discrete-Time System Characteristic Equation

The transfer function of a closed-loop discrete-time system, similar to the one in Fig. 6.2 with $D(z) = 1$, can be written as

$$T(z) = \frac{C(z)}{R(z)} = \frac{G(z)}{1 + \overline{GH}(z)} \tag{6.10}$$

The transfer function $T(z)$ can be expressed as

$$T(z) = K \frac{\prod_{i=1}^{m} (z - z_i)}{\prod_{j=1}^{n} (z - p_j)} \tag{6.11}$$

where z_i, $i = 1, \ldots, m$ and p_j, $j = 1, \ldots, n$ are zeros and poles of the closed-loop system transfer function and K is the gain. The z-transform of the system output can be expressed using Eq. (6.11) and partial fraction expansion, as

$$C(z) = \sum_{j=1}^{n} k_i \frac{z}{z - p_j} + C_R(z) \tag{6.12}$$

where the first n terms are related to the natural or characteristic response of the system and $C_R(z)$ represents response terms related to the input $R(z)$. For a stable system, the first n terms lead to transient response. The inverse z-transform of the jth of the first n terms can be expressed as

$$Z^{-1}\left[k_j \frac{z}{z - p_j}\right] = k_j\, p_j^{\,k}, k \geq 0 \tag{6.13}$$

The system transient response is represented by these n terms with the poles p_j, $j = 1, \ldots, n$ as the roots of the closed-loop system characteristic equation:

$$1 + \overline{GH}(z) = 0 \tag{6.14}$$

6.4 Mapping of the s-Plane in the z-Plane

The relationship between the z-domain and the s-domain is expressed in Chap. 2 as

$$z = e^{sT} \tag{6.15}$$

The mapping of the s-plane in the z-plane can be studied using the general expression for complex s consisting of real (σ) and imaginary ($j\omega$) components, $j = \sqrt{-1}$, as

$$s = \sigma + j\omega, \ -\infty < \sigma < \infty, \ -\infty < \omega < \infty \tag{6.16}$$

$$z = e^{sT} = e^{(\sigma + j\omega)T} = e^{\sigma T} e^{j\omega T} = r\angle\theta \tag{6.17}$$

where

$$r = e^{\sigma T}, \quad 0 \leq r < \infty \tag{6.18}$$

$$\theta = \omega T, \quad -\infty < \theta < \infty \tag{6.19}$$

Left-Half of s-Plane The entire left half of the s-plane can be characterized by $-\infty < \sigma < 0$, i.e., with a negative real part. Each point on the left half of the s-plane can be mapped in the z-plane as

$$z = e^{sT} = e^{(\sigma + j\omega)T} = r\angle\theta, \quad 0 \leq r < 1, \ -\infty < \theta < \infty \tag{6.20}$$

The entire left half of the s-plane is mapped in the z-plane inside the unit circle (circle with a radius of 1 and the center at the origin (0, 0) of the z-plane). It means any point on the left half of the s-plane is mapped within the unit circle ($0 \leq r < 1$) in the z-plane.

Right-Half of s-Plane The entire right half of the s-plane can be characterized by $0 < \sigma < \infty$, i.e., with a positive real part. Each point on the right half of the s-plane can be mapped in the z-plane as

$$z = e^{sT} = e^{(\sigma + j\omega)T} = r\angle\theta, \quad 1 < r < \infty, \quad -\infty < \theta < \infty \tag{6.21}$$

The entire right half of the s-plane is mapped in the z-plane outside the unit circle (with $r > 1$). It means any point with $\sigma > 0$ in the s-plane is mapped outside the circumference of the unit circle in the z-plane ($r > 1$).

Imaginary Axis in the s-Plane The imaginary axis in the s-plane can be expressed as $\sigma = 0$, $-\infty < \omega < \infty$. Each point on the imaginary axis in the s-plane can be mapped in the z-plane as

$$z = e^{sT} = e^{(\sigma + j\omega)T} = e^{0.T} e^{j\omega T} = 1\angle\omega T \tag{6.22}$$

It means each point on the imaginary axis in the s-plane is mapped onto the circumference of the unit circle (radius $r = 1$) at an angle $\theta = \omega T$ in the z-plane. In the s-plane, the imaginary axis ($\sigma = 0$) is the boundary separating the stable left half ($-\infty < \sigma < 0$) and the unstable right half ($0 < \sigma < \infty$) regions. In the z-plane, the circumference of the unit circle ($r = 1$) separates the stable interior ($r < 1$) and unstable exterior ($r > 1$) regions. Stability analysis is covered in Chap. 7.

Primary and Complementary Strips It is instructive to study the mapping of a point on the imaginary axis in the s-plane, $s = j\omega, -\infty < \omega < \infty$, and into the z-plane, $z = e^{sT} = 1 \angle \omega T$. A point, $s = j\omega$, in the frequency range $-\frac{\omega_s}{2} \le \omega \le \frac{\omega_s}{2}$ ($\omega_s = \frac{2\pi}{T}$, ω_s is the sampling frequency and T is the sampling time), is mapped onto the circumference of the unit circle with an angle θ, $-\pi \le \theta \le \pi$ in the z-plane. Similarly, a point $s = j\omega, \frac{\omega_s}{2} \le \omega \le \frac{3\omega_s}{2}$, in the s-plane is mapped on the circumference of the unit circle with $\pi \le \theta \le 3\pi$. It implies that when a point on the imaginary axis in the s-plane moves from $-\infty$ to ∞, the corresponding point mapped onto the circumference of the unit circle in the z-plane completes an infinite number of counter-clockwise rotations. It means each strip of width ω_s in the left half of the s-plane is mapped into the inside of the unit circle in the z-plane. This leads to the division of the left half of the s-plane into an infinite number of periodic strips of width ω_s as shown in Fig. 6.6. The primary strip is defined in the frequency range $-\frac{\omega_s}{2} \le \omega \le \frac{\omega_s}{2}$. The complementary strips are defined in the frequency ranges $\pm m\frac{\omega_s}{2} \le \omega \le \pm (m+2)\frac{\omega_s}{2}, m = 1, 3, 5, \ldots$ and same signs appear together, i.e., +/+ for $\omega > 0$ and $-/-$ for $\omega < 0$.

A path with the sequence of points A-B-C-D-E-A in the primary strip in the s-plane is mapped on the unit circle in the z-plane in the same sequence A-B-C-D-E-A as shown in Fig. 6.7.

The area enclosed within any of the complementary strips is mapped into the same unit circle in the z-plane implying that a point in the z-plane corresponds to an infinite number of points in the s-plane. However, each point in the s-plane corresponds to a point in the z-plane.

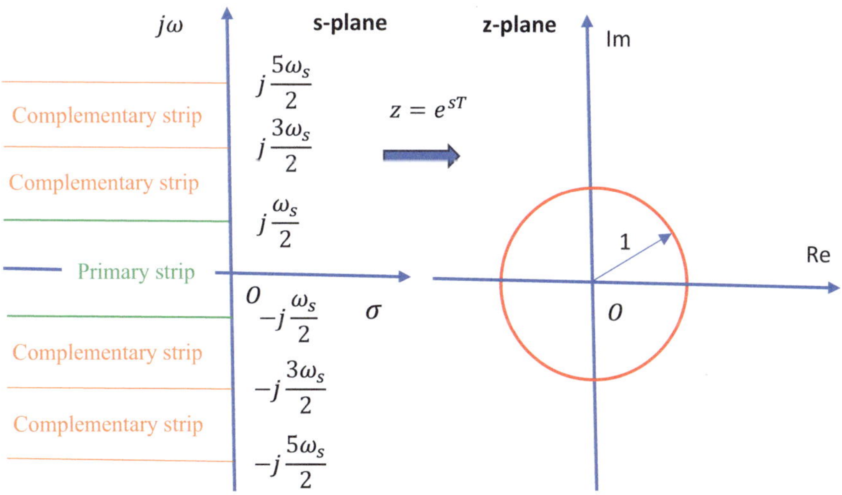

Fig. 6.6 Periodic strips in the s-plane mapped in the z-plane as the unit circle

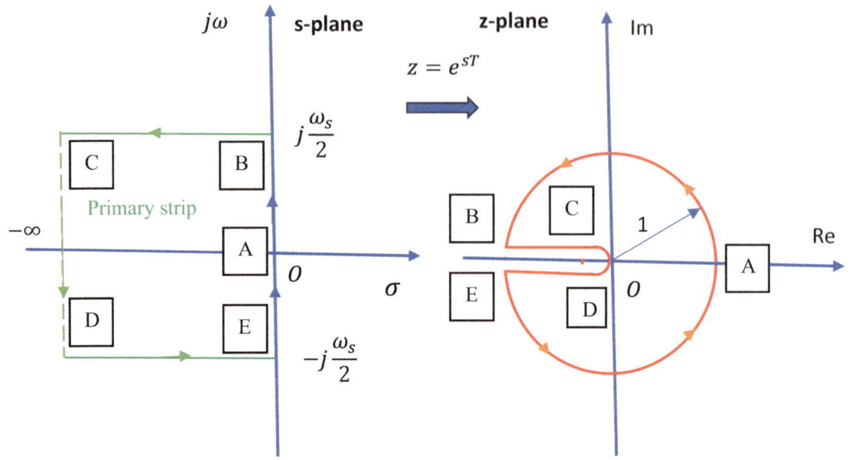

Fig. 6.7 Primary strip in the s-plane mapped in the z-plane

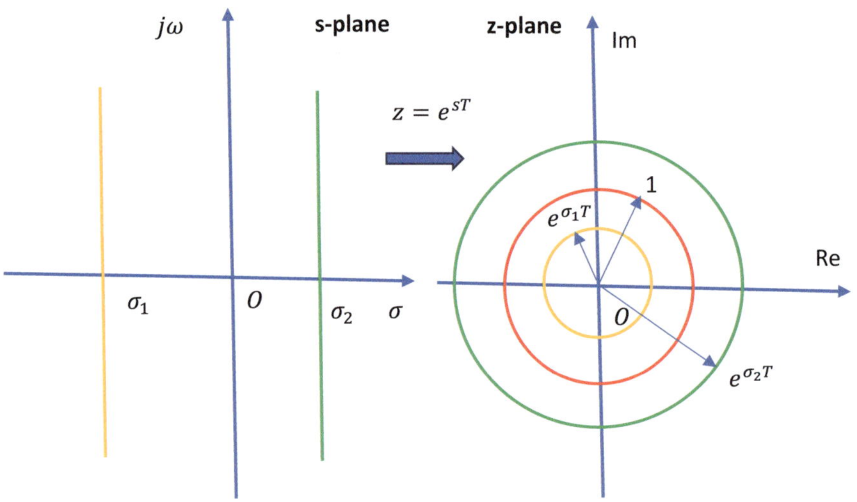

Fig. 6.8 Constant damping lines in the s-plane mapped in the z-plane as concentric circles

Next mapping of some of the commonly used lines characterizing constant damping, constant frequency, and constant damping ratio from the s-plane to the z-plane is covered. Readers are referred to Ogata [19] for further details.

Vertical Lines or Constant Damping Lines in s-Plane (with Constant Real Part)
Constant real part (σ) lines in the s-plane can be represented in the z-plane as concentric circles with the center at the origin as shown in Fig. 6.8. The radius of the concentric circles depends on the real part (σ), $r = e^{\sigma T}$, for $\sigma < 0,\ r < 1$ (inside the unit circle), and for $\sigma > 0,\ r > 1$ (outside the unit circle).

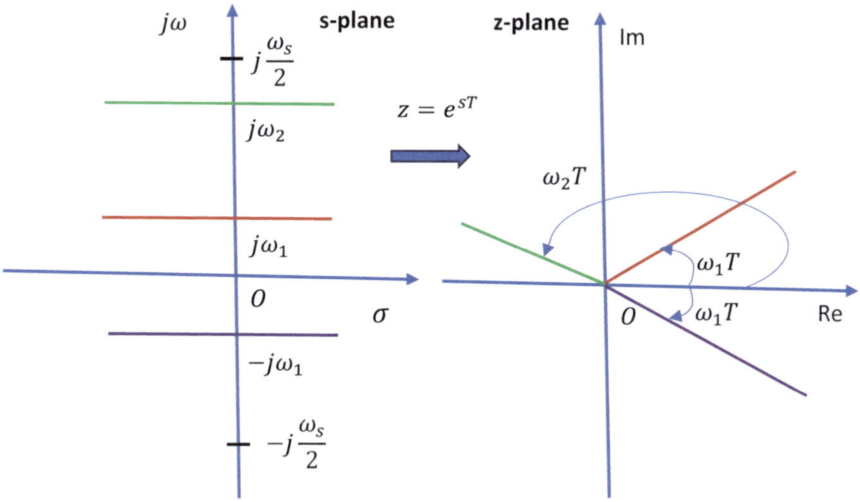

Fig. 6.9 Constant frequency loci in the s-plane mapped in the z-plane as radial lines

Horizontal Lines or Constant Frequency Lines in s-Plane (with Constant Imaginary Part) Constant imaginary part ($j\omega$) lines in the s-plane can be represented in the z-plane as radial lines passing through the origin and at an angle $\theta = \omega T$ as shown in Fig. 6.9. Constant frequency lines at $\omega = \pm\frac{\omega_s}{2}$ are mapped into the negative real axis ($\omega T = \pm\pi$) in the z-plane as follows: the left half in the s-plane ($-\infty < \sigma < 0$) to $0 > z > -1$ in the z-plane and the right half in the s-plane ($0 < \sigma < \infty$), $-1 > z > -\infty$ in the z-plane. The negative real axis in the s-plane corresponds to the positive real axis between 1 and 0 ($1 \geq z \geq 0$) in the z-plane. Constant frequency lines at $\omega = \pm n\omega_s$, $n = 0, 1, 2, \ldots$ in the right half of the s-plane are mapped into the positive real axis between 1 and ∞ ($1 \leq z < \infty$) in the z-plane.

Radial Lines or Constant Damping Ratio Lines in s-Plane (Constant Damping Ratio with $\frac{\omega}{\sigma} = \tan\beta$) The constant damping ratio lines in the s-plane can be represented in the z-plane as

$$z = e^{sT} = e^{\sigma T} e^{j\omega T} = e^{\sigma T} \angle \sigma T \tan\beta \tag{6.23}$$

Assuming $\sigma < 0$, the locus can be represented as a logarithmic spiral with reducing magnitude and increasing angle with increasing magnitude of σ as shown approximately in Fig. 6.10.

An analysis for the constant damping ratio line in the s-plane can be made using the definition of the complex pole in terms of damping ratio (ζ) and undamped natural frequency (ω_n) as

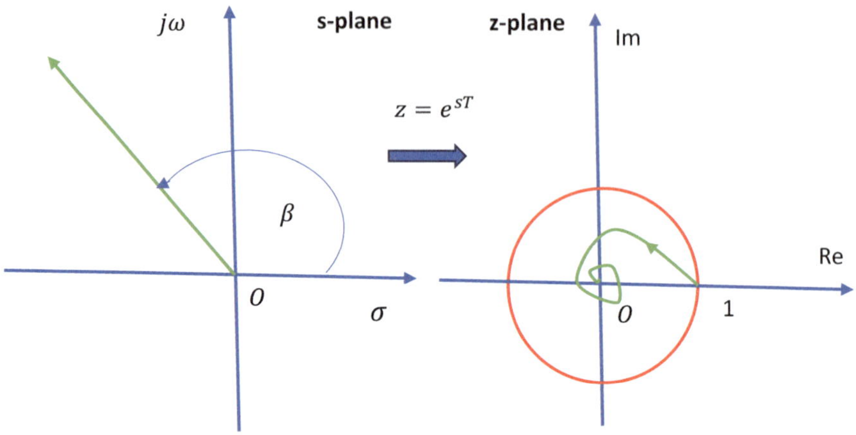

Fig. 6.10 Constant damping ratio line in the s-plane mapped in the z-plane as a logarithmic spiral

$$s = \sigma + j\omega = -\zeta\omega_n + j\omega_n\sqrt{1-\zeta^2} \tag{6.24}$$

$$\tan\beta = \frac{\sqrt{1-\zeta^2}}{-\zeta} \tag{6.25}$$

The constant damping ratio line in the s-plane line is mapped in the z-plane as

$$z = e^{sT} = e^{-\zeta\omega_n T + j\omega_n T\sqrt{1-\zeta^2}} \tag{6.26}$$

Equation (6.26) can be expressed in terms of damping ratio and frequency ratio $\frac{\omega}{\omega_s}$ as

$$z = e^{\left(-\frac{2\pi\zeta}{\sqrt{1-\zeta^2}}\frac{\omega}{\omega_s} + j2\pi\frac{\omega}{\omega_s}\right)} = r\angle\theta \tag{6.27}$$

where

$$r = e^{\left(-\frac{2\pi\zeta}{\sqrt{1-\zeta^2}}\frac{\omega}{\omega_s}\right)} \tag{6.28}$$

$$\theta = 2\pi\frac{\omega}{\omega_s} \tag{6.29}$$

Equations (6.28) and (6.29) can be used to plot the logarithmic spiral curve in the z-plane with a decreasing magnitude and an increasing angle as the frequency ω

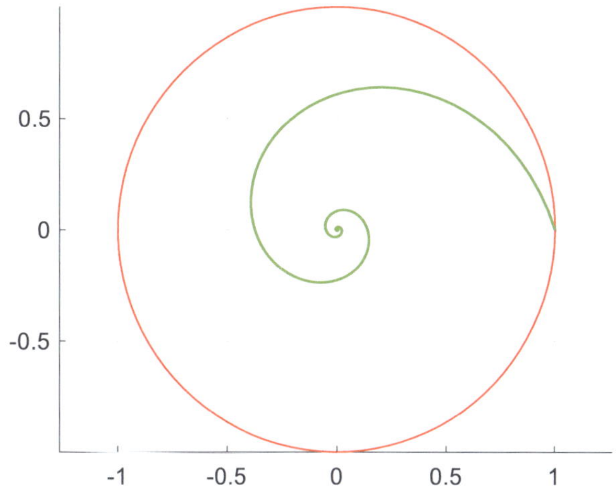

Fig. 6.11 Constant damping ratio line in the s-plane mapped as a logarithmic spiral in the z-plane

increases for a constant damping ratio (ζ) and a specified sampling time T or sampling frequency (ω_s). An example of mapping a constant damping ratio line with $\zeta = 0.30$ and $T = 0.10$ s in the z-plane is shown in Fig. 6.11.

***Complex Conjugate Poles in s-Plane* (s = σ ± jω)** The complex conjugate poles in the s-plane lead to the transient response $Ae^{\sigma t}\cos(\omega t + \phi)$, and the corresponding poles in z-plane can be represented as

$$z = e^{sT} = e^{(\sigma \pm j\omega)T} = e^{\sigma T} e^{\pm j\omega T} = r\angle \pm \theta \tag{6.30}$$

These poles in the z-plane lead to the transient response of the discrete-time systems as $Ae^{\sigma kT}\cos(k\omega T + \phi)$ or $Ar^k\cos(k\theta + \phi)$.

Relationship of Poles in s- and z-Domains A second-order system transfer function in s-domain in standard form is written as

$$G(s) = \frac{\omega_n^2}{s^2 + 2\zeta\omega_n s + \omega_n^2} \tag{6.31}$$

where ω_n and ζ are known as undamped natural frequency and damping ratio, respectively.

The poles are expressed in s-plane, assuming $\zeta < 1$ (underdamped system), as

$$s_{1,2} = -\zeta\omega_n \pm j\omega_n\sqrt{1-\zeta^2}, \ \ j = \sqrt{-1} \tag{6.32}$$

The corresponding poles in the z-plane can be obtained as

$$z_{1,2} = e^{sT}\Big|_{s=s_{1,2}} = e^{\left(-\zeta\omega_n \pm j\omega_n\sqrt{1-\zeta^2}\right)T} = e^{-\zeta\omega_n T} e^{\pm j\omega_{nT}\sqrt{1-\zeta^2}} = r\angle \pm\theta \tag{6.33}$$

where

$$r = e^{-\zeta\omega_n T} \tag{6.34}$$

$$\theta = T\omega_n\sqrt{1-\zeta^2} \tag{6.35}$$

From (6.34) and (6.35), the damping ratio and natural frequency can be expressed as

$$\zeta = \frac{-\ln r}{\sqrt{(\ln r)^2 + \theta^2}} \tag{6.36}$$

$$\omega_n = \frac{1}{T}\sqrt{(\ln r)^2 + \theta^2} \tag{6.37}$$

The time constant of the poles is expressed as

$$\tau = \frac{1}{\zeta\omega_n} = -\frac{T}{\ln r} \tag{6.38}$$

The second-order system step response has two features: overshoot and settling time. The overshoot for a step response can be expressed as

$$\alpha = e^{\frac{-\pi\zeta}{\sqrt{1-\zeta^2}}} \tag{6.39}$$

Settling time can be expressed as

$$t_{\text{set}} = \frac{4}{\zeta\omega_n} \tag{6.40}$$

Example 6.3 Obtain closed-loop zero and pole for the discrete-time closed-loop system of Example 6.1. Estimate the time constant, rise time, and settling time for the closed-loop system step response.

Solution The closed-loop discrete-time system transfer function is obtained in Example 6.1 as

$$T(z) = \frac{G(z)}{1+G(z)} = \frac{0.2654}{z-0.5134}$$

The closed-loop system pole in the z-domain is at $z = 0.5134$ and no zero.

The closed-loop system pole in s-domain can be obtained from $z = e^{sT}$ or $s = \ln(z)/T = -6.667$.

Time constant $\tau = \frac{1}{6.667} = 0.15$ s

Rise time $t_R = 2.303\tau = 2.303 * 0.15 \text{ s} = 0.3454$ s

Settling time $t_{set} = 4\tau = 4 * 0.15 \text{ s} = 0.6$ s

Matlab code and results are presented next. The closed-loop system step response characteristics are shown in Fig. 6.12.

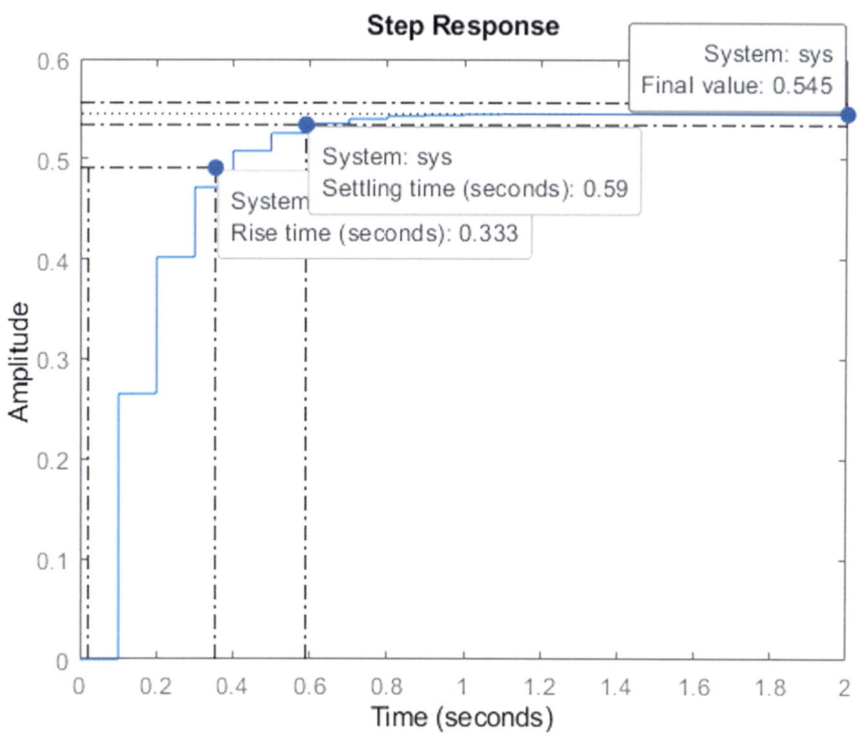

Fig. 6.12 Closed-loop discrete-time system step response characteristics for Example 6.1

```
% Ch6 Ex 6_1 Time response of a closed-loop discrete-time system
% Gp(s)=3/(s+2.5) First order system
T=0.1 ; % Sampling time
Gp = tf(3, [1 2.5]) %Gp(s)

% R(s)=1/s R(z)= z/(z-1)
Rz = tf([1 0], [1 -1], T) % R(z)

% G(z) = Z(Gho(s)Gp(s))
Gz =c2d(Gp,T) % G(z)

% Closed loop Transfer function
Tz = feedback(Gz, 1) % Closed loop TF with unity feedback
css = dcgain(Tz) % steady-state value of c(kT) with unit step input

% C(z) =T(z)*R(z)
Cz = Tz * Rz

% zpk for of C(z)
Cz = zpk(Cz) % zero-pole-gain form of C(z)

% Closed-loop step response
dstep(Tz,1)

% Inverse z-transform of C(z) using iztrans
syms z k
Cz =0.26544*z/(z^2-1.5134*z +0.5134);
ck = vpa(iztrans(Cz, k),4)

% Example 6.3

[ntz, dtz] = tfdata(Tz, 'v'); %  T(z)=ntz(z)/dtz(z)
[z,p,k] = tf2zp(ntz, dtz) % closed loop TF T(z) zero-pole-gain
ps = log(p)/T  % closed loop pole in s-plane
tau = 1/abs(ps) % time constant (s)
tR = 2.303*tau % rise time
tset = 4*tau  % settling time

Results:

T(z) =
     0.2654
  ----------
  z - 0.5134

z =  0×1 empty double column vector

p =    0.5134

k =    0.2654

ps =   -6.6677

tau =    0.1500 s,  tR =    0.3454 s, tset =    0.5999 s
```

Example 6.4 Obtain closed-loop zero(s) and pole(s) for the discrete-time closed-loop system of Example 6.2 with $G_p(s) = \frac{3}{s(s+2)}$, a zero-order hold (ZOH), and a sampling time $T = 0.1$ s. Estimate time constant, rise time, settling time, damping ratio, undamped natural frequency, and overshoot for the closed-loop system step response.

Solution The closed-loop discrete-time system transfer function is obtained in Example 6.1 as

$$T(z) = \frac{G(z)}{1 + G(z)} = \frac{0.01405z + 0.01314}{z^2 - 1.8047z + 0.8319}$$

The closed-loop system zero in z-domain is at $z = -\frac{0.01314}{0.01405} = -0.9355$.

The closed-loop system poles in the z-domain are

$$0.9023 + 0.1329i$$
$$0.9023 - 0.1329i$$

Magnitude of the poles $r = \sqrt{(0.9023\ )^2 + (0.1329)^2} = 0.9121$
Angle of poles $\theta = \pm \tan^{-1}\left(\frac{0.1329}{0.9023}\right) = \pm 0.1462$ rad
Time constant $\tau = -\frac{T}{\ln(r)} = 1.0865$ s
Rise time $t_R = 2.303\tau = 2.303 * 1.0865$ s $= 2.5022$ s
Settling time $t_{set} = 4\tau = 4 * 1.0865$ s $= 4.3460$ s
Undamped natural frequency $\omega_n = \frac{1}{T}\sqrt{(\ln r)^2 + \theta^2} = 1.7275$ rad/s
Damping ratio $\zeta = \frac{-\ln r}{\sqrt{(\ln r)^2 + \theta^2}} = 0.5328$
Overshoot $\alpha = e^{\frac{-\pi\zeta}{\sqrt{1-\zeta^2}}} = 0.1384$

Matlab code and results are presented next. The closed-loop system step response characteristics are shown in Fig. 6.13.

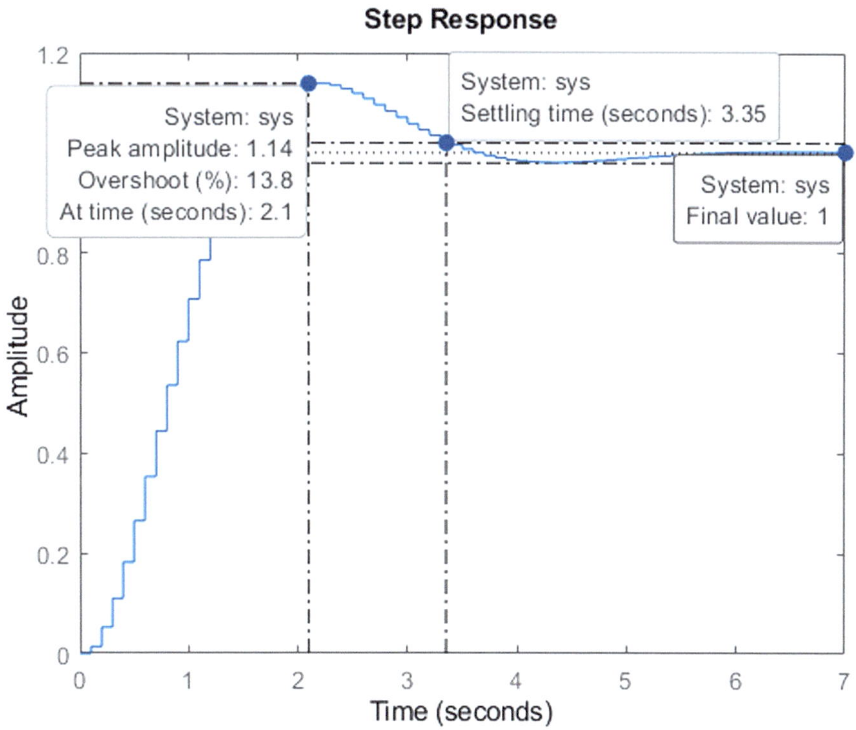

Fig. 6.13 Closed-loop discrete-time system step response characteristics for Example 6.2

```
% Ch6 Ex 6_4 Time response of a 2nd order closed-loop discrete-time system
% Gp(s)=3/s(s+2) First order system
clear all; close all; clc
T=0.1 ; % Sampling time
Gp = tf(3, [1 2 0]) %Gp(s)

% R(s)=1/s R(z)= z/(z-1)
Rz = tf([1 0], [1 -1], T) % R(z)

% G(z) = Z(Gho(s)Gp(s))
Gz =c2d(Gp,T) % G(z)

% Closed loop Transfer function
Tz = feedback(Gz, 1) % Closed loop TF with unity feedback
css = dcgain(Tz) % steady-state value of c(kT) with unit step input

% C(z) =T(z)*R(z)
Cz = Tz * Rz

% zpk for of C(z)
Cz = zpk(Cz) % zero-pole-gain form of C(z)

% Closed-loop step response
figure(1)
dstep(Tz,1)

% % comparison with continuous-time system
% Ts = feedback(Gp, 1) % closed-loop system transfer function
% figure(2)
% stepplot(Ts) % closed-loop system step response

% Example 6.4
% closed-loop system transient response characteristics
[ntz, dtz] = tfdata(Tz, 'v') % T(z)=ntz(z)/dtz(z)
[z,p,k]=tf2zp(ntz,dtz) % zero-pole-gain of T(z)
r=abs(p) % closed-loop pole magnitude
th=angle(p) % closed-loop pole angle
tau=-T./log(r) % time constant
tR=2.303*tau    % rise time
tset=4*tau    % settling time
A=sqrt(log(r(1))^2+th(1)^2)
wn=A/T  % natural frequency of the closed-loop system
zeta=-log(r(1))/A % damping ratio
x=-pi*zeta/sqrt(1-zeta^2)
alpha=exp(x) % percentage overshoot
```

```
Results:

T(z) =
   0.01405 z + 0.01314
  ----------------------
  z^2 - 1.805 z + 0.8319

z =   -0.9355

p =
   0.9023 + 0.1329i
   0.9023 - 0.1329i

k =    0.0140

r =    0.9121,    th =    0.1462 rad,   tau =    1.0865 s,    tR =    2.5022 s

tset =    4.3460 s,   wn =    1.7275 rad/s,   zeta =    0.5328,   alpha =  0.1384
```

The closed-loop step response characteristics are shown in Fig. 6.13.

6.5 Discrete Equivalents

The approach of analysis and design of digital control systems in the discrete-time domain using z-transform is adopted throughout the book. However, to take advantage of existing continuous-time controllers (filters), it is necessary to obtain discrete equivalents of the continuous-time controllers (filters) before implementation in the discrete-time domain. The indirect method of digital controller design through discrete equivalents is known as emulation. In this section, three methods of finding discrete equivalents from the transfer functions of continuous-time filters are briefly presented for completeness. Readers are referred to Franklin et al. [7] for further details.

The first method (numerical integration) is based on the numerical integration of the differential equation representing the transfer function in the continuous-time domain. Among different methods of numerical integration, rectangular (Euler) and trapezoid rules are considered. The second method (zero-pole matching) is based on the matching of zeros and poles of the filter transfer function in continuous-time with those of the approximating discrete-time filter. The final method (hold equivalence) is based on sampling the input signal, approximating the sampled signal between samples through extrapolation, and passing the approximated signal through the filter.

6.5.1 *Discrete Equivalents via Numerical Integration*

Three rules of numerical integration, namely, forward rectangular (Euler), backward rectangular, and trapezoid (Tustin), are considered along with an extended version of Tustin's method (bilinear with prewarping).

Forward Rectangular (Euler) Integration A discrete-time forward rectangular (Euler) integration rule is presented in Fig. 6.14. The area under the curve in the time interval $(k-1)T \leq t \leq kT$ can be approximated by the rectangle with a height of $e((k-1)T)$ and width of T.

In the s-domain, the transfer function for the integrator, with the input $E(s)$ and the output $M(s)$, is given as

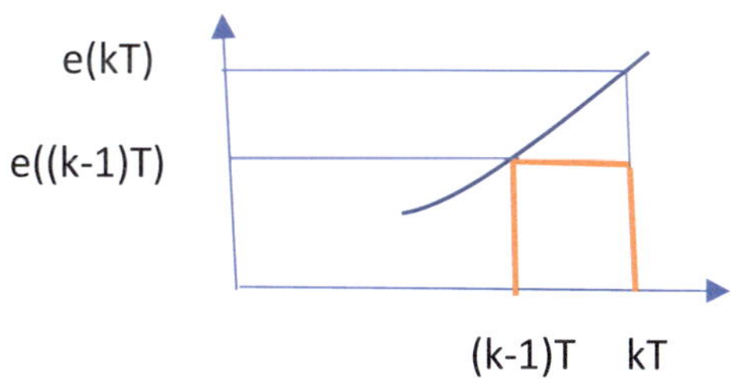

Fig. 6.14 Discrete-time forward rectangular integration

$$\frac{M(s)}{E(s)} = \frac{1}{s} \tag{6.41}$$

The output $m(kT)$ can be approximated using the forward rectangular (Euler) integration rule of the input $e(kT)$ as

$$m(kT) = m((k-1)T) + e((k-1)T)T \tag{6.42}$$

Taking z-transform, Eq. (6.41) can be expressed as

$$\left(1 - z^{-1}\right)M(z) = z^{-1}E(z)T \tag{6.43}$$

After some rearrangement, the transfer function for the discrete-time integration filter, using the forward Euler rule, can be expressed as

$$\frac{M(z)}{E(z)} = \frac{T\,z^{-1}}{1 - z^{-1}} = \frac{T}{z-1} \tag{6.44}$$

Comparing transfer functions in the s- and z-domain, the approximation for the frequency variable s, with the forward Euler rule, can be obtained as

$$s \leftarrow \frac{z-1}{T} \tag{6.45}$$

Backward Rectangular (Euler) Integration A discrete-time backward rectangular (Euler) integration rule is presented in Fig. 6.15. The area under the curve in the time interval $(k-1)T \leq t \leq kT$ can be approximated by the rectangle with a height of $e(kT)$ and width of T.

The output $m(kT)$ is obtained using the backward rectangular (Euler) integration rule of the input $e(kT)$ as

$$m(kT) = m((k-1)T) + e(kT)T \tag{6.46}$$

Taking z-transform, Eq. (6.46) can be expressed as

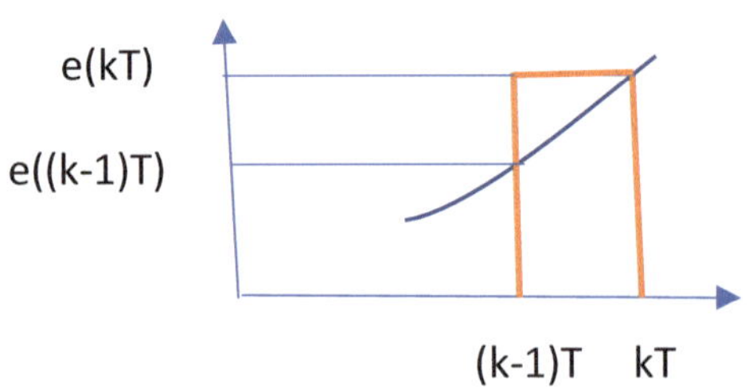

Fig. 6.15 Discrete-time backward rectangular integration

$$\left(1-z^{-1}\right)M(z)=E(z)T \tag{6.47}$$

After some rearrangement, the transfer function for the discrete-time integration filter, using the backward Euler rule, can be expressed as

$$\frac{M(z)}{E(z)}=\frac{T}{1-z^{-1}}=\frac{Tz}{z-1} \tag{6.48}$$

Comparing transfer functions in the s- and z-domain, the approximation for the frequency variable s, with the backward Euler rule, can be obtained as

$$s\leftarrow\frac{z-1}{Tz} \tag{6.49}$$

Trapezoidal Integration A discrete-time trapezoidal integration rule is presented in Fig. 6.16. The area under the curve in the time interval $(k-1)T\leq t\leq kT$ can be approximated by the trapezoid with the average height of $\frac{(e((k-1)T)+e(kT))}{2}$ and width of T.

The output $m(kT)$ is obtained using the trapezoidal rule for numerical integration of the input $e(kT)$ as

$$m(kT)=m((k-1)T)+\frac{T}{2}(\,e((k-1)T)+e(kT)) \tag{6.50}$$

Taking z-transform, Eq. (6.50) can be expressed as

$$M(z)=z^{-1}M(z)+\frac{T}{2}\left(z^{-1}E(z)+E(z)\right) \tag{6.51}$$

After some rearrangement, the transfer function for the discrete-time integral filter, using the trapezoidal rule, can be expressed as

$$\frac{M(z)}{E(z)}=\frac{T}{2}\frac{1+z^{-1}}{1-z^{-1}}=\frac{T}{2}\left(\frac{z+1}{z-1}\right) \tag{6.52}$$

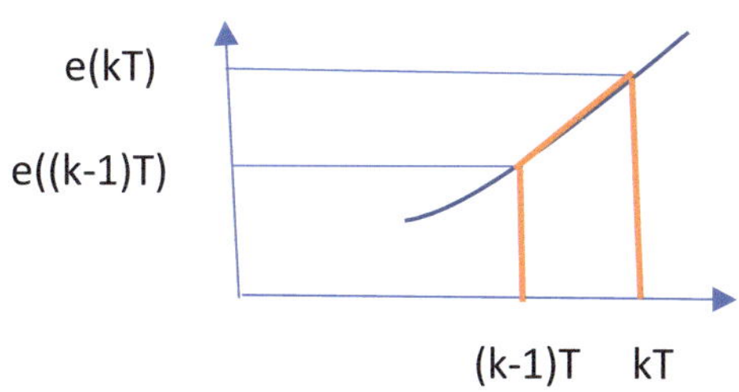

Fig. 6.16 Trapezoidal rule for numerical integration

Comparing transfer functions in the s- and z-domain, the approximation for the frequency variable s, with the trapezoidal rule, can be obtained as

$$s \leftarrow \frac{2}{T}\left(\frac{z-1}{z+1}\right) \tag{6.53}$$

This is also known as Tustin's rule or bilinear transformation.

Mapping of Left Half of the s-Plane in the z-Plane Relations (6.45), (6.49), and (6.53) can be used to obtain discrete-time transfer function $G(z)$ from the s-domain transfer function $G(s)$ for forward Euler, backward Euler, and trapezoid rule of numerical integration, respectively. These relations are also considered as the mapping from the s-plane to the z-plane. The imaginary axis in the s-plane, $s = j\omega$, is the boundary separating the stable system poles (in the left half of the s-plane) and the unstable system poles (in the right half of the s-plane). It is interesting to study the mapping of the imaginary axis in the s-plane, $s = j\omega$, and the left half of the s-plane to the z-plane for these three rules of numerical integration.

Relations (6.45), (6.49), and (6.53) can be used to express z in terms of s as

$$z = 1 + Ts \tag{6.54}$$

$$z = \frac{1}{1 - Ts} \tag{6.55}$$

$$z = \frac{1 + Ts/2}{1 - Ts/2} \tag{6.56}$$

Relation (6.54) can be written with $s = j\omega$ as

$$z = 1 + j\omega T = \sqrt{1 + \omega^2 T^2} \angle \tan^{-1}(\omega T) \tag{6.57}$$

Relation (6.57) can be used to map the imaginary axis and the left half of the s-plane in the z-plane as the shaded area in Fig. 6.17. The right side of the mapped (shaded) region in the z-plane is bounded by vertical line $z = 1$. From Fig. 6.17, it is possible that with the forward Euler rule, a stable continuous-time filter in the s-plane may be mapped outside the unit circle in the z-plane leading to an unstable discrete-time (digital) filter.

Similarly, relation (6.55) can be written with $s = j\omega$ as

$$z = \frac{1}{1 - j\,\omega T} = \frac{1}{2} + \frac{1}{1 - j\,\omega T} - \frac{1}{2} = \frac{1}{2} + \frac{1}{2}\left(\frac{1 + j\,\omega T}{1 - j\,\omega T}\right) \tag{6.58}$$

Relation (6.58) can be written as

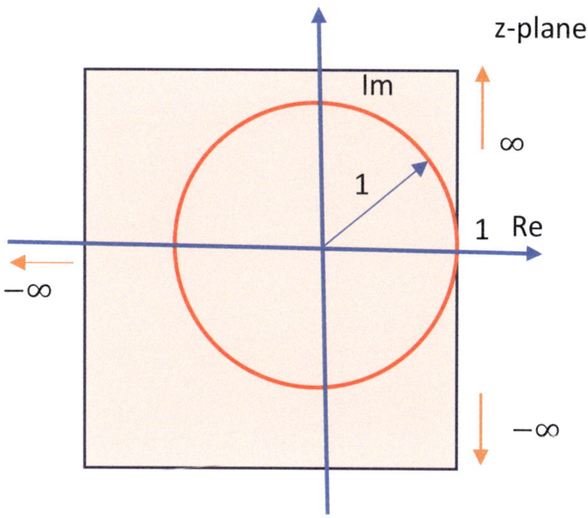

Fig. 6.17 Left half of the s-plane mapped in the z-plane for forward Euler rule

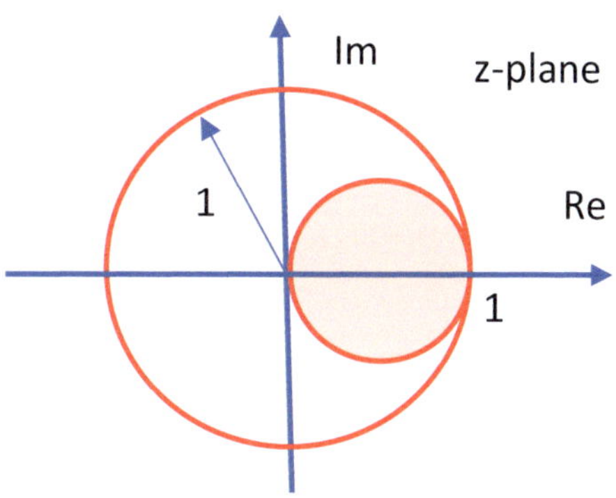

Fig. 6.18 Left half of the s-plane mapped in the z-plane for backward Euler rule

$$z - \frac{1}{2} = \frac{1}{2}\left(\frac{1 + j\,\omega T}{1 - j\,\omega T}\right) \tag{6.59}$$

Relation (6.59) can be expressed as

$$\left|z - \frac{1}{2}\right| = \frac{1}{2} \tag{6.60}$$

Relation (6.60) can be represented as a circle with a radius of (1/2) and the center at (1/2, 0) in the z-plane as shown in Fig. 6.18. Relation (6.60) can be used to map the imaginary axis and the left half of the s-plane in the z-plane as the shaded region in Fig. 6.18.

From Fig. 6.18, it is possible to state that with the backward Euler rule, a stable continuous-time filter is always mapped inside the unit circle in the z-plane leading to a stable digital filter. However, it is possible that with the backward Euler rule, an

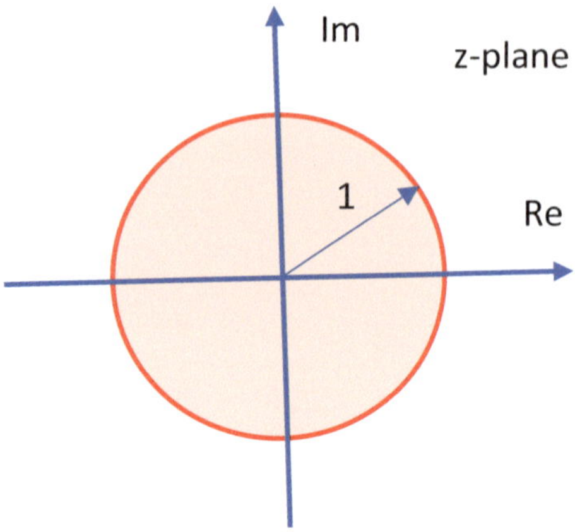

Fig. 6.19 Left half of the s-plane mapped in the z-plane for the trapezoid rule

unstable continuous-time filter may be mapped inside the unit circle but outside the smaller circle leading to a stable digital filter.

Similarly, relation (6.56) can be written with $s = j\omega$ as

$$z = \left(\frac{1 + j\,\omega T/2}{1 - j\,\omega T/2} \right) \tag{6.61}$$

Relation (6.61) can be expressed as

$$|z| = 1 \tag{6.62}$$

Relation (6.61) represents the unit circle with a radius of 1 and the center (0, 0) in the z-plane as shown in Fig. 6.19. For the trapezoid rule, the stable region of the s-plane is exactly mapped in the stable region of the z-plane. However, the entire imaginary axis ($s = j\omega$) is compressed into the unit circle circumference (of length 2π) leading to substantial distortion (warping) of mapped frequencies.

Bilinear Transformation with Prewarping To correct for the distortion of mapped frequencies in Tustin's rule, relation (6.53) is modified with prewarping as

$$s \leftarrow \frac{\omega_1}{\tan\left(\frac{\omega_1 T}{2}\right)} \left(\frac{z - 1}{z + 1} \right) \tag{6.63}$$

where ω_1 is the critical frequency at which the transfer functions in continuous time and discrete time are matched [7]. It is worth noting that for small $\omega_1 T$, $\tan\left(\frac{\omega_1 T}{2}\right) \approx \frac{\omega_1 T}{2}$, and (6.63) is the same as (6.53).

The continuous-time filter transfer function $G(s)$ can be approximated to the discrete-time filter transfer function using forward Euler ($G_F(z)$), backward Euler ($G_B(z)$), Tustin ($G_T(z)$), and bilinear transformation with prewarping ($G_P(z)$), respectively, as

$$G_F(z) = G(s)|_{s=\frac{z-1}{T}} \tag{6.64}$$

$$G_B(z) = G(s)|_{s=\frac{z-1}{Tz}} \tag{6.65}$$

$$G_T(z) = G(s)|_{s=\frac{2}{T}\left(\frac{z-1}{z+1}\right)} \tag{6.66}$$

$$G_P(z) = G(s)|_{s=\frac{\omega_1}{\tan\left(\frac{\omega_1 T}{2}\right)}\left(\frac{z-1}{z+1}\right)} \tag{6.67}$$

6.5.2 Zero-Pole Matching Equivalents

The discrete-time equivalent transfer function ($G_{zp}(z)$) can be approximated from the continuous-time transfer function $G(s)$ through zero-pole matching. In this method, all poles and finite zeros of $G(s)$ are converted to discrete-time poles and zeros of $G_{zp}(z)$ using $z = e^{sT}$; all infinite zeros of $G(s)$, $s = \infty$, are mapped in $G_{zp}(z)$, at $z = -1$; and the steady-state gains of $G_{zp}(z)$ and $G(s)$ are matched.

6.5.3 Hold Equivalents

Zero-order hold (ZOH) and first-order hold (FOH), discussed in Chap. 3 (Sect. 3.4), can also be used to obtain discrete-time equivalent transfer functions ($G_{h0}(z)$, $G_{h1}(z)$) from the continuous-time transfer function ($G(s)$).

In Matlab, continuous-to-discrete (c2d) command like $Gz = \text{c2d}(Gs, T, \text{'method'})$ can be used to find the discrete-time transfer function (Gz) from the continuous-time transfer function (Gs) for a given sampling time (T) and "method." The methods among those discussed in this section include "tustin" (Tustin's rule), "prewrap" (bilinear with prewarping), "matched" (zero-pole matching), "zoh" (zero-order hold), and "foh" (first-order hold). For "prewarp," the critical frequency ($w1$) for the matched condition has to be provided as $Gz = \text{c2d}(Gs, T, \text{'prewarp'}, w1)$. The process of finding discrete equivalents using the methods discussed in this section is illustrated using a simple example along with the Matlab code.

Example 6.5 Obtain discrete-time transfer function of a continuous-time transfer function $G(s) = \frac{a}{s+a}$ with $a = 1$ for a sampling time $T = 0.10$ s using forward Euler, backward Euler, Tustin, bilinear with prewarping, zero-order hold, and first-order hold.

Solution Discrete-time equivalent transfer function using forward and backward Euler rules can be obtained manually. Results for other methods can be obtained using the Matlab command c2d() as listed.

Discrete-time equivalent transfer function with forward Euler rule can be obtained from (6.63) as

$$G_F(z) = G(s)|_{s=\frac{z-1}{T}} = \frac{1}{\frac{z-1}{0.10}+1} = \frac{1}{10z-9}$$

Discrete-time equivalent transfer function with backward Euler rule can be obtained from (6.64) as

$$G_B(z) = G(s)|_{s=\frac{z-1}{Tz}} = \frac{1}{\frac{z-1}{0.10z}+1} = \frac{z}{11z-10}$$

Matlab code and results for other methods are presented.

```
% Example 6.5 Discrete Equivalents
clear all; clc
% Sampling time
T=0.1
a=1 % pole -a
% Ssytem TF G(s)=a/(s+a)
G=tf(a,[1 a])
% Discrete Equivalents
% Tustin
Gt=c2d(G,T,'tustin')
% Tustin with prewarping
Gtp=c2d(G,T,'prewarp',1)
% Zero-Ploe Matching
Gzp=c2d(G,T,'matched')
% Zero-order Hold (ZOH)
Gh0=c2d(G,T,'zoh')
% First-order Hold (FOH)
Gh1=c2d(G,T,'foh')

Results:
Gt =

  0.04762 z + 0.04762
  -------------------
      z - 0.9048

Gtp =

  0.04766 z + 0.04766
  -------------------
      z - 0.9047

Gzp =

   0.09516
  ----------
  z - 0.9048


Gh0 =

   0.09516
  ----------
  z - 0.9048

Gh1 =

  0.04837 z + 0.04679
  -------------------
      z - 0.9048
```

6.6 Discrete-Time System Steady-State Accuracy

The open-loop system transfer function is represented for the discrete-time system of Fig. 6.2 as

$$G_o(z) = D(z)G(z) \tag{6.68}$$

The open system transfer function $G_o(z)$ can be represented in zero-pole form as

$$G_o(z) = K \frac{\prod_{i=1}^{m} (z - z_i)}{(z-1)^t \prod_{j=1}^{n-t} (z - p_j)} \tag{6.69}$$

where K is the static gain, z_i, $i = 1, \ldots, m$ are zeros, p_j, $j = 1, \ldots, (n - t)$ are non-unity poles, and t is the number of poles at $z = 1$ of the open-loop system. The order and the type of the open-loop system are n and t, respectively.

The closed-loop system transfer function $T(z)$ can be obtained from the schematic of the closed-loop discrete-time control system, Fig. 6.2, as

$$T(z) = \frac{C(z)}{R(z)} = \frac{D(z)G(z)}{1 + D(z)G(z)} \tag{6.70}$$

The denominator of the closed-loop system transfer function $T(z)$ leads to the characteristic equation of the discrete-time closed-loop system. The roots of the characteristic equation are the closed-loop poles of the system.

$$1 + D(z)G(z) = 0 \tag{6.71}$$

Steady-State Errors

For a closed-loop discrete-time system, the error, $E(z)$, the output of the comparator, can be obtained as

$$E(z) = R(z) - C(z) = \frac{1}{1 + D(z)G(z)} R(z) \tag{6.72}$$

Steady-state error can be obtained by applying the final value theorem to Eq. (6.72) with the reference input represented as the z-transform of the sampled input function $r(t)$.

Step Input For example, in the case of a unit step input, $r(t) = u(t), R(s) = \frac{1}{s}$, $R(z) = \frac{z}{z-1}$, the steady-state error can be expressed as

$$\begin{aligned} e_{\text{ss,step}} &= \lim_{z\to 1}(z-1)E(z) = \lim_{z\to 1}(z-1)\frac{1}{1+D(z)G(z)}\left(\frac{z}{z-1}\right) \\ &= \lim_{z\to 1}\frac{1}{1+D(z)G(z)} = \frac{1}{1+k_p} \end{aligned} \tag{6.73}$$

where

$$k_p = \lim_{z\to 1}(D(z)G(z)) \tag{6.74}$$

k_p is termed as position error constant.

For a type zero ($t = 0$) system, k_p is a finite value, and the steady-state error due to a unit step input is a finite value. For type 1 ($t = 1$) and type 2 ($t = 2$) systems, k_p approaches infinity making the steady-state error due to a step input as zero.

Ramp Input For a unit ramp input, $r(t) = tu(t), R(s) = \frac{1}{s^2}, R(z) = \frac{Tz}{(Z-1)^2}$, the steady-state error can be expressed as

$$\begin{aligned} e_{\text{ss,ramp}} &= \lim_{z\to 1}(z-1)E(z) = \lim_{z\to 1}(z-1)\frac{1}{1+D(z)G(z)}\left(\frac{Tz}{(Z-1)^2}\right) \\ &= \lim_{z\to 1}\frac{T}{(z-1)(1+D(z)G(z))} = \frac{1}{k_v} \end{aligned} \tag{6.75}$$

where

$$k_v = \frac{1}{T}\lim_{z\to 1}((z-1)D(z)G(z)) \tag{6.76}$$

k_v is termed as velocity (or rate) error constant. For a type zero ($t = 0$) system, k_v is zero, and the steady-state error due to a unit ramp input is infinite. For a type 1 ($t = 1$) system, k_v is a finite value, and the steady-state error due to a ramp input is a finite value. For type 2 ($t = 2$) systems, k_v approaches infinity making the steady-state error due to a ramp input as zero.

Parabolic Input For a unit parabolic input, $r(t) = \frac{t^2}{2}u(t), R(s) = \frac{1}{s^3}, R(z) = \frac{T^2 z(z+1)}{2(z-1)^3}$, the steady-state error can be expressed as

$$\begin{aligned} e_{\text{ss,par}} &= \lim_{z\to 1}(z-1)E(z) = \lim_{z\to 1}(z-1)\frac{1}{1+D(z)G(z)}\left(\frac{T^2 z(z+1)}{2(z-1)^3}\right) \\ &= \lim_{z\to 1}\frac{T^2}{(z-1)^2(1+D(z)G(z))} = \frac{1}{k_a} \end{aligned} \tag{6.77}$$

where

Table 6.1 Steady-state error depending on type of system and input signal

	Type 0 ($t = 0$)	Type 1 ($t = 1$)	Type 2 ($t = 2$)
Step, $r(t) = u(t), R(s) = \frac{1}{s}, R(z) = \frac{z}{z-1}$	$\frac{1}{1+k_p}$	0	0
Ramp, $r(t) = tu(t)$, $R(s) = \frac{1}{s^2}, R(z) = \frac{Tz}{(z-1)^2}$	∞	$\frac{1}{k_v}$	0
Parabolic, $r(t) = \frac{t^2}{2}u(t)$, $R(s) = \frac{1}{s^3}, R(z) = \frac{T^2z(z+1)}{2(z-1)^3}$	∞	∞	$\frac{1}{k_a}$

$$k_a = \frac{1}{T^2}\lim_{z\to 1}\left((z-1)^2D(z)G(z)\right) \tag{6.78}$$

k_a is termed as acceleration (or rate of rate) error constant. For type zero ($t = 0$) and type 1 ($t = 1$) systems, k_a is zero, and the steady-state error due to a unit parabolic input is infinite. For a type 2 ($t = 2$) system, k_a is a finite value, and the steady-state error due to a unit parabolic input is a finite value. The steady-state error for different types ($t = 0$, 1, 2) of discrete-time systems and different inputs (step, ramp, and parabolic) is summarized in Table 6.1. The desired situation should be on the upper diagonal of Table 6.1 or on the main diagonal, and the lower diagonal should be avoided as much as possible.

Example 6.6 Obtain the system type of the discrete-time open-loop system for Example 6.1 and estimate steady-state errors due to unit step, ramp, and parabolic inputs with the digital controller transfer function (a) $D(z) = 1$ and (b) $D(z) = 1 + \frac{0.1z}{z-1}$. Use sampling time $T = 0.10$ s.

Solution

(a) With $D(z) = 1$, the open-loop discrete-time system transfer function

$$D(z)G(z) = \frac{0.2654}{z-0.7788}$$

Type of the system $t = 0$, as there is no pole at $z = 1$.
Position error constant

$$k_p = \lim_{z\to 1}(D(z)G(z)) = \lim_{z\to 1}\left(\frac{0.2654}{z-0.7788}\right) = 1.2000$$

Steady-state error due to unit step input

$$e_{ss,step} = \frac{1}{1+k_p} = \frac{1}{1+1.2000} = 0.4545$$

Velocity error constant

$$k_v = \frac{1}{T} \lim_{z \to 1} ((z-1)D(z)G(z)) = \frac{1}{T} \lim_{z \to 1} \left[(z-1) \left(\frac{0.2654}{z-0.7788} \right) \right] = 0$$

Steady-state error due to a unit ramp input

$$e_{\text{ss,ramp}} = \frac{1}{k_v} = \infty$$

Acceleration error constant

$$k_a = \frac{1}{T^2} \lim_{z \to 1} \left((z-1)^2 D(z)G(z) \right) = \frac{1}{T^2} \lim_{z \to 1} \left[(z-1)^2 \left(\frac{0.2654}{z-0.7788} \right) \right] = 0$$

Steady-state error due to a unit parabolic input

$$e_{\text{ss,par}} = \frac{1}{k_a} = \infty$$

(b) With $D(z) = 1 + \frac{0.1z}{z-1} = \frac{1.1z-1}{z-1}$, the open-loop discrete-time system transfer function

$$D(z)G(z) = \left(\frac{0.2654}{z-0.7788} \right) \left(\frac{1.1z-1}{z-1} \right)$$

Type of the system $t = 1$, as there is one pole at $z = 1$.

Position error constant

$$k_p = \lim_{z \to 1} (D(z)G(z)) = \lim_{z \to 1} \left[\left(\frac{0.2654}{z-0.7788} \right) \left(\frac{1.1z-1}{z-1} \right) \right] = \infty$$

Steady-state error due to unit step input

$$e_{\text{ss,step}} = \frac{1}{1+k_p} = \frac{1}{1+\infty} = 0.$$

Velocity error constant

$$k_v = \frac{1}{T} \lim_{z \to 1} ((z-1)D(z)G(z)) = \frac{1}{T} \lim_{z \to 1} \left[(z-1) \left(\frac{0.2654}{z-0.7788} \right) \left(\frac{1.1z-1}{z-1} \right) \right] = 1.200$$

Steady-state error due to a unit ramp input

$$e_{\text{ss,ramp}} = \frac{1}{k_v} = 0.8333$$

Acceleration error constant

$$k_a = \frac{1}{T^2} \lim_{z\to 1}\left((z-1)^2 D(z)G(z)\right) = \frac{1}{T^2} \lim_{z\to 1}\left[(z-1)^2\left(\frac{0.2654}{z-0.7788}\right)\left(\frac{1.1z-1}{z-1}\right)\right] = 0$$

Steady-state error due to a unit parabolic input

$$e_{\text{ss,par}} = \frac{1}{k_a} = \infty$$

Matlab code and results are presented next. The closed-loop system step response is shown in Fig. 6.20.

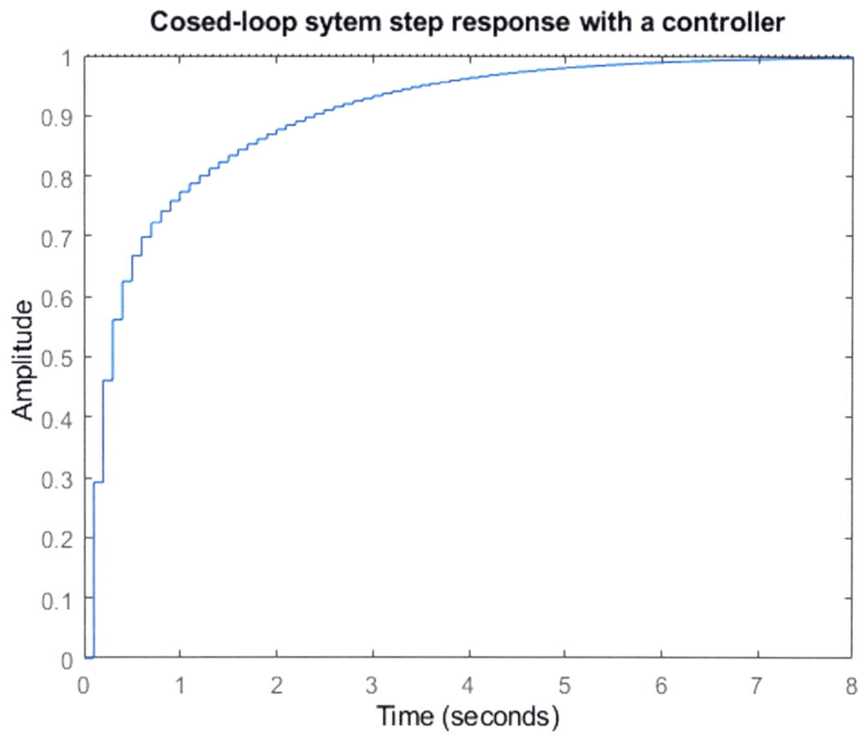

Fig. 6.20 Closed-loop system step response with a controller

```
% Ch6 Ex 6_ 1 Time response of a closed-loop discrete-time system
% Gp(s)=3/(s+2.5) First order system
clear all; close all; clc
T=0.1 ; % Sampling time
Gp = tf(3, [1 2.5]) %Gp(s)=3/(s+2.5)

% R(s)=1/s R(z)= z/(z-1)
Rz = tf([1 0], [1 -1], T) % R(z)

% G(z) = Z(Gho(s)Gp(s))
Gz =c2d(Gp,T) % G(z)

% Closed loop Transfer function
Tz = feedback(Gz, 1) % Closed loop TF with unity feedback
css = dcgain(Tz) % steady-state value of c(kT) with unit step input

% C(z) =T(z)*R(z)
Cz = Tz * Rz

% zpk for of C(z)
Cz = zpk(Cz) % zero-pole-gain form of C(z)

% Closed-loop step response
dstep(Tz,1)

% Inverse z-transform of C(z) using iztrans
syms z k
Cz =0.26544*z/(z^2-1.5134*z +0.5134);
ck = vpa(iztrans(Cz, k),4)

% Example 6.3 Time response characteristics

[ntz, dtz] = tfdata(Tz, 'v'); %  T(z)=ntz(z)/dtz(z)
[z,p,k] = tf2zp(ntz, dtz) % closed loop TF T(z) zero-pole-gain
ps = log(p)/T  % closed loop pole in s-plane
tau = 1/abs(ps) % time constant (s)
tR = 2.303*tau % rise time
tss = 4*tau  % settling time

% Example 6.5 Steady state errors
% System type and steady-state errors with D(z)=1
Dz  = tf(1,1,T); % D(z)=1
DGz = Dz*Gz % D(z)*G(z)
zm1 = tf([1 -1],1, T); % (z-1)
zm1DGz = zm1*DGz;  % (z-1)*D(z)G(z)
zm1DGz = minreal(zm1DGz); % min realization after pole zero cancellation
zm1DGz = zpk(zm1DGz)
zm2DGz = zm1*zm1DGz;  % (z-1)*(z-1)*D(z)G(z)
zm2DGz = minreal(zm2DGz); % min realization after pole zero cancellation
zm2DGz = zpk(zm2DGz)
Kp = dcgain(DGz) % DGz at z=1
Kv = (1/T)*dcgain(zm1DGz)
```

```
Ka = (1/T)^2*dcgain(zm2DGz)
essStep = 1/(1+Kp)
if (abs(essStep) < 1e-10)
    essStep = 0
end

essRamp = 1/Kv
if (abs(essRamp) < 1e-10)
    essRamp = 0
end

essPar  = 1/Ka
if (abs(essPar) < 1e-10)
    essPar = 0
end
% with controller D(z)=kp+ki*z/(z-1)= ((kp+ki)*z-kp)/(z-1)
kp=1; ki=0.1; % kp and ki gains of PI controller
Dz = tf([kp+ki -kp], [1 -1],T) % PI controller D(z)
DGz = Dz * Gz; % D(z)G(z)
DGz = zpk(DGz)
zm1DGz = zm1*DGz;  % (z-1)*D(z)G(z)
zm1DGz = minreal(zm1DGz); % min realization after pole zero cancellation
zpk(zm1DGz)
zm2DGz = zm1*zm1DGz;  % (z-1)*(z-1)*D(z)G(z)
zm2DGz = minreal(zm2DGz); % min realization after pole zero cancellation
zpk(zm2DGz)
Kp = dcgain(DGz)
Kv = (1/T)*dcgain(zm1DGz)
Ka = (1/T)^2*dcgain(zm2DGz)
essStep = 1/(1+Kp)
if (abs(essStep) < 1e-10)
    essStep = 0
end

essRamp = 1/Kv
if (abs(essRamp) < 1e-10)
    essRamp = 0
end

essPar  = 1/Ka
if (abs(essPar) < 1e-10)
    essPar = 0
end

Results:
D(z) = 1
D(z)G(z) =

    0.2654
  ----------
  z - 0.7788
```

```
(z-1)*D(z)G(z) =

  0.26544 (z-1)
  -------------
   (z-0.7788)

(z-1)^2*D(z)G(z) =

  0.26544 (z-1)^2
  ---------------
    (z-0.7788)

Kp =     1.2000   Kv =      0   Ka =      0

essStep =     0.4545    essRamp =    Inf     essPar =    Inf
```

The steady-state error for a unit step input is verified from the step response of the closed-loop system presented in Fig. 6.3, `essStep=1-css=1-0.5455=0.4545`.

```
D(z)=

 1.1 z - 1
 ---------
 z - 1

D(z)G(z) =

 0.29198 (z-0.9091)
 ------------------
 (z-1) (z-0.7788)

(z-1)*D(z)G(z) =

 0.29198 (z-0.9091)
 ------------------
 (z-0.7788)

(z-1)^2*D(z)G(z) =

 0.29198 (z-1) (z-0.9091)
 ------------------------
 (z-0.7788)

Kp = abs(-2.7022e+14) =Inf Kv = 1.2000 Ka = 0

essStep = 0 essRamp = 0.8333 essPar = Inf
```

The steady-state error for a unit step input is verified from the step response of the closed-loop system with a controller as shown in Fig. 6.20, `essStep=1-css=1-1=0`.

Example 6.7 Obtain the system type of the discrete-time open-loop system for Example 6.2 with $G_p(s) = \frac{3}{s(s+2)}$, a zero-order hold (ZOH), and a sampling time $T = 0.1$ s. Estimate steady-state errors due to unit step, ramp, and parabolic inputs with the digital controller transfer function (a) $D(z) = 1$ and (b) $D(z) = 1 + \frac{0.1z}{z-1}$. Use sampling time $T = 0.10$ s.

Solution

(a) With $D(z) = 1$, the open-loop discrete-time system transfer function

$$D(z)G(z) = \frac{0.01405z + 0.01314}{(z-1)(z-0.8187)}$$

Type of the system $t = 1$, as there is one pole at $z = 1$.
Position error constant

$$k_p = \lim_{z\to 1}(D(z)G(z)) = \lim_{z\to 1}\left[\frac{0.01405z + 0.01314}{(z-1)(z-0.8187)}\right] = \infty$$

Steady-state error due to unit step input

$$e_{\text{ss,step}} = \frac{1}{1+k_p} = \frac{1}{1+\infty} = 0$$

Velocity error constant

$$k_v = \frac{1}{T}\lim_{z\to 1}((z-1)D(z)G(z)) = \frac{1}{T}\lim_{z\to 1}\left[(z-1)\left(\frac{0.01405z + 0.01314}{(z-1)(z-0.8187)}\right)\right] = 1.5000$$

Steady-state error due to a unit ramp input

$$e_{\text{ss,ramp}} = \frac{1}{k_v} = 0.6667$$

Acceleration error constant

$$k_a = \frac{1}{T^2}\lim_{z\to 1}\left((z-1)^2 D(z)G(z)\right) = \frac{1}{T^2}\lim_{z\to 1}\left[(z-1)^2\left(\frac{0.01405z + 0.01314}{(z-1)(z-0.8187)}\right)\right] = 0$$

Steady-state error due to a unit ramp input

$$e_{\text{ss,par}} = \frac{1}{k_a} = \infty$$

(b) With $D(z) = 1 + \frac{0.1z}{z-1} = \frac{1.1z-1}{z-1}$, the open-loop discrete-time system transfer function

$$D(z)G(z) = \left(\frac{0.01405z + 0.01314}{(z-1)(z-0.8187)}\right)\left(\frac{1.1z-1}{z-1}\right)$$

Type of the system $t = 2$, as there are two poles at $z = 1$.

Position error constant

$$k_p = \lim_{z\to1}(D(z)G(z)) = \lim_{z\to1}\left[\left(\frac{0.01405z + 0.01314}{(z-1)(z-0.8187)}\right)\left(\frac{1.1z-1}{z-1}\right)\right] = \infty$$

Steady-state error due to unit step input

$$e_{\text{ss,step}} = \frac{1}{1+k_p} = \frac{1}{1+\infty} = 0.$$

Velocity error constant

$$\begin{aligned} k_v &= \frac{1}{T}\lim_{z\to1}((z-1)D(z)G(z)) \\ &= \frac{1}{T}\lim_{z\to1}\left[(z-1)\left(\frac{0.01405z + 0.01314}{(z-1)(z-0.8187)}\right)\left(\frac{1.1z-1}{z-1}\right)\right] = \infty \end{aligned}$$

Steady-state error due to a unit ramp input

$$e_{\text{ss,ramp}} = \frac{1}{k_v} = 0$$

Acceleration error constant

$$\begin{aligned} k_a &= \frac{1}{T^2}\lim_{z\to1}\left((z-1)^2 D(z)G(z)\right) \\ &= \frac{1}{T^2}\lim_{z\to1}\left[(z-1)^2\left(\frac{0.01405z + 0.01314}{(z-1)(z-0.8187)}\right)\left(\frac{1.1z-1}{z-1}\right)\right] = 1.5000 \end{aligned}$$

Steady-state error due to a unit parabolic input

$$e_{\text{ss,par}} = \frac{1}{k_a} = 0.6667$$

Matlab code and results are presented next.

```
% Ch6 Ex 6_4 Time response of a 2nd order closed-loop discrete-time system
% Gp(s)=3/s(s+2) First order system
clear all; close all; clc
T=0.1 ; % Sampling time
Gp = tf(3, [1 2 0]) %Gp(s)

% R(s)=1/s R(z)= z/(z-1)
Rz = tf([1 0], [1 -1], T) % R(z)

% G(z) = Z(Gho(s)Gp(s))
Gz =c2d(Gp,T) % G(z)

% Closed loop Transfer function
Tz = feedback(Gz, 1) % Closed loop TF with unity feedback
css = dcgain(Tz) % steady-state value of c(kT) with unit step input

% C(z) =T(z)*R(z)
Cz = Tz * Rz

% zpk for of C(z)
Cz = zpk(Cz) % zero-pole-gain form of C(z)

% Closed-loop step response
figure(1)
dstep(Tz,1)

% % comparison with continuous-time system
% Ts = feedback(Gp, 1) % closed-loop system transfer function
% figure(2)
% stepplot(Ts) % closed-loop system step response

% Example 6.4
% closed-loop system transient response characteristics
[ntz, dtz] = tfdata(Tz, 'v') % T(z)=ntz(z)/dtz(z)
[z,p,k]=tf2zp(ntz,dtz) % zero-pole-gain of T(z)
r=abs(p) % closed-loop pole magintude
th=angle(p) % closed-loop pole angle
tau=-T./log(r) % timeconstant
tR=2.303*tau   % rise time
tss=4*tau     % sttling time
A=sqrt(log(r(1))^2+th(1)^2)
wn=A/T  % natural frequency of the closed-loop system
zeta=-log(r(1))/A % damping ratio
x=-pi*zeta/sqrt(1-zeta^2)
alpha=exp(x) % percentage overshoot

% Example 6.6 Steady-State Errors
% System type and steady-state errors with D(z)=1
Dz  = tf(1,1,T); % D(z)=1
DGz = Dz*Gz % D(z)*G(z)
zm1 = tf([1 -1],1, T); % (z-1)
zm1DGz = zm1*DGz;  % (z-1)*D(z)G(z)
zm1DGz = minreal(zm1DGz); % min realization after pole zero cancellation
```

```
zm1DGz = zpk(zm1DGz)
zm2DGz = zm1*zm1DGz;  % (z-1)*(z-1)*D(z)G(z)
zm2DGz = minreal(zm2DGz); % min realization after pole zero cancellation
zm2DGz = zpk(zm2DGz)
Kp = dcgain(DGz)
Kv = (1/T)*dcgain(zm1DGz)
Ka = (1/T)^2*dcgain(zm2DGz)
essStep = 1/(1+Kp)
if (abs(essStep) < 1e-10)
    essStep = 0
end

essRamp = 1/Kv
if (abs(essRamp) < 1e-10)
    essRamp = 0
end

essPar  = 1/Ka
if (abs(essPar) < 1e-10)
    essPar = 0
end

% with controller D(z)=kp+ki*z/(z-1)= ((kp+ki)*z-kp)/(z-1)
kp=1; ki=0.1; % kp and ki gains of PI controller
Dz = tf([kp+ki -kp], [1 -1],T) % PI controller D(z)
DGz = Dz * Gz; % D(z)G(z)
DGz = zpk(DGz)
zm1DGz = zm1*DGz;  % (z-1)*D(z)G(z)
zm1DGz = minreal(minreal(zm1DGz,0.01),0.01); % min realization after pole
zero cancellation
zpk(zm1DGz)
zm2DGz = zm1*zm1DGz;  % (z-1)*(z-1)*D(z)G(z)
zm2DGz = minreal(minreal(zm2DGz,0.01),0.01); % min realization after pole
zero cancellation
zpk(zm2DGz)
Kp = dcgain(DGz);
Kv = (1/T)*dcgain(zm1DGz);
Ka = (1/T)^2*dcgain(zm2DGz);
essStep = 1/(1+Kp)
if (abs(essStep) < 1e-10)
    essStep = 0
end

essRamp = 1/Kv
if (abs(essRamp) < 1e-10)
    essRamp = 0
end

essPar  = 1/Ka
if (abs(essPar) < 1e-10)
    essPar = 0
end
```

```
Results:
D(z) =1

D(z)G(z) =

   0.01405 z + 0.01314
  ----------------------
  z^2 - 1.819 z + 0.8187

(z-1)D(z)G(z) =

  0.014048 (z+0.9355)

  -------------------
      (z-0.8187)


(z-1)^2*D(z)G(z) =

  0.014048 (z-1) (z+0.9355)
  -------------------------
         (z-0.8187)

Kp =   2.4491e+14   Kv =    1.5000    Ka =    0
essStep =      0     essRamp =     0.6667   essPar =     Inf


D(z) =

  1.1 z - 1
  ---------
    z - 1
D(z)G(z) =

  0.015453 (z+0.9355) (z-0.9091)
  ------------------------------
        (z-1)^2 (z-0.8187)

(z-1)*D(z)G(z)=

  0.015453 (z+0.9355) (z-0.9091)
  ------------------------------
         (z-0.8187) (z-1)

(z-1)^2*D(z)G(z)=

  0.015453 (z+0.9355) (z-0.9091)
  ------------------------------
            (z 0.8187)

essStep =      0     essRamp =     0     essPar =     0.6667
```

Closed-loop step response of the system with (a) $D(z) = 1$, (b) $D(z) = 1 + 0.1z/(z - 1)$ are shown in Figs. 6.21 and 6.22.

Example 6.8 A closed-loop system with a digital filter and a disturbance input is shown in Fig. 6.23. The digital filter transfer function is given as $D(z) = \frac{0.0855K}{z-0.7261}$. Sampling time $T = 0.10$ s. Obtain the expression for the output $C(z)$ with (a) unit step

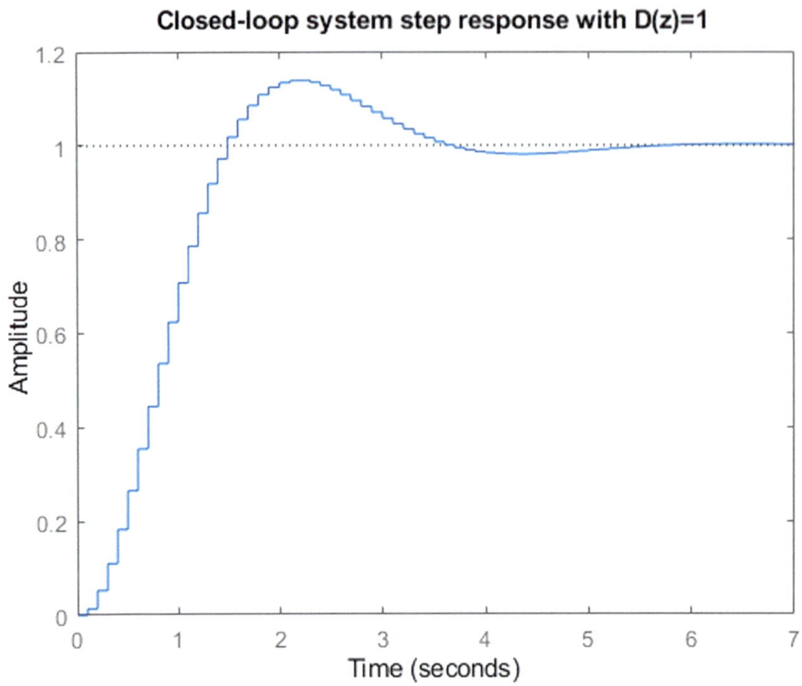

Fig. 6.21 Closed-loop system step response with $D(z) = 1$

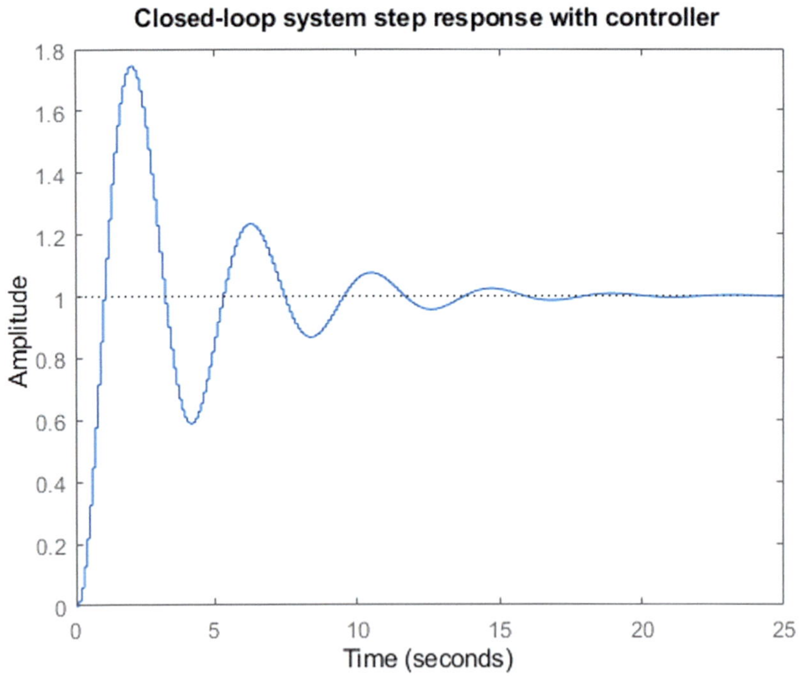

Fig. 6.22 Closed-loop system step response with controller

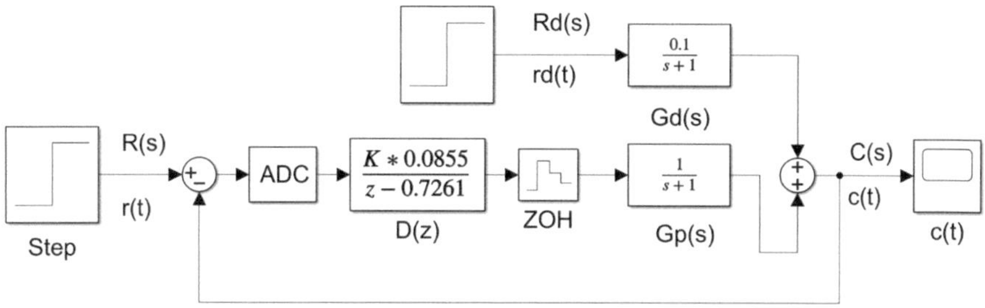

Fig. 6.23 Model of closed-loop system with digital filter and a disturbance input (Example 6.8)

reference input $R(s) = \frac{1}{s}$ only, (b) unit step disturbance input $R_d(s) = \frac{1}{s}$ only, and (c) both unit step reference and step disturbance inputs, $R(s) = \frac{1}{s}$ and $R_d(s) = \frac{1}{s}$. Find K such that the steady-state error due to the unit step disturbance input only is limited to 2%. Use this K to find the steady-state error due to unit step reference input only. Simulate the system response, and verify these steady-state errors.

Solution In the z-domain, the open-loop transfer functions for the plant and disturbance with ZOH can be obtained as

$$G_p(z) = Z\left[\frac{1 - e^{-sT}}{s} \frac{1}{s+1}\right]_{e^{sT}=z} = \frac{0.0952}{z - 0.9048}$$

$$G_d(z) = Z\left[\frac{1 - e^{-sT}}{s} \frac{0.1}{s+1}\right]_{e^{sT}=z} = \frac{0.00952}{z - 0.9048}$$

(a) The output due to reference input only can be obtained (with $R(z) = \frac{z}{z-1}$ and $R_d(z) = 0$) as

$$C_R(z) = \frac{D(z)G_p(z)}{1 + D(z)G_p(z)} R(z) = \frac{\left(\frac{0.0855K}{z - 0.7261}\right)\left(\frac{0.0952}{z - 0.9048}\right)}{1 + \left(\frac{0.0855K}{z - 0.7261}\right)\left(\frac{0.0952}{z - 0.9048}\right)} R(z)$$

$$= \left(\frac{0.00814K}{z^2 - 1.6309z + 0.6570 + 0.00814K}\right)\left(\frac{z}{z-1}\right)$$

(b) The output due to disturbance input only can be obtained (with $R(z) = 0$ and $R_d(z) = \frac{z}{z-1}$) as

$$C_D(z) = \frac{G_d(z)}{1 + D(z)G_p(z)} R_d(z) = \frac{\left(\frac{0.00952}{z - 0.9048}\right)}{1 + \left(\frac{0.0855K}{z - 0.7261}\right)\left(\frac{0.0952}{z - 0.9048}\right)} R_d(z)$$

$$= \left(\frac{0.00952(z - 0.7261)}{z^2 - 1.6309z + 0.6570 + 0.00814K}\right)\left(\frac{z}{z - 1}\right)$$

(c) The output due to both reference and disturbance unit step inputs is obtained as

$$C(z) = C_R(z) + C_D(z)$$

Error due to disturbance input is the same as $C_D(z)$

$$e_D(z) = C_D(z) = \left(\frac{0.00952(z - 0.7261)}{z^2 - 1.6309z + 0.6570 + 0.00814K}\right)\left(\frac{z}{z - 1}\right)$$

Steady-state error due to step disturbance input can be obtained as

$$e_{\text{ssStep},D} = \lim_{z \to 1} (z - 1)C_D(z) = \frac{0.002608}{0.0261 + 0.00814K}$$

$$\frac{0.002608}{0.0261 + 0.00814K} = 0.02$$

$$K = \frac{\left(\frac{0.00260}{0.02} - 0.0261\right)}{0.00814} = 12.7641$$

Steady-state error due to step reference input can be obtained as

$$e_{\text{ssStep},R} = \lim_{z \to 1} (z - 1)(R(z) - C_R(z)) = \frac{1}{1 + k_p}$$

where

$$k_p = \lim_{z \to 1} \left[D(z)G_p(z)\right] = \lim_{z \to 1} \left[\left(\frac{0.0855K}{z - 0.7261}\right)\left(\frac{0.0952}{z - 0.9048}\right)\right] = 3.9844$$

$$e_{\text{ssStep},R} = \frac{1}{1 + 3.9844} = 0.2006 \text{ or } 20.06\%.$$

Matlab code and results are presented next. The closed-loop system step response with reference input is shown in Fig. 6.24 and with disturbance input in Fig. 6.25.

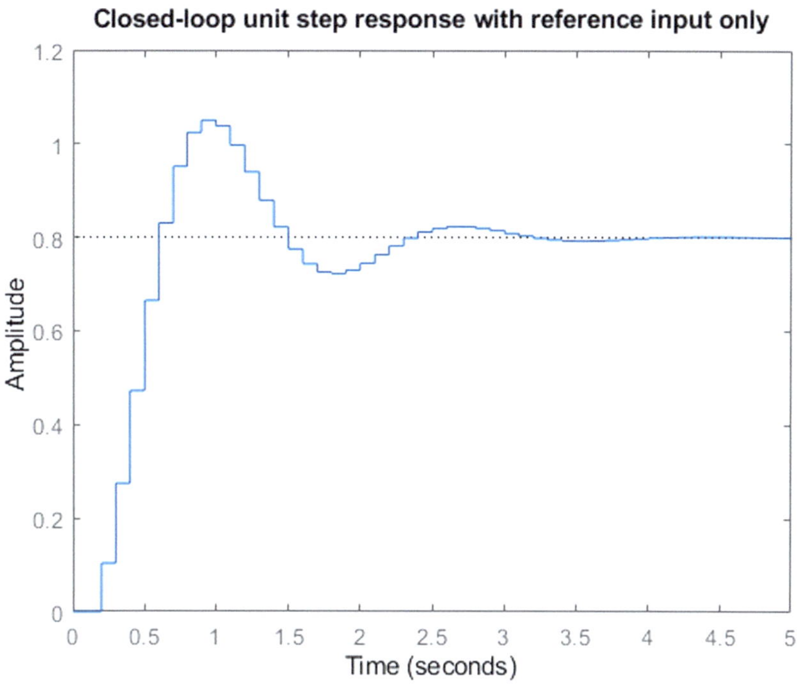

Fig. 6.24 Closed-loop unit step response with $R(s) = 1/s$ only (Example 6.7)

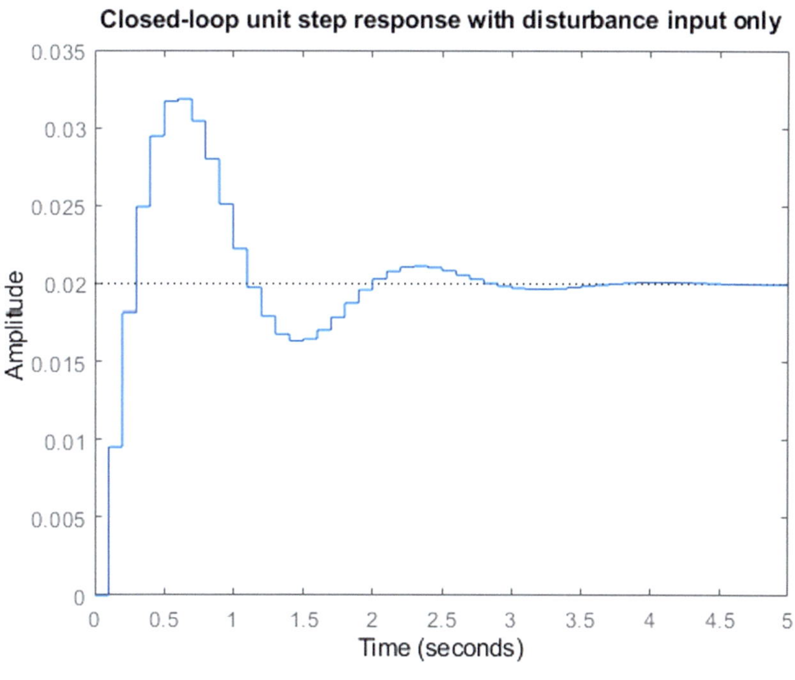

Fig. 6.25 Closed-loop unit step response with $R_d(s) = 1/s$ only (Example 6.7)

```
close all; clear all; clc
T=0.1;
K=12.7641;
Gp=tf(1,[1 1]) ;   % Gp(s)
Gd=tf(0.1,[1 1]); % Gd(s)
Dz=tf(0.0855*K, [1 -0.7261],T) % D(z)
% conversion to z-domain
Gpz=c2d(Gp,T)    % Gp(z)
Gdz=c2d(Gd,T)    % Gd(z)
% Closed-loop transfer function with R only
DGpz=Dz*Gpz;  % D(z)*Gp(z)
TR = feedback(DGpz,1) % TR(z)
% Closed-loop transfer function with Rd only
TD =Gdz/(1+DGpz);
TD=minreal(TD)
figure
dstep(TR,1)
title('Closed-loop unit step response with reference input only')
figure
dstep(TD,1)
title('Closed-loop unit step response with disturbance input only')
```

```
Results
Dz =

    1.091
  ----------
  z - 0.7261

Gpz =

   0.09516
  ----------
  z - 0.9048

Gdz =

   0.009516
  ----------
  z - 0.9048

TR =

          0.1039
  ----------------------
  z^2 - 1.631 z + 0.7609

TD =

   0.009516 z - 0.00691
  ----------------------
  z^2 - 1.631 z + 0.7609
```

Exercises

6.1 Obtain step response of a closed-loop discrete-time system, shown in Fig. P6.1, with the plant transfer function $G_p(s) = \frac{4}{s+4.5}$ and a zero-order hold (ZOH) for a sampling time $T = 0.1$ s. Obtain the output $c(kT)$ using inverse z-transform.

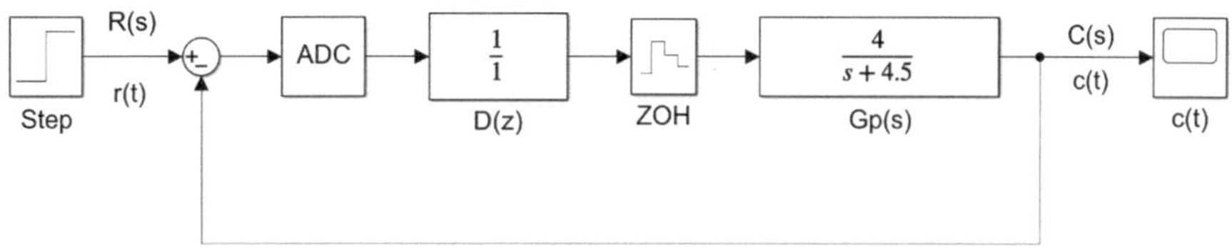

Fig. P6.1 Closed-loop control system of Problem 6.1

6.2 Obtain step response of a closed-loop discrete-time system shown in Fig. P6.2, with the plant transfer function $G_p(s) = \frac{2}{(s+1.5)(s+3.5)}$ and a zero-order hold (ZOH) for a sampling time $T = 0.05$ s.

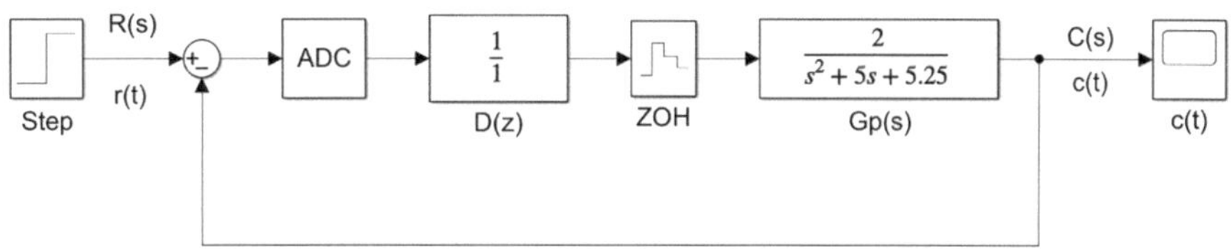

Fig. P6.2 Closed-loop control system of Problem 6.2

6.3 Obtain step response of a closed-loop discrete-time system, shown in Fig. P6.3, with the plant transfer function $G_p(s) = \frac{2}{s(s+4)}$ and a zero-order hold (ZOH) for a sampling time $T = 0.05$ s. Compare the discrete-time closed-loop step response with the corresponding continuous-time closed-loop system response.

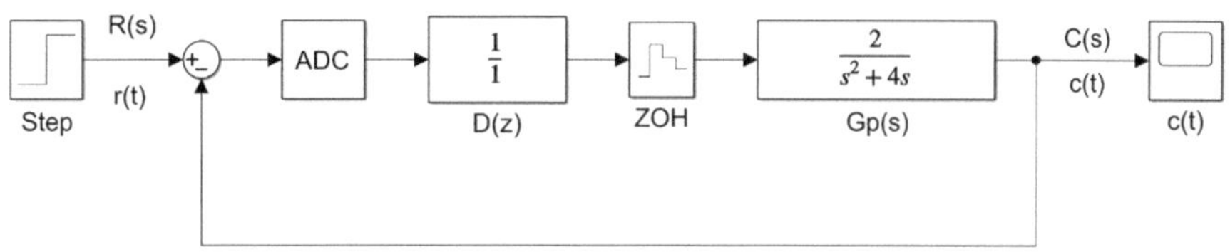

Fig. P6.3 Closed-loop control system of Problem 6.3

6.4 Obtain step response of a closed-loop discrete-time system, shown in Fig. P6.4, with the plant transfer function $G_p(s) = \frac{4}{s(s+2.5)(s+3.5)}$ and a zero-order hold (ZOH) for a sampling time $T = 0.05$ s. Compare the discrete-time closed-loop step response with the corresponding continuous-time closed-loop system response.

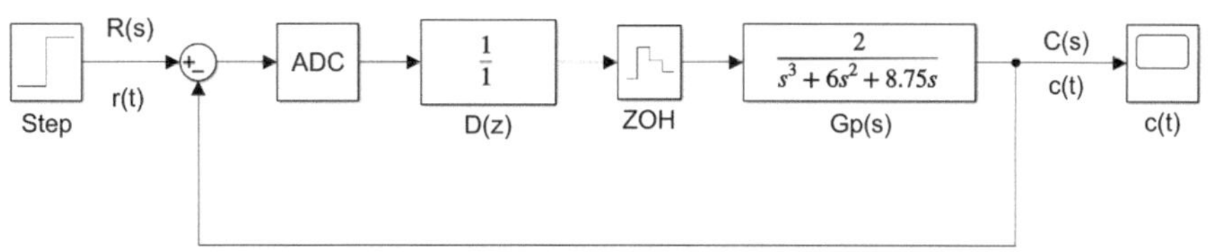

Fig. P6.4 Closed-loop control system of Problem 6.4

6.5 Obtain closed-loop zero and poles for the discrete-time closed-loop system shown in Fig. P6.5, with the plant transfer function $G_p(s) = \frac{10}{s^2+4s+5.5}$ and a zero-order hold (ZOH) for a sampling time $T = 0.1$ s. Estimate the time constant, rise time, settling time, and overshoot for the closed-loop system step response. Verify these values from the closed-loop system step response.

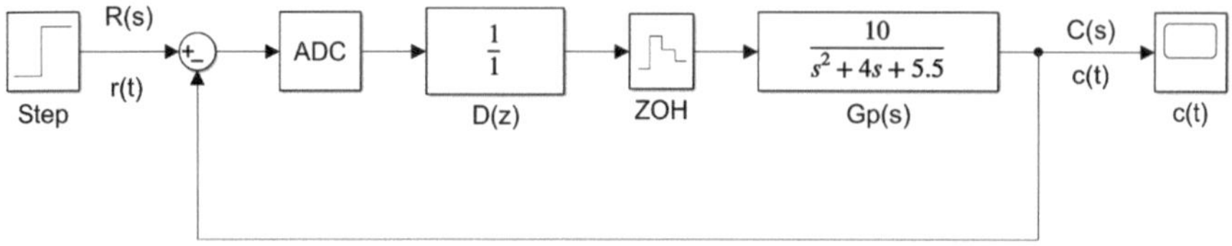

Fig. P6.5 Closed-loop control system of Problem 6.5

6.6 Obtain closed-loop zero and pole for the discrete-time closed-loop system shown in Fig. P6.6, with the plant transfer function $G_p(s) = \frac{10}{s(s^2+6s+6.5)}$ and a zero-order hold (ZOH) for a sampling time $T = 0.10$ s. Estimate the time constant, rise time, settling time, and overshoot for the closed-loop system step response. Verify these values from the closed-loop system step response.

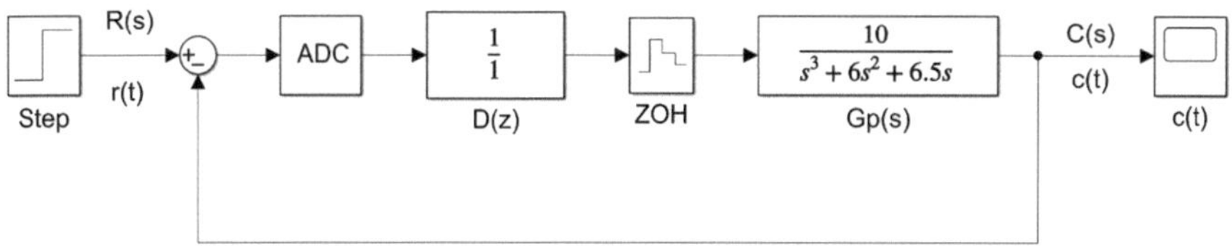

Fig. P6.6 Closed-loop control system of Problem 6.6

6.7 Obtain closed-loop zero and pole for the discrete-time closed-loop system shown in Fig. P6.7, with the plant transfer function $G_p(s) = \frac{18}{s^2+5.5s+5}$ and a zero-order hold (ZOH) for a sampling time $T = 0.05$ s. Estimate the time constant, rise time, settling time, and overshoot for the closed-loop system step response. Verify these values from the closed-loop system step response.

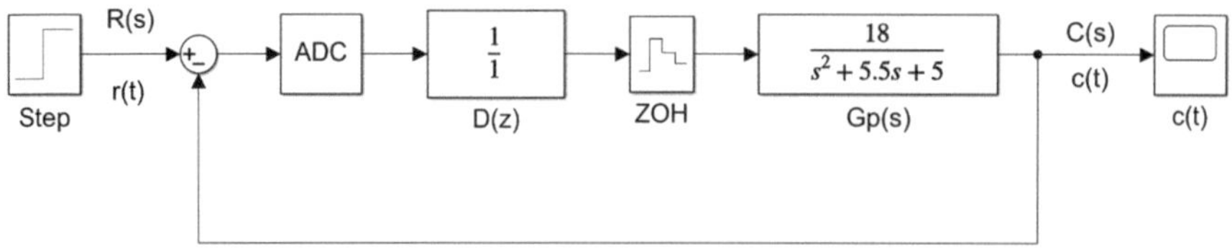

Fig. P6.7 Closed-loop control system of Problem 6.7

6.8 Obtain closed-loop zero and pole for the discrete-time closed-loop system shown in Fig. P6.8, with the plant transfer function $G_p(s) = \frac{15}{s^3+3.5s^2+8.5s+6}$ and a zero-order hold (ZOH) for a sampling time $T = 0.05$ s. Estimate the time constant, rise time, settling time, and overshoot for the closed-loop system step response. Verify these values from the closed-loop system step response.

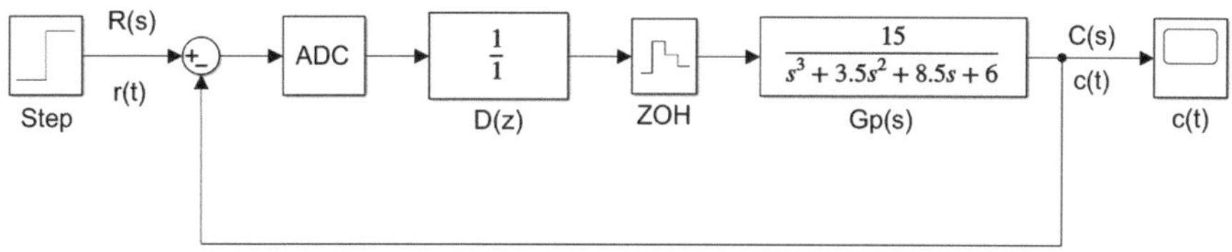

Fig. P6.8 Closed-loop control system of Problem 6.8

6.9 Obtain the system type of the discrete-time open-loop system shown in Fig. P6.9, with the plant transfer function $G_p(s) = \frac{10}{s^2+3.5s+8.5}$ and a zero-order hold (ZOH) for a sampling time $T = 0.05$ s. Estimate steady-state errors due to unit step, ramp, and parabolic inputs with the digital controller transfer function (a) $D(z) = 1$ and (b) $D(z) = 1 + 0.1z/(z - 1)$.

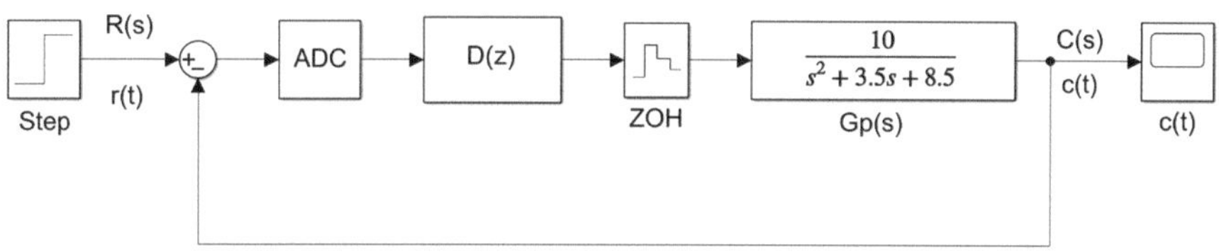

Fig. P6.9 Closed-loop control system of Problem 6.9

6.10 Obtain the system type of the discrete-time open-loop system shown in Fig. P6.10, with the plant transfer function $G_p(s) = \frac{2}{s^3+4.5s^2+7.5s+3}$ and a zero-order hold (ZOH) for a sampling time $T = 0.05$ s. Estimate steady-state errors due to unit step, ramp, and parabolic inputs with the digital controller transfer function (a) $D(z) = 1$ and (b) $D(z) = 1 + 0.05z/(z - 1)$.

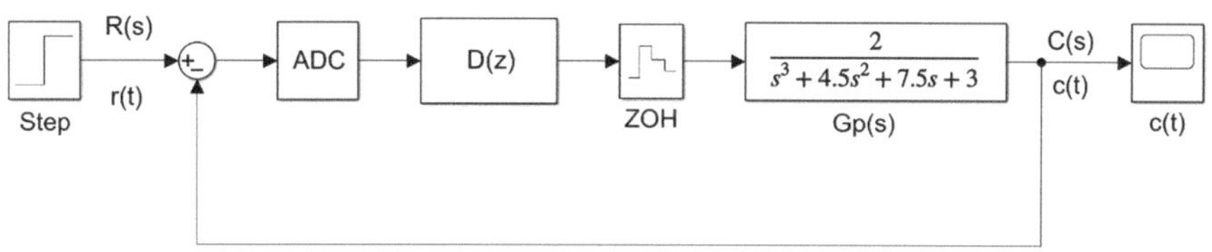

Fig. P6.10 Closed-loop control system of Problem 6.10

6.11 Obtain the system type of the discrete-time open-loop system shown in Fig. P6.11, with the plant transfer function $G_p(s) = \frac{K}{s^2+2.5s+3.5}$ and a zero-order hold (ZOH) for a sampling time $T = 0.02$ s. Obtain gain K such that the steady-state error with $D(z) = 1$ is 10% for a step input. Estimate steady-state errors due to unit step, ramp, and parabolic inputs with the digital controller transfer function $D(z) = 1 + 0.02z/(z - 1)$.

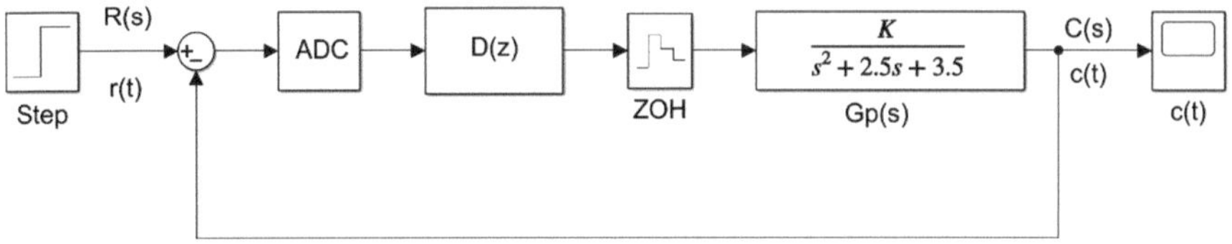

Fig. P6.11 Closed-loop control system of Problem 6.11

6.12 Obtain the system type of the discrete-time open-loop system shown in Fig. P6.12, with the plant transfer function $G_p(s) = \frac{K}{s^3+3.5s^2+6.5s+5}$ and a zero-order hold (ZOH) for a sampling time $T = 0.01$ s. Obtain gain K such that the steady-state error with $D(z) = 1$ is 40% for a step input. Estimate steady-state errors due to unit step, ramp, and parabolic inputs with the digital controller transfer function $D(z) = 1 + 0.01z/(z - 1)$.

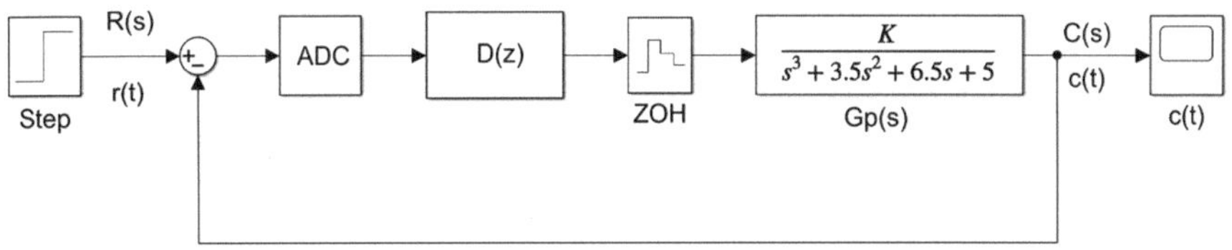

Fig. P6.12 Closed-loop control system of Problem 6.12

6.13 A closed-loop system with a digital filter and a disturbance input is shown in Fig. P6.13. The digital filter transfer function is given as $D(z) = \frac{0.0906K}{z-0.8187}$. Sampling time $T = 0.10$ s. Obtain the expression for the output $C(z)$ with (a) unit step reference input $R(s) = \frac{1}{s}$ only, (b) unit step disturbance input $R_d(s) = \frac{1}{s}$ only, and (c) both unit step reference and step disturbance inputs: $R(s) = \frac{1}{s}$ and $R_d(s) = \frac{1}{s}$. Find K such that the steady-state error due to the unit step disturbance input only is limited to 1.25%. Use this K to find the steady-state error due to unit step reference input only. Simulate the system response, and verify these steady-state errors.

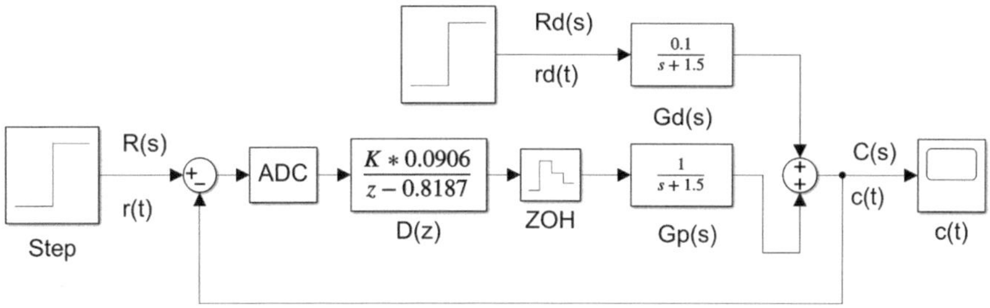

Fig. P6.13 Closed-loop control system of Problem 6.13

Further Reading

1. Asadi F (2023) Signals and systems with MATLAB® and Simulink®. Springer, Cham
2. Astrom KJ, Murray RM (2021) Feedback systems: an introduction for scientists and engineers, 2nd edn. Princeton University Press, Princeton
3. Chen C-T (2006) Analog and digital control system design: transfer-function, state-space, and algebraic methods. Oxford University Press, New York
4. Dorf R, Bishop R (2016) Modern control systems, 13th edn. Pearson, Essex
5. Fadali MS, Visioli A (2020) Digital control engineering analysis and design, 3rd edn. Academic Press, London
6. Franklin GF, Powell JD, Emami-Naeini A (2019) Feedback control of dynamic systems, 8th edn. Pearson, Upper Saddle River
7. Franklin GF, Powell JD, Workman M (1998) Digital control of dynamic systems, 3rd edn. Addison-Wesley, Boston
8. Isermann R (1981) Digital control systems. Springer, Berlin
9. Jalili N, Candelino NW (2023) Dynamic systems and control engineering. Cambridge University Press, Cambridge
10. Keviczky L, Bars R, Hetthéssy J, Bányász C (2019) Control engineering. Springer, Singapore
11. Koeing DM (2009) Practical control engineering: a guide for engineers, managers, and practitioners. McGraw-Hill, New York
12. Kuo BC (1995) Digital control systems, 2nd edn. Oxford University Press, Oxford
13. Landau ID, Gito G (2006) Digital control systems: design, identification and implementation. Springer, London
14. Leigh JR (2006) Applied digital control: theory, design and implementation, 2nd edn. Dover, Mineola
15. Mathworks (2023) Control System Toolbox. https://www.mathworks.com/products/control.html
16. Mathworks (2023) Matlab. https://www.mathworks.com/products/matlab.html
17. Nise NS (2019) Control systems engineering, 8th edn. Wiley, Hoboken
18. Nof SY (ed) (2009) Springer handbook of automation. Springer, Berlin
19. Ogata K (1995) Discrete-time control systems, 2nd edn. Pearson, Hoboken
20. Phillips CL, Nagle HT, Chakrabortty A (2015) Digital control system analysis and design, 4th edn. Pearson, Essex
21. Rabbath CA, Lechevin N (2014) Discrete-time control system design with applications. Springer, New York
22. Samanta B (2023) Introduction to mechatronics – an integrated approach. Springer, Cham
23. Shertukde HM (2015) Digital control applications illustrated with MATLAB®. CRC Press, Boca Raton

24. Starr G (2020) Introduction to applied digital controls. Springer, Cham
25. Veloni A, Miridakis NI (2018) Digital control systems-theoretical problems and simulation tools. CRC Press, Boca Raton
26. Vilanova R, Visioli A (2012) PID control in the third millennium – lessons learned and new approaches. Springer, London
27. Visioli A (2006) Practical PID control. Springer, London

Chapter 7
Stability Analysis of Discrete-Time Systems

7.1 Introduction

Stability analysis of discrete-time systems without explicit solution of the closed-loop system characteristic equation is covered in this chapter. Jury's test and Nyquist criterion are presented to analyze the stability of closed-loop discrete-time systems in the z-domain. Bilinear transformation is presented to transform discrete-time system transfer functions from the z-domain to equivalent continuous-time system transfer functions in the w-domain enabling the application of s-domain techniques in the w-domain. Finally, Routh-Hurwitz criterion is presented to analyze the stability of discrete-time systems in the w-domain. The procedures are illustrated with suitable examples and corresponding Matlab codes in each section. End-of-chapter exercises are provided to enhance understanding of the concepts covered and consolidate problem-solving skills.

7.2 Stability of Discrete-Time Systems

A closed-loop discrete-time system is shown in Fig. 7.1.

The closed-loop discrete-time system transfer function can be expressed as

$$T(z) = \frac{C(z)}{R(z)} = \frac{G(z)D(z)}{1 + \overline{GH}(z)D(z)} \tag{7.1}$$

where

$$\overline{GH}(z) = Z\left[G_{\mathrm{h}}(s)G_{\mathrm{p}}(s)H(s)\right] \tag{7.2}$$

B. Samanta, *Introduction to Digital Control*,
https://doi.org/10.1007/978-3-031-66830-2_7

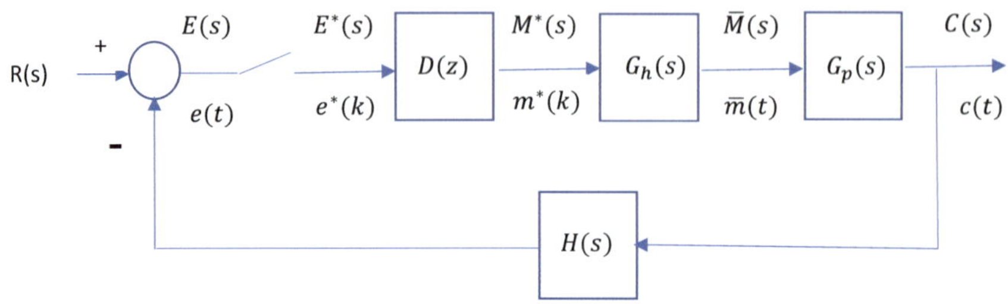

Fig. 7.1 Closed-loop discrete-time system with a digital filter and a sensor

The characteristic equation for the closed-loop discrete-time system can be written as

$$1 + \overline{GH}(z)D(z) = 0 \tag{7.3}$$

The characteristic equation (7.3) can be expressed in terms of its roots as

$$P(z) = \prod_{i=1}^{n} (z - p_i) = 0 \tag{7.4}$$

where roots p_i, $i = 1, \ldots, n$ are termed as closed-loop poles of the discrete-time system and n is the order of the system. For stability, all closed-loop poles (p_i) must be within the unit circle in z-plane.

The stability of a closed-loop discrete-time system can be analyzed by directly finding the roots of the characteristic equation (7.3) and examining each pole location in the z-plane. However, there are situations where it would be required to analyze the stability without directly solving the characteristic equation. In the following sections, stability analysis using these indirect methods, without the need for an explicit solution of the characteristic equation, is presented.

7.3 Jury's Stability Test

The characteristic polynomial is written as

$$P(z) = a_n z^n + a_{n-1} z^{n-1} + a_{n-2} z^{n-2} + \cdots + a_1 z + a_0 = 0, \quad a_n > 0 \tag{7.5}$$

The roots of the characteristic polynomial are within the unit circle in the z-plane if and only if the following $(n + 1)$ conditions are satisfied:

$$
\begin{array}{ll}
(1) & P(1) > 0 \\
(2) & (-1)^n P(-1) > 0 \\
(3) & |a_0| < a_n \\
(4) & |b_0| > |b_{n-1}| \\
(5) & |c_0| > |c_{n-2}| \\
 & \vdots \\
(n+1) & |q_0| > q_2
\end{array}
\tag{7.6}
$$

Where the terms in $(n + 1)$ conditions are calculated from Table 7.1. The entries in Jury's array for stability analysis of Table 7.1 are calculated as

Table 7.1 Jury's array for stability analysis

Row	z^0	z^1	z^2	…	…	z^{n-1}	z^n
1	a_0	a_1	a_2	…	…	a_{n-1}	a_n
2	a_n	a_{n-1}	a_{n-2}	…	…	a_1	a_0
3	b_0	b_1	b_2	…	…	b_{n-1}	
4	b_{n-1}	b_{n-2}	b_{n-3}	…	…	b_0	
5	c_0	c_1	c_2	…	c_{n-2}		
6	c_{n-2}	c_{n-3}	c_{n-4}	…	c_0		
$2n-5$	p_0	p_1	p_2	p_3			
$2n-4$	p_3	p_2	p_1	p_0			
$2n-3$	q_0	q_1	q_2				

$$b_k = \begin{vmatrix} a_0 & a_{n-k} \\ a_n & a_k \end{vmatrix}, \quad k = 0, 1, \ldots, n-1$$

$$c_k = \begin{vmatrix} b_0 & b_{n-k-1} \\ b_{n-1} & b_k \end{vmatrix}, \quad k = 0, 1, \ldots, n-2 \tag{7.7}$$

$$\vdots$$

$$q_0 = \begin{vmatrix} p_0 & p_3 \\ p_3 & p_0 \end{vmatrix}, \quad q_1 = \begin{vmatrix} p_0 & p_2 \\ p_3 & p_1 \end{vmatrix}, \quad q_3 = \begin{vmatrix} p_0 & p_1 \\ p_3 & p_2 \end{vmatrix}$$

For an nth order system, the number of rows in Jury array is given as $(2n - 3)$, and number of conditions (constraints) to satisfy is $(n + 1)$.

It is worth noting that the first three conditions in Eq. (7.6) can be checked without constructing the Jury's array. If any of the first three conditions is not satisfied, there will be a least one root of the characteristic polynomial on or outside the unit circle in the z-plane, making the closed-loop discrete-time system unstable. In that case, there is no need to construct the Jury's array of Table 7.1. If all first three conditions are satisfied, the procedure to construct Jury's array needs to be completed, and rest of the conditions in Eq. (7.6) should be tested to assess stability of the closed-loop discrete-time system.

The procedure is illustrated using examples next.

Example 7.1 A closed-loop discrete-time system, similar to Fig. 7.1, is with the plant transfer function $G_\text{p}(s) = \frac{3}{s(s+2)}$, a zero-order hold (ZOH), a digital controller with a simple gain, $D(z) = K$, a sensor with a unity gain, $H(s) = 1$. Use a sampling time $T = 0.1$ s. Use Jury's test to find the range of the controller gain K for the stability of the closed-loop discrete-time system.

Solution The open-loop transfer function of the discrete-time system with ZOH is obtained (in Example 6.2) as

$$\begin{aligned} G(z) &= \overline{G_{\text{h}0}G_\text{p}}(z) = Z\left[\frac{1-e^{-sT}}{s}\,G_\text{p}(s)\right]_{e^{sT}=z} \\ &= \frac{0.01405z + 0.01314}{z^2 - 1.8187z + 0.8187} \end{aligned}$$

The closed-loop discrete-time system characteristic equation can be expressed as

$$1 + D(z)G(z) = 1 + K\frac{0.01405z + 0.01314}{z^2 - 1.8187z + 0.8187} = 0$$

Or

$$z^2 - 1.8187z + 0.8187 + K(0.01405z + 0.01314) = 0$$

Or

$$z^2 + (0.01405K - 1.8187)z + (0.8187 + 0.01314K) = 0$$

For the second-order system $n = 2$, there is only one row $(2n - 3)$ in Jury's array as

	z^0	z^1	z^2
	$0.8187 + 0.01314K$	$0.01405K - 1.8187$	1

Applying condition (1) $P(1) > 0$

$$1 + (0.01405K - 1.8187)1 + (0.8187 + 0.01314K) = 0.02719K > 0 \rightarrow K > 0$$

Applying condition (2) $(-1)^2 P(-1) > 0$

$$1 + (0.01405K - 1.8187)(-1) + (0.8187 + 0.01314K) > 0$$
$$3.6374 - 0.910 \times 10^{-3}K > 0 \rightarrow K < 3997$$

Applying condition (3) $|a_0| < a_2$

$$(0.8187 + 0.01314K) < 1 \rightarrow K < \frac{1 - 0.8187}{0.01314} = 13.798$$

For the stability of the closed-loop discrete-time system, the range of controller gain K is obtained as

$$0 < K < 13.798$$

The closed-loop discrete-time system is marginally stable at $K = 13.798$. At this critical value of K, the characteristic equation is

$$z^2 + (0.01405K - 1.8187)z + (0.8187 + 0.01314K) = 0$$
$$z^2 - 1.6248\,z + 1 = 0$$

The roots of the characteristic equation are obtained as

$$z = 0.8124 \pm j\,0.5831 = 1\angle \pm 0.6225 \text{ rad} = 1\angle \pm \omega T$$

For the sampling time $T = 0.1$ s, the closed-loop system is expected to have an oscillatory response with a frequency $\omega = 6.225$ rad/s.

Matlab code and results are presented to confirm the stability analysis. The closed-loop system step response with $K < K_{\text{critical}}$, $K = K_{\text{critical}}$, and $K > K_{\text{critical}}$ are shown in Figs. 7.2, 7.3, and 7.4, respectively.

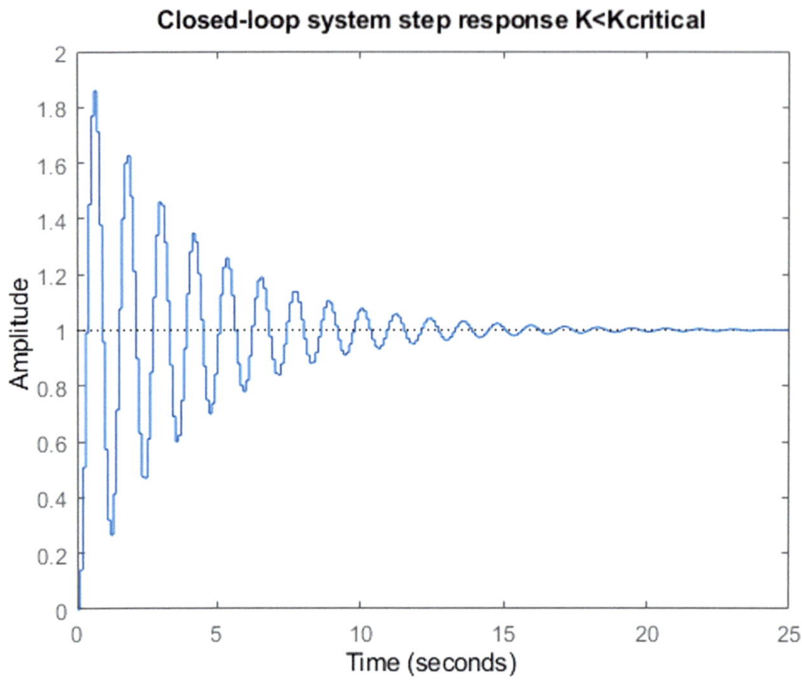

Fig. 7.2 Closed-loop system step response at gain $K = 10$ ($<K_{\text{crtical}}$)

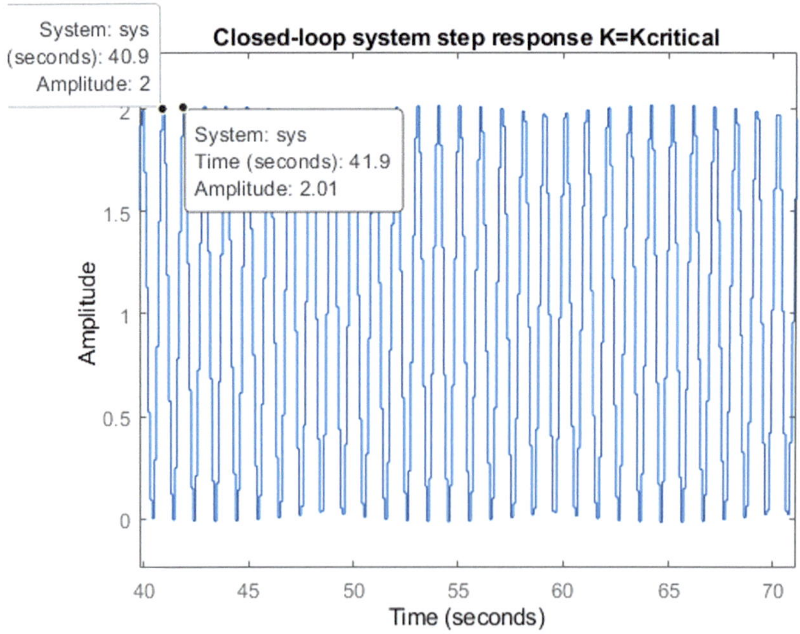

Fig. 7.3 Closed-loop system step response at gain $K = 13.798$ ($=K_{\text{crtical}}$) (zoomed in)

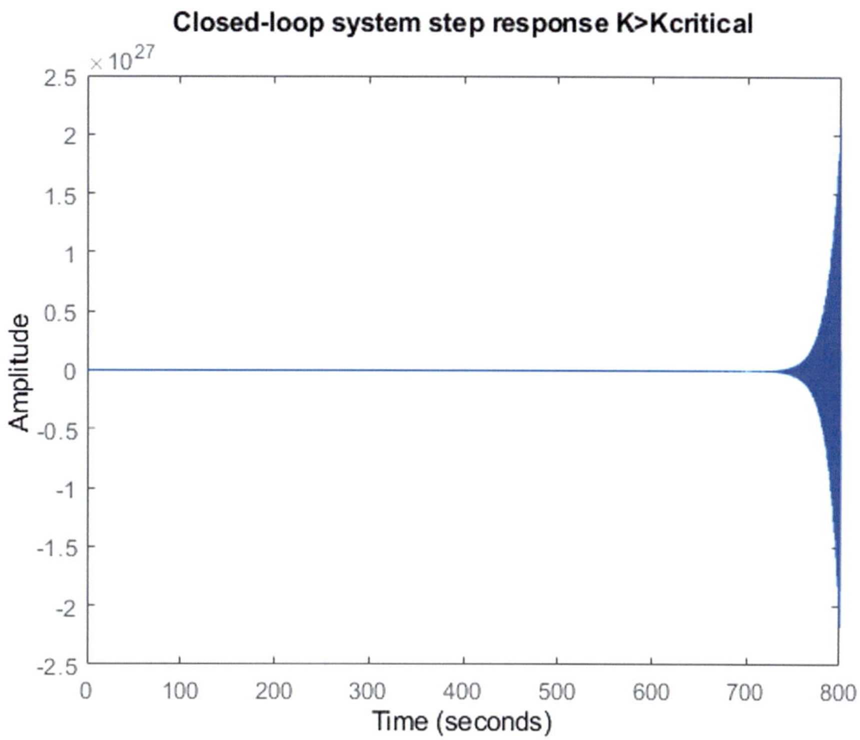

Fig. 7.4 Closed-loop system step response at gain $K = 15$ ($>K_{\mathrm{crtical}}$)

```
%Jury Test
% Ch7 Ex 7_1 Stability Analysis of a 2nd order closed-loop discrete-time
system
% Gp(s)=3/s(s+2), H(s)=1, D(z)=K
clear all; close all; clc
T=0.1 ; % Sampling time
Gp = tf(3, [1 2 0]) %Gp(s)

% C(z) = Z(Cho(s)Gp(s))
Gz =c2d(Gp,T) % G(z)
Gz=zpk(Gz)
[ngz,dgz]=tfdata(Gz,'v')
[z,p,k]=tf2zp(ngz,dgz)
n=length(dgz)-1 % order of the system (number of open-loop poles)
m=length(ngz)-1 % number of open-loop zeros
syms K z
a2=(dgz(1)+K*ngz(1));
a1=(dgz(2)+K*ngz(2));
a0=(dgz(3)+K*ngz(3));
P=a2*z^n+a1*z^(n-1)+a0*z^(n-2);
P=vpa(simplify(P),4)
```

```
% condition 1 P(1)>0
c1=vpa(subs(P,z,1),4)
% condition 2 P(-1)>0
c2=vpa(subs(P,z,-1),4)
% condition 3 |a0|<a2
c3=vpa(a0-a2,4)
K1=double(solve(c1==0,K))
K2=double(solve(c2==0,K))
K3=double(solve(c3==0,K))
% P at Kcritical
Pc=vpa(subs(P,K,K3),4) % characteristic ploynomial at K=Kcrtical
zc=double(solve(Pc==0, z)) % closed-loop poles at K=Kcrtical
r=abs(zc)  % magnitude of closed-loop poles
th=angle(zc) % angles of closed-loop poles
w=th/T % frequency of oscillation

% Closed-loop Step Response
% Gp(s)=3/s(s+2), H(s)=1, D(z)=K
clear all; close all; clc
T=0.1 ; % Sampling time
Gp = tf(3, [1 2 0]) %Gp(s)

% G(z) = Z(Gho(s)Gp(s))
Gz =c2d(Gp,T) % G(z)

% K<13.798
K=10
Tz = feedback(K*Gz, 1)
figure(1)
dstep(Tz,1)
title('Closed-loop system step response K<Kcritical')

% K=13.798
K=13.798; % critical gain
% Closed loop Transfer function
Tz = feedback(K*Gz, 1) % Closed loop TF with unity feedback
figure(2)
dstep(Tz,1)
title('Closed-loop system step response K=Kcritical')
% closed loop poles at Kcritical
[ntz,dtz] = tfdata(Tz, 'v')
[z,p,k]=tf2zp(ntz,dtz)
r=abs(p(1))
th=angle(p(1))
wosc = th/T  % frequency of oscillation at Kcritical

% K>13.797
K=15; % > Kcritical
Tz = feedback(K*Gz, 1) % Closed loop TF with unity feedback
figure(3)
dstep(Tz,1)
title('Closed-loop system step response K>Kcritical')
```

```
Results:
G(z) =

 0.01405 z + 0.01314
 ----------------------
 z^2 - 1.819 z + 0.8187

P(z) = z^2 + (0.01405*K - 1.819)*z + 0.01314*K + 0.8187

c1: 0.02719*K > 0 K> 0

c2: 3.637 - 0.0009057*K > 0 K < 4016

c3: 0.01314*K - 0.1813 < 0 K < 13.7928

P(z) = z^2 - 1.625*z + 1.0

z = 0.8125 - 0.5830i; 0.8125 + 0.5830i

r = 1.0000; 1.0000
th = -0.6224; 0.6224

w = 6.2240 rad/s

step response K< Kcritical
K = 10
T(z) =
 0.1405 z + 0.1314
 ----------------------------
 z^2 - 1.678 z + 0.9502



At critical gain K=13.798
T(z) =
 0.1938 z + 0.1813
 -------------------------
 z^2 - 1.625 z + 1

closed-loop poles are p = 0.8125 + 0.5831i, 0.8125 - 0.5831i
r = 1.0000 th= 0.6225 rad, wosc = 6.2248 rad/s
```

Frequency of oscillation $\omega = \frac{2\pi}{(41.9 - 40.9)} = 6.2832 \text{ rad/s}$.

```
At K=15 (<Kcritical)

T(z) =
 0.2107 z + 0.1971
 ------------------------------
 z^2 - 1.608 z + 1.016
```

Example 7.2 A closed-loop discrete-time system, similar to Fig. 7.1, is with the plant transfer function $G_p(s) = \frac{3}{s(s+2)}$, a zero-order hold (ZOH), a PI controller $D(z) = K_P\left(1 + \frac{K_I}{K_P}\frac{T}{2}\left(\frac{z+1}{z-1}\right)\right)$, a sensor with a unity gain, $H(s) = 1$. Use a sampling time $T = 0.1$ s, $\frac{K_I}{K_P} = 0.2$. Use Jury's test to find the range of the PI controller proportional gain K_P for the stability of the closed-loop discrete-time system.

Solution The open-loop transfer function of the discrete-time system with ZOH is obtained (in Example 6.2) as

$$\begin{aligned} G(z) &= \overline{G_{h0}G_p}(z) = Z\left[\frac{1-e^{-sT}}{s}\, G_p(s)\right]_{e^{sT}=z} \\ &= \frac{0.01405z + 0.01314}{z^2 - 1.8187z + 0.8187} \end{aligned}$$

The closed-loop discrete-time system characteristic equation can be expressed as

$$1 + D(z)G(z) = 1 + K_P\left(1 + 0.01\left(\frac{z+1}{z-1}\right)\right)\frac{0.01405z + 0.01314}{z^2 - 1.8187z + 0.8187} = 0$$

Or

$$(z-1)\left(z^2 - 1.8187z + 0.8187\right) + K_P(1.01z - 0.99)(0.01405z + 0.01314) = 0$$

Or

$$\begin{aligned} &z^3 + (0.01419K_P - 2.819)z^2 + (2.637 - 0.000638K_P)z \\ &- (0.8187 + 0.01301K_P) = 0 \end{aligned}$$

For the third-order system $n = 3$, there are $(2n - 3)$ rows in Jury's array as

Row	z^0	z^1	z^2	z^3
1	$-(0.8187 + 0.01301K_P)$	$2.637 - 0.000638K_P$	$0.01419K_P - 2.819$	1
2	1	$0.01419K_P - 2.819$	$2.637 - 0.000638K_P$	$-(0.8187 + 0.01301K_P)$
3	b_0	b_1	b_2	

Applying condition (1) $P(1) > 0$

$$1 + (0.01419K_P - 2.819)1 + (2.637 - 0.000638K_P)1 - (0.8187 + 0.01301K_P) > 0$$
$$0.0005438K_P - 3.638e - 12 > 0\ K_P > 6.6899e - 09$$

Applying condition (2) $(-1)^3 P(-1) > 0$

$$1 - (0.01419K_P - 2.819)1 + (2.637 - 0.000638K_P)1 + (0.8187 + 0.01301K_P) > 0$$
$$7.275 - 0.00182K_P > 0 \rightarrow K_P < 3997$$

Applying condition (3) $|a_0| < a_3$

$$(0.8187 + 0.01301K_P) < 1 \rightarrow K_P < \frac{1 - 0.8187}{0.01301} = 13.937$$

$$b_0 = \begin{vmatrix} -(0.8187 + 0.01301K_P) & 1 \\ 1 & -(0.8187 + 0.01301K_P) \end{vmatrix}$$
$$= (0.8187 + 0.01301K_P)^2 - 1$$

$$b_1 = \begin{vmatrix} -(0.8187 + 0.01301K_P) & 0.01419K_P - 2.819 \\ 1 & 2.637 - 0.000638K_P \end{vmatrix}$$
$$= 8.301e - 6\ K_P^2 - 0.04798\ K_P + 0.6595$$

$$b_2 = \begin{vmatrix} -(0.8187 + 0.01301K_P) & 2.637 - 0.000638K_P \\ 1 & 0.01419K_P - 2.819 \end{vmatrix}$$
$$= -0.0001846K_P^2 + 0.02569K_P - 0.3297$$

Applying condition (4) $|b_0| > |b_{n-1}|$

$$\left|(0.8187 + 0.01301K_P)^2 - 1\right| > -0.0001846K_P^2 + 0.02653K_P - 0.3322$$
$$K_P(0.004387 - 0.0003538K_P) + 4.829 \times 10^{-13} > 0$$

Numerically solving condition (4), one can obtain $0 < K_P < 12.40$.

For the stability of the closed-loop discrete-time system, the range of controller gain K_P is obtained as

$$0 < K_P < 12.40$$

The closed-loop discrete-time system is marginally stable at $K_P = 12.40$. At this critical value of K_P, the characteristic equation is

$$z^3 - 2.643\, z^2 + 2.629\, z - 0.9800 = 0$$
$$(z - 0.9800)(z - 0.8314 - j0.5557)(z - 0.8314 + j0.5557) = 0$$

The complex roots of the characteristic equation are obtained as

$$z = 0.8314 \pm j\, 0.5557 = 1\angle \pm 0.5892 \text{ rad} = 1\angle \pm \omega T$$

For the sampling time $T = 0.1$ s, the closed-loop system is expected to have an oscillatory response with a frequency $\omega = 5.892$ rad/s.

Matlab code and results are presented to confirm the stability analysis. The closed-loop system step response with $K = K_{critical}$ is shown in Fig. 7.5.

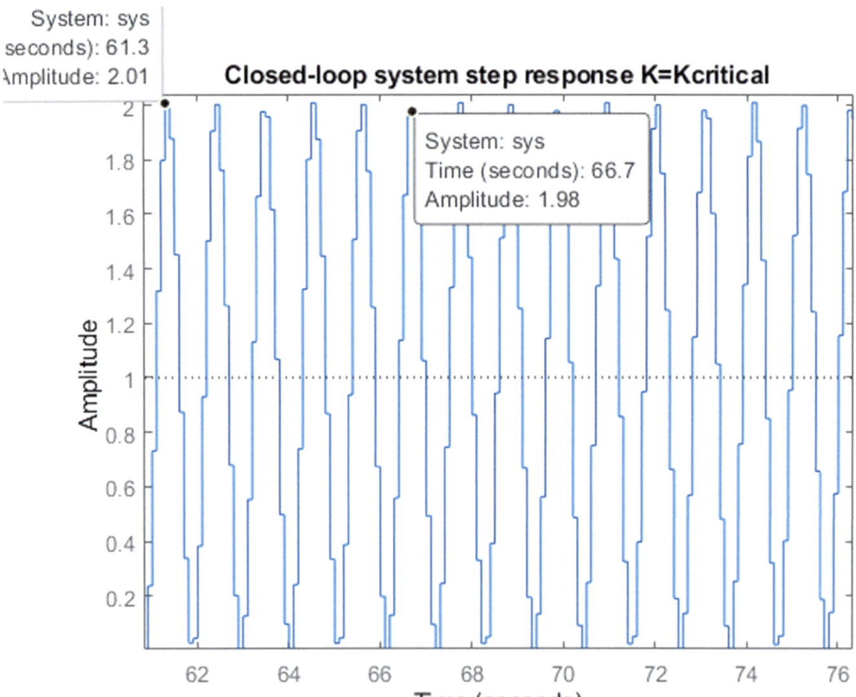

Fig. 7.5 Closed-loop system step response at gain $K = 12.4$ ($=K_{crtical}$) (zoomed in)

```
% Gp(s)=3/s(s+1), H(s)=1, D(z)=K(1+0.01(z+1)/(z-1))

clear all; close all;clc
T=0.1 ; % Sampling time
Gp = tf(3, [1  2  0 ]) %Gp(s)

% G(z) = Z(Gho(s)Gp(s))
Gz =c2d(Gp,T) % G(z)
[ngz, dgz]=tfdata(Gz,'v')

syms K z
KI=0.01
Ng=vpa(0.01405*z+0.01314,4)
Dg=vpa(z^2-1.8187*z+0.8187,4)
Ngp=vpa(simplify(Ng*((KI+1)*z+(KI-1))),4)
Dgp=vpa(simplify(Dg*(z-1)),4)
ba_bi = diff(Ngp,z)*Dgp-diff(Dgp,z)*Ngp;
babi=double(solve(ba_bi==0, z))

% Generalized G(z)

ngzp=[0 0.01405*1.01 (1.01*0.01314-0.99*0.01405) -0.01314*0.99]
dgzp=[1 -(1.8187+1) (1.8187+0.8187) -0.8187]
Gpz=tf(ngzp, dgzp, T)
n=length(dgzp)-1 % order of the system (number of open-loop poles)
m=length(ngzp)-1 % number of open-loop zeros

a3=(dgzp(1)+K*ngzp(1));
a2=(dgzp(2)+K*ngzp(2));
a1=(dgzp(3)+K*ngzp(3));
a0=(dgzp(4)+K*ngzp(4));
P=a3*z^n+a2*z^(n-1)+a1*z^(n-2)+a0*z^(n-3);
P=vpa(simplify(P),4)
% condition 1 P(1)>0
c1=vpa(subs(P,z,1),4)
% condition 2 P(-1)>0
c2=vpa(subs((-1)^n*P,z,-1),4)
% condition 3 |a0|<a3
c3=vpa(a3-abs(a0),4)

%condition 4
b0=vpa(simplify(a0^2-a3^2),4)
b1=vpa(simplify(a0*a1-a2*a3),4)
b2=vpa(simplify(a0*a2-a1*a3),4)
c4=vpa(simplify(b2-b0),4)

K1=double(solve(c1==0,K))
K2=double(solve(c2==0,K))
K3=double(solve(c3==0,K))
K4=double(solve(c4==0,K))

% P at Kcritical
KK =[ abs(K2), abs(K3), K4(1),K4(2)]
Km=min(KK)
Km=12.4



 Pc=vpa(subs(P,K,Km),4) % characteristic ploynomial at K=Kcrtical
 zc=double(solve(Pc==0, z)) % closed-loop poles at K=Kcrtical
 r=abs(zc)  % magnitude of closed-loop poles
 th=angle(zc) % angles of closed-loop poles
 w=th/T % frequency of oscillation


 % step response
 Tz=feedback(Km*Gpz,1)
 figure(1)
 dstep(Tz,1)
 title('Closed-loop system step response K=Kcritical')
```

```
Results:

G(z) =

    0.01405 z + 0.01314
  ----------------------
  z^2 - 1.819 z + 0.8187



D(z)G(z) =

   0.01419 z^2 - 0.0006381 z - 0.01301

  ----------------------------------------------

  z^3 - 2.819 z^2 + 2.637 z - 0.8187
```

$P(z) = z^3 + (0.01419\,K_P - 2.819)z^2 - (0.0006381\,K_P - 2.637)z - (0.01301\,K_P + 0.8187)$

c1 = 0.0005438 K_P - 3.638e-12 K_P > 6.6899e-09

c2 = 7.275 - 0.00182 K_P K_P < 3997

c3 = 1.0 - (0.01301*K + 0.8187) K_P < 13.9369

b0 = (0.01301 K_P + 0.8187)^2 - 1.0

b1 = 8.301e-6 $K_P{}^2 -$ 0.04798 K_P + 0.6595

b2 = - 0.0001846 $K_P{}^2$ + 0.02569 K_P - 0.3297

c4 = - 0.0003538 $K_P{}^2$ + 0.004387 K_P + 4.829e-13 $0 < K_P < 12.4$

$P(z) = z^3 - 2.643z^2 + 2.629\,z - 0.98$

```
 zc =    0.9800 + 0.0000i;     0.8314 - 0.5557i;     0.8314 + 0.5557i

r =     0.9800;       1.0000;     1.0000

th =          0;     -0.5892;     0.5892
w =          0;    -5.8924;     5.8924

T(z) =

   0.176 z^2 - 0.007912 z - 0.1613

  ------------------------------------------

  z^3 - 2.643 z^2 + 2.629 z - 0.98
```

Frequency of oscillation $\omega = \frac{2\pi}{(66.7 - 61.3)/5} = 5.8178$ rad/s that is close to the value obtained from closed-loop poles (5.892 rad/s).

7.4 Nyquist Criterion for Stability Analysis

The characteristic equation for the closed-loop discrete-time system, Eq. (7.3) can be written as

$$1+\overline{GH}(z)D(z)=1+\frac{R(z)}{Q(z)}=\frac{R(z)+Q(z)}{Q(z)}=0 \tag{7.8}$$

$$\overline{GH}(z)D(z)=\frac{R(z)}{Q(z)} \tag{7.9}$$

Zeros of the characteristic equation are the poles of the closed-loop discrete-time system. The poles of the open-loop discrete-time system are the poles of the characteristic equation. For stability of the closed-loop discrete-time system, all poles of the closed-loop system or all zeros of the characteristic equation must be within the unit circle in the z-plane. In other words, for stability, there must not be any zero of the characteristic equation outside the unit circle in the z-plane.

The stability of a closed-loop discrete-time system can be analyzed using Nyquist criterion for the open-loop system transfer function $\overline{GH}(z)D(z)$ of Eq. (7.9) in relation with the characteristic equation (7.8) on the unit circle as the Nyquist path. Nyquist criterion can be derived for analyzing stability of a closed-loop discrete-time system in the z-plane as

$$Z=N+P \tag{7.10}$$

where Z is the number of zeros of the characteristic equation outside the unit circle, P is the number of poles of the open-loop discrete-time system that are outside the unit circle, and N is the number of clockwise encirclements of the point $-1+j0$ by the Nyquist diagram in the z-plane. The Nyquist diagram can be obtained by plotting the open-loop discrete-time transfer function $\overline{GH}(z)D(z)$ for $z=e^{j\omega T}$ (unit circle) as

$$\overline{GH}(z)D(z)\big|_{z=e^{j\omega T}} \tag{7.11}$$

For stability, there must not be a zero of the characteristic equation (7.8) outside the unit circle in the z-plane, i.e., $Z=0$, and if there is no open-loop discrete-time system pole outside the unit circle, i.e., $P=0$, then the number of clockwise encirclements of the point $-1+j0$ by the Nyquist diagram must be zero, i.e., $N=0$.

Nyquist criterion can also be used to find the maximum gain of a controller for the stability of a closed-loop discrete-time system.

Example 7.3 A closed-loop discrete-time system, similar to Fig. 7.1, is with the plant transfer function $G_{\mathrm{p}}(s)=\frac{3}{s(s+2)}$, a zero-order hold (ZOH), a digital controller with a simple gain, $D(z)=K$, a sensor with a unity gain, $H(s)=1$. Use a sampling time $T=0.1$ s. Use Nyquist criterion to find the maximum controller gain K for the stability of the closed-loop discrete-time system.

Solution The open-loop transfer function of the discrete-time system with ZOH is obtained (in Example 6.2) as

$$\begin{aligned} G(z) &= \overline{G_{\mathrm{h0}}G_{\mathrm{p}}}(z) = Z\left[\tfrac{1-e^{-sT}}{s}\,G_{\mathrm{p}}(s)\right]_{e^{sT}=z} \\ &= \frac{0.01405z+0.01314}{z^2-1.8187z+0.8187} = \frac{0.01405z+0.01314}{(z-1)(z-0.8187)} \end{aligned}$$

The closed-loop discrete-time system characteristic equation can be expressed as

$$1 + D(z)G(z) = 1 + K\frac{0.01405z+0.01314}{(z-1)(z-0.8187)} = 0$$

There is an open-loop pole at $z = 1$, so the Nyquist path will be a small detour around $z = 1$ using $z = 1 + \rho e^{j\omega T}$, $\rho \ll 1$ resulting in a large circle of diameter $(1/\rho)$

$$D(z)G(z) = K\frac{0.01405+0.01314}{\rho e^{j\omega T}(1-0.8187)}$$

For z on the rest of the upper half of the unit circle the Nyquist diagram can be obtained as

$$K\frac{0.01405e^{j\omega T}+0.01314}{(e^{j\omega T}-1)(e^{j\omega T}-0.8187)}.$$

With $K = 1$, the Nyquist diagram can be obtained using Matlab command nyquist () as shown in Fig. 7.6. The Nyquist diagram crosses the real axis at -0.00268 and -0.072 and does not encircle the point $-1 + j0$, i.e., $N = 0$. There is no open-loop system pole outside the unit circle, i.e., $P = 0$. It means the number of zeros of the characteristic equation, $Z = N + P = 0 + 0 = 0$. In other words, there is no closed-loop system pole outside the unit circle in the z-plane meaning the closed-loop system is stable for the controller gain $K = 1$.

The minimum controller gain to make the closed-loop discrete-time system unstable can be obtained as $K = \frac{1}{0.072} = 13.89$. It means the closed-loop discrete-time system will remain stable for the controller gain

$$K < 13.89.$$

This confirms the range of gain for stability obtained using Jury's stability test in Example 7.1.

Matlab code and results are presented next.

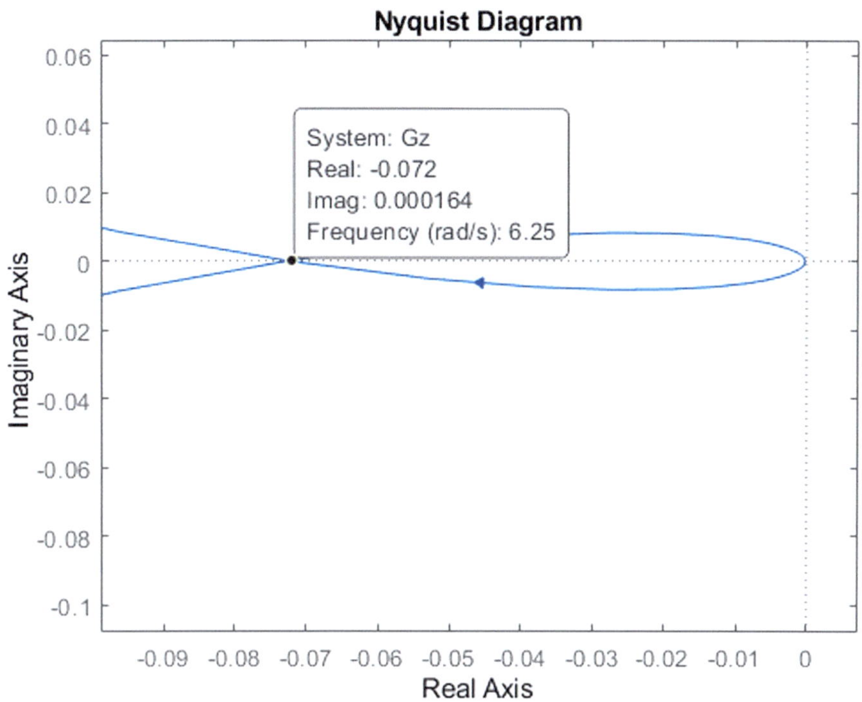

Fig. 7.6 Nyquist diagram of the open-loop transfer function of Example 7.3 ($K = 1$)

```
% Ch7 Ex 7_3 Stability Analysis of a 2nd order closed-loop discrete-time system
% Gp(s)=3/s(s+2), H(s)=1, D(z)=K
clear all; close all; clc
T=0.1 ; % Sampling time
Gp = tf(3, [1 2 0]) %Gp(s)

% G(z) = Z(Gho(s)Gp(s))
Gz =c2d(Gp,T) % G(z)

% Nyquist plot
nyquist(Gz)
axis([-1.2 0.2 -0.5 0.5])


% K=13.798
K=13.798; % critical gain
% Closed loop Transfer function
Tz = feedback(K*Gz, 1) % Closed loop TF with unity feedback
figure(2)
dstep(Tz,1)
title('Closed-loop system step response K=Kcritical')
% closed loop poles at Kcritical
[ntz,dtz] = tfdata(Tz, 'v')
[z,p,k]=tf2zp(ntz,dtz)
r=abs(p(1))
th=angle(p(1))
wosc = th/T  % frequency of oscillation at Kcritical
```

Example 7.4 A closed-loop discrete-time system, similar to Fig. 7.1, is with the plant transfer function $G_{\rm P}(s) = \frac{3}{s(s+2)}$, a zero-order hold (ZOH), a PI controller $D(z) = K_{\rm P}\left(1 + \frac{K_{\rm I}}{K_{\rm P}}\frac{T}{2}\left(\frac{z+1}{z-1}\right)\right)$, a sensor with a unity gain, $H(s) = 1$. Use a sampling time $T = 0.1$ s, $\frac{K_{\rm I}}{K_{\rm P}} = 0.2$. Use Nyquist criterion to confirm the PI controller proportional gain $K_{\rm P}$ for the stability of the closed-loop discrete-time system.

Solution The open-loop transfer function of the discrete-time system with ZOH is obtained (in Example 6.2) as

$$\begin{aligned} G(z) &= \overline{G_{\rm h0}G_{\rm p}}(z) = Z\left[\frac{1-e^{-sT}}{s}G_{\rm p}(s)\right]_{e^{sT}=z} \\ &= \frac{0.01405z + 0.01314}{(z-1)(z-0.8187)} \end{aligned}$$

The closed-loop discrete-time system characteristic equation can be expressed as

$$1 + D(z)G(z) = 1 + K_{\rm P}\left(1 + 0.01\left(\frac{z+1}{z-1}\right)\right)\frac{0.01405z + 0.01314}{z^2 - 1.8187z + 0.8187} = 0$$

There are two open-loop poles at $z = 1$, so the Nyquist path will be a small detour around $z = 1$ using $z = 1 + \rho e^{j\omega T}$, $\rho \ll 1$ resulting in a large circle of diameter $(1/\rho)$

$$D(z)G(z) = K_{\rm P}\left(\frac{\rho e^{j\omega T} + 0.02}{\rho e^{j\omega T}}\right)\left(\frac{0.01405 + 0.01314}{\rho e^{j\omega T}(1 - 0.8187)}\right)$$

For z on the rest of the upper half of the unit circle the Nyquist diagram can be obtained as

$$K_{\rm P}\left(\frac{e^{j\omega T} - 1 + 0.01(1 + e^{j\omega T})}{e^{j\omega T} - 1}\right)\frac{0.01405e^{j\omega T} + 0.01314}{(e^{j\omega T} - 1)(e^{j\omega T} - 0.8187)}.$$

With $K_{\rm P} = 1$, the Nyquist diagram can be obtained using Matlab command nyquist() as shown in Fig. 7.7. The Nyquist diagram crosses the real axis at -0.0803. It means the Nyquist diagram does not encircle the point $-1 + j0$, i.e., $N = 0$ for $K_{\rm P} = 1$. There is no open-loop system pole outside the unit circle, i.e., $P = 0$. It means the number of zeros of the characteristic equation, $Z = N + P = 0$. In other words, there is no closed-loop system pole outside the unit circle in the z-plane meaning the closed-loop system is stable for the controller gain $K_{\rm P} = 1$.

The PI controller proportional gain to make the closed-loop discrete-time system unstable can be obtained as $K_{\rm P} = \frac{1}{0.0803} = 12.45$. It means the closed-loop discrete-time system will remain stable for the PI controller proportional gain

$$K_{\rm P} < 12.45.$$

This confirms the range of gain for stability obtained using Jury's stability test in Example 7.2.

The closed-loop system becomes marginally stable at $K_{\rm P} = 12.4$. This is confirmed by the closed-loop step response of the system as shown in Fig. 7.8.

Matlab code and results are presented next.

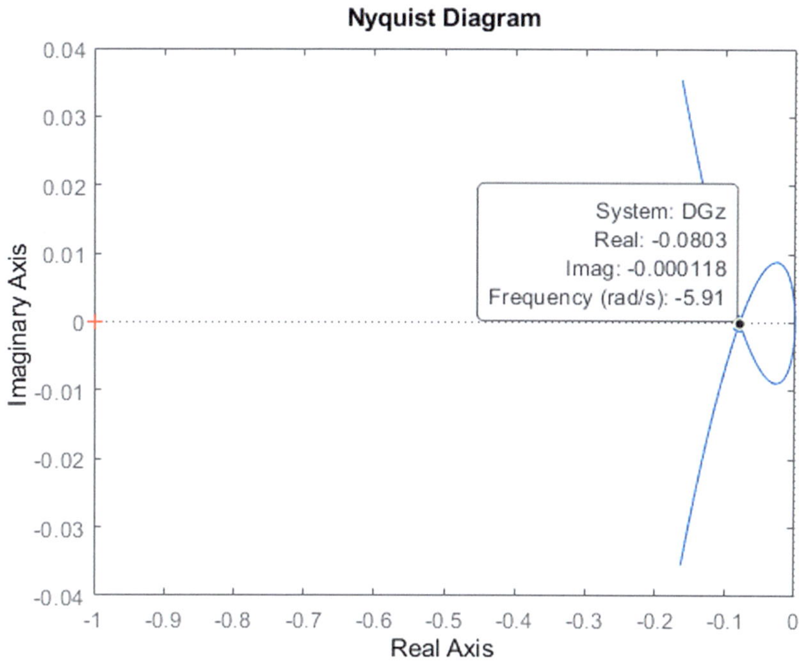

Fig. 7.7 Nyquist diagram of the open-loop transfer function of Example 7.4

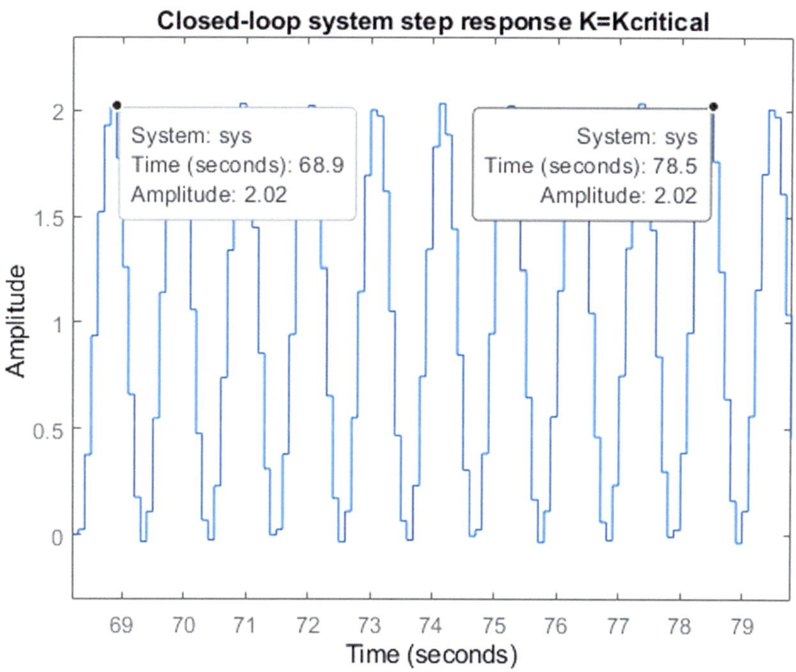

Fig. 7.8 Closed-loop system step response at gain $K_P = 12.4$ ($=K_{crtical}$) (zoomed in)

```
% Ch7 Ex 7_4 Stability Analysis of a 2nd order closed-loop discrete-time system
% Gp(s)=3/s(s+2), H(s)=1, D(z)=K
clear all; close all; clc
T=0.1 ; % Sampling time
Gp = tf(3, [1 2 0]) %Gp(s)

% G(z) = Z(Gho(s)Gp(s))
Gz =c2d(Gp,T) % G(z)
% D(z)=Kp+0.1z/(z-1)
Dz = tf([1.01 -0.99], [1 -1], T)
DGz =Dz*Gz
% Nyquist plot
figure
nyquist(DGz, {4,40})
Km=12.4
% Closed-loop step response
Tz= feedback(Km*DGz, 1)
[ntz, dtz]=tfdata(Tz, 'v')
[z,p,k]=tf2zp(ntz, dtz)
r=abs(p(2))
th=angle(p(2))
wosc=th/T

figure
dstep(Tz,1)
title('Closed-loop system step response K=Kcritical')

ω = 2*pi*9/(78.5-68.9)=5.8905 rad/s
```

7.5 Bilinear Transformation

The techniques for analyzing continuous-time linear time-invariant (LTI) systems in the s-domain cannot be directly applied for discrete-time LTI systems in the z-domain as the stability boundary of the imaginary axis in the s-plane is changed to the unit circle in the z-plane. Instead, if the discrete-time systems can be converted to an equivalent continuous-time w-domain using the bilinear transformation such that the stability boundary of the unit circle in the z-plane is changed to the imaginary axis in the w-plane, the techniques like Routh-Hurwitz criterion and Bode diagrams can be used for discrete-time systems in the w-domain. The bilinear transformation is defined as

$$z = \frac{1 + \left(\frac{T}{2}\right)w}{1 - \left(\frac{T}{2}\right)w} \tag{7.12}$$

Alternatively, the relationship between z and w can be obtained as

$$w = \frac{2}{T}\frac{z - 1}{z + 1} \tag{7.13}$$

The unit circle in the z-plane can be transformed to the imaginary axis in the w-plane as

$$w=\frac{2}{T}\frac{e^{j\omega T}-1}{e^{j\omega T}+1}=\frac{2}{T}\frac{e^{\frac{j\omega T}{2}}-e^{\frac{-j\omega T}{2}}}{e^{\frac{j\omega T}{2}}+e^{\frac{-j\omega T}{2}}}=j\frac{2}{T}\tan\left(\frac{\omega T}{2}\right) \tag{7.14}$$

The w-plane frequency (ω_w) can be defined in terms of the s-plane frequency ω as

$$\omega_w=\frac{2}{T}\tan\left(\frac{\omega T}{2}\right) \tag{7.15}$$

Equation (7.15) can also be written to obtain the s-plane frequency ω in terms of the w-plane frequency ω_w as

$$\omega=\frac{2}{T}\tan^{-1}\left(\frac{\omega_w T}{2}\right) \tag{7.16}$$

For small $\frac{\omega T}{2}$, such that $\tan\left(\frac{\omega T}{2}\right)=\frac{\omega T}{2}$, Eq. (7.15) can be written as

$$\omega_w=\frac{2}{T}\tan\left(\frac{\omega T}{2}\right)=\frac{2}{T}\left(\frac{\omega T}{2}\right)=\omega \tag{7.17}$$

It means that for small $\frac{\omega T}{2}$, such that $\tan\left(\frac{\omega T}{2}\right)=\frac{\omega T}{2}$, the w-plane frequency is very close to the s-plane frequency.

The transfer function of a discrete-time system, similar to Fig. 7.1, in the z-domain can be transformed to the w-domain using the bilinear transformation as

$$\overline{GH}(w)=Z\left[G_{\mathrm{h}}(s)G_{\mathrm{p}}(s)H(s)\right]\Big|_{z=\frac{1+\left(\frac{T}{2}\right)w}{1-\left(\frac{T}{2}\right)w}} \tag{7.18}$$

The closed-loop discrete-time system transfer function in the w-domain can be expressed as

$$T(w)=\frac{C(w)}{R(w)}=\frac{G(w)D(w)}{1+\overline{GH}(w)D(w)} \tag{7.19}$$

The characteristic equation for the closed-loop discrete-time system in the w-domain can be written as

$$1+\overline{GH}(w)D(w)=0 \tag{7.20}$$

Example 7.5 A closed-loop discrete-time system, similar to Fig. 7.1, is with the plant transfer function $G_p(s)=\frac{3}{s(s+2)}$, a zero-order hold (ZOH), a digital controller with a simple gain, $D(z)=K$, a sensor with a unity gain, $H(s)=1$. Use a sampling time $T=0.1$ s.Obtain the transfer function of the open-loop system in the z-domain, $G(z)$ and in the w-domain, $G(w)$. Obtain the zeros and poles of open-loop system in the w-plane and compare these with the ones for the original open-loop continuous-time system in the s-plane.

Solution The open-loop transfer function of the discrete-time system with ZOH is obtained (in Example 6.2) as

$$\begin{aligned} G(z) &= \overline{G_{h0}G_p}(z)=Z\left[\frac{1-e^{-sT}}{s}G_p(s)\right]_{e^{sT}=z} \\ &= \frac{0.01405z+0.01314}{(z-1)(z-0.8187)} \end{aligned}$$

The open-loop transfer function of the discrete-time system in the w-domain using bilinear transformation can be obtained as

$$G(w)=\left.\frac{0.01405z+0.01314}{(z-1)(z-0.8187)}\right|_{z=\frac{1+\frac{Tw}{2}}{1-\frac{Tw}{2}}}$$

$$\begin{aligned} G(w) &= \frac{0.01405\left(\frac{1+0.05w}{1-0.05w}\right)+0.01314}{\left(\frac{1+0.05w}{1-0.05w}-1\right)\left(\frac{1+0.05w}{1-0.05w}-0.8187\right)} \\ &= 2.5014\times10^{-4}\frac{(w-20)(w+597.5824)}{w(w+1.9934)} \end{aligned}$$

The effect of sampling can be observed in the introduction of two zeros (at $w=-597.58, 20$) and the shift of the nonzero pole in the w-plane to $w=-1.9934$ from its original location in the s-plane at $s=-2$. These values are close to that obtained using Matlab command *d2c(Gz, 'tustin')* as shown in Matlab results.

Matlab code and results are presented next.

```
% Ch7 Ex 7_5 w-plane transfer function of a 2nd order discrete-time system

% Gp(s)=3/s(s+2), H(s)=1, D(z)=K
clear all; close all; clc
T=0.1 ; % Sampling time
Gp = tf(3, [1 2 0]) %Gp(s)

% G(z) = Z(Gho(s)Gp(s))
Gz =c2d(Gp,T) % G(z)

% G(w) = W(G(z))
Gw=d2c(Gz,'tustin'); % G(w) using Tustin bilinear transformation
[ngw, dgw]=tfdata(Gw, 'v') % num/den form of G(w)
printsys(ngw, dgw, 'w')
[ngg,dgg,k]=tf2zp(ngw,dgw) % G(w) in zero-pole form

Results:
Gp(s) =
      3
  ---------
  s^2 + 2 s

poles: s=0, -2, no finite zero.

G(z) =

   0.01405 z + 0.01314
  ---------------------
  z^2 - 1.819 z + 0.8187
zero  in the z-plane zo =   -0.9355
poles in the z-plane po =    1.0000,    0.8187
k =    0.0140

The effect of sampling can be observed in the introduction of a zero in the
z-plane at z= -0.9355.

G(w) =

   -0.000249 w^2 - 0.14452 w + 2.99
   --------------------------------
      w^2 + 1.9934 w + 1.1065e-14

zeros in the w-plane zw = -600.3999, 20.0000
poles in the w-plane  pw = -1.9934, -0.0000
k = -2.4900e-04
```

The effect of sampling can be observed in the introduction of two zeros (at $w = -600.399, 20$) and the shift of the non-zero pole in the w-plane to $w = -1.9931$ from its original location in the s-plane at $s = -2$.

The effect of sampling can be observed in the introduction of two zeros (at $w = -600.399$, 20) and the shift of the nonzero pole in the w-plane to $w = -1.9934$ from its original location in the s-plane at $s = -2$.

Example 7.6 A closed-loop discrete-time system, similar to Fig. 7.1, is with the plant transfer function $G_\mathrm{p}(s) = \frac{3}{s(s+2)}$, a zero-order hold (ZOH), a PI controller $D(z) = K_\mathrm{P}\left(1 + \frac{K_\mathrm{I}}{K_\mathrm{P}} \frac{T}{2}\left(\frac{z+1}{z-1}\right)\right)$, a sensor with a unity gain, $H(s) = 1$. Use a sampling time $T = 0.1$ s, $\frac{K_\mathrm{I}}{K_\mathrm{P}} = 0.2$. Obtain the zeros and poles of open-loop system in the w-

plane and compare these with the ones for the original open-loop continuous-time system in the s-plane.

Solution The open-loop transfer function of the discrete-time system with ZOH is obtained (in Example 6.2) as

$$
\begin{aligned}
G(z) &= \overline{G_{h0}G_p}(z) = Z\left[\frac{1-e^{-sT}}{s}G_p(s)\right]_{e^{sT}=z} \\
&= \frac{0.01405z+0.01314}{(z-1)(z-0.8187)}
\end{aligned}
$$

The open-loop discrete-time system with the controller transfer function can be expressed as

$$
D(z)G(z) = K_P\left(1+0.01\left(\frac{z+1}{z-1}\right)\right)\frac{0.01405z+0.01314}{(z-1)(z-0.8187)}
$$

The open-loop discrete-time system with the controller transfer function in the w-domain can be expressed using bilinear transformation as

$$
DG(w) = K_P\left(\frac{1.01z-0.99}{z-1}\right)\frac{0.01405z+0.01314}{(z-1)(z-0.8187)}\Bigg|_{z=\frac{1+\frac{Tw}{2}}{1-\frac{Tw}{2}}}
$$

$$
DG(w) = K_P\left(\frac{1.01\left(\frac{1+0.05w}{1-0.05w}\right)-0.99}{\left(\frac{1+0.05w}{1-0.05w}\right)-1}\right)\left(\frac{0.01405\left(\frac{1+0.05w}{1-0.05w}\right)+0.01314}{\left(\frac{1+0.05w}{1-0.05w}-1\right)\left(\frac{1+0.05w}{1-0.05w}-0.8187\right)}\right)
$$

$$
DG(w) = K_P\left(\frac{0.02+0.05(2.0)w}{0.10w}\right)\left(2.49\times10^{-4}\frac{(w-20)(w+600.4)}{w(w+1.9934)}\right)
$$

$$
DG(w) = K_P\left(2.49\times10^{-4}\right)\left(\frac{w+0.2}{w}\right)\left(\frac{(w-20)(w+600.4)}{w(w+1.9934)}\right)
$$

In the w-plane, open-loop zeros: $w = -600.4, -0.2, 20$, open-loop poles: $w = -1.9934, 0, 0$. These are close to the ones obtained using Matlab as shown in Matlab results.

Matlab code and results are presented next.

```
% Ch7 Ex 7_6 w-plane transfer function of a 2nd order discrete-time system
% with a digital controller
% Gp(s)=3/s(s+2), H(s)=1, D(z)=Kp(1+0.1z/(z-1))
clear all; close all; clc
T=0.1 ; % Sampling time
Gp = tf(3, [1 2 0]) %Gp(s)

% G(z) = Z(Gho(s)Gp(s))
Gz =c2d(Gp,T) % G(z)
Kp = 1
% D(z)=Kp((1.01z-0.99)/(z-1))
Dz = tf(Kp*[1.01 -0.99], [1 -1], T)
DGz =Dz*Gz
[ngz, dgz]=tfdata(DGz, 'v') % num/den form of D(z)G(z)
[zo,po,k]=tf2zp(ngz,dgz) % zero-pole-gain of D(z)G(z)

% DG(w) W(D(z)G(z))
DGw = d2c(DGz,'tustin');  % DG(w)
[ngw, dgw]=tfdata(DGw, 'v') % num/den form of G(w)
printsys(ngw, dgw, 'w')
[zw,pw,k]=tf2zp(ngw,dgw) % zero-pole-gain of DG(w)

Results:
Gp(s) =
       3
  ---------
  s^2 + 2 s

G(z) =

   0.01405 z + 0.01314
  ----------------------
  z^2 - 1.819 z + 0.8187

D(z) =

  1.01 z - 0.99
  --------------
    z - 1




D(z)G(z) =

  0.01419 z^2 - 0.0006338 z - 0.01301
  -----------------------------------
  z^3 - 2.819 z^2 + 2.637 z - 0.8187

zeros in the z-plane zo =   -0.9355, 0.9802
poles in the z-plane po =   1.0000 + 0.0000i, 1.0000 - 0.0000i, 0.8187 + 0.0000i
k =    0.0142

DG(w) =

   -0.000249 w^3 - 0.14457 w^2 + 2.9611 w + 0.59801
   -----------------------------------------------
      w^3 + 1.9934 w^2 + 6.616e-14 w + 2.7491e-13

zeros in the w-plane zw = -600.3999, 20.0000, -0.2
poles in the w-plane pw =-1.9934 + 0.0000i, 0.0000 + 0.0000i, 0.0000 -
0.0000i
k = 2.49e-04
```

7.6 Routh-Hurwitz Criterion for Discrete-Time Systems

Procedure for applying Routh-Hurwitz criterion for the stability analysis of discrete-time systems in the w-plane is discussed in this section.

Let the characteristic equation, Eq. (7.20), of a closed-loop discrete-time system in the w-domain be expressed as

$$P(w) = a_n w^n + a_{n-1} w^{n-1} + a_{n-2} w^{n-2} + \ldots + a_1 w + a_0 = 0, \quad a_n > 0 \qquad (7.21)$$

First two rows (rows for w^n and w^{n-1}) of Routh array can be constructed by putting alternately the coefficients of the characteristic polynomial $P(w)$ as shown in Table 7.2. The remaining rows of the Routh array, up to the last row (for w^0) of Table 7.2, are calculated as shown in Eq. (7.22).

$$b_1 = \frac{a_{n-1}a_{n-2} - a_n a_{n-3}}{a_{n-1}} \quad b_2 = \frac{a_{n-1}a_{n-4} - a_n a_{n-5}}{a_{n-1}} \quad b_3 = \frac{a_{n-1}a_{n-6} - a_n a_{n-7}}{a_{n-1}}$$
$$c_1 = \frac{b_1 a_{n-3} - a_{n-1} b_2}{b_1} \quad c_2 = \frac{b_1 a_{n-5} - a_{n-1} b_3}{b_1} \quad c_3 = \frac{b_1 a_{n-7} - a_{n-1} b_4}{b_1} \qquad (7.22)$$

Table 7.2 Routh array for stability analysis

w^n	a_n	a_{n-2}	a_{n-4}	...
w^{n-1}	a_{n-1}	a_{n-3}	a_{n-5}	...
w^{n-2}	b_1	b_2	b_3	...
w^{n-3}	c_1	c_2	c_3	...
w^{n-4}	d_1	d_2	d_3	...
w^2	k_1	k_2		
w^1	l_1			
w^0	m_1			

It should be noted that Routh array can be constructed with the assumption that the first element of each row is not zero. Once the Routh array is completed to the row of w^0, the number of sign change in the first column of Routh array is equal to the number of closed loop poles in the right half of the w-plane.

Special Case I: The First Element of a Row Is Zero

In case the first element of a row is zero, the element is replaced with a small number $\in$, $|\in| \rightarrow 0$. Routh array can be constructed with $\in$ and the number of sign change in the first element of the rows is equal to the number of closed-loop poles in the right half of the w-plane.

Special Case II: All Elements of a Row Are Zero

In case all elements of a row (say, w^r) are zero, an auxiliary polynomial $A(w)$ needs to be constructed from the row above the row with all zero, i.e., row w^{r+1} as

$$A(w) = g_{r+1}w^{r+1} + g_{r-1}w^{r-1} + g_{r-3}w^{r-3} + \dots \tag{7.23}$$

where $g_{r+1}, g_{r-1}, g_{r-3}, \dots$ are the corresponding coefficients of the auxiliary polynomial. Differentiating the auxiliary polynomial $A(w)$ with respect to w, the polynomial is obtained as

$$\frac{dA(w)}{dw} = (g_{r+1})(r+1)w^r + (g_{r-1})(r-1)w^{r-2} + (g_{r-3})(r-3)w^{r-4} + \dots \tag{7.24}$$

The zeros of row w^r are replaced with the corresponding coefficients of the differentiated auxiliary polynomial $A_r(w)$ of Eq. (7.25) as

$$A_r(w) = (g_{r+1})(r+1)w^r + (g_{r-1})(r-1)w^{r-2} + (g_{r-3})(r-3)w^{r-4} + \dots \tag{7.25}$$

The entries in the new row w^r are updated as shown in Table 7.3 and Routh array is completed till the last row w^0 as usual. The number of sign change in the first column of Routh array is equal to the number of closed-loop poles in the right half of the w-plane.

The existence of a row(w^r) with all elements as zero implies that the auxiliary polynomial $A(w)$ is a factor of the characteristic polynomial $P(w)$, and the imaginary roots of the auxiliary polynomial $A(w) = 0$ are indicative of oscillatory response of the closed-loop system step response. The magnitudes of these imaginary roots are the corresponding frequencies of oscillating step response.

Table 7.3 Updated row w^r in Routh array

w^r	$(g_{r+1})(r+1)$	$(g_{r-1})(r-1)$	$(g_{r-3})(r-3)$	...

Example 7.7 Consider Example 7.5 with the digital controller as a simple gain, $D(z) = K$ with everything else remaining the same. Use Routh-Hurwitz criterion to analyze stability of the closed-loop discrete-time system. Obtain the range of gain K for stability of the closed-loop system. Obtain the critical value of the gain K_{critical} at which the closed-loop step response becomes oscillatory. Obtain the corresponding frequency of oscillation. Verify these using closed-loop step response of the discrete-time system using Matlab.

Solution The open-loop transfer function in the w-domain of the discrete-time system with the digital controller can be obtained as

$$\overline{DGH}(w) = K\,\frac{-0.000249w^2 - 0.14452w + 2.99}{w^2 + 1.9934w}$$

The characteristic equation for the closed-loop discrete-time system in the w-domain can be expressed as

$$1 + \overline{DGH}(w) = 1 + K\,\frac{-0.000249w^2 - 0.14452w + 2.99}{w^2 + 1.9934w} = 0$$

The characteristic polynomial $P(w)$ can be written as

$$(1 - 0.000249K)w^2 + (1.9934 - 0.14452K)w + 2.99K = 0$$

Routh array can be constructed as shown in Table 7.4

Examining the first column of Routh array for stability

w^2 row: $1 - 0.000249\text{K} > 0, \rightarrow K < \frac{1}{0.0000249} = 4016.1$
w^1 row: $1.9934 - 0.14452K > 0, \rightarrow K < \frac{1.9934}{0.14452} = 13.7932$
w^0 row: $2.99K > 0, \rightarrow K > 0$

For stability of the closed-loop discrete-time system, the range of controller gain can be obtained as

$$0 < K < 13.7932$$

At $K = 13.7932$, all elements of w^1 row become zero. The auxiliary polynomial can be obtained from the w^2 row as

Table 7.4 Routh array for Example 7.7

w^2	$1 - 0.000249K$	$2.99K$
w^1	$1.9934 - 0.14452K$	
w^0	$2.99K$	

$$A(w) = (1 - 0.000249K)w^2 + 2.99K$$

Substituting $K = 13.7932$

$$A(w) = 0.9966w^2 + 41.2417 = 0.9966\left(w^2 + 41.3824\right) = 0.9966(w + j6.4329) \times (w - j6.4329)$$

At $K = 13.7932 = K_{\text{critical}}$, the closed-loop step response is expected to be oscillatory and the frequency of oscillation in the w-domain is $\omega_w = 6.4329$. The frequency oscillation (in s-plane) is obtained, using Eq. (7.16) as

$$\omega = \frac{2}{T} \tan^{-1}\left(\frac{\omega_w T}{2}\right) = \frac{2}{0.1} \tan^{-1}\left(\frac{6.4329 \times 0.1}{2}\right) = 6.2239 \text{ rad/s}$$

Matlab code and results are presented next. The closed-loop system step response with $K = K_{\text{critical}}$ is shown in Fig. 7.9.

```
% Ch7 Ex 7_7 R-H criterion in w-plane
% Gp(s)=3/s(s+2), H(s)=1, D(z)=K
clear all; close all; clc
T=0.1 ; % Sampling time
Gp = tf(3, [1 2 0]) %Gp(s)

% G(z) = Z(Gho(s)Gp(s))
Gz =c2d(Gp,T) % G(z)
[ngz, dgz]=tfdata(Gz, 'v') % num/den form of G(z)
[zo,po,k]=tf2zp(ngz,dgz) % zero-pole-gain of G(z)

% G(w) = W(G(z))
Gw=d2c(Gz,'tustin'); % G(w) using Tustin bilinear transformation
[ngw, dgw]=tfdata(Gw, 'v') % num/den form of G(w)
printsys(ngw, dgw, 'w')
% Routh-Hurwitz Array
syms K
a2 = 1-0.000249*K
a0 = 2.99*K
a1 = 1.9934-0.14452*K
b1 = a1
c1 = a0
K1 = double(solve(a2--0, K)) % 1-0.000249K>0 K<1/0.000249=4016.1
K2 = double(solve(a1==0, K)) % 1.9934-0.14452*K>0 K<1.9934/0.14452 =13.7932
%K3 = double(a0==0, K)  % 2.99K > 0,  K > 0,    0<K<13.7932
K=1.9934/0.14452
% Closed-loop TF T(z)
Tz = feedback(K*Gz, 1)
dstep(Tz, 1)
title('Closed-loop step response at K=13.7932')
```

Frequency of oscillation $\omega = \frac{2\pi \text{ rad}}{(94.5 - 93.5) \text{ s}} = 6.2832$ rad/s. This is close to the value obtained from calculation using Routh array.

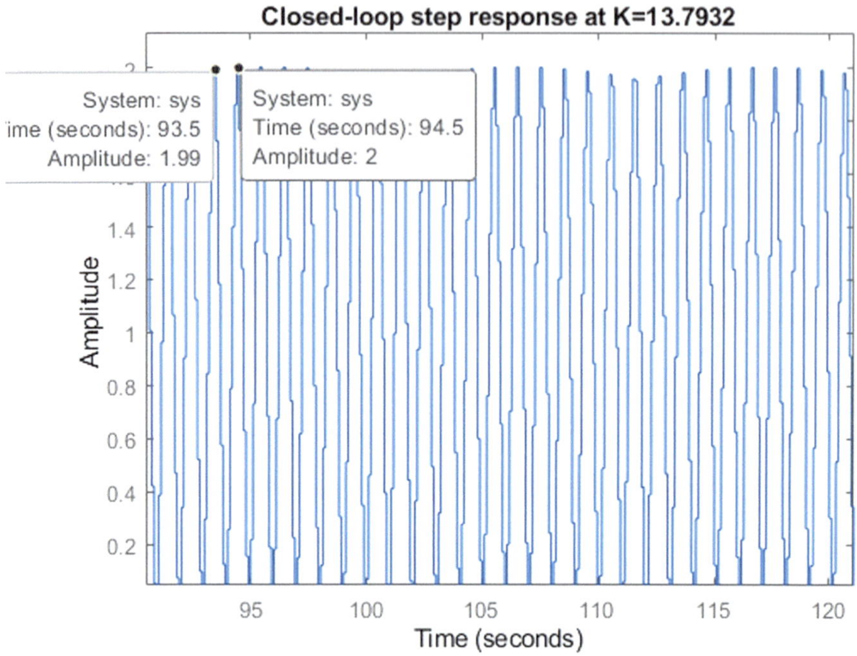

Fig. 7.9 Closed-loop system step response at gain $K_P = K_{crtical}$ (zoomed in)

Example 7.8 Consider Example 7.6 with the digital controller as a PI controller, $D(z) = K_P\left(1 + 0.01\frac{z+1}{z-1}\right)$. Use Routh-Hurwitz criterion to analyze stability of the closed-loop discrete-time system. Obtain the range of gain K for stability of the closed-loop system. Obtain the critical value of the gain $K_{critical}$ at which the closed-loop step response becomes oscillatory. Obtain the corresponding frequency of oscillation. Verify these using closed-loop step response of the discrete-time system using Matlab.

Solution The open-loop discrete-time system with the controller transfer function in the w-domain can be expressed using bilinear transformation, in Example 7.6, as

$$D\overline{GH}(w) = K_P \frac{-0.000249w^3 - 0.14457w^2 + 2.9611w + 0.59801}{w^3 + 1.9934w^2}$$

The characteristic equation for the closed-loop discrete-time system in the w-domain can be expressed as

Table 7.5 Routh array for Example 7.8

w^3	$1 - 0.000249K_P$	$2.9611K_P$
w^2	$1.9934 - 0.14457K_P$	$0.59801K_P$
w^1	b_1	
w^0	$0.59801K_P$	

$$1 + \overline{DGH}(w) = 1 + K_P \frac{-0.000249w^3 - 0.14457w^2 + 2.9611w + 0.59801}{w^3 + 1.9934w^2} = 0$$

The characteristic polynomial $P(w)$ can be written as

$$(1 - 0.000249K_P)w^3 + (1.9934 - 0.14457K_P)w^2 + 2.9611K_P w + 0.59801K_P = 0$$

Routh array can bc constructed as shown in Table 7.5

Examining the first column of Routh array for stability

w^3 row: $1 - 0.000249K_P > 0, \quad \rightarrow \quad K_P < \frac{1}{0.0000249} = 4016$

w^2 row: $1.9934 - 0.14457K_P > 0, \rightarrow K_P < \frac{1.9934}{0.14457} = 13.7880$

w^1 row: $b_1 = (1.9934 - 0.14457K_P)2.9611K_P - (1 - 0.000249K_P)0.59801K_P > 0,$

$\rightarrow (\ 5.305 - 0.4279K_P\ )K_P > 0 \rightarrow \quad K_P > 0, \ \ K_P < \frac{5.305}{0.4279} = 12.397$

w^0 row: $0.59801K_P > 0, \rightarrow K_P > 0$

For stability of the closed-loop discrete-time system, the range of controller gain can be obtained as

$$0 < K_P < 12.397$$

At $K_P = 12.397$, all elements of w^1 row become zero. The auxiliary polynomial can be obtained from the w^2 row as

$$A(w) = (\ 1.9934 - 0.14457K_P)w^2 + 0.59801K_P$$

Substituting $K_P = 12.397$

$$A(w) = 0.2012w^2 + 7.4135 = 0.2012\left(w^2 + 36.8466\right) = 0.2012(w + j6.0701) \times (w - j6.0701)$$

At $K_P = 12.397 = K_{Pcritical}$, the closed-loop step response is expected to be oscillatory and the frequency of oscillation in the w-domain is $\omega_w = 6.0701$. The frequency oscillation (in s-plane) is obtained, using Eq. (7.16) as

$$\omega = \frac{2}{T}\tan^{-1}\left(\frac{\omega_w T}{2}\right) = \frac{2}{0.1}\tan^{-1}\left(\frac{6.0701 \times 0.1}{2}\right) = 5.8923 \text{ rad/s}$$

These values are close to the ones found in Examples 7.2 and 7.4.

Matlab code and results are presented next.

```
% Ch7 Ex 7_8 w-plane transfer function of a 2nd order discrete-time system
% with a digital controller
% Gp(s)=3/s(s+2), H(s)=1, D(z)=Kp(1+0.1z/(z-1))
clear all; close all; clc
T=0.1 ; % Sampling time
Gp = tf(3, [1 2 0]) %Gp(s)

% G(z) = Z(Gho(s)Gp(s))
Gz =c2d(Gp,T) % G(z)
Kp = 1
% D(z)=Kp((1.01z-0.99)/(z-1))
Dz = tf(Kp*[1.01 -0.99], [1 -1], T)
DGz =Dz*Gz
[ngz, dgz]=tfdata(DGz, 'v') % num/den form of D(z)G(z)
[zo,po,k]=tf2zp(ngz,dgz) % zero-pole-gain of D(z)G(z)

% DG(w) W(D(z)G(z))
DGw = d2c(DGz,'tustin'); % DG(w)
[ngw, dgw]=tfdata(DGw, 'v') % num/den form of G(w)
printsys(ngw, dgw, 'w')
[zw,pw,k]=tf2zp(ngw,dgw) % zero-pole-gain of DG(w)

% RH criterion
syms Kp w
a3=vpa(dgw(1)+Kp*ngw(1),4)
a2=vpa(dgw(2)+Kp*ngw(2),4)
a1=vpa(Kp*ngw(3),4)
a0=vpa(Kp*ngw(4),4)
%RH
K1=double(solve(a3==0, Kp))
K2=double(solve(a2==0,Kp))
b1=vpa(a2*a1-a3*a0,4)
K3=double(solve(b1==0,Kp))
K4=double(solve(a0==0, Kp))
a3 =  1.0 - 0.000249*Kp >0 Kp<4016
a2 =  1.993 - 0.1446*Kp >0 Kp<13.7889
a1 =  2.961*Kp
a0 =  0.598*Kp

b1 =  - 0.4279*Kp^2 + 5.305*Kp >0   0<Kp<12.3954
        0 <K < 12.3954
```

Exercises

7.1. A closed-loop discrete-time system, Fig. P7.1, is with the plant transfer function $G_p(s) = \frac{3}{(s+1)(s+2)}$, a zero-order hold (ZOH), a digital controller with a simple gain, $D(z) = K$, a sensor with a unity gain, $H(s) = 1$. Use a sampling time $T = 0.1$ s. Use Jury's test to find the range of the controller gain K for the stability of the closed-loop discrete-time system.

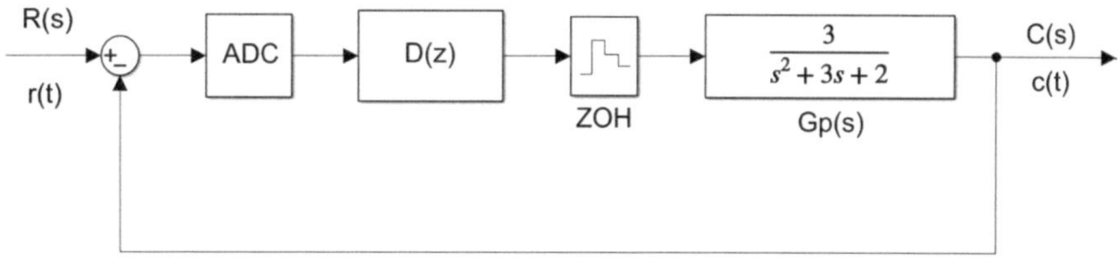

Fig. P7.1 Closed-loop discrete-time system for Problem 7.1

7.2. A closed-loop discrete-time system, Fig. P7.2, is with the plant transfer function $G_P(s) = \frac{2}{(s+1)(s+2)}$, a zero-order hold (ZOH), a PI controller $D(z) = K_P\left(1 + \frac{K_I}{K_P}\frac{T}{2}\left(\frac{z+1}{z-1}\right)\right)$, a sensor with a unity gain, $H(s) = 1$. Use a sampling time $T = 0.1$ s, $\frac{K_I}{K_P} = 0.2$. Use Jury's test to find the range of the PI controller proportional gain K_P for the stability of the closed-loop discrete-time system.

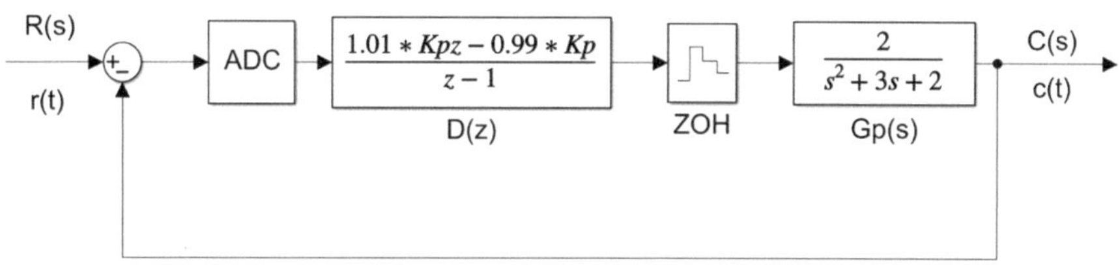

Fig. P7.2 Closed-loop discrete-time system for Problem 7.2

7.3. A closed-loop discrete-time system, Fig. P7.1, is with the plant transfer function $G_P(s) = \frac{3}{(s+1)(s+2)}$, a zero-order hold (ZOH), a digital controller with a simple gain, $D(z) = K$, a sensor with a unity gain, $H(s) = 1$. Use a sampling time $T = 0.1$ s. Use Nyquist criterion to find the maximum controller gain K for the stability of the closed-loop discrete-time system.

7.4. A closed-loop discrete-time system, Fig. P7.2, is with the plant transfer function $G_P(s) = \frac{2}{(s+1)(s+2)}$, a zero-order hold (ZOH), a PI controller $D(z) = K_P\left(1 + \frac{K_I}{K_P}\frac{T}{2}\left(\frac{z+1}{z-1}\right)\right)$, a sensor with a unity gain, $H(s) = 1$. Use a sampling time $T = 0.1$ s, $\frac{K_I}{K_P} = 0.2$. Use Nyquist criterion to confirm the PI controller proportional gain K_P for the stability of the closed-loop discrete-time system.

7.5. A closed-loop discrete-time system, Fig. P7.1, is with the plant transfer function $G_P(s) = \frac{3}{(s+1)(s+2)}$, a zero-order hold (ZOH), a digital controller with a simple gain, $D(z) = K$, a sensor with a unity gain, $H(s) = 1$. Use a sampling time $T = 0.1$ s.Obtain the transfer function of the open-loop system in the z-domain, $G(z)$ and in the w-domain, $G(w)$. Obtain the zeros and poles of open-loop system in the w-plane and compare these with the ones for the original open-loop continuous-time system in the s-plane.

7.6. A closed-loop discrete-time system, Fig. P7.2, is with the plant transfer function $G_{\mathrm{P}}(s) = \frac{2}{(s+1)(s+2)}$, a zero-order hold (ZOH), a PI controller $D(z) = K_{\mathrm{P}}\left(1 + \frac{K_{\mathrm{I}}}{K_{\mathrm{P}}}\frac{T}{2}\left(\frac{z+1}{z-1}\right)\right)$, a sensor with a unity gain, $H(s) = 1$. Use a sampling time $T = 0.1$ s, $\frac{K_{\mathrm{I}}}{K_{\mathrm{P}}} = 0.2$. Obtain the zeros and poles of open-loop system in the w-plane and compare these with the ones for the original open-loop continuous-time system in the s-plane.

7.7. A closed-loop discrete-time system, Fig. P7.1, is with the plant transfer function $G_{\mathrm{P}}(s) = \frac{3}{(s+1)(s+2)}$, a zero-order hold (ZOH), a digital controller with a simple gain, $D(z) = K$, a sensor with a unity gain, $H(s) = 1$. Use a sampling time $T = 0.1$ s. Use Routh-Hurwitz criterion to analyze stability of the closed-loop discrete-time system. Obtain the range of gain K for stability of the closed-loop system. Obtain the critical value of the gain K_{critical} at which the closed-loop step response becomes oscillatory. Obtain the corresponding frequency of oscillation. Verify these using closed-loop step response of the discrete-time system using Matlab.

7.8. A closed-loop discrete-time system, Fig. P7.2, is with the plant transfer function $G_{\mathrm{P}}(s) = \frac{2}{(s+1)(s+2)}$, a zero-order hold (ZOH), a PI controller $D(z) = K_{\mathrm{P}}\left(1 + \frac{K_I}{K_P}\frac{T}{2}\left(\frac{z+1}{z-1}\right)\right)$, a sensor with a unity gain, $H(s) = 1$. Use a sampling time $T = 0.1$ s, $\frac{K_{\mathrm{I}}}{K_{\mathrm{P}}} = 0.2$. Use Routh-Hurwitz criterion to analyze stability of the closed-loop discrete-time system. Obtain the range of gain K for stability of the closed-loop system. Obtain the critical value of the gain K_{critical} at which the closed-loop step response becomes oscillatory. Obtain the corresponding frequency of oscillation. Verify these using closed-loop step response of the discrete-time system using Matlab.

7.9. The closed-loop vehicle distance control system for a traffic control system can be represented as shown in Fig. P7.3. The plant transfer function of the distance control system can be represented in the s-domain as $G_{\mathrm{P}}(s) = \frac{1.5}{s^2+12s+18}$. The digital controller is a simple gain, $D(z) = K$. The sampling time is given as $T = 0.10$ s. Use Jury's test to find the range of the controller gain K for the stability of the closed-loop discrete-time system.

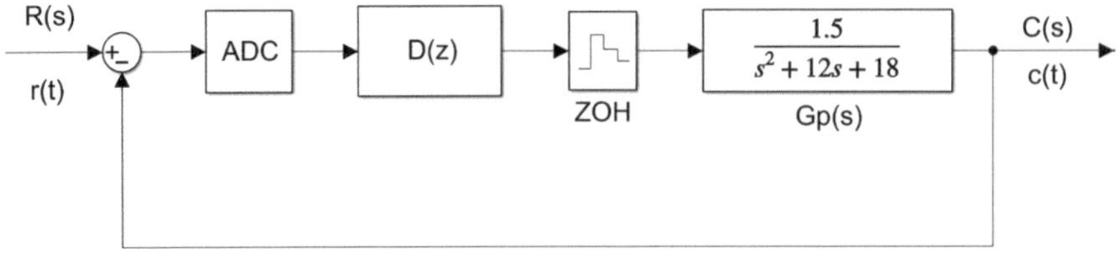

Fig. P7.3 Closed-loop discrete-time system for Problem 7.9

7.10. The closed-loop vehicle distance control system for a traffic control system can be represented as shown in Fig. P7.3. The plant transfer function of the distance control system can be represented in the s-domain as $G_p(s) = \frac{1.5}{s^2+12s+18}$. The digital controller is a simple gain, $D(z) = K$. The sampling time is given as $T = 0.10$ s. Use Nyquist criterion to find the range of the controller gain K for the stability of the closed-loop discrete-time system.

7.11. The closed-loop vehicle distance control system for a traffic control system can be represented as shown in Fig. P7.3. The plant transfer function of the distance control system can be represented in the s-domain as $G_p(s) = \frac{1.5}{s^2+12s+18}$. The digital controller is a simple gain, $D(z) = K$. The sampling time is given as $T = 0.10$ s. Use Routh-Hurwitz criterion to analyze stability of the closed-loop discrete-time system. Obtain the range of gain K for stability of the closed-loop system. Obtain the critical value of the gain K_{critical} at which the closed-loop step response becomes oscillatory. Obtain the corresponding frequency of oscillation. Verify these using closed-loop step response of the discrete-time system using Matlab.

7.12. The closed-loop roll angle control system for a jet airplane can be represented as shown in Fig. P7.4. The plant transfer function of roll control system is represented in the s-domain as $G_p(s) = \frac{10}{(s+10)(s^2+2.5s+25)}$. The sampling time is given as $T = 0.05$ s. Use Jury's test to find the range of the controller gain K for the stability of the closed-loop discrete-time system.

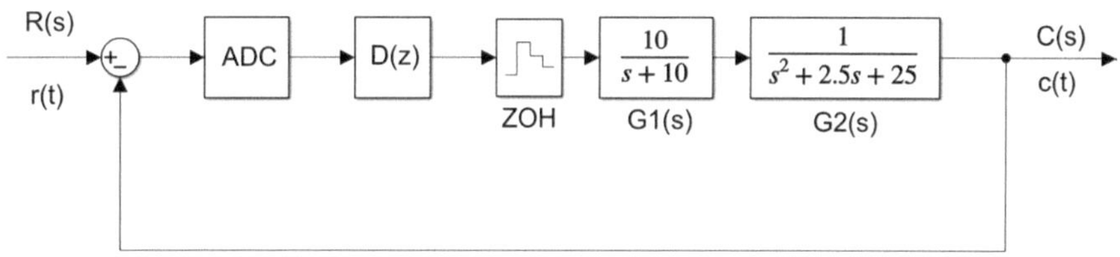

Fig. P7.4 Closed-loop discrete-time system for Problem 7.12

7.13. The closed-loop roll angle control system for a jet airplane can be represented as shown in Fig. P7.4. The plant transfer function of roll control system is represented in the s-domain as $G_p(s) = \frac{10}{(s+10)(s^2+2.5s+25)}$. The sampling time is given as $T = 0.05$ s. Use Nyquist criterion to find the range of the controller gain K for the stability of the closed-loop discrete-time system.

7.14. The closed-loop roll angle control system for a jet airplane can be represented as shown in Fig. P7.4. The plant transfer function of roll control system is represented in the s-domain as $G_p(s) = \frac{10}{(s+10)(s^2+2.5s+25)}$. The sampling time is given as $T = 0.05$ s. Use Routh-Hurwitz criterion to analyze stability of the closed-loop discrete-time system. Obtain the range of gain K for stability of the closed-loop system. Obtain the critical value of the gain K_{critical} at which the closed-loop step response becomes oscillatory. Obtain the corresponding frequency of oscillation. Verify these using closed-loop step response of the discrete-time system using Matlab.

Further Reading

1. Asadi F (2023) Signals and systems with MATLAB® and Simulink®. Springer, Cham
2. Astrom KJ, Murray RM (2021) Feedback systems: an introduction for scientists and engineers, 2nd edn. Princeton University Press, Princeton
3. Chen C-T (2006) Analog and digital control system design: transfer-function, state-space, and algebraic methods. Oxford University Press, New York
4. Dorf R, Bishop R (2016) Modern control systems, 13th edn. Pearson, Essex
5. Fadali MS, Visioli A (2020) Digital control engineering analysis and design, 3rd edn. Academic Press, London
6. Franklin GF, Powell JD, Workman M (1998) Digital control of dynamic systems, 3rd edn. Addison-Wesley, Boston
7. Franklin GF, Powell JD, Emami-Naeini A (2019) Feedback control of dynamic systems, 8th edn. Pearson, New Jersey
8. Isermann R (1981) Digital control systems. Springer-Verlag, Berlin
9. Jalili N, Candelino NW (2023) Dynamic systems and control engineering. Cambridge University Press, Cambridge
10. Keviczky L, Bars R, Hetthéssy J, Bányász C (2019) Control engineering. Springer, Singapore
11. Koeing DM (2009) Practical control engineering: a guide for engineers, managers, and practitioners. McGraw-Hill, New York
12. Kuo BC (1995) Digital control systems, 2nd edn. Oxford University Press, Oxford
13. Landau ID, Gito G (2006) Digital control systems: design, identification and implementation. Springer-Verlag, London
14. Leigh JR (2006) Applied digital control: theory, design and implementation, 2nd edn. Dover, Mineola
15. Mathworks (2023) Control System Toolbox. https://www.mathworks.com/products/control.html
16. Mathworks (2023) Matlab. https://www.mathworks.com/products/matlab.html
17. Nise NS (2019) Control systems engineering, 8th edn. Wiley, Hoboken
18. Nof SY (ed) (2009) Springer handbook of automation. Springer-Verlag, Berlin
19. Ogata K (1995) Discrete-time control systems, 2nd edn. Pearson, New Jersey
20. Phillips CL, Nagle HT, Chakrabortty A (2015) Digital control system analysis and design, 4th edn. Pearson, Essex
21. Rabbath CA, Lechevin N (2014) Discrete-time control system design with applications. Springer, New York
22. Samanta B (2023) Introduction to mechatronics: an integrated approach. Springer, Cham
23. Shertukde HM (2015) Digital control applications illustrated with MATLAB®. CRC Press, Boca Raton
24. Starr G (2020) Introduction to applied digital controls. Springer, Cham
25. Veloni A, Miridakis NI (2018) Digital control systems-theoretical problems and simulation tools. CRC Press, Boca Raton
26. Vilanova R, Visioli A (2012) PID control in the third millennium – lessons learned and new approaches. Springer-Verlag, London
27. Visioli A (2006) Practical PID control. Springer, London

Chapter 8
Root Locus for Discrete-Time Systems

8.1 Introduction

Stability analysis of discrete-time closed-loop systems using root locus techniques in the z-domain is presented in this chapter. Root locus in z-domain is introduced to analyze loci of closed-loop discrete-time system poles from the open-loop system transfer functions. Stability analysis of discrete-time closed-loop systems in standard and generalized forms is covered. The procedures are illustrated through examples along with Matlab codes. A laboratory experiment for stability analysis using z-domain techniques is presented for a DC motor module. End-of-chapter exercises are provided to enhance understanding of the concepts covered and consolidate problem-solving skills.

8.2 Root Locus in z-Plane

A closed-loop discrete-time system is shown in Fig. 8.1.

The closed-loop discrete-time system transfer function can be expressed as

$$T(z) = \frac{C(z)}{R(z)} = \frac{G(z)D(z)}{1 + \overline{GH}(z)D(z)} \tag{8.1}$$

where

$$\overline{GH}(z) = Z\left[G_h(s)G_p(s)H(s)\right] \tag{8.2}$$

The characteristic equation for the closed-loop discrete-time system can be written as

B. Samanta, *Introduction to Digital Control*,
https://doi.org/10.1007/978-3-031-66830-2_8

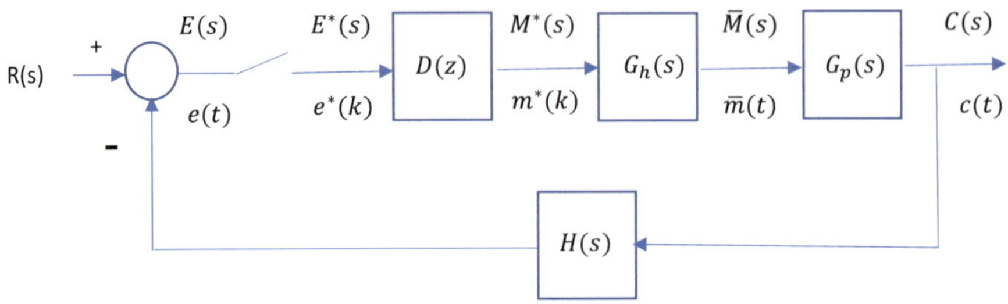

Fig. 8.1 Closed-loop discrete-time system with a digital filter and a sensor

$$1 + \overline{GH}(z)D(z) = 0 \tag{8.3}$$

If the controller is a simple gain, i.e., $D(z) = K$, the characteristic Eq. (8.3) can be expressed as

$$1 + K\,\overline{GH}(z) = 0 \tag{8.4}$$

It should be noted that each root of the characteristic equation is a pole of the closed-loop system. It is important to analyze the locus of each root of the characteristic equation or closed-loop system pole for positive gain $0 < K < \infty$ in the z-plane without explicitly solving for roots of the characteristic equation. The technique to study the locus of each root of the characteristic equation is known as the root locus technique. The root locus technique is applicable for continuous-time systems in the s-plane and for discrete-time systems in the z-plane with almost similar processes. The main difference is in the region of consideration for stability analysis: for continuous-time systems, the closed-loop poles have to be in the left half of the s-plane, whereas, for discrete-time systems, the closed-loop poles have to be within the unit circle in the z-plane for system stability. The root locus technique is applied to study the closed-loop system poles using the open-loop system transfer function for continuous-time and discrete-time systems.

The process of constructing root locus for discrete-time systems in the z-plane is considered in this section.

The characteristic equation of a closed-loop discrete-time system, Eq. (8.4), can be written in terms of the open-loop system zeros and poles as

$$1 + Kk\frac{\prod_{i=1}^{m}(z - z_i)}{\prod_{j=1}^{n}(z - p_j)} = 0 \tag{8.5}$$

where z_i, $i = 1, \ldots, m$ and p_j, $j = 1, \ldots, n$, $m \leq n$ are zeros and poles, respectively, and k is the gain of the open-loop system, $\overline{GH}(z)$. The gains K and k can be combined into a single gain $\boldsymbol{K} = K * k$. The root locus technique is used to analyze the locus of the roots Eq. (8.5) using the open-loop system zeros and poles for $0 < K < \infty$.

Equation (8.4) can also be written as

$$K\,\overline{GH}(z) = -1 \tag{8.6}$$

It implies that for a point z in the z-plane to be on the root locus, the magnitude, and the angle criteria, Eqs. (8.7) and (8.8), respectively, must be satisfied.

$$\left|\overline{GH}(z)\right| = k\frac{\left|\prod_{i=1}^{m}(z-z_i)\right|}{\left|\prod_{j=1}^{n}(z-p_j)\right|} = 1/K \tag{8.7}$$

$$\angle\overline{GH}(z) = \angle\prod_{i=1}^{m}(z-z_i) - \angle\prod_{j=1}^{n}(z-p_j) = (2q+1)\pi, q=0,1,2,\ \ldots \tag{8.8}$$

1. Root locus starting points ($K = 0$):

   Equation (8.5) can be rewritten as

$$\prod_{j=1}^{n}(z-p_j) + \boldsymbol{K}\prod_{i=1}^{m}(z-z_i) = 0 \tag{8.9}$$

   At $K = 0$, Eq. (8.6) can be written as

$$\prod_{j=1}^{n}(z-p_j) = 0 \tag{8.10}$$

   It means that the root locus originates at the open-loop system poles.

2. Root locus terminating points ($K \to \infty$):

   Equation (8.6) open-loop system transfer can be rewritten as

$$\frac{1}{\boldsymbol{K}}\prod_{j=1}^{n}(z-p_j) + \prod_{i=1}^{m}(z-z_i) = 0 \tag{8.11}$$

   As $K \to \infty$, Eq. (8.8) can be reduced to

$$\prod_{i=1}^{m}(z-z_i)=0 \tag{8.12}$$

It means that the root locus terminates at the open-loop system zeros. In general, number of open-loop zeros (m) is less than or equal to the number of open-loop poles (n). If the number of finite zeros is less than the number of poles, there are (n – m) asymptotes going toward $\pm\infty$.

3. Root locus symmetry about real axis:

Since the coefficients of the open-loop system transfer function are real numbers, the roots of the characteristics equation or the closed-loop system poles are either real or complex conjugates. It implies that the root locus is symmetric about the real axis.

4. Root locus on real axis:

From Eq. (8.5), applying the angle criterion, the root locus lies on the real axis to the left of an odd number of poles and zeros on the real axis to make the net contributions of the poles and zeros on the real axis as an odd multiple of π.

5. Asymptotes

The number (N) of asymptotes is equal to the difference in number of open-loop poles (n) and zeros (m):

$$N=n-m, \quad n>m \tag{8.13}$$

The intercept of asymptotes ($N \geq 1$) on the real axis can be expressed as

$$\sigma=\frac{\sum_{j=1}^{n}p_j-\sum_{i=1}^{m}z_i}{n-m}, \quad n>m \tag{8.14}$$

The angle of asymptotes with real axis can be expressed as

$$\theta_k=\frac{(2k+1)\pi}{n-m}, k=0, 1, \ldots, (n-m-1) \tag{8.15}$$

6. Break-in and break-away points:

The break-in and break-away points can be obtained as

$$\frac{d\left[\overline{GH}(z)\right]}{dz}=0 \tag{8.16}$$

8.3 Stability Analysis Using Root Locus in z-Plane

Root locus techniques are useful in analyzing the stability of closed-loop discrete-time systems in the z-plane. For stability of discrete-time systems, the closed-loop poles must be within the unit circle in the z-plane. It means that the root locus has to be within the unit circle in the z-plane for the stability of the discrete-time systems. The range of positive gain for which the root locus stays within the unit circle indicates the range of gain for stability of the closed-loop discrete-time systems. The gain at which the root locus intersects with the unit circle in the z-plane is the critical gain for the oscillatory response of the closed-loop systems with step inputs. In other words, at critical gains, the closed-loop systems become marginally stable. The closed-loop poles at critical gains include the imaginary poles that give the frequency of oscillation for the oscillatory response.

The procedure is illustrated through two examples.

Example 8.1 A closed-loop discrete-time system, similar to Fig. 8.1, is with the plant transfer function $G_p(s) = \frac{3}{s(s+2)}$; a zero-order hold (ZOH); a digital controller with a simple gain, $D(z) = K$; and a sensor with a unity gain, $H(s) = 1$. Use a sampling time $T = 0.1$ s. Analyze the system stability using root locus. Obtain the range of the controller gain K for the stability of the closed-loop discrete-time system. Find the critical gain at which the closed-loop step response becomes oscillatory. Verify the frequency of oscillation from the closed-loop step response of the discrete-time system.

Solution The open-loop transfer function of the discrete-time system with ZOH is obtained (in Example 6.2) as

$$\begin{aligned}\overline{GH}(z) &= \overline{G_{h0}G_pH}(z) = Z\left[\frac{1-e^{-sT}}{s}\,G_p(s)H(s)\right]_{e^{sT}=z} \\ &= \frac{0.01405z + 0.01314}{z^2 - 1.8187z + 0.8187}\end{aligned}$$

The closed-loop discrete-time system characteristic equation can be expressed as

$$1 + D(z)\overline{GH}(z) = 1 + K\frac{0.01405z + 0.01314}{z^2 - 1.8187z + 0.8187} = 0$$

The root locus of the system can be obtained as follows:

Open-loop zero of the system

$$0.01405z + 0.01314 = 0$$

$$z = -\frac{0.01314}{0.01405} = -0.9355, \quad m = 1$$

Open-loop poles

$$z^2 - 1.8187z + 0.8187 = 0$$

$$z = 0.8187, \quad 1.000, \quad n = 2$$

Root locus starts at the open-loop poles $z = 0.8187, 1.000$, one ends at the open-loop zero at $z = -0.9355$, and the other one $(n - m = 1)$ goes to $-\infty$.

The root locus on the real axis is to the left of the open-loop pole at $z = 1$ till the second pole at $z = 0.8187$, i.e., $0.8187 \leq z \leq 1$. There is no root locus on the real axis between the $-0.9355 < z < 0.8187$.

Number of asymptotes $N = n - m1 = 2 - 1 = 1$, no intercept of the real axis

The angle of asymptote $\theta_k = \frac{(2k+1)\pi}{n-m}, k = 0, 1, \ldots, (n - m - 1), \theta = \pi$

Break-away and break-in points can be obtained as

$$\frac{d\left[\overline{GH}(z)\right]}{dz} = 0$$

$$\frac{d}{dz}\left[\frac{0.01405z + 0.01314}{z^2 - 1.8187z + 0.8187}\right] = 0$$

$$0.01405\left(z^2 - 1.8187z + 0.8187\right) - (2z - 1.8187)(0.01405z + 0.01314) = 0$$

$$-0.01405z^2 - 0.02628z + 0.0354 = 0$$

$$z = -2.7776, \quad 0.9071$$

Root locus can be obtained using Matlab command rlocus() as shown in Fig. 8.2. The value of critical gain at which the root locus crosse the unit circle is found to be 13.9 which is close to the value (13.8) obtained in Chap. 7, Example 7.1. The frequency of oscillation is 6.25 rad/s which is close to the value (6.23 rad/s) obtained in Example 7.1.

The root locus starting points (open-loop poles), end point (open-loop zero), and break-away points are shown on the root locus in Fig. 8.3. The break-in/break-away points are close to the calculated (−2.7776, 0.9071).

The closed-loop step response is oscillatory at the critical gain as shown in Fig. 8.4. The frequency of oscillation is close to 6.22 rad/s which is also close to the calculated value.

Example 8.2 A closed-loop discrete-time system, similar to Fig. 8.1, is with the plant transfer function $G_p(s) = \frac{3}{s(s+2)}$, a zero-order hold (ZOH), a PI controller $D(z) = K_P\left(1 + \frac{K_I}{K_P}\frac{T}{2}\left(\frac{z+1}{z-1}\right)\right)$, and a sensor with a unity gain, $H(s) = 1$. Use a sampling time $T = 0.1$ s, $\frac{K_I}{K_P} = 0.2$. Analyze the system stability using root locus. Obtain the range of the controller gain K_P for the stability of the closed-loop discrete-time system. Find the critical gain at which the closed-loop step response becomes

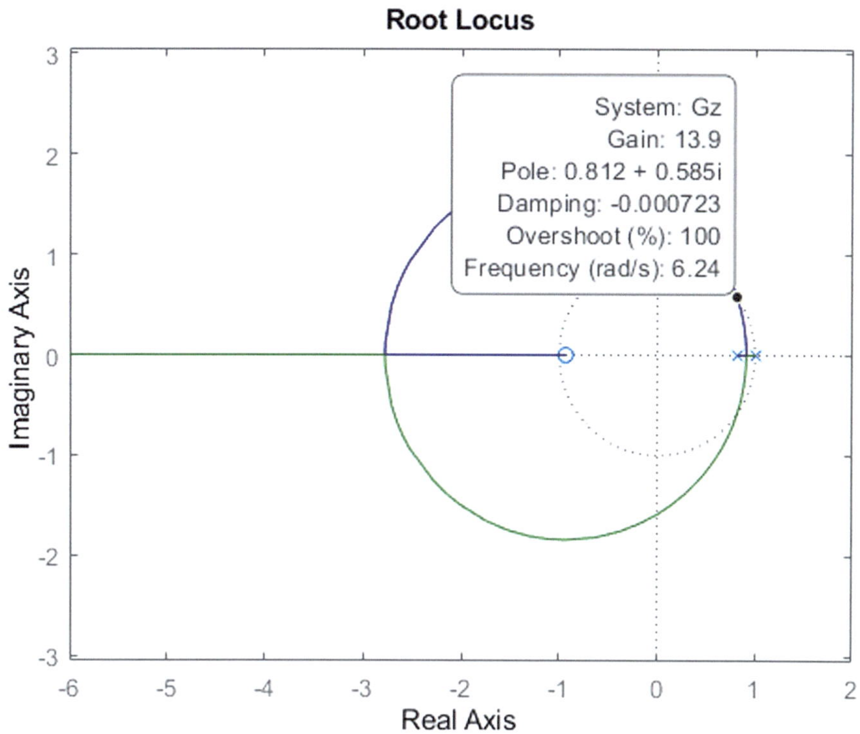

Fig. 8.2 Root locus for the discrete-time system of Example 8.1

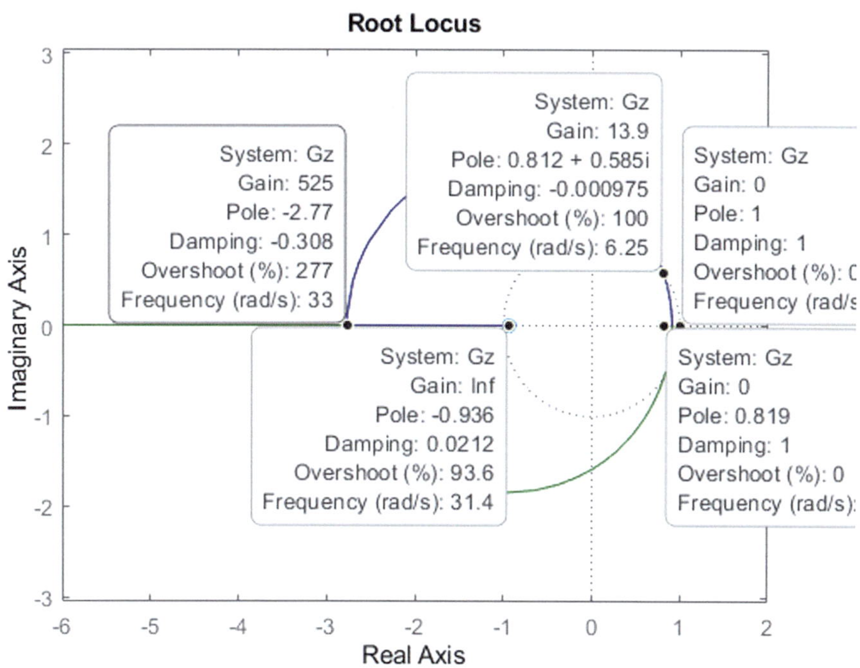

Fig. 8.3 Root locus for the discrete-time system of Example 8.1 with details

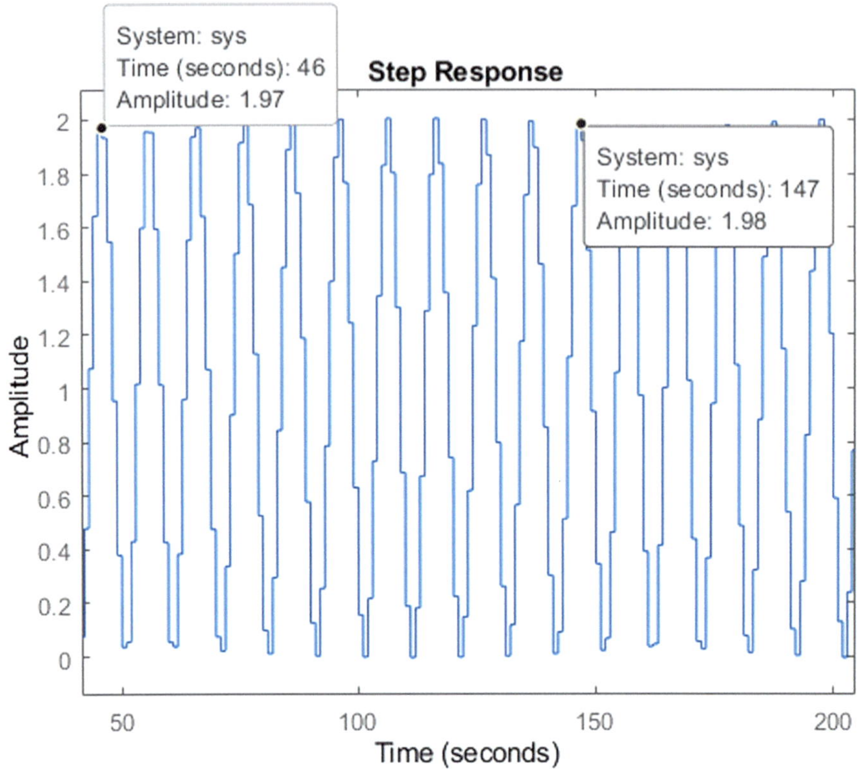

Fig. 8.4 Closed-loop system step response of Example 8.1 at K_{critical}

oscillatory. Verify the frequency of oscillation from the closed-loop step response of the discrete-time system.

Solution The open-loop transfer function of the discrete-time system with ZOH is obtained (in Example 6.2) as

$$\overline{GH}(z) = \overline{G_{h0}G_p}(z) = Z\left[\frac{1-e^{-sT}}{s}\,G_p(s)\right]_{e^{sT}=z}$$
$$= \frac{0.01405z + 0.01314}{(z-1)(z-0.8187)}$$

The open-loop discrete-time system with the controller transfer function can be expressed as

$$D(z)\overline{GH}(z) = K_P\left(1 + 0.01\left(\frac{z+1}{z-1}\right)\right)\left(\frac{0.01405z + 0.01314}{(z-1)(z-0.8187)}\right)$$
$$= K_P\left(\frac{1.01z - 0.99}{z-1}\right)\left(\frac{0.01405z + 0.01314}{(z-1)(z-0.8187)}\right)$$

Open-loop zeros of the system

$$(1.01z - 0.99)(0.01405z + 0.01314) = 0$$
$$z = \frac{0.99}{1.01} = 0.9802, \quad z = -\frac{0.01314}{0.01405} = -0.9355, \quad m = 2$$

Open-loop poles

$$(z-1)(z-1)(z-0.8187) = 0$$
$$z = 0.8187, 1.000, 1.000, \quad n = 3$$

Root locus starts at the open-loop poles $z = 0.8187, 1.000, 1.000$, two end at the open-loop zeros at $z = 0.9802, -0.9355$, and the other one ($n - m = 1$) goes to $-\infty$.

The root locus on the real axis is to the left of the open-loop zero at $z = 0.9802$, between the pole at $z = 0.8187$ and zero at $z = 0.9802$, i.e., $0.8187 \le z \le 0.9802$. There is no root locus on the real axis between $0.9802 < z < 1$ and $-0.9355 < z < 0.8187$.

Number of asymptotes $N = n - m = 3 - 2 = 1$, no intercept of the real axis

The angle of asymptote $\theta_k = \frac{(2k+1)\pi}{n-m}, k = 0, 1, \ldots, (n-m-1), \theta = \pi$

Break-away and break-in points can be obtained as

$$\frac{d\left[\overline{GH}(z)\right]}{dz} = 0$$
$$\frac{d}{dz}\left[\frac{(1.01z - 0.99)}{(z-1)}\frac{0.01405z + 0.01314}{z^2 - 1.8187z + 0.8187}\right] = 0$$

Roots are obtained as $z = -2.7871, 0.9279, 0.9491, 1.0000$. These are root locus break-away and break-in points. Root locus starting at the double pole at $z = 1.0$ break away and break in at $z = 0.9491$. One locus terminates at zero $z = 0.9802$, and the other end moves to the left and meets another locus originating from the pole at $z = 0.8187$, breaking away at $z = 0.9279$. These two loci break in at $z = -2.7871$: one of the loci moves on the real axis to the right of $z = -2.7871$ and terminates at zero at $z = -0.9355$, while the other locus moves on the real axis to the left of $z = -2.7871$ toward $-\infty$.

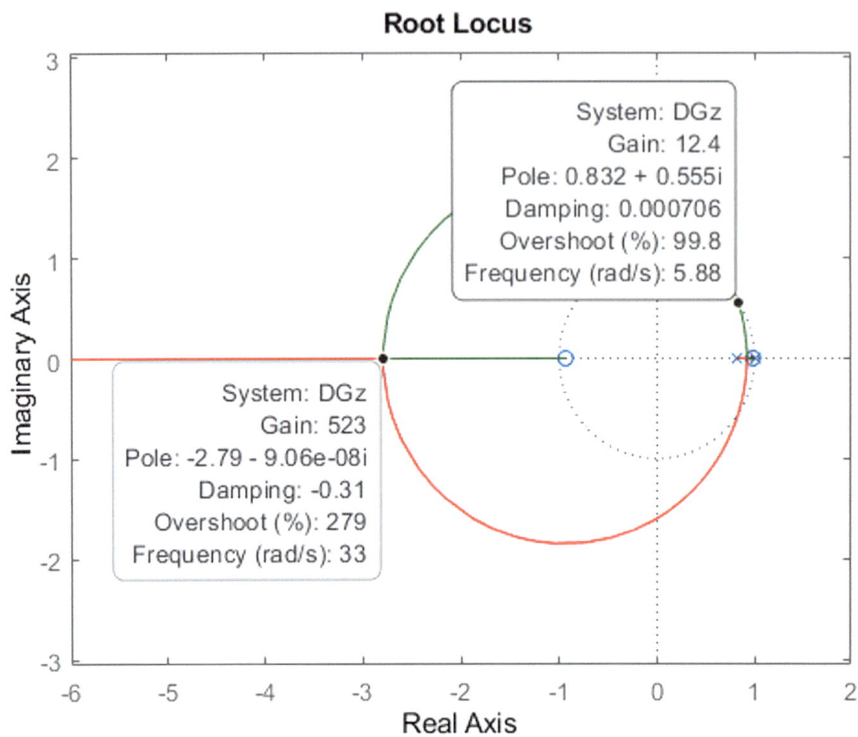

Fig. 8.5 Root locus for the discrete-time system of Example 8.2

Root locus can be obtained using Matlab command rlocus() as shown in Fig. 8.5. The value of critical gain at which the root locus crosses the unit circle is found to be 12.4 which is close to the value obtained in Chap. 7, Example 7.8. The frequency of oscillation is 5.88 rad/s which is close to the value (5.89 rad/s) obtained in Example 7.8.

The root locus starting points (open-loop poles), end point (open-loop zero), and break-away points are shown on the root locus in Fig. 8.5. The break-in/break-away points are close to the calculated values. A zoomed-in view of the root locus is presented in Fig. 8.6.

The closed-loop step response is oscillatory at the critical gain as shown in Fig. 8.7. The frequency of oscillation is $2\pi \frac{18}{116 - 96.6} = 5.8298\ \text{rad/s}$ which is close to the value obtained in Example 7.8.

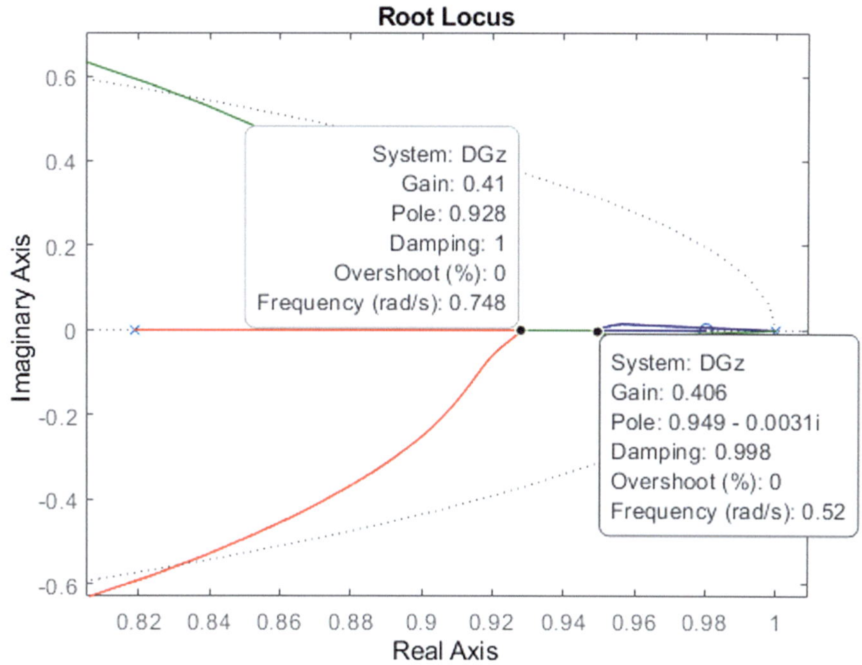

Fig. 8.6 Zoomed in view of root locus for the discrete-time system of Example 8.2

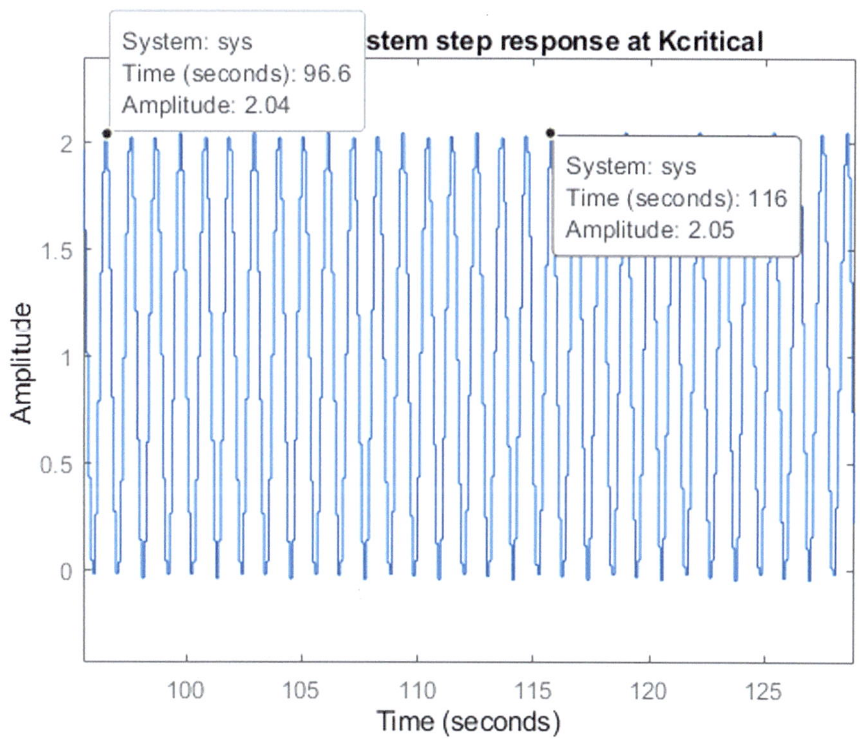

Fig. 8.7 Closed-loop system step response of Example 8.2 at K_{critical}

Matlab code and results are presented next.

```
%Ch8_ex8-2 Root Locus of a 2nd order discrete-time system
% with a digital controller
% Gp(s)=3/s(s+2), H(s)=1, D(z)=Kp(1+0.01(z+1)/(z-1))
clear all; close all; clc
T=0.1 ; % Sampling time
Gp = tf(3, [1 2 0]) %Gp(s)

% G(z) = Z(Gho(s)Gp(s))
Gz =c2d(Gp,T) % G(z)
Kp = 1
% D(z)=Kp((1.01z-0.99)/(z-1))
Dz = tf(Kp*[1.01 -0.99], [1 -1], T)
DGz =Dz*Gz
[ngz, dgz]=tfdata(DGz, 'v') % num/den form of D(z)G(z)
[zo,po,k]=tf2zp(ngz,dgz) % zero-pole-gain of D(z)G(z)

%root locus
figure(1)
rlocus(DGz)
axis('equal')

% break-away break-in points
syms z
u=vpa((1.01*z-0.99)/(z-1),4)
up = diff(u,z);
v = vpa((0.01405*z+0.01314)/(z^2-1.8187*z+0.8187),4)
vp=diff(v,z);
P=up*v+vp*u;
P=vpa(simplify(P),4)
zp=double(solve(P==0,z))

% closed-loop system step response
K=12.4
Tz=feedback(K*DGz,1)
[ntz,dtz]=tfdata(Tz,'v')
[z, p, k]=tf2zp(ntz,dtz)
r=abs(p)
th=angle(p)
w=th/T
figure(2)
dstep(Tz,1)
title('Closed-loop system step response at Kcritical')
```

```
G(z) =

   0.01405 z + 0.01314
  ----------------------
  z^2 - 1.819 z + 0.8187
D(z) =

  1.01 z - 0.99
  -------------
      z - 1

DGz =

  0.01419 z^2 - 0.0006338 z - 0.01301
  -----------------------------------
  z^3 - 2.819 z^2 + 2.637 z - 0.8187

open-loop zeros zo = 0.9802, -0.9355
open-loop poles po = 1.0000 + 0.0000i, 1.0000 - 0.0000i, 0.8187 + 0.0000i

Break-in break-away points
u = (1.01*z - 0.99)/(z - 1.0)
v = (0.01405*z + 0.01314)/(z^2 - 1.819*z + 0.8187)
P = (1.25e-34*(- 7.095e+96*z^4 + 6.381e+95*z^3 + 3.733e+97*z^2 - 4.829e+97*z
+ 1.742e+97))/(2.5e+32*z^3 - 7.047e+32*z^2 + 6.593e+32*z - 2.047e+32)^2

Break-in/break-away points zp = -2.7871, 0.9279, 0.9491, 1.0000

K = 12.4000

Tz =

   0.1759 z^2 - 0.00786 z - 0.1613
  ---------------------------------
  z^3 - 2.643 z^2 + 2.63 z - 0.9801

closed-loop system zeros z = 0.9802,  -0.9355
closed-loop system poles p = 0.8314 + 0.5557i,0.8314 -0.5557i, 0.9800 +
0.0000i

closed-loop pole magnitude r = 1.0000, 1.0000,  0.9800
closed-loop pole angles th =  0.5892,  -0.5892,  0 rad
frequencies                w =  5.8923,  -5.8923,  0 rad/s
```

8.4 Generalized Root Locus in z-Plane

The characteristic equation, Eq. (8.4), of a closed-loop discrete-time system similar to Fig. 8.1 with the digital controller $D(z) = K$ is generally used for stability analysis using root locus techniques for positive gain $0 < K' < \infty$. However, quite often, it is necessary to apply root locus techniques for the analysis of closed-loop systems in terms of parameters other than the simple gain K. In such cases, the characteristic equation needs to be rearranged in the form of the standard characteristic Eq. (8.4) with the parameter K and the open-loop system transfer function $\overline{GH}(z)$ redefined as K' and $\overline{GH}'(z)$ in Eq. (8.17):

$$1 + K'\,\overline{GH}'(z) = 0 \tag{8.17}$$

This modified characteristic equation is used for root locus analysis with the modified open-loop system transfer function $\overline{GH}'(z)$ for positive gain $0 < K' < \infty$. This is called the generalized root locus.

As an example, let us consider a system with the digital controller $D(z)$ of a simple gain which is replaced with a proportional-integral (PI) controller as

$$D(z) = \left(K_P + K_I \frac{T}{2} \frac{z+1}{z-1}\right) \tag{8.18}$$

where K_P and K_I are the proportional and the integral gains, respectively. The characteristic equation for the closed-loop discrete-time system is written, with $K_I' = K_I \frac{T}{2}$, as

$$1 + \left(K_P + K_I' \frac{z+1}{z-1}\right) \overline{GH}(z) = 0 \tag{8.19}$$

It is necessary to apply the root locus technique for positive proportional gain $0 < K_P < \infty$ for given K_I. The characteristic equation can be rearranged in the form of Eq. (8.20) as

$$1 + K_I' \frac{z+1}{z-1} \overline{GH}(z) + K_P \overline{GH}(z) = 0 \tag{8.20}$$

$$1 + K_P \frac{\overline{GH}(z)}{1 + K_I' \frac{z+1}{z-1} \overline{GH}(z)} = 0 \tag{8.21}$$

$\overline{GH}(z)$ can be expressed in terms of numerator and denominator polynomials, $N_G(z)$, $D_G(z)$, as

$$\overline{GH}(z) = \frac{N_G(z)}{D_G(z)} \tag{8.22}$$

Equation (8.21) can be rewritten, using Eq. (8.22) and after some rearrangement, as

$$1 + K_P G'(z) = 0 \tag{8.23}$$

where the modified open-loop system transfer function for root locus analysis can be expressed as

$$G'(z) = \frac{(z-1)N_G(z)}{(z-1)D_G(z) + K_I'(z+1)N_G(z)} \tag{8.24}$$

Example 8.3 A closed-loop discrete-time system, similar to Fig. 8.1, is with the plant transfer function $G_p(s) = \frac{3}{(s+1)(s+2)}$, a zero-order hold (ZOH), a PI controller $D(z) = \left(K_P + K_I \frac{T}{2} \frac{z+1}{z-1}\right)$, and a sensor with a unity gain, $H(s) = 1$. Use a sampling time $T = 0.1$, $K_I = 0.2$. Analyze the system stability using generalized root locus for the positive proportional gain $0 < K_P < \infty$. Obtain the range of the controller gain K_P for the stability of the closed-loop discrete-time system. Find the critical gain at which the closed-loop step response becomes oscillatory. Verify the frequency of oscillation from the closed-loop step response of the discrete-time system.

Solution The open-loop transfer function of the discrete-time system with ZOH is obtained (in Example 6.2) as

$$\overline{GH}(z) = \overline{G_{h0}G_p}(z) = Z\left[\frac{1-e^{-sT}}{s}G_p(s)\right]_{e^{sT}=z}$$
$$= \frac{0.01358z + 0.01229}{z^2 - 1.7236z + 0.7408}$$

The open-loop discrete-time system with the controller transfer function can be expressed as

$$D(z)\overline{GH}(z) = \left(K_P + K_I\frac{T}{2}\frac{z+1}{z-1}\right)\frac{0.01358z + 0.01229}{z^2 - 1.7236z + 0.7408}$$
$$= \left(K_P + 0.01\frac{z+1}{z-1}\right)\frac{0.01358z + 0.01229}{z^2 - 1.7236z + 0.7408}$$

Using $N_G(z) = 0.01358z + 0.01229$, $D_G(z) = z^2 - 1.7236z + 0.7408$, the modified open-loop system transfer function as

$$G'(z) = \frac{(z-1)N_G(z)}{(z-1)D_G(z) + K_I'(z+1)N_G(z)}$$

$$G'(z) = \frac{0.01358\,z^2 - 0.00129\,z - 0.01229}{z^3 - 2.723\,z^2 + 2.465\,\text{z} - 0.7407}$$

Open-loop zeros of the system

$$0.01358\,z^2 - 0.00129\,z - 0.01229 = 0$$
$$z = -0.9050, \quad z = 1, \quad m = 2$$

Open-loop poles

$$z^3 - 2.723\,z^2 + 2.465\,\text{z} - 0.7407 = 0$$
$$z = 0.9700 + j0.0714, \quad 0.9700 - j0.0714, \quad 0.7829, \quad n = 3$$

One root locus starts at the open-loop pole $z = 0.7829$, moves on the real axis to the right, and terminates at the open-loop zero at $z = 1$. Two other root loci start at the open-loop poles $z = 0.9700 + j0.0714$, $0.9700 - j0.0714$. The root locus on the real axis is to the left of the open-loop zero at $z = 1.000$, starting at the pole $z = 0.7829$ and ending zero at $z = 1$, i.e., $0.7829 \le z \le 1$. There is no root locus on the real axis between $-0.9050 < z < 0.7829$.

Number of asymptotes $N = n - m = 3 - 2 = 1$, no intercept of the real axis
The angle of asymptote $\theta_k = \frac{(2k+1)\pi}{n-m}, k = 0, 1, \ldots, (n-m-1), \theta = \pi$
Break-away and break-in points can be obtained as

$$\frac{d[G'(z)]}{dz} = 0$$
$$\frac{d}{dz}\left[\frac{0.01358\, z^2 - 0.00129\, z - 0.01229}{z^3 - 2.723\, z^2 + 2.465\, z - 0.7407}\right] = 0$$

Roots are obtained as $z = -2.6713, 0.8775, 0.9444, 1.0394$. Of these, $z = -2.6713$ is the valid root locus break-in and break-away point. Root locus starts at pole-pair at $z = 0.9700 \pm j0.0714$ break in and break away at $z = -2.6713$. One locus terminates at zero $z = -0.9050$, while the other locus moves on the real axis to the left of $z = -2.6713$ toward $-\infty$.

Root locus can be obtained using Matlab command rlocus() as shown in Fig. 8.8. A zoomed-in view of root locus is shown in Fig. 8.9. Two loci start at the open-loop pole-pair at $z = 0.9700 \pm j0.0714$. The value of critical gain at which the root locus crosses the unit circle is found to be 21.1. The frequency of oscillation is 7.7 rad/s.

The closed-loop step response is oscillatory at the critical gain as shown in Fig. 8.10. The frequency of oscillation is close to 7.85 rad/s which is also close to the value of 7.7 rad/s obtained from the root locus.

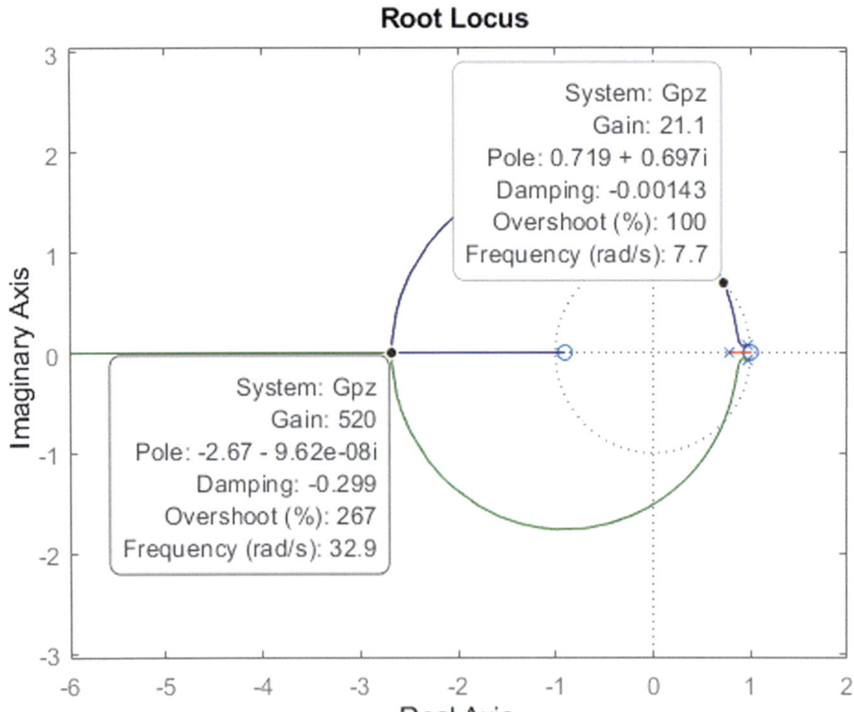

Fig. 8.8 Root locus for the discrete-time system of Example 8.3

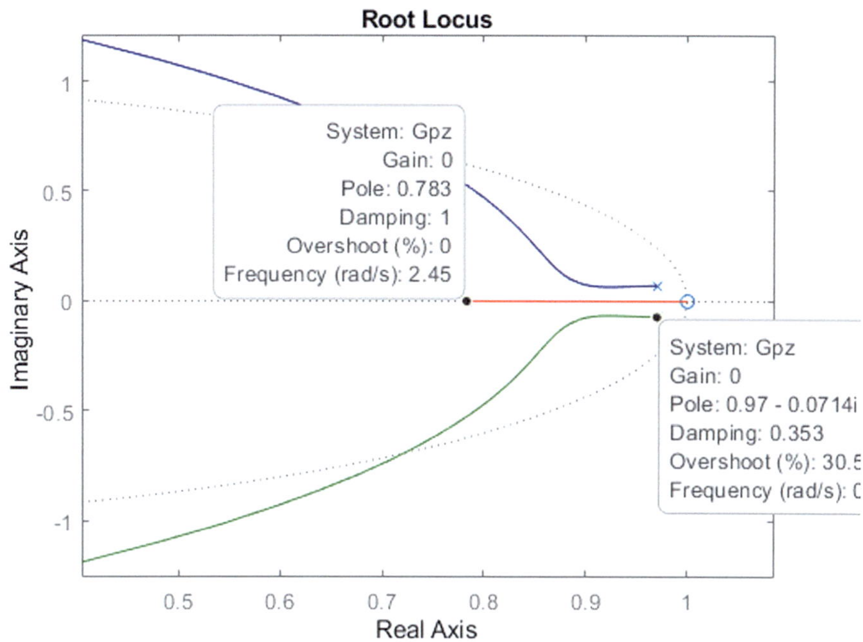

Fig. 8.9 Zoomed in view of root locus for the discrete-time system of Example 8.3

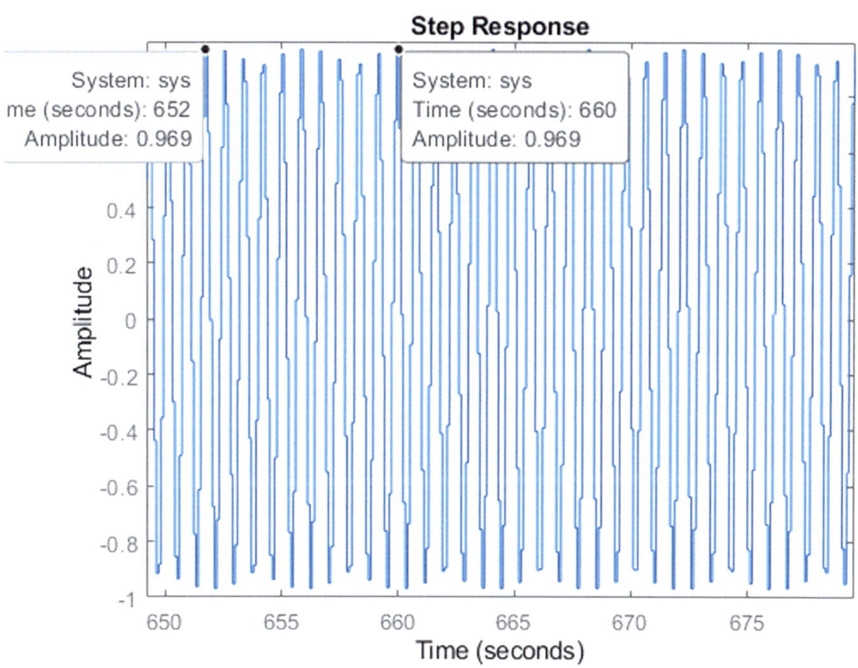

Fig. 8.10 Closed-loop system step response of Example 8.3 at $K_{critical}$

Matlab code for Jury's test for stability analysis is presented along with the code for root locus and closed-loop step response.

```
%Ch8 Ex8_3 Generalized Root Locus
$ Gp(s)3/(s+1)(s+2) H(s)=1, D(z)=Kp+KI(T/2)(z+1)/(z-1)
clear all; close all;clc
T=0.1 ; % Sampling time
Gp = tf(3, [1  3  2 ]) %Gp(s)

% G(z) = Z(Gho(s)Gp(s))
Gz =c2d(Gp,T) % G(z)
[ngz, dgz]=tfdata(Gz,'v')
% symbolic operation
syms z
KI=0.01
Ng=vpa(0.01358*z+0.01229,4)
Dg=vpa(z^2-1.7236*z+0.7408,4)
Ngp=vpa(simplify(Ng*(z-1)),4)
Dgp=vpa(simplify(Dg*(z-1)+KI*(z+1)*Ng),4)

% open-loop TF DG(z)
ngzp=[0.01358 -0.00129 -0.01229]
dgzp=[1 -2.723 2.465 -0.7407]
zp=roots(ngzp)
pp=roots(dgzp)
r=abs(pp)
Gpz=tf(ngzp, dgzp, T)

% break-in break-away points
ba_bi = diff(Ngp,z)*Dgp-diff(Dgp,z)*Ngp;
babi=double(solve(ba_bi==0, z))

%root locus
figure(1)
rlocus(Gpz)
axis('equal')

% closed-loop system step response
k=20.9091
Tz=feedback(k*Gpz,1)
[ntz,dtz]=tfdata(Tz,'v')
[z,p,k]=tf2zp(ntz,dtz)
r=abs(p)
th=angle(p)
w=th/T
figure(2)
dstep(Tz,1)

Results:
G(z) =
```

```
    0.01358 z + 0.01229
  ----------------------
  z^2 - 1.724 z + 0.7408
G'(z) =
  0.01358 z^2 - 0.00129 z - 0.01229
  ----------------------------------
  z^3 - 2.723 z^2 + 2.465 z - 0.7407
open-loop zeros zp = 1.0000, -0.9050
open-loop poles pp = 0.9700 + 0.0714i, 0.9700 - 0.0714i, 0.7829 + 0.0000i
open-loop pole magnitude r = 0.9726, 0.9726, 0.7829

closed-loop system transfer function at k = 20.9091
T(z) =

    0.2839 z^2 - 0.02697 z - 0.257
  ----------------------------------
  z^3 - 2.439 z^2 + 2.438 z - 0.9977

closed-loop zeros z = 1.0000, -0.9050
closed-loop poles p = 0.9977 + 0.0000i, 0.7207 + 0.6933i, 0.7207 - 0.6933i
closed-loop pole magnitude r = 0.9977, 1.0000, 1.0000
closed-loop pole angles th = 0, 0.7660, -0.7660
frequency w = 0, 7.6600, -7.6600 rad/s
```

Jury Test for PI controller proportional gain K_p

```
% Ch8 Ex 8_3 Stability Analysis of a 2nd order closed-loop discrete-time
system
% Gp(s)=3/(s+1)(s+2), H(s)=1, D(z)=Kp+KI(T/2)(z+1)/(z-1)
clear all; close all; clc
T=0.1 ; % Sampling time
Gp = tf(3, [1  3  2 ]) %Gp(s)

% G(z) = Z(Gho(s)Gp(s))
Gz =c2d(Gp,T) % G(z)
[ngz, dgz]=tfdata(Gz,'v')

% symbolic operations to get G'(z) 1+K'G'(z)=0
syms K z
KI=0.01
Ng=vpa(0.01358*z+0.01229,4)
Dg=vpa(z^2-1.7236*z+0.7408,4)
Ngp=vpa(simplify(Ng*(z-1)),4)
Dgp=vpa(simplify(Dg*(z-1)+KI*(z+1)*Ng),4)

% break-in break-away points
ba_bi = diff(Ngp,z)*Dgp-diff(Dgp,z)*Ngp;
babi=double(solve(ba_bi==0, z))

% Generalized G(z)
ngzp=[0 0.01358 -0.00129 -0.01229]
dgzp=[1 -2.723 2.465 -0.7407]
Gpz=tf(ngzp, dgzp, T)
n=length(dgzp)-1 % order of the system (number of open-loop poles)
m=length(ngzp)-1 % number of open-loop zeros
```

```
% Jusy Test
a3=(dgzp(1)+K*ngzp(1));
a2=(dgzp(2)+K*ngzp(2));
a1=(dgzp(3)+K*ngzp(3));
a0=(dgzp(4)+K*ngzp(4));
P=a3*z^n+a2*z^(n-1)+a1*z^(n-2)+a0*z^(n-3);
P=vpa(simplify(P),4)
% condition 1 P(1)>0
c1=vpa(subs(P,z,1),4)
% condition 2 (-1)^n *P(-1)>0
c2=vpa(subs((-1)^n*P,z,-1),4)
% condition 3 |a0|<a3
c3=vpa(a3-abs(a0),4)

%condition 4
b0=vpa(simplify(a0^2-a3^2),4)
b1=vpa(simplify(a0*a1-a2*a3),4)
b2=vpa(simplify(a0*a2-a1*a3),4)
c4=vpa(simplify(b2-b0),4)

K1=double(solve(c1==0,K))
K2=double(solve(c2==0,K))
K3=double(solve(c3==0,K))
K4=double(solve(c4==0,K))

% P at Kcritical
Pc=vpa(subs(P,K,K4(2)),4) % characteristic ploynomial at K=Kcrtical
zc=double(solve(Pc==0, z)) % closed-loop poles at K=Kcrtical
r=abs(zc)  % magnitude of closed-loop poles
th=angle(zc) % angles of closed-loop poles
w=th/T % frequency of oscillation

Results:

G(z) =

   0.01358 z + 0.01229
  ----------------------
  z^2 - 1.724 z + 0.7408

Ng = 0.01358*z + 0.01229
Dg = z^2 - 1.724*z + 0.7408
Ngp = (0.01358*z + 0.01229)*(z - 1.0)
Dgp = z^3 - 2.723*z^2 + 2.465*z - 0.7407

G'(z) =

  0.01358 z^2 - 0.00129 z - 0.01229
  ----------------------------------
  z^3 - 2.723 z^2 + 2.465 z - 0.7407
```

```
Break-away-break-in point
babi = -2.6713, 0.8775, 0.9444, 1.0394

P =z^3 + (0.01358*K - 2.723)z^2 -(0.00129*K - 2.465)z - 0.01229*K - 0.7407 = 0

c1 = 0.0013 - 1.776e-14*K > 0  K < 7.3183e+10
c2 = 6.929 - 0.00258*K > 0     K < 2.6855e+03
c3 = 1.0 - 1.0*abs(0.01229*K + 0.7407) > 0 K < 21.0985

b0 = (0.01229*K + 0.7407)^2 - 1.0
b1 = 1.585e-5*K^2 - 0.04292*K + 0.8972
b2 = - 0.0001669*K^2 + 0.0247*K - 0.4481
c4 = - 0.0003179*K^2 + 0.006491*K + 0.00329 > 0 -0.4948 < K < 20.9091
Pc = z^3 - 2.439*z^2 + 2.438*z - 0.9977

Range of proportional gain Kp for stability -0.4948 < K < 20.9091

Closed-loop poles  0.9977 + 0.0000i, 0.7207 - 0.6933i, 0.7207 + 0.6933i

Closed-loop pole magnitude r =  0.9977, 1.0000, 1.0000

Closed-loop pole angles th =  0, -0.7660, 0.7660

Frequency w =  0, -7.6600, 7.6600 rad/s
```

8.5 Laboratory Experiment for Stability Analysis of Discrete-Time Systems Using *z*-Domain Techniques

Objectives The objectives are to determine the range of proportional gain for stability using *z*-domain techniques and validate through an experiment on a physical module of a DC motor.

Introduction There are different methods of analyzing the stability of a dynamic system. In this lab, some of these techniques in the *z*-domain, namely, Jury's test for stability and root locus techniques, are studied using the identified model parameters for the DC motor. The results are validated through simulation and actual experimentation with the DC motor module.

I. *Stability Analysis and Simulation*

   A. *Open-Loop System Transfer Function in z-Domain*

      1. The transfer function $G(s)$ for the speed dynamics of the DC motor using the identified parameters, static gain K and time constant τ, can be expressed as

$$G(s) = \frac{K}{\tau s + 1} \tag{8.25}$$

2. Using the speed transfer function, the position transfer function $G_p(s)$ is obtained as

$$G_p(s) = \frac{K}{s(\tau s + 1)} \tag{8.26}$$

3. Use $G_p(s)$ and a proper sampling time ($T_s \approx \tau/10$) to find $G(z)$, the z-domain transfer function for the position of the DC motor. For the DC motor control trainer (DCMCT) module, a sampling time of 0.01 s is selected.

B. *Jury's Test for Stability Analysis*

1. For a proportional controller with transfer function $D(z) = K_P$, the closed-loop transfer function can be expressed as

$$T(z) = \frac{D(z)G(z)}{1 + D(z)G(z)} = \frac{K_P G(z)}{1 + K_P G(z)} \tag{8.27}$$

2. The characteristic equation for the closed-loop discrete-time system can be expressed as

$$1 + K_P G(z) = 0 \tag{8.28}$$

3. Obtain the range of gain K_P through Jury's test for stability analysis for the closed-loop system.
4. Obtain the upper limit of the positive gain $K_P = K_u$ for stability and the corresponding closed-loop poles. Estimate the frequency of oscillation (ω_{osc}) from the closed-loop poles.
5. Verify that the closed-loop system step response is oscillatory for $K_P = K_u$. Estimate the frequency of oscillation ω_{osc} (rad/s).
6. Complete the corresponding column in Table 8.1.

Table 8.1 Summary of results for stability analysis in z-domain

			Closed-loop system step response	
	Jury's test	Root locus	Simulation	Experiment
K_u				
Closed-loop poles				
ω_{osc} (rad/s)				

C. *Stability Analysis Using Root Locus*

1. Obtain the root locus of the system in the z-domain, using the open-loop system transfer function $G(z)$.
2. Obtain the poles (p) and limiting gain K_u where the root locus crosses the unit circle.
3. Use gain K_u to calculate the closed-loop poles of the system by setting $D(z) = K_u$ in the closed-loop transfer function.
4. Compare the calculated poles with the poles found from the root locus in step 2.
5. From the closed-loop poles, estimate the frequency ω and period T of oscillation at the limiting gain K_u.
6. Plot the step response of the closed-loop system with $D(z) = K_u$. From the plot, estimate the frequency ω and period T of oscillation, and verify these values with those estimated in step 5.
7. Complete the corresponding column in Table 8.1.

II. *Experimental Validation*

1. Set up NI ELVIS II board and DC motor module. Select the device, and set the sampling rate. For the DC motor control trainer (DCMCT) module, a sampling rate of 100 Hz is used.
2. Study the block diagram for the position control experiment (*01-QNET_DCMCT_Position_Control.vi*).
3. Identify the different components of the control system, including the input, error, controller, plant, sensor, and output.
4. Set the controller parameters to $K_p = K_u$, $K_d = 0$, $K_i = 0$, and change the input waveform frequency to a small value so that it behaves like a step input.
5. Select your input parameters. For the DCMCT module, typical values for the input square wave voltage are as follows: amplitude of 1.0 V, frequency of 0.40 Hz, and offset and 3.0 V.
6. Run VI, and observe the system response. If the actual system response does not perfectly oscillate, slowly adjust the gain K_p until it does so.
7. Make sure to set the number of significant figures to at least 5 for the display format of each graph on the front panel VI (right-click on the graph → display format → decimal → 6). Export the time response graph, motor rotation $\theta(t)$, to Excel by right-clicking on the graph and selecting the Export to Excel option (right-click → Export to Excel). Compare it to the simulated response.
8. Estimate the frequency ω and period T of the closed-loop response. Compare these values to the ones found in Section I.
9. Complete the Corresponding column in Table 8.1.

III. *Deliverables*

1. Complete Table 8.1.
2. Include the closed-loop step response graph for $K_p = K_u$ from simulation.
3. Include the closed-loop response graph for $K_p = K_u$ from experiment.

Lab Report
Present all deliverables in the report and discuss results.

Exercises

8.1 A closed-loop discrete-time system, Fig. P8.1, is with the plant transfer function $G_p(s) = \frac{3}{(s+1)(s+2)}$; a zero-order hold (ZOH); a digital controller with a simple gain, $D(z) = K$; and a sensor with a unity gain, $H(s) = 1$. Use a sampling time $T = 0.1$ s. Analyze the system stability using root locus. Obtain the range of the controller gain K for the stability of the closed-loop discrete-time system. Find the critical gain at which the closed-loop step response becomes oscillatory. Verify the frequency of oscillation from the closed-loop step response of the discrete-time system.

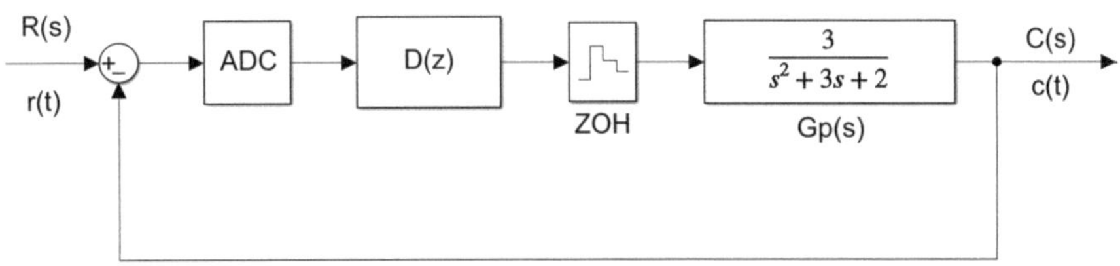

Fig. P8.1 Closed-loop discrete-time system for Problem 8.1

8.2 A closed-loop discrete-time system, Fig. P8.2, is with the plant transfer function $G_p(s) = \frac{2}{(s+1)(s+2)}$, a zero-order hold (ZOH), a PI controller $D(z) = K_P\left(1 + \frac{K_I}{K_P}\frac{T}{2}\left(\frac{z+1}{z-1}\right)\right)$, and a sensor with a unity gain, $H(s) = 1$. Use a sampling time $T = 0.1$ s, $\frac{K_I}{K_P} = 0.2$. Analyze the system stability using root locus. Obtain the range of the controller gain K_P for the stability of the closed-loop discrete-time system. Find the critical gain at which the closed-loop step response becomes oscillatory. Verify the frequency of oscillation from the closed-loop step response of the discrete-time system.

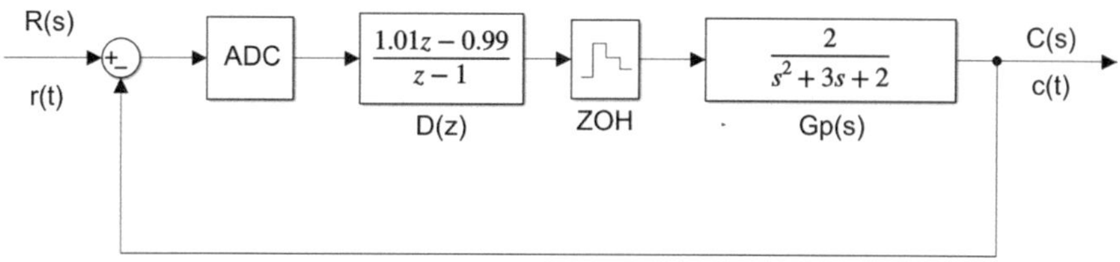

Fig. P8.2 Closed-loop discrete-time system for Problem 8.2

8.3 A closed-loop discrete-time system, Fig. P8.3, is with the plant transfer function $G_p(s) = \frac{2}{(s+3)(s+4)}$, a zero-order hold (ZOH), a PI controller $D(z) = \left(K_P + K_I \frac{T}{2}\frac{z+1}{z-1}\right)$, and a sensor with a unity gain, $H(s) = 1$. Use a sampling time $T = 0.1$, $K_I = 0.2$. Analyze the system stability using generalized root locus for the positive proportional gain $0 < K_P < \infty$. Obtain the range of the controller gain K_P for the stability of the closed-loop discrete-time system. Find the critical gain at which the closed-loop step response becomes oscillatory. Verify the frequency of oscillation from the closed-loop step response of the discrete-time system.

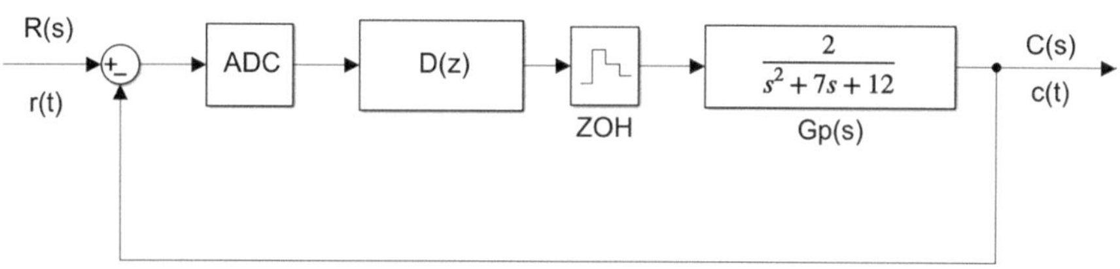

Fig. P8.3 Closed-loop discrete-time system for Problem 8.3

8.4 A closed-loop discrete-time system, Fig. P8.4, is with the plant transfer function $G_p(s) = \frac{14}{(s+3)(s+4)}$, a zero-order hold (ZOH), a PI controller $D(z) = \left(K_P + K_I \frac{T}{2}\frac{z+1}{z-1}\right)$, and a sensor with a unity gain, $H(s) = 1$. Use a sampling time $T = 0.1$, $K_I = 0.2$. Analyze the system stability using generalized root locus for the positive proportional gain $0 < K_P < \infty$. Obtain the range of the controller gain K_P for the stability of the closed-loop discrete-time system. Find the critical gain at which the closed-loop step response becomes oscillatory. Verify the frequency of oscillation from the closed-loop step response of the discrete-time system.

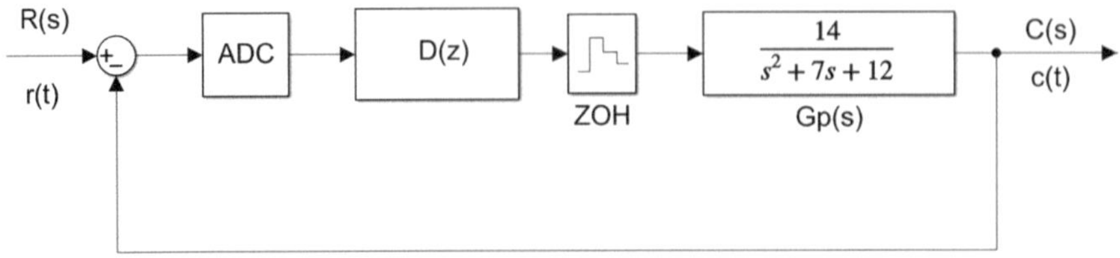

Fig. P8.4 Closed-loop discrete-time system for Problem 8.4

8.5 The closed-loop vehicle distance control system for a traffic control system can be represented as shown in Fig. P8.5. The plant transfer function of the distance control system can be represented in s-domain as $G_p(s) = \frac{1.5}{s^2+12s+18}$. The digital controller is a simple gain, $D(z) = K$. The sampling time is given as $T = 0.10$ s. Analyze the system stability using root locus. Obtain the range of the controller gain K for the stability of the closed-loop discrete-time system. Find the critical gain at which the closed-loop step response becomes oscillatory. Verify the frequency of oscillation from the closed-loop step response of the discrete-time system. Obtain the closed-loop poles and the closed-loop step response for $K = \frac{K_{\text{critical}}}{2}$.

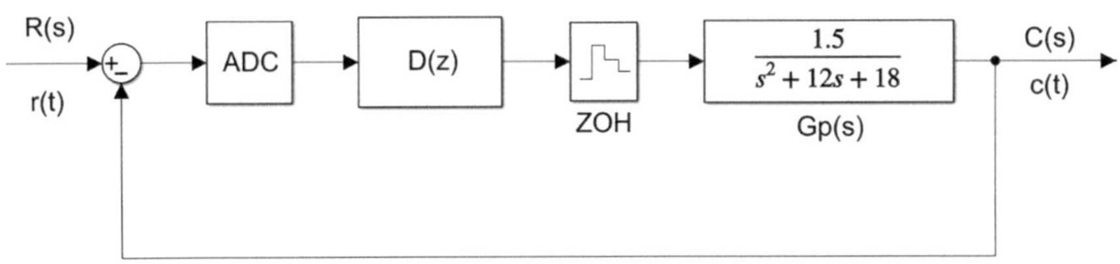

Fig. P8.5 Closed-loop discrete-time system for Problem 8.5

8.6 The closed-loop roll angle control system for a jet airplane can be represented as shown in Fig. P8.6. The plant transfer function of the roll control system is represented in the s-domain as $G_p(s) = \frac{10}{(s+10)(s^2+2.5s+25)}$. The digital controller is a simple gain, $D(z) = K$. The sampling time is given as $T = 0.05$ s. Analyze the system stability using root locus. Obtain the range of the controller gain K for the stability of the closed-loop discrete-time system. Find the critical gain at which the closed-loop step response becomes oscillatory. Verify the frequency of oscillation from the closed-loop step response of the discrete-time system. Obtain the closed-loop poles and the closed-loop step response for $K = \frac{K_{\text{critical}}}{3}$.

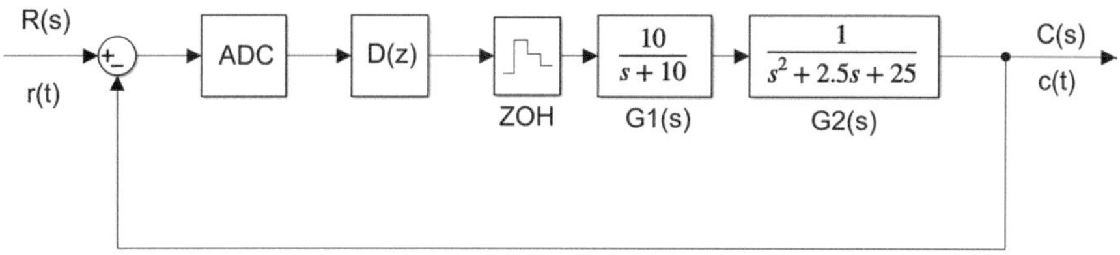

Fig. P8.6 Closed-loop discrete-time system for Problem 8.6

8.7 The closed-loop steering control system for a second-order system with a time delay can be represented as shown in Fig. P8.7, $G_p(s) = \frac{1}{(s+1)(s+3)} e^{-sT}$. The digital controller is a simple gain, $D(z) = K$. The sampling time is given as $T = 0.2$ s. Analyze the system stability using root locus. Obtain the range of the controller gain K for the stability of the closed-loop discrete-time system. Find the critical gain at which the closed-loop step response becomes oscillatory. Verify the frequency of oscillation from the closed-loop step response of the discrete-time system. Obtain the closed-loop poles and the closed-loop step response for $K = \frac{K_{\text{critical}}}{2}$.

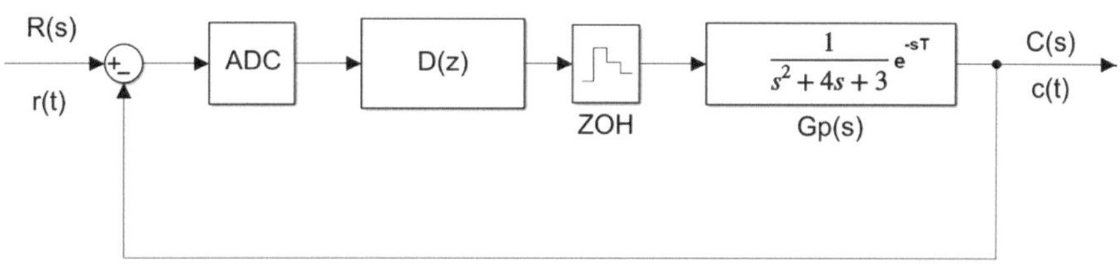

Fig. P8.7 Closed-loop discrete-time system for Problem 8.7

8.8 The closed-loop thickness control system in a metal processing plant can be represented as shown in Fig. P8.8. The plant transfer function of the thickness control system is represented in the s-domain as $G_p(s) = \frac{1}{s(s+2)(s^2+s+10)}$. The digital controller is a simple gain, $D(z) = K$. The sampling time is given as $T = 0.10$ s. Analyze the system stability using root locus. Obtain the range of the controller gain K for the stability of the closed-loop discrete-time system. Find the critical gain at which the closed-loop step response becomes oscillatory. Verify the frequency of oscillation from the closed-loop step response of the discrete-time system. Obtain the closed-loop poles and the closed-loop step response for $K = \frac{K_{\text{critical}}}{2}$.

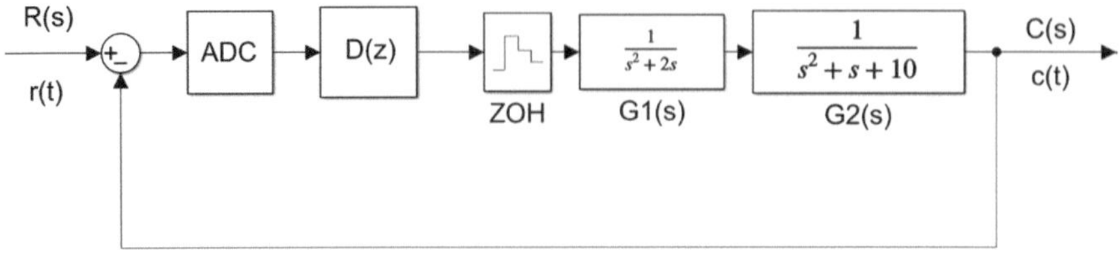

Fig. P8.8 Closed-loop discrete-time system for Problem 8.8

8.9 The closed-loop tilt control system in a high-speed train for passenger comfort can be represented as shown in Fig. P8.9. The plant transfer function of the train tilt control system is represented in the *s*-domain as $G_p(s) = \frac{1}{s(s+2)(s+3)}$. The digital controller is a simple gain, $D(z) = K$. The sampling time is given as $T = 0.10$ s. Analyze the system stability using root locus. Obtain the range of the controller gain K for the stability of the closed-loop discrete-time system. Find the critical gain at which the closed-loop step response becomes oscillatory. Verify the frequency of oscillation from the closed-loop step response of the discrete-time system. Obtain the closed-loop poles and the closed-loop step response for $K = \frac{K_{\text{critical}}}{3}$.

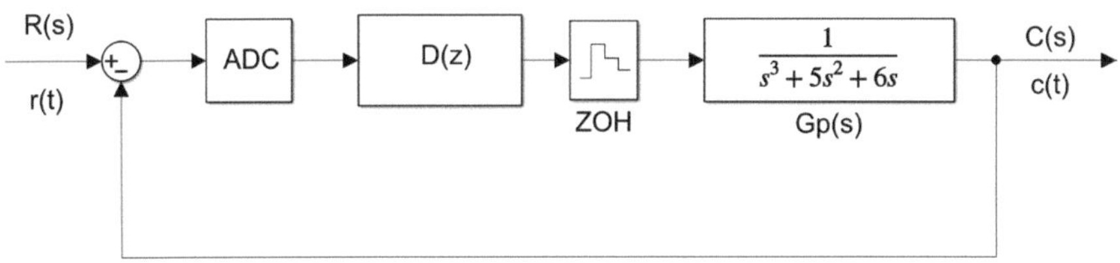

Fig. P8.9 Closed-loop discrete-time system for Problem 8.9

8.10 The closed-loop position control system for one of the axes of an experimental automated surface-grinding wheel can be represented as shown in Fig. P8.10. The plant transfer function of the axis position control system can be represented in *s*-domain as [4] $G_p(s) = \frac{1}{s(0.2s+1)(0.3s+1)}$. The digital controller is a simple gain, $D(z) = K$. The sampling time is given as $T = 0.05$ s. Analyze the system stability using root locus. Obtain the range of the controller gain K for the stability of the closed-loop discrete-time system. Find the critical gain (K_{critical}) at which the closed-loop step response becomes oscillatory. Verify the frequency of oscillation from the closed-loop step response of the discrete-time system. Obtain the closed-loop poles and the closed-loop step response for $K = \frac{K_{\text{critical}}}{2}$.

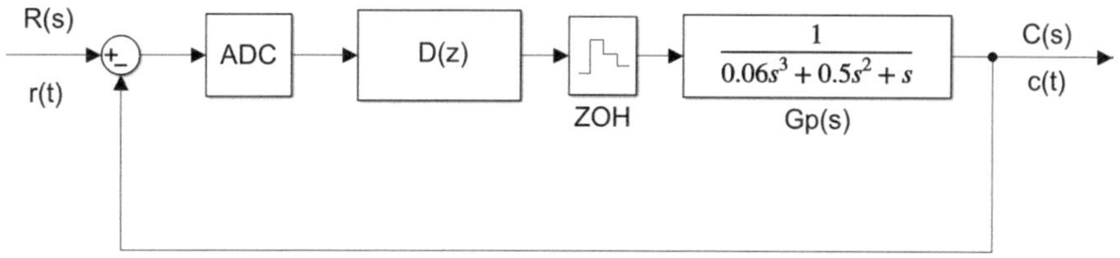

Fig. P8.10 Closed-loop discrete-time system for Problem 8.10

Further Reading

1. Asadi F (2023) Signals and systems with MATLAB® and Simulink®. Springer, Cham
2. Astrom KJ, Murray RM (2021) Feedback systems: an introduction for scientists and engineers, 2nd edn. Princeton University Press, Princeton
3. Chen C-T (2006) Analog and digital control system design: transfer-function, state-space, and algebraic methods. Oxford University Press, New York
4. Dorf R, Bishop R (2016) Modern control systems, 13th edn. Pearson, Essex
5. Fadali MS, Visioli A (2020) Digital control engineering analysis and design, 3rd edn. Academic Press, London
6. Franklin GF, Powell JD, Emami-Naeini A (2019) Feedback control of dynamic systems, 8th edn. Pearson, Upper Saddle River
7. Franklin GF, Powell JD, Workman M (1998) Digital control of dynamic systems, 3rd edn. Addison-Wesley, Boston
8. Isermann R (1981) Digital control systems. Springer, Berlin
9. Jalili N, Candelino NW (2023) Dynamic systems and control engineering. Cambridge University Press, Cambridge
10. Keviczky L, Bars R, Hetthéssy J, Bányász C (2019) Control engineering. Springer, Singapore
11. Koeing DM (2009) Practical control engineering: a guide for engineers, managers, and practitioners. McGraw-Hill, New York
12. Kuo BC (1995) Digital control systems, 2nd edn. Oxford University Press, Oxford
13. Landau ID, Gito G (2006) Digital control systems: design, identification and implementation. Springer, London
14. Leigh JR (2006) Applied digital control: theory, design and implementation, 2nd edn. Dover, Mineola
15. Mathworks (2024) Control System Toolbox. https://www.mathworks.com/products/control.html
16. Mathworks (2024) Matlab. https://www.mathworks.com/products/matlab.html
17. National Instruments (2024) LabVIEW. https://www.ni.com/en/support/downloads/software-products/download.labview.html#487445
18. National Instruments (2024) NI-ELVIS-II. https://www.ni.com/en-us/support/model.ni-elvis-ii.html
19. National Instruments (2024) Labs for Quanser QNET DC Motor Add-On Board. https://education.ni.com/teach/resources/180/labs-for-quanser-qnet-dc-motor-add-on-board
20. Nise NS (2019) Control systems engineering, 8th edn. Wiley, Hoboken
21. Nof SY (ed) (2009) Springer handbook of automation. Springer, Berlin
22. Ogata K (1995) Discrete-time control systems, 2nd edn. Pearson, Hoboken
23. Phillips CL, Nagle HT, Chakrabortty A (2015) Digital control system analysis and design, 4th edn. Pearson, Essex
24. Quanser (2024) QUBE Servo 2. https://www.quanser.com/products/qube-servo-2/

25. Rabbath CA, Lechevin N (2014) Discrete-time control system design with applications. Springer, New York
26. Samanta B (2023) Introduction to mechatronics – an integrated approach. Springer, Cham
27. Shertukde HM (2015) Digital control applications illustrated with MATLAB®. CRC Press, Boca Raton
28. Starr G (2020) Introduction to applied digital controls. Springer, Cham
29. Veloni A, Miridakis NI (2018) Digital control systems-theoretical problems and simulation tools. CRC Press, Boca Raton
30. Vilanova R, Visioli A (2012) PID control in the third millennium – lessons learned and new approaches. Springer, London
31. Visioli A (2006) Practical PID control. Springer, London

Chapter 9
Frequency Domain Analysis of Discrete-Time Systems

9.1 Introduction

Two frequency domain techniques, namely, Bode plots and Nyquist diagrams, for stability analysis of discrete-time systems are presented in this chapter. Discrete-time systems represented in the w-domain are analyzed for stability. Gain margin, phase margin, gain cross-over frequency, and phase cross-frequency are defined using both Bode plots and Nyquist diagrams. Generalized Bode plots are presented for stability analysis of discrete-time systems. Examples along with corresponding Matlab codes are presented to illustrate the techniques. End-of-chapter exercises are provided to enhance understanding of the concepts covered and consolidate problem-solving skills.

9.2 Frequency Domain Analysis of Discrete-Time Systems

An open-loop discrete-time system is shown in Fig. 9.1 where the input, the output, and the transfer function are represented in the z-domain as

$$G(z) = \frac{Y(z)}{X(z)} \tag{9.1}$$

The open-loop system can be represented in the w-domain using the bilinear transformation introduced in Chap. 7, Eq. (7.12):

$$z = \frac{1 + \left(\frac{T}{2}\right)w}{1 - \left(\frac{T}{2}\right)w} \tag{9.2}$$

where T is the sampling time.

B. Samanta, *Introduction to Digital Control*,
https://doi.org/10.1007/978-3-031-66830-2_9

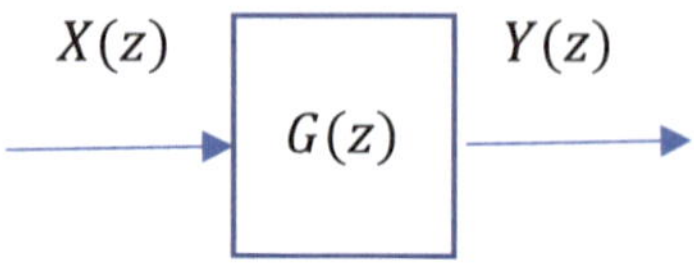

Fig. 9.1 Open-loop discrete-time system in *z*-domain

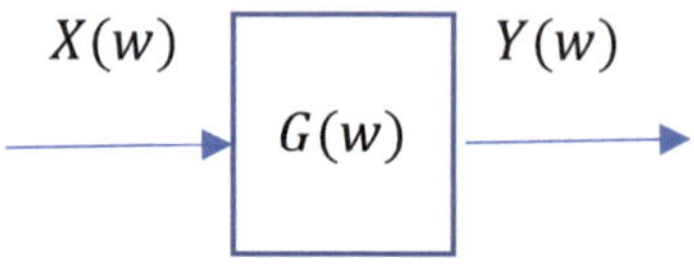

Fig. 9.2 Open-loop discrete-time system in *w*-domain

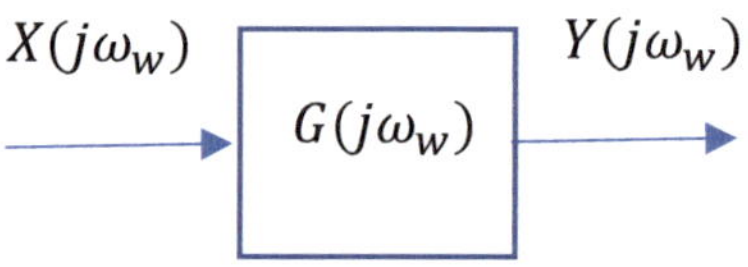

Fig. 9.3 Open-loop discrete-time system in *w*-domain

The *w*-domain representation of the open-loop system of Fig. 9.1 is shown in Fig. 9.2.

Where

$$G(w) = G(z)\Big|_{z=\frac{1+\left(\frac{T}{2}\right)w}{1-\left(\frac{T}{2}\right)w}} \tag{9.3}$$

The open-loop system can be analyzed in the frequency domain replacing $w = j\omega_w$ where ω_w is termed as the frequency in the *w*-domain. The system of Fig. 9.2 can be represented for frequency response analysis as in Fig. 9.3:

Where the input, the output, and the transfer function are represented in the frequency domain in terms of the corresponding magnitudes (X, Y, M) and phase angles (ϕ_X, ϕ_Y, ϕ) corresponding to the frequency ω_w as

$$X(j\omega_w) = Xe^{j\phi_X} \tag{9.4}$$

$$Y(j\omega_w) = Ye^{j\phi_Y} \tag{9.5}$$

$$G(j\omega_w) = Me^{j\phi} \tag{9.6}$$

$$G(j\omega_w) = \frac{Y(j\omega_w)}{X(j\omega_w)} = \left(\frac{Y}{X}\right)e^{j(\phi_Y - \phi_X)} \tag{9.7}$$

$$M = \frac{Y}{X} \tag{9.8}$$

$$\phi = \phi_Y - \phi_X \tag{9.9}$$

It should be emphasized that magnitudes (X, Y, M) and phase angles (ϕ_X, ϕ_Y, ϕ) are functions of frequency ω_w. It is often necessary to analyze the frequency response of the system, $G(j\omega_w)$, both magnitude $M(\omega_w)$ and phase $\phi(\omega_w)$, with varying frequency ω_w. One of the common methods used for frequency response analysis is to plot magnitude $M(\omega_w)$ and phase $\phi(\omega_w)$ versus frequency ω_w. These plots are known as Bode plots. Another method is to plot $Me^{j\phi}(\omega_w)$ in the complex plane, and the plot is known as the Nyquist diagram.

9.3 Bode Plots for Discrete-Time Systems

A closed-loop discrete-time system is shown in Fig. 9.4.

The closed-loop discrete-time system transfer function in w-domain can be expressed as

$$T(w) = \frac{C(w)}{R(w)} = \frac{G(w)D(w)}{1 + \overline{GH}(w)D(w)} \tag{9.10}$$

where the open-loop system transfer function and discrete-time controller are presented in w-domain, using bilinear transformation, as

$$G(w) = G(z)\rfloor_{z=\frac{1+\frac{Tw}{2}}{1-\frac{Tw}{2}}} \tag{9.11}$$

$$D(w) = D(z)|_{z=\frac{1+\frac{Tw}{2}}{1-\frac{Tw}{2}}} \tag{9.12}$$

$$\overline{GH}(w) = \overline{GH}(z)\rfloor_{z=\frac{1+\frac{Tw}{2}}{1-\frac{Tw}{2}}} \tag{9.13}$$

The characteristic equation for the closed-loop system in w-domain can be written as

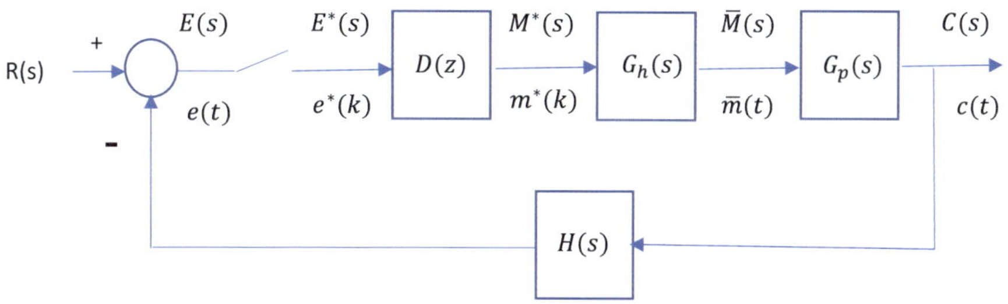

Fig. 9.4 Closed-loop discrete-time system with a digital filter and a sensor

$$1+\overline{GH}(w)D(w)=0 \tag{9.14}$$

For the characteristic Eq. (9.14), the magnitude and phase angle relations can be represented as

$$\left|\overline{GH}(j\omega_w)D(j\omega_w)\right|=1 \tag{9.15}$$

$$\angle\overline{GH}(j\omega_w)D(j\omega_w)=180\,^\circ \tag{9.16}$$

In Eq. (9.16), the phase angle is presented in degrees. There are two frequencies of interest in the analysis of Bode plots. The frequency at which the magnitude is equal to 1, i.e., Eq. (9.15), is called gain cross-over frequency, ω_{wgc}:

$$\left|\overline{GH}(j\omega_{wgc})D(j\omega_{wgc})\right|=1 \tag{9.17}$$

The phase margin (ϕ_M) is defined as the difference of 180° and phase angle at gain cross-over frequency ω_{wgc}:

$$\phi_M=180\,^\circ+\angle\overline{GH}(j\omega_{wgc})D(j\omega_{wgc}) \tag{9.18}$$

The frequency at which the phase is equal to 180°, i.e., Eq. (9.16), is called phase cross-over frequency, ω_{wpc}:

$$\angle\overline{GH}(j\omega_{wpc})D(j\omega_{wpc})=180\,^\circ \tag{9.19}$$

The gain margin (GM) is defined as the reciprocal of the magnitude at phase cross-over frequency ω_{wpc}:

$$GM=1/\left|\overline{GH}(j\omega_{wpc})D(j\omega_{wpc})\right| \tag{9.20}$$

In dB scale, at gain cross-over frequency, $\omega_{wgc.}$

$$20\log_{10}\left|\overline{GH}(j\omega_{wgc})D(j\omega_{wgc})\right|=0\text{ dB} \tag{9.21}$$

The gain margin in the dB scale can be represented as

$$GM=-20\log_{10}\left|\overline{GH}(j\omega_{wpc})D(j\omega_{wpc})\right|\text{ dB} \tag{9.22}$$

For stability, both the gain margin and phase margin need to be positive.

9.4 Stability Analysis of Discrete-Time Systems Using Bode Plots

In this section, the process of analyzing the stability of discrete-time systems is illustrated through two examples.

Example 9.1 A closed-loop discrete-time system, similar to Fig. 9.1, is with the plant transfer function $G_p(s) = \frac{3}{s(s+2)}$; a zero-order hold (ZOH); a digital controller with a simple gain, $D(z) = K$; and a sensor with a unity gain, $H(s) = 1$. Use a sampling time $T = 0.1$ s. Obtain the transfer function of the open-loop system in the z-domain, $G(z)$, and in the w-domain, $G(w)$. Obtain Bode plots, obtain gain and phase margins, gain cross-over frequency, and phase cross-over frequency.

Solution The open-loop transfer function of the discrete-time system with ZOH is obtained (in Example 6.2) as

$$\begin{aligned} G(z) &= \overline{G_{h0}G_p}(z) = Z\left[\frac{1-e^{-sT}}{s}\,G_p(s)\right]_{e^{sT}=z} \\ &= \frac{0.01405z + 0.01314}{(z-1)(z-0.8187)} \end{aligned}$$

The open-loop transfer function of the discrete-time system in the w-domain using bilinear transformation can be obtained as

$$G(w) = \left.\frac{0.01405z + 0.01314}{(z-1)(z-0.8187)}\right|_{z=\frac{1+\frac{Tw}{2}}{1-\frac{Tw}{2}}}$$

$$\begin{aligned} G(w) &= \frac{0.01405\left(\dfrac{1+0.05w}{1-0.05w}\right) + 0.01314}{\left(\dfrac{1+0.05w}{1-0.05w} - 1\right)\left(\dfrac{1+0.05w}{1-0.05w} - 0.8187\right)} \\ &= 2.5014 \times 10^{-4}\frac{(w-20)(w+597.5824)}{w(w+1.9934)} \end{aligned}$$

Bode plots can be obtained using the Matlab command bode(). The gain and phase margins and gain and phase cross-over frequencies can be obtained using the Matlab command margin(). Bode plots are presented in Fig. 9.5.

Gain margin gm = 13.7928
Phase margin pm = 54.029°
Phase cross-over frequency $\omega_{wpc} = 6.4330$ $\omega_{pc} = (\frac{2}{T})atan(\frac{wT}{2}) = 6.224$ rad/s

Gain cross-over frequency $\omega_{wgc} = 1.2681$ $\omega_{gc} = (\frac{2}{T})atan(\frac{wT}{2}) = 1.266$ rad/s

The gain margin is close to the critical gain obtained in Examples 7.1, 7.3, and 7.7. The frequency of oscillation at the critical gain is close to the phase cross-over frequency. This is also close to the values obtained in Examples 7.1, 7.3, and 7.7.

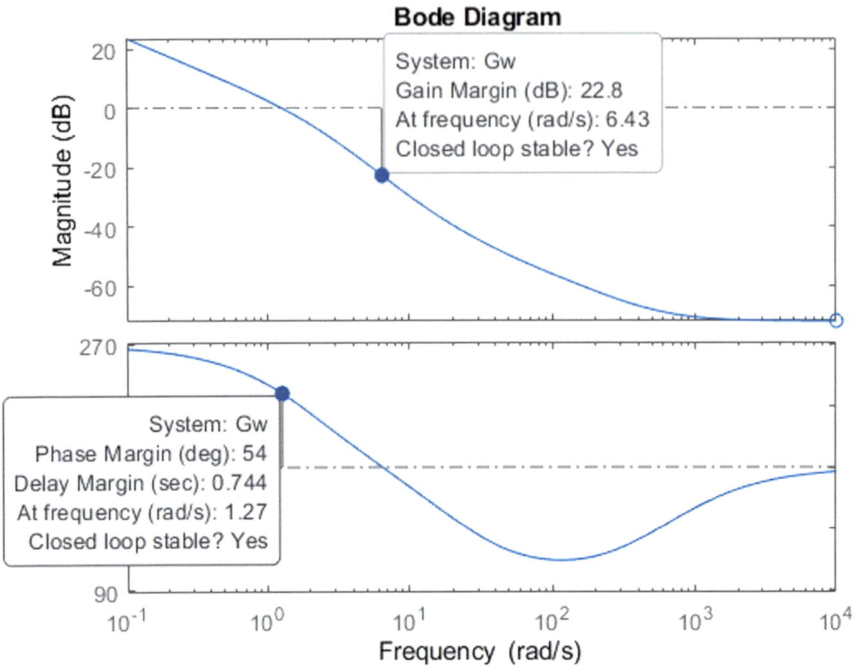

Fig. 9.5 Bode plots for Example 9.1

Matlab code and results are presented next.

```
%Ch 9_ex9-1 Bode of a 2nd order discrete-time system
% Gp(s)=3/s(s+2), H(s)=1, D(z)=K
clear all; close all; clc
T=0.1 ; % Sampling time
Gp = tf(3, [1 2 0]) %Gp(s)

% G(z) = Z(Gho(s)Gp(s))
Gz =c2d(Gp,T) % G(z)
Gz=zpk(Gz)
[ngz,dgz]=tfdata(Gz,'v');
[z,p,k]=tf2zp(ngz,dgz)

% G(w)
Gw = d2c(Gz,'tustin')
[ngw,dgw] = tfdata(Gw, 'v')
Gw = zpk(Gw)

% Bode plot
figure(1)
bode(Gw)
[gm,pm,wpc,wgc]=margin(Gw)
Results:
G(z) =

   0.01405 z + 0.01314
  ----------------------
  z^2 - 1.819 z + 0.8187

G(w)
   -0.000249 w^2 - 0.14452 w + 2.99
   --------------------------------
      w^2 + 1.9934 w + 1.1065e-14
```

Example 9.2 A closed-loop discrete-time system, similar to Fig. 8.1, is with the plant transfer function $G_p(s) = \frac{3}{s(s+2)}$, a zero-order hold (ZOH), a PI controller $D(z) = K_P\left(1 + \frac{K_I}{K_P}\frac{T}{2}\left(\frac{z+1}{z-1}\right)\right)$, and a sensor with a unity gain, $H(s) = 1$. Use a sampling time $T = 0.1$ s, $\frac{K_I}{K_P} = 0.2$. Analyze the system stability using Bode plots. Obtain gain and phase margins, gain cross-over frequency, and phase cross-over frequency.

Solution The open-loop transfer function of the discrete-time system with ZOH is obtained (in Example 6.2) as

$$\begin{aligned}\overline{GH}(z) &= \overline{G_{h0}G_p}(z) = Z\left[\frac{1-e^{-sT}}{s}G_p(s)\right]_{e^{sT}=z} \\ &= \frac{0.01405z + 0.01314}{(z-1)(z-0.8187)}\end{aligned}$$

The open-loop discrete-time system with the controller transfer function can be expressed as

$$\begin{aligned}D(z)\overline{GH}(z) &= K_P\left(1 + 0.01\left(\frac{z+1}{z-1}\right)\right)\left(\frac{0.01405z + 0.01314}{(z-1)(z-0.8187)}\right) \\ &= K_P\left(\frac{1.01z - 0.99}{z-1}\right)\left(\frac{0.01405z + 0.01314}{(z-1)(z-0.8187)}\right)\end{aligned}$$

The open-loop discrete-time system with the controller transfer function in the w-domain can be expressed using bilinear transformation as

$$DG(w) = K_P\left(\frac{1.01z - 0.99}{z-1}\right)\frac{0.01405z + 0.01314}{(z-1)(z-0.8187)}\bigg|_{z=\frac{1+\frac{Tw}{2}}{1-\frac{Tw}{2}}}$$

$$DG(w) = K_P(2.49 \times 10^{-4})\left(\frac{w+0.2}{w}\right)\left(\frac{(w-20)(w+600.4)}{w(w+1.9934)}\right)$$

Bode plots, obtained using Matlab, are presented in Fig. 9.6.

Gain margin gm = 12.3954
Phase margin pm = 44.8724°
Phase cross-over frequency $\omega_{wpc} = 6.0678$ $\omega_{pc} = 5.8913$ rad/s
Gain cross-over frequency $\omega_{wgc} = 1.2801$ $\omega_{gc} = 1.2783$ rad/s

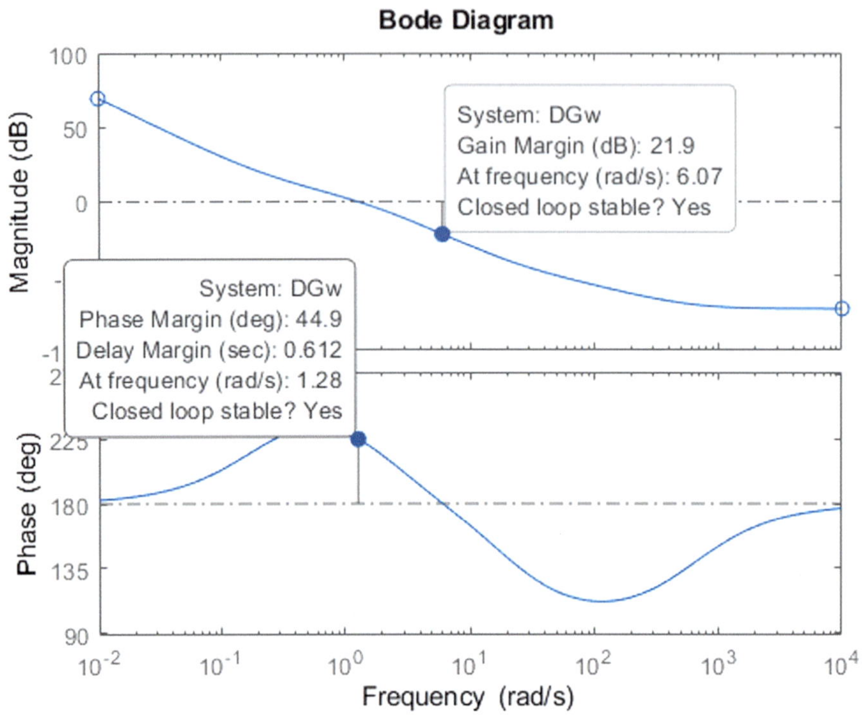

Fig. 9.6 Bode plots for Example 9.2

The gain margin is close to the critical gain obtained in Examples 7.2, 7.4, and 7.8. The frequency of oscillation at the critical gain is close to the phase cross-over frequency. This is also close to the values obtained in Examples 7.2, 7.4, and 7.8.

Matlab code and results are presented next.

```
% Ch9 Ex 9_26 w-plane transfer function of a 2nd order discrete-time system
% with a digital controller
% Gp(s)=3/s(s+2), H(s)=1, D(z)=Kp(1+0.1z/(z-1))
clear all; close all; clc
T=0.1 ; % Sampling time
Gp = tf(3, [1 2 0]) %Gp(s)

% G(z) = Z(Gho(s)Gp(s))
Gz =c2d(Gp,T) % G(z)
Kp = 1
% D(z)=Kp((1.01z-0.99)/(z-1))
Dz = tf(Kp*[1.01 -0.99], [1 -1], T)
DGz =Dz*Gz
[ngz, dgz]=tfdata(DGz, 'v') % num/den form of D(z)G(z)
[zo,po,k]=tf2zp(ngz,dgz) % zero-pole-gain of D(z)G(z)

% DG(w) W(D(z)G(z))
DGw = d2c(DGz,'tustin');  % DG(w)
[ngw, dgw]=tfdata(DGw, 'v') % num/den form of G(w)
printsys(ngw, dgw, 'w')
[zw,pw,k]=tf2zp(ngw,dgw) % zero-pole-gain of DG(w)
```

```
% Bode plot
figure(1)
bode(DGw)
[gm,pm,wpc,wgc]=margin(DGw)
wpcs=(2/T)*atan(wpc*T/2)
wgcs=(2/T)*atan(wgc*T/2)

Results:

DG(z)=

  0.01419 z^2 - 0.0006338 z - 0.01301
  -----------------------------------
  z^3 - 2.819 z^2 + 2.637 z - 0.8187

DG(w)=

   -0.000249 w^3 - 0.14457 w^2 + 2.9611 w + 0.59801
   ------------------------------------------------
      w^3 + 1.9934 w^2 + 6.616e-14 w + 2.7491e-13
```

9.5 Stability Analysis of Discrete-Time Systems Using Nyquist Diagrams

The Nyquist diagram of an open-loop discrete-time system, represented in w-domain, $\overline{GH}(w)$, can be obtained by plotting $\overline{GH}(j\omega_w)$ in the complex plane as shown in Fig. 9.7. Like Bode plots, there are two frequencies of interest.

The gain cross-over frequency, ω_{wgc}, is the frequency at which the magnitude is equal to 1, $\left|\overline{GH}\left(j\omega_{wgc}\right)\right| = 1$, similar to Eq. (9.17). This is the point of intersection of the Nyquist diagram with the unit circle (radius = 1). The corresponding real (α) and imaginary (β) parts of $\overline{GH}\left(j\omega_{wgc}\right)$ can be used to obtain phase angle (ϕ) and phase margin (ϕ_M) (in degrees) similar to Eq. (9.18), as

$$\varphi = \angle\overline{GH}(j\omega_{wgc}) = atan2d(\beta, \alpha) \tag{9.23}$$

$$\phi_M = 180\,° + \angle\overline{GH}\left(j\omega_{wgc}\right) \tag{9.24}$$

The phase cross-over frequency, ω_{wpc}, is the frequency at which the Nyquist plot intersects the negative real axis (180° line), $\angle\overline{GH}\left(j\omega_{wgc}\right) = 180\,°$, similar to Eq. (9.19). The reciprocal of the magnitude $\left|\overline{GH}\left(j\omega_{wpc}\right)\right|$ is known as gain margin, similar to Eq. (9.20), as

$$GM = 1/\left|\overline{GH}\left(j\omega_{wpc}\right)\right| \tag{9.25}$$

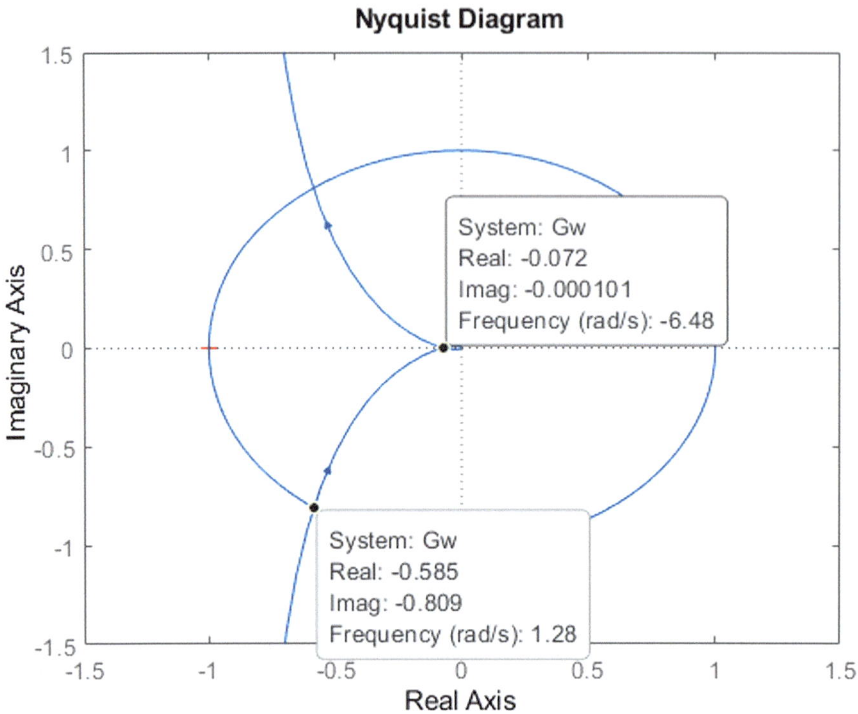

Fig. 9.7 Nyquist diagram for Example 9.3

Example 9.3 A closed-loop discrete-time system, similar to Fig. 9.1, is with the plant transfer function $G_p(s) = \frac{3}{s(s+2)}$, a zero-order hold (ZOH), a digital controller with a simple gain, $D(z) = K$, and a sensor with a unity gain, $H(s) = 1$. Use a sampling time $T = 0.1$ s. Obtain the Nyquist diagram, and find the gain margin, phase margin, gain cross-over frequency, and phase cross-over frequency for stability analysis of the closed-loop discrete-time system.

Solution The open-loop transfer function of the discrete-time system with ZOH is obtained (in Example 6.2) as

$$G(z) = \overline{G_{h0}G_p}(z) = Z\left[\frac{1-e^{-sT}}{s}\,G_p(s)\right]_{e^{sT}=z}$$
$$= \frac{0.01405z + 0.01314}{(z-1)(z-0.8187)}$$

The open-loop transfer function of the discrete-time system in the w-domain using bilinear transformation can be obtained as

$$G(w) = \frac{0.01405z + 0.01314}{(z-1)(z-0.8187)}\Big|_{z=\frac{1+\frac{Tw}{2}}{1-\frac{Tw}{2}}}$$

$$G(w) = \frac{0.01405\left(\frac{1+0.05w}{1-0.05w}\right) + 0.01314}{\left(\frac{1+0.05w}{1-0.05w} - 1\right)\left(\frac{1+0.05w}{1-0.05w} - 0.8187\right)}$$

$$= 2.5014 \times 10^{-4} \frac{(w-20)(w+597.5824)}{w(w+1.9934)}$$

The Nyquist diagram obtained using the Matlab command nyquist() is shown in Fig. 9.7.

Gain cross-over frequency $\omega_{wgc} = 1.28$, $\omega_{gc} = 1.2783$ rad/s
Phase cross-over frequency $\omega_{wpc} = 6.48$, $\omega_{gc} = 6.2665$ rad/s
Gain margin $GM = \frac{1}{|G(j\omega_{wpc})|} = \frac{1}{0.072} = 13.8889$
Phase margin $\phi_M = 180\ ° + a\tan2d(-0.809, -0.585) = 54.1287°$

```
%Ch 9_ex9-3 Nyquist diagram of a 2nd order discrete-time system
% Gp(s)=3/s(s+2), H(s)=1, D(z)=K
clear all; close all; clc
T=0.1 ; % Sampling time
Gp = tf(3, [1 2 0]) %Gp(s)

% G(z) = Z(Gho(s)Gp(s))
Gz =c2d(Gp,T) % G(z)
Gz=zpk(Gz)
[ngz,dgz]=tfdata(Gz,'v');
[z,p,k]=tf2zp(ngz,dgz)

% G(w)
Gw = d2c(Gz,'tustin');
[ngw,dgw] = tfdata(Gw, 'v');
Gw = zpk(Gw);
printsys(ngw,dgw,'w')

%nyquist plot
figure
nyquist(Gw)
hold on
axis('equal')
t=linspace(0,2*pi,100);
x=sin(t); y=cos(t);
plot(x,y,'-')
axis([-1.5 1.5 -1.5 1.5])
wpc =6.48
wpcs=(2/T)*atan(wpc*T/2)
wgc =1.28
wpcs=(2/T)*atan(wpc*T/2)
gm=1/0.072
pm=atand(0.809/0.585)
```

Example 9.4 A closed-loop discrete-time system, similar to Fig. 7.1, is with the plant transfer function $G_p(s) = \frac{3}{s(s+2)}$, a zero-order hold (ZOH), a PI controller $D(z) = K_P\left(1 + \frac{K_I}{K_P}\frac{T}{2}\left(\frac{z+1}{z-1}\right)\right)$, and a sensor with a unity gain, $H(s) = 1$. Use a sampling time $T = 0.1$ s, $\frac{K_I}{K_P} = 0.2$. Obtain the Nyquist diagram, and find the gain margin, phase margin, gain cross-over frequency, and phase cross-over frequency for stability analysis of the closed-loop discrete-time system.

Solution The open-loop transfer function of the discrete-time system with ZOH is obtained (in Example 6.2) as

$$\begin{aligned} G(z) &= \overline{G_{h0}G_p}(z) = Z\left[\frac{1-e^{-sT}}{s}\,G_p(s)\right]_{e^{sT}=z} \\ &= \frac{0.01405z + 0.01314}{(z-1)(z-0.8187)} \end{aligned}$$

The open-loop discrete-time system with the controller transfer function in the w-domain can be expressed using bilinear transformation as

$$DG(w) = K_P\left(\frac{1.01z - 0.99}{z-1}\right)\frac{0.01405z + 0.01314}{(z-1)(z-0.8187)}\Bigg|_{z=\frac{1+\frac{Tw}{2}}{1-\frac{Tw}{2}}}$$

$$DG(w) = K_P\left(2.49\times 10^{-4}\right)\left(\frac{w+0.2}{w}\right)\left(\frac{(w-20)(w+600.4)}{w(w+1.9934)}\right)$$

The Nyquist diagram obtained using the Matlab command nyquist() is shown in Fig. 9.8.

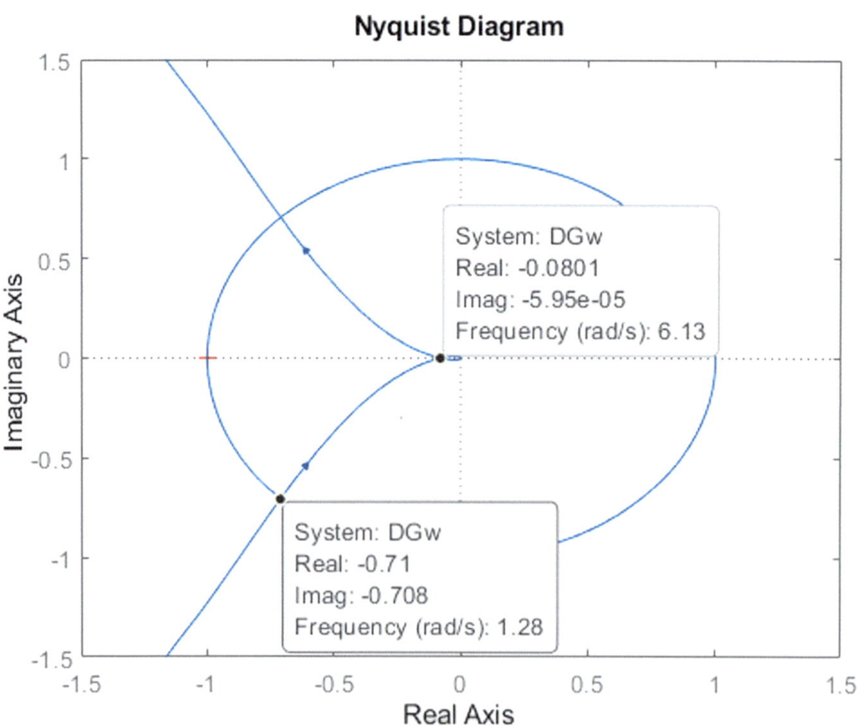

Fig. 9.8 Nyquist diagram for Example 9.4

Gain cross-over frequency $\omega_{wgc} = 1.28$, $\omega_{gc} = 1.2783$ rad/s
Phase cross-over frequency $\omega_{wpc} = 6.13$, $\omega_{gc} = 5.9482$ rad/s
Gain margin $GM = \frac{1}{\left|G\left(j\omega_{wpc}\right)\right|} = \frac{1}{0.0801} = 12.4844$
Phase margin $\phi_M = 180\ ^\circ + a\tan 2d(-0.708, -0.71) = 44.9192°$

```
% Ch9 Ex 9_4 w-plane transfer function of a 2nd order discrete-time system
% with a digital controller
% Gp(s)=3/s(s+2), H(s)=1, D(z)=Kp(1+0.1z/(z-1))
clear all; close all; clc
T=0.1 ; % Sampling time
Gp = tf(3, [1 2 0]) %Gp(s)

% G(z) = Z(Gho(s)Gp(s))
Gz =c2d(Gp,T) % G(z)
Kp = 1
% D(z)=Kp((1.01z-0.99)/(z-1))
Dz = tf(Kp*[1.01 -0.99], [1 -1], T)
DGz =Dz*Gz
[ngz, dgz]=tfdata(DGz, 'v') % num/den form of D(z)G(z)
[zo,po,k]=tf2zp(ngz,dgz) % zero-pole-gain of D(z)G(z)

% DG(w) W(D(z)G(z))
DGw = d2c(DGz,'tustin');  % DG(w)
[ngw, dgw]=tfdata(DGw, 'v') % num/den form of G(w)
printsys(ngw, dgw, 'w')
[zw,pw,k]=tf2zp(ngw,dgw) % zero-pole-gain of DG(w)

% Bode plot
figure(1)
bode(DGw)
[gm,pm,wpc,wgc]=margin(DGw)
wpcs=(2/T)*atan(wpc*T/2)
wgcs=(2/T)*atan(wgc*T/2)%nyquist plot
%nyquist
figure(2)
nyquist(DGw)
hold on
axis('equal')
axis([-1.5 1.5 -1.5 1.5])
t=linspace(0,2*pi,100);
x=sin(t); y=cos(t);
plot(x,y)
wpc =6.13
wpcs=(2/T)*atan(wpc*T/2)
wgc =1.28
wgcs=(2/T)*atan(wgc*T/2)
gm=1/0.0801
pm=atand(0.708/0.71)
```

9.6 Generalized Bode Plots for Discrete-Time Systems

Similar to the generalized root locus presented in Chap. 8, Sect. 8.4, a generalized bode plot analysis can be done in the w-domain representing the generalized characteristic equation for the closed-loop system in terms of redefined gain K' and open-loop system transfer function $\overline{GH}'(w)$:

$$1 + K'\overline{GH}'(w) = 0 \tag{9.26}$$

This modified characteristic equation is used for frequency analysis through Bode plots with the modified open-loop system transfer function $\overline{GH}'(w)$ for positive gain $0 < K' < \infty$. This is called generalized Bode plot analysis.

As an example, let us consider a system with the digital controller $D(z)$ of a simple gain which is replaced with a proportional-integral (PI) controller as

$$D(z) = \left(K_P + K_I \frac{T}{2}\frac{z+1}{z-1}\right) \tag{9.27}$$

where K_P and K_I are the proportional and the integral gains, respectively. The characteristic equation for the closed-loop discrete-time system can be written in the w-domain as

$$1 + \left(K_P + K_I \frac{T}{2}\frac{z+1}{z-1}\right)\overline{GH}(z)\bigg|_{z=\frac{1+\left(\frac{T}{2}\right)w}{1-\left(\frac{T}{2}\right)w}} = 0 \tag{9.28}$$

It is necessary to apply the root locus technique for positive proportional gain $0 < K_P < \infty$ for given K_I. The characteristic equation can be rearranged in the form of Eq. (9.29) as

$$1 + \frac{K_I}{w}\overline{GH}(w) + K_P\overline{GH}(w) = 0 \tag{9.29}$$

$$1 + K_P \frac{\overline{GH}(w)}{1 + \frac{K_I}{w}\overline{GH}(w)} = 0 \tag{9.30}$$

$\overline{GH}(w)$ can be expressed in terms of numerator and denominator polynomials, $N_G(w)$, $D_G(w)$, as

$$\overline{GH}(w) = \frac{N_G(w)}{D_G(w)} \tag{9.31}$$

Equation (9.30) can be rewritten, using Eq. (9.31) and after some rearrangement, as

$$1 + K_P G'(w) = 0 \tag{9.32}$$

where the modified open-loop system transfer function for bode analysis can be expressed as

$$G'(w) = \frac{w N_G(w)}{w D_G(w) + K_I N_G(w)} \tag{9.33}$$

Alternatively, the characteristic Eq. (9.28) can be analyzed in the z-domain to obtain the modified open-loop transfer function in the z-domain as

$$1 + K_p G_z{}'(z) = 0 \tag{9.34}$$

where

$$G_z{}'(z) = \frac{(z-1)N_{Gz}(z)}{(z-1)D_{Gz}(z) + (K_I T/2)(z+1)N_{Gz}(z)} \tag{9.35}$$

$\overline{GH}(z)$ can be expressed in terms of numerator and denominator polynomials, $N_{Gz}(z)$, $D_{Gz}(z)$ as

$$\overline{GH}(z) = \frac{N_{Gz}(z)}{D_{Gz}(z)} \tag{9.36}$$

Next, for stability analysis in the w-domain, the modified open-loop transfer $G_z^{'}(z)$ can be converted to $G^{'}(w)$ using bilinear transformation as

$$G'(w) = G_z{}'(z)\Big|_{z=\frac{1+\left(\frac{T}{2}\right)w}{1-\left(\frac{T}{2}\right)w}} \tag{9.37}$$

The procedure is illustrated using Examples 9.5 and 9.6.

Example 9.5 A closed-loop discrete-time system, similar to Fig 9.1, is with the plant transfer function $G_p(s) = \frac{3}{(s+1)(s+2)}$, a zero-order hold (ZOH), a PI controller $D(z) = \left(K_P + K_I \frac{T}{2} \frac{z+1}{z-1}\right)$, and a sensor with a unity gain, $H(s) = 1$. Use a sampling time $T = 0.1$, $K_I = 0.2$. Analyze the system stability using generalized Bode plots for proportional gain K_P. Obtain gain margin, phase margin, gain cross-over frequency, and phase cross-over frequency for stability analysis of the closed-loop discrete-time system.

Solution The open-loop transfer function of the discrete-time system with ZOH is obtained (in Example 6.2) as

$$\begin{aligned}\overline{GH}(z) &= \overline{G_{h0}G_p}(z) = Z\left[\tfrac{1-e^{-sT}}{s}\,G_p(s)\right]_{e^{sT}=z} \\ &= \frac{0.01358z + 0.01229}{z^2 - 1.7236z + 0.7408}\end{aligned}$$

The open-loop discrete-time system with the controller transfer function can be expressed in the z-domain, as

$$\begin{aligned}D(z)\overline{GH}(z) &= \left(K_P + K_I\frac{T}{2}\frac{z+1}{z-1}\right)\frac{0.01358z + 0.01229}{z^2 - 1.7236z + 0.7408} \\ &= \left(K_P + 0.01\frac{z+1}{z-1}\right)\frac{0.01358z + 0.01229}{z^2 - 1.7236z + 0.7408}\end{aligned}$$

The open-loop discrete-time system with the controller transfer function can be expressed in the w-domain, as

$$D(w)\overline{GH}(w) = (K_P + K_I/w)\left.\frac{0.01358z + 0.01229}{z^2 - 1.7236z + 0.7408}\right|_{z=\frac{1+\left(\frac{T}{2}\right)w}{1-\left(\frac{T}{2}\right)w}}$$

$$\overline{GH}(w) = \frac{-0.00037313\, w^2 - 0.14191\, w + 2.9876}{w^2 + 2.9925w + 1.9917}$$

Using $N_G(w) = -0.00037313\ w^2 - 0.14191\ w + 2.9876$ and $D_G(w) = w^2 + 2.9925w + 1.9917$, the modified transfer function can be expressed as

$$G'(w) = \frac{wN_G(w)}{wD_G(w) + K_I N_G(w)} = \frac{-0.00037313\, w^3 - 0.14191\, w^2 + 2.9876w}{w^3 + 2.992\ w^2 + 1.963\, w + 0.5975}$$

Bode plots obtained using Matlab are presented in Fig. 9.9.

Gain margin gm = 21.0218 (26.5 dB)
Phase margin pm = 99.2171°

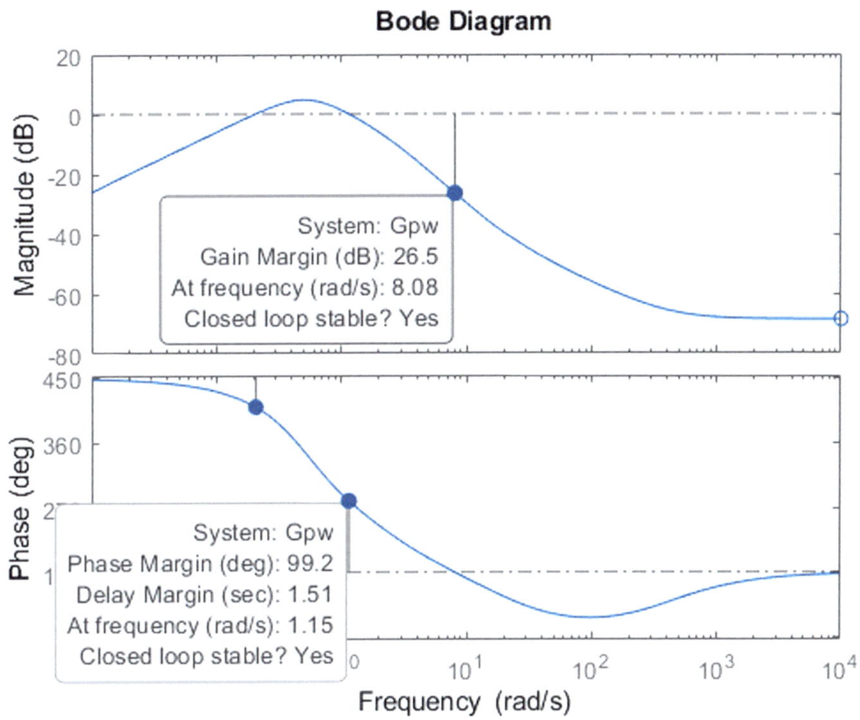

Fig. 9.9 Bode plots for Example 9.5

Phase cross-over frequency $\omega_{wpc} = 8.0795$ $\omega_{pc} = 7.6786$ rad/s
Gain cross-over frequency $\omega_{wgc} = 1.1505$ $\omega_{gc} = 1.1492$ rad/s

If the modified z-domain transfer function $G'(z)$ is converted to $G'(w)$ directly using bilinear transformation instead of converting $D(z)$ and $\overline{GH}(z)$ individually and getting the modified transfer function $G'(w)$, the results are the same.

The results are close to the values obtained in Example 8.3 using the root locus technique.

Matlab code and results are presented next.

```
%Ch9 Ex9_5 Generalized Bode plot
clear all; close all;clc
T=0.1 ; % Sampling time
Gp = tf(3, [1  3  2 ]) %Gp(s)

% G(z) = Z(Gho(s)Gp(s))
Gz =c2d(Gp,T) % G(z)
[ngz, dgz]=tfdata(Gz,'v')
%G(w)
Gw = d2c(Gz,'tustin')
[ngw, dgw]=tfdata(Gw,'v')
printsys(ngw, dgw, 'w')

KI=0.2

% open-loop TF DG(z)
ngwp=[ngw 0]
dgwp=[dgw 0] +KI*[0 ngw]
zp=roots(ngwp)
pp=roots(dgwp)
r=abs(pp)
Gpw=tf(ngwp, dgwp)

figure
bode(Gpw)
[gm,pm,wpc,wgc]=margin(Gpw)
wgcs=(2/T)*atan(wgc*T/2)
wpcs=(2/T)*atan(wpc*T/2)
%
figure
nyquist(Gpw)

% open-loop TF DG(z)
zm1=[1 -1]; zp1=[1 1]
ngzp=conv(ngz, zm1)
dgzp=conv(dgz,zm1)+(KI*T/2)*conv(ngz,zp1)

zp=roots(ngzp)
pp=roots(dgzp)
r=abs(pp)
Gpz=tf(ngzp, dgzp, T)
% bode plot
Gpw = d2c(Gpz,'tustin');
[ngwp, dgwp]=tfdata(Gpw,'v');
printsys(ngwp,dgwp,'w')
figure
bode(Gpw)
[gm,pm,wpc,wgc]=margin(Gpw)
wgcs=(2/T)*atan(wgc*T/2)
wpcs=(2/T)*atan(wpc*T/2)
% step response
Kp=gm
Tz=feedback(Kp*Gpz, 1)

[ntz,dtz]=tfdata(Tz,'v')
Figure
Nyquist(Gpw)
dstep(Tz,1)
% % Nyquist plot
figure
nyquist(Gpw)
```

Example 9.6 A closed-loop discrete-time system, similar to Fig. 9.1, is with the plant transfer function $G_p(s) = \frac{3}{(s+1)(s+2)}$, a zero-order hold (ZOH), a PI controller $D(z) = \left(K_P + K_I \frac{T}{2} \frac{z+1}{z-1}\right)$, and a sensor with a unity gain, $H(s) = 1$. Use a sampling time $T = 0.1$, $K_I = 0.2$. Analyze the system stability using a generalized Nyquist diagram for proportional gain K_P. Obtain gain margin, phase margin, gain cross-over frequency, and phase cross-over frequency for stability analysis of the closed-loop discrete-time system.

Solution The modified transfer function in the w-domain obtained in Example 9.5 is

$$G'(w) = \frac{-0.00037313\, w^3 - 0.14191\, w^2 + 2.9876w}{w^3 + 2.992\ w^2 + 1.963\, w + 0.5975}$$

The Nyquist diagram obtained using the Matlab command *nyquist()* is shown in Fig. 9.10.

Gain cross-over frequency $\omega_{wgc} = 1.15$, $\omega_{gc} = 1.149$ rad/s
Phase cross-over frequency $\omega_{wpc} = 8.53$, $\omega_{gc} = 8.063$ rad/s
Gain margin $GM = \frac{1}{|G(j\omega_{wpc})|} = \frac{1}{0.0437} = 22.88$
Phase margin $\phi_M = 180\ ° + a\tan 2d(-0.987, 0.176) = 100.1°$

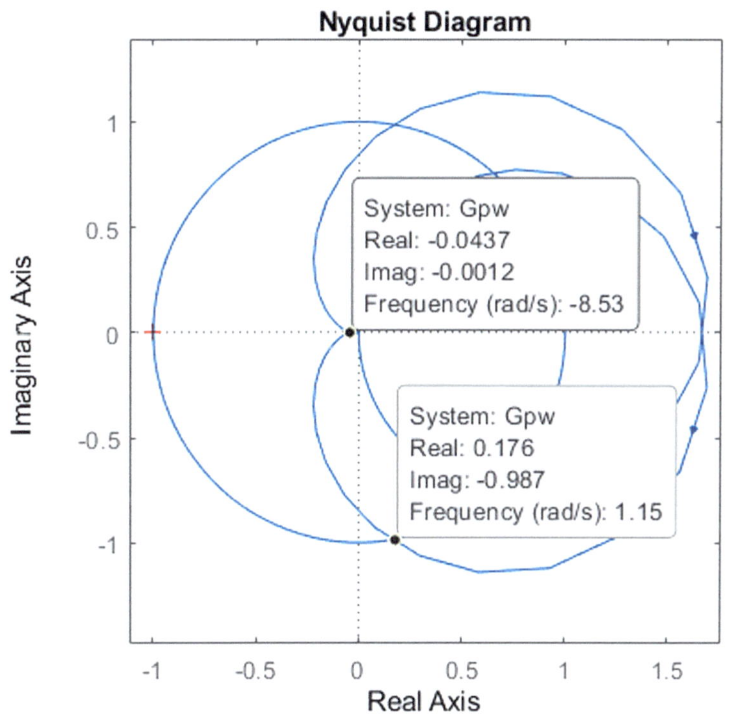

Fig. 9.10 Nyquist diagram for Example 9.6

9.7 Laboratory Experiment for Stability Analysis of Discrete-Time Systems Using Frequency Domain Techniques

Objectives The objectives are to determine the range of proportional gain for stability using frequency domain techniques and validate through an experiment on a physical DC motor module.

Introduction There are different methods of analyzing the stability of a dynamic system. In this lab, some of these techniques in the frequency domain (w-domain), namely, the Routh-Hurwitz criterion, Nyquist criterion, and Bode plots for stability, are studied using the model parameters for the DC motor. These results are validated through simulation and actual experimentation with the DC motor module.

I. *Stability Analysis and Simulation*

A. *Open-Loop System Transfer Function in w-Domain*

1. The transfer function $G(s)$ for the speed dynamics of the DC motor using the identified parameters, static gain K and time constant τ, can be expressed as

$$G(s) = \frac{K}{\tau s + 1} \tag{9.38}$$

2. Using the speed transfer function, the position transfer function $G_p(s)$ is obtained as

$$G_p(s) = \frac{K}{s(\tau s + 1)} \tag{9.39}$$

3. Use $G_p(s)$ and a proper sampling time ($t_s \approx \tau/10$) to find $G(z)$, the z-domain transfer function for the position of the DC motor.
4. Convert discrete-time transfer function for the DC motor position model $G(z)$ to $G(w)$ in w-domain.

B. *Routh-Hurwitz Criterion*

1. For a proportional controller with transfer function $D(w) = K_P$, the closed-loop transfer function can be expressed as

$$T(w) = \frac{D(w)G(w)}{1 + D(w)G(w)} = \frac{K_P G(w)}{1 + K_P G(w)} \tag{9.40}$$

2. The characteristic equation for the closed-loop discrete-time system can be expressed as

$$1 + K_P G(w) = 0 \tag{9.41}$$

3. Obtain a range of gain K_P using the Routh-Hurwitz criterion for stability analysis for the closed-loop system.
4. Obtain the upper limit of the positive gain $K_P = K_u$ for stability and the corresponding closed-loop poles in the w-domain. Estimate the frequency of oscillation from the closed-loop poles.
5. Complete the corresponding columns in Tables 9.1 and 9.2.

C. *Nyquist Diagram*

1. Obtain Nyquist plot for the system in w-domain, using the open-loop system transfer function $G(w)$.
2. Locate the gain cross-over point (point of intersection of Nyquist plot with the unit circle). Note down the real part (α) and the frequency (ω_{wgc}) at the gain cross-over point. Use these values to obtain gain margin ($GM = 1/|\alpha|$) and gain cross-over frequency in s-domain (ω_{gc}).
3. Locate phase cross-over point (point of intersection of Nyquist plot with the negative real axis: 180° line). Note down the real (α) and imaginary

Table 9.1 Summary of results for stability analysis in w-domain

Routh-Hurwitz criterion		Nyquist criterion				Bode plots			
K_u	Closed-loop poles	GM	ϕ_M	ω_{gc} (rad/s)	ω_{pc} (rad/s)	GM	ϕ_M	ω_{gc} (rad/s)	ω_{pc} (rad/s)

Table 9.2 Summary of results for stability analysis in w-domain

	Routh-Hurwitz criterion	Nyquist criterion	Bode plots	Closed-loop system step response: Simulation	Closed-loop system step response: Experimental
K_u					
ω_{osc} (rad/s)					

(β) parts and the frequency (ω_{wpc}). Use these values to obtain phase margin $\varphi = \angle\overline{GH}(j\omega_{wgc}) = atan2d(\beta, \alpha)$,

$$\phi_M = 180^\circ + \angle\overline{GH}(j\omega_{wgc})$$

4. Complete the corresponding columns in Tables 9.1 and 9.2.

D. *Bode Plots*

1. Obtain Bode plots using the open-loop transfer function in w-domain $G(w)$.
2. Obtain stability margins. In Matlab, this can be done by right-clicking on any of the Bode plots → characteristics → all stability margins and placing the cursor at the gain and phase margin points.
3. Note down gain margin (GM, dB), phase margin (ϕ_M), gain cross-over frequency (ω_{wgc}), and phase cross-over frequency (ω_{wpc}). Use these frequencies to obtain gain and phase cross-over frequencies in s-domain (ω_{gc}, ω_{pc}).
4. Complete the corresponding columns in Tables 9.1 and 9.2.

II. *Experimental Validation*

1. Set up NI ELVIS II board and DC motor module. Select the device and set the sampling rate. For the DC motor control trainer (DCMCT) module, a sampling rate of 100 Hz is used.
2. Study the block diagram for the position control experiment (*01-QNET_DCMCT_Position_Control.vi*).
3. Identify the different components of the control system, including the input, error, controller, plant, sensor, and output.
4. Set the controller parameters to $K_p = K_u$, $K_d = 0$, $K_i = 0$, and change the input waveform frequency to a small value so that it behaves like a step input. Alternatively, change the block diagram to use a step input.
5. Select your input parameters. For the DCMCT module, typical values for the input square wave voltage are as follows: amplitude of 1.0 V, frequency of 0.40 Hz, and offset of 3.0 V.
6. Run VI, and observe the system response.
7. Make sure to set the number of significant figures to at least 5 for the display format of each graph on the front panel VI (right-click on the graph → display format → decimal → 6). Export the time response graph, motor rotation $\theta(t)$, to Excel by right-clicking on the graph and selecting the Export to Excel option (right-click → Export to Excel). Compare it to the simulated response.
8. Estimate the frequency ω and period T of the response, and compare these values to the ones found in Section I.
9. Complete the corresponding columns in Tables 9.1 and 9.2.

III. *Deliverables*

1. Complete Tables 9.1 and 9.2.
2. Include a Nyquist diagram showing all relevant data tips.
3. Include Bode plots showing all stability margins, gain, and phase cross-over frequencies.
4. Include a closed-loop response graph for $K_p = K_u$ from simulation.
5. Include a closed-loop response graph for $K_p = K_u$ from experiment.

Lab Report
Present all deliverables in the report and discuss results.

Exercises

9.1 A closed-loop discrete-time system, Fig. P9.1, is with the plant transfer function $G_p(s) = \frac{3}{(s+1)(s+2)}$; a zero-order hold (ZOH); a digital controller with a simple gain, $D(z) = K$; and a sensor with a unity gain, $H(s) = 1$. Use a sampling time $T = 0.1$ s. Obtain the transfer function of the open-loop system in the z-domain, $G(z)$, and in the w-domain, $G(w)$. Obtain Bode plots, obtain gain and phase margins, gain cross-over frequency, and phase cross-over frequency.

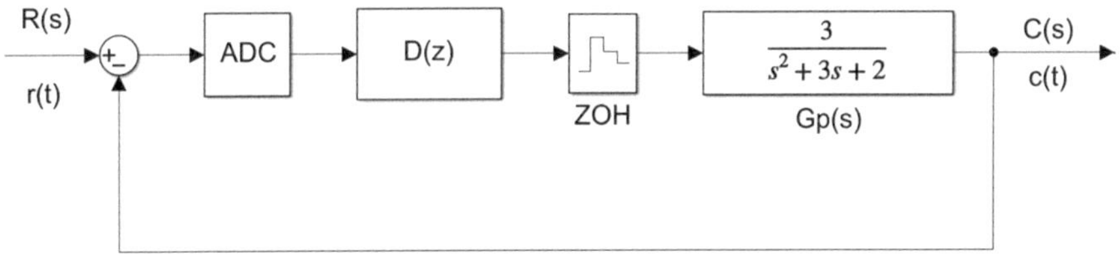

Fig. P9.1 Closed-loop discrete-time system for Problem 9.1

9.2 A closed-loop discrete-time system, Fig. P9.2, is with the plant transfer function $G_p(s) = \frac{2}{(s+1)(s+2)}$, a zero-order hold (ZOH), a PI controller $D(z) = K_P\left(1 + \frac{K_I}{K_P}\frac{T}{2}\left(\frac{z+1}{z-1}\right)\right)$, and a sensor with a unity gain, $H(s) = 1$. Use a sampling time $T = 0.1$ s, $\frac{K_I}{K_P} = 0.2$. Analyze the system stability using Bode plots. Obtain gain and phase margins, gain cross-over frequency, and phase cross-over frequency.

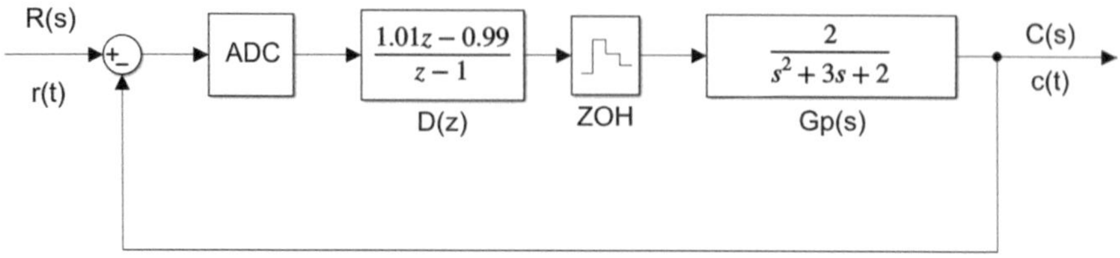

Fig. P9.2 Closed-loop discrete-time system for Problem 9.2

9.3 A closed-loop discrete-time system, Fig. P9.1, is with the plant transfer function $G_p(s) = \frac{3}{(s+1)(s+2)}$; a zero-order hold (ZOH); a digital controller with a simple gain, $D(z) = K$; and a sensor with a unity gain, $H(s) = 1$. Use a sampling time $T = 0.1$ s. Obtain the Nyquist diagram, and find the gain margin, phase margin, gain cross-over frequency, and phase cross-over frequency for stability analysis of the closed-loop discrete-time system.

9.4 A closed-loop discrete-time system, similar to Fig. P9.2, is with the plant transfer function $G_p(s) = \frac{2}{(s+1)(s+2)}$, a zero-order hold (ZOH), a PI controller $D(z) = K_P\left(1 + \frac{K_I}{K_P}\frac{T}{2}\left(\frac{z+1}{z-1}\right)\right)$, and a sensor with a unity gain, $H(s) = 1$. Use a sampling time $T = 0.1$ s, $\frac{K_I}{K_P} = 0.2$. Obtain the Nyquist diagram, and find the gain margin, phase margin, gain cross-over frequency, and phase cross-over frequency for stability analysis of the closed-loop discrete-time system.

9.5 A closed-loop discrete-time system, similar to Fig. P9.3, is with the plant transfer function $G_p(s) = \frac{14}{(s+3)(s+4)}$, a zero-order hold (ZOH), a PI controller $D(z) = \left(K_P + K_I\frac{T}{2}\frac{z+1}{z-1}\right)$, and a sensor with a unity gain, $H(s) = 1$. Use a sampling time $T = 0.1$, $K_I = 0.2$. Analyze the system stability using generalized Bode plots and Nyquist diagrams. Obtain gain and phase margins, gain cross-over frequency, and phase cross-over frequency.

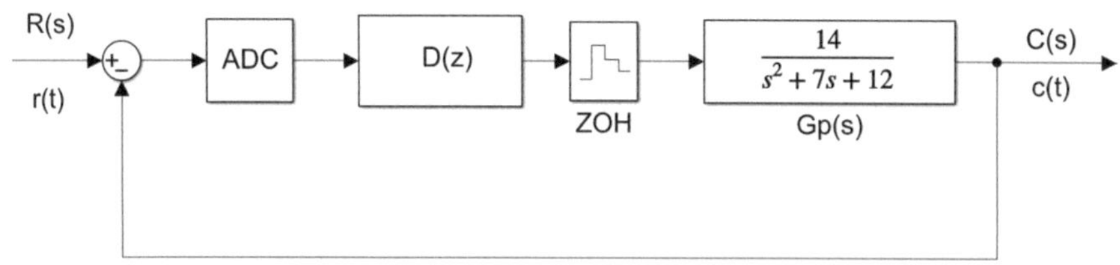

Fig. P9.3 Closed-loop discrete-time system for Problem 9.5

9.6 A closed-loop discrete-time system, Fig. P9.4, is with the plant transfer function $G_p(s) = \frac{14}{(s+3)(s+4)}$, a zero-order hold (ZOH), a PI controller $D(z) = \left(K_P + K_I\frac{T}{2}\frac{z+1}{z-1}\right)$, and a sensor with a unity gain, $H(s) = 1$. Use a sampling time $T = 0.1$, $K_I = 0.4$. Analyze the system stability using the Nyquist diagram. Obtain gain and phase margins, gain cross-over frequency, and phase cross-over frequency.

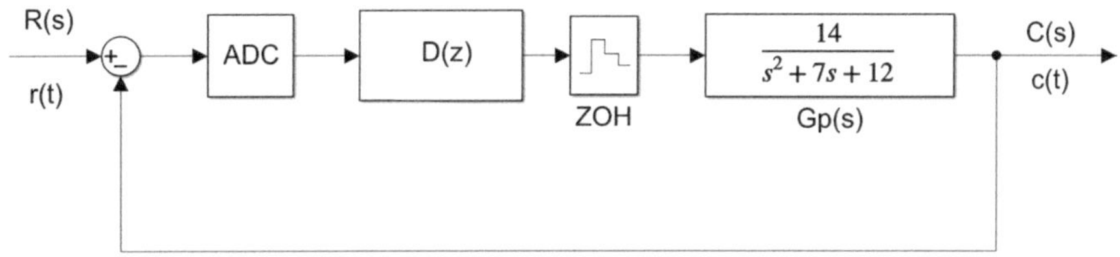

Fig. P9.4 Closed-loop discrete-time system for Problem 9.6

9.7 The closed-loop vehicle distance control system for a traffic control system can be represented as shown in Fig. P9.5. The plant transfer function of the distance control system can be represented in s-domain as $G_p(s) = \frac{1.5}{s^2+12s+18}$. The digital controller is a simple gain, $D(z) = K$. The sampling time is given as $T = 0.10$ s. Analyze the system stability using Bode plots and Nyquist diagrams. Obtain gain and phase margins, gain cross-over frequency, phase cross-over frequency, and the closed-loop step response for K equal to half of the original gain margin.

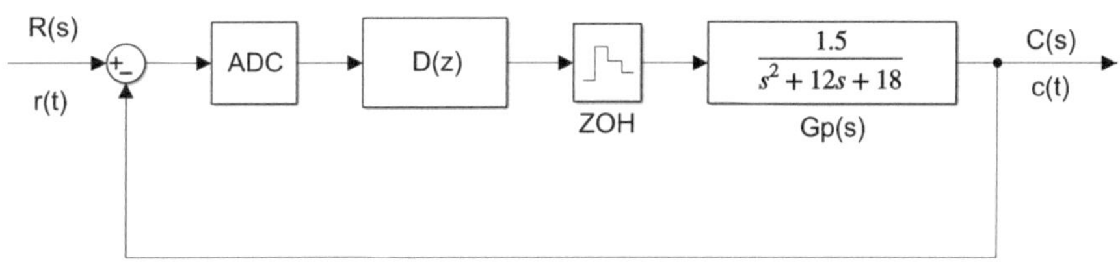

Fig. P9.5 Closed-loop discrete-time system for Problem 9.7

9.8 The closed-loop roll angle control system for a jet airplane can be represented as shown in Fig. P9.6. The plant transfer function of the roll control system is represented in the s-domain as $G_p(s) = \frac{10}{(s+10)(s^2+2.5s+25)}$. The digital controller is a simple gain, $D(z) = K$. The sampling time is given as $T = 0.05$ s. Analyze the system stability using Bode plots and Nyquist diagrams. Obtain gain and phase margins, gain cross-over frequency, phase cross-over frequency, and the closed-loop step response for K equal to one-third of the original gain margin.

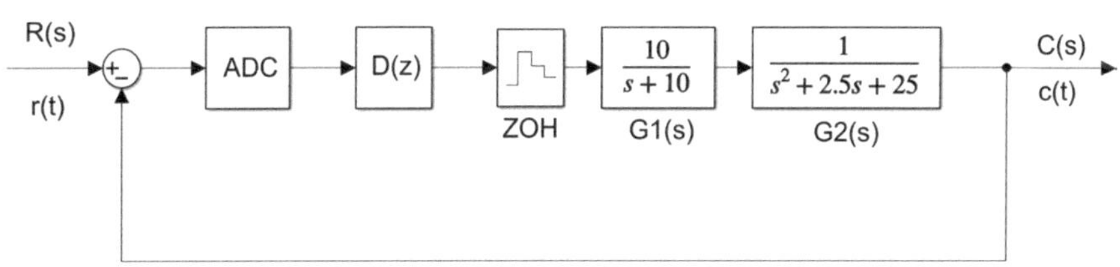

Fig. P9.6 Closed-loop discrete-time system for Problem 9.8

9.9 The closed-loop steering control system for a second-order system with a time delay can be represented as shown in Fig. P9.7, $G_p(s) = \frac{1}{(s+1)(s+3)}e^{-sT}$. The digital controller is a simple gain, $D(z) = K$. The sampling time is given as $T = 0.02$ s. Analyze the system stability using Bode plots and Nyquist diagrams. Obtain gain and phase margins, gain cross-over frequency, phase cross-over frequency, and the closed-loop step response for K equal to half of the original gain margin.

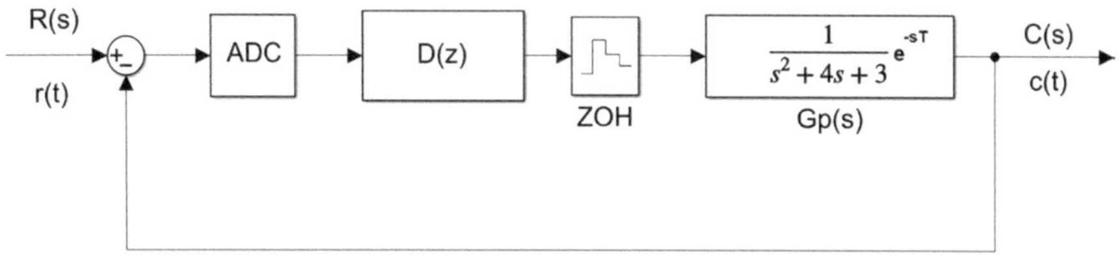

Fig. P9.7 Closed-loop discrete-time system for Problem 9.9

9.10 The closed-loop thickness control system in a metal processing plant can be represented as shown in Fig. P9.8. The plant transfer function of the thickness control system is represented in the s-domain as $G_p(s) = \frac{5}{s(s+2)(s^2+s+10)}$. The digital controller is a simple gain, $D(z) = K$. The sampling time is given as $T = 0.10$ s. Analyze the system stability using Bode plots and Nyquist diagrams. Obtain gain and phase margins, gain cross-over frequency, phase cross-over frequency, and the closed-loop step response for K equal to one-third of the original gain margin.

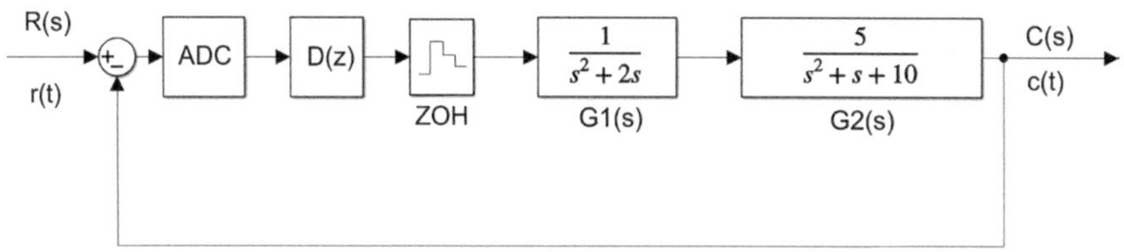

Fig. P9.8 Closed-loop discrete-time system for Problem 9.10

9.11 The closed-loop tilt control system in a high-speed train for passenger comfort can be represented as shown in Fig. P9.9. The plant transfer function of the train tilt control system is represented in the s-domain as $G_p(s) = \frac{1}{s(s+2)(s+3)}$. The digital controller is a simple gain, $D(z) = K$. The sampling time is given as $T = 0.10$ s. Analyze the system stability using Bode plots and Nyquist diagrams. Obtain gain and phase margins, gain cross-over frequency, phase cross-over frequency, and the closed-loop step response for K equal to half of the original gain margin.

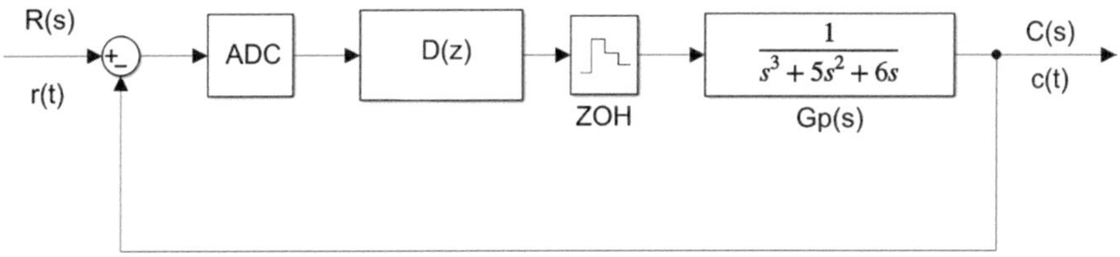

Fig. P9.9 Closed-loop discrete-time system for Problem 9.11

9.12 The closed-loop position control system for one of the axes of an experimental automated surface-grinding wheel can be represented as shown in Fig. P9.10. The plant transfer function of the axis position control system can be represented in the s-domain as $G_p(s) = \frac{5}{s(0.2s+1)(0.3s+1)}$ [4]. The digital controller is a simple gain, $D(z) = K$. The sampling time is given as $T = 0.05$ s. Analyze the system stability using Bode plots and Nyquist diagrams. Obtain gain and phase margins, gain cross-over frequency, phase cross-over frequency, and the closed-loop step response for K equal to one-third of the original gain margin.

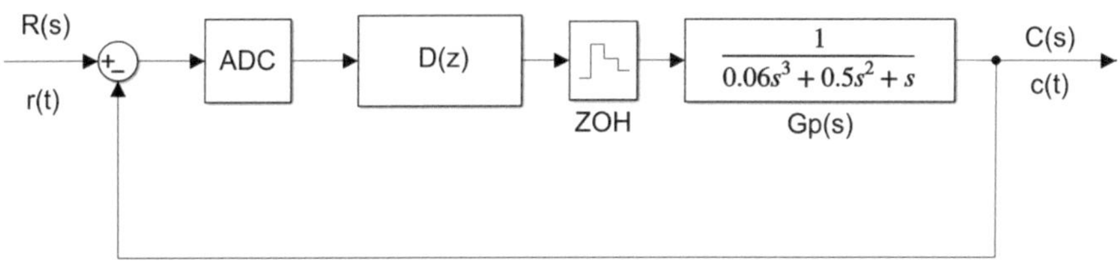

Fig. P9.10 Closed-loop discrete-time system for Problem 9.12

Further Reading

1. Asadi F (2023) Signals and systems with MATLAB® and Simulink®. Springer, Cham
2. Astrom KJ, Murray RM (2021) Feedback systems: an introduction for scientists and engineers, 2nd edn. Princeton University Press, Princeton
3. Chen C-T (2006) Analog and digital control system design: transfer-function, state-space, and algebraic methods. Oxford University Press, New York
4. Dorf R, Bishop R (2016) Modern control systems, 13th edn. Pearson, Essex
5. Fadali MS, Visioli A (2020) Digital control engineering analysis and design, 3rd edn. Academic Press, London
6. Franklin GF, Powell JD, Emami-Naeini A (2019) Feedback control of dynamic systems, 8th edn. Pearson, Upper Saddle River
7. Franklin GF, Powell JD, Workman M (1998) Digital control of dynamic systems, 3rd edn. Addison-Wesley, Boston
8. Isermann R (1981) Digital control systems. Springer, Berlin
9. Jalili N, Candelino NW (2023) Dynamic systems and control engineering. Cambridge University Press, Cambridge
10. Keviczky L, Bars R, Hetthéssy J, Bányász C (2019) Control engineering. Springer, Singapore

11. Koeing DM (2009) Practical control engineering: a guide for engineers, managers, and practitioners. McGraw-Hill, New York
12. Kuo BC (1995) Digital control systems, 2nd edn. Oxford University Press, Oxford
13. Landau ID, Gito G (2006) Digital control systems: design, identification and implementation. Springer, London
14. Leigh JR (2006) Applied digital control: theory, design and implementation, 2nd edn. Dover, Mineola
15. Mathworks (2024) Control System Toolbox. https://www.mathworks.com/products/control.html
16. Mathworks (2024) Matlab. https://www.mathworks.com/products/matlab.html
17. National Instruments (2024) LabVIEW. https://www.ni.com/en/support/downloads/software-products/download.labview.html#487445
18. National Instruments (2024) NI-ELVIS-II. https://www.ni.com/en-us/support/model.ni-elvis-ii.html
19. National Instruments (2024) Labs for Quanser QNET DC Motor Add-On Board. https://education.ni.com/teach/resources/180/labs-for-quanser-qnet-dc-motor-add-on-board
20. Nise NS (2019) Control systems engineering, 8th edn. Wiley, Hoboken
21. Nof SY (ed) (2009) Springer handbook of automation. Springer, Berlin
22. Ogata K (1995) Discrete-time control systems, 2nd edn. Pearson, Hoboken
23. Phillips CL, Nagle HT, Chakrabortty A (2015) Digital control system analysis and design, 4th edn. Pearson, Essex
24. Quanser (2024) QUBE Servo 2. https://www.quanser.com/products/qube-servo-2/
25. Rabbath CA, Lechevin N (2014) Discrete-time control system design with applications. Springer, New York
26. Samanta B (2023) Introduction to mechatronics – an integrated approach. Springer, Cham
27. Shertukde HM (2015) Digital control applications illustrated with MATLAB®. CRC Press, Boca Raton
28. Starr G (2020) Introduction to applied digital controls. Springer, Cham
29. Veloni A, Miridakis NI (2018) Digital control systems-theoretical problems and simulation tools. CRC Press, Boca Raton
30. Vilanova R, Visioli A (2012) PID control in the third millennium – lessons learned and new approaches. Springer, London
31. Visioli A (2006) Practical PID control. Springer, London

Chapter 10
Discrete-Time System Performance Specifications for Controller Design

10.1 Introduction

Discrete-time system performance specifications for controller design are presented in this chapter. Control system performance specifications of steady-state accuracy, transient response characteristics of overshoot and settling time, and frequency domain specifications of stability margins are covered. The relationships between time domain and frequency domain specifications are discussed. Examples are presented along with Matlab codes to illustrate time and frequency domain specifications. End-of-chapter exercises are provided to enhance understanding of covered topics and consolidate problem-solving skills.

10.2 Steady-State Accuracy

The steady-state accuracy of a closed-loop discrete-time system depends on the type of the system and the type of input, as discussed in Sect. 6.5 and summarized in Table 6.1. The steady-state accuracy specification for step, ramp, or parabolic inputs can be used to get information about system gain.

For a system with a specification on the steady-state error due to a step input, the following relation can be used:

$$e_{ss,\text{step}} = \frac{1}{1 + K_p} \tag{10.1}$$

where

B. Samanta, *Introduction to Digital Control*,
https://doi.org/10.1007/978-3-031-66830-2_10

$$k_p = \lim_{z\to 1}(D(z)G(z)) \tag{10.2}$$

Similarly for a specification on the steady-state error due to a ramp input, the corresponding relations from Sect. 6.5 can be used:

$$e_{ss,\text{ramp}} = \frac{1}{k_v} \tag{10.3}$$

where

$$k_v = \frac{1}{T}\lim_{z\to 1}((z-1)D(z)G(z)) \tag{10.4}$$

Similarly for a specification on the steady-state error due to a parabolic input, the corresponding relations from Sect. 6.5 can be used:

$$e_{ss,\text{par}} = \frac{1}{k_a} \tag{10.5}$$

where

$$k_a = \frac{1}{T^2}\lim_{z\to 1}\left((z-1)^2D(z)G(z)\right) \tag{10.6}$$

10.3 Time Domain Specifications

Desired time domain transient response characteristics of a closed-loop discrete-time system can be specified assuming the system response can be approximated by a second-order system using two dominant poles. The second-order system approximation can be expressed in the s-domain, using closed-loop transfer function *T(s)*, as

$$T(s) = \frac{\omega_n^2}{s^2 + 2\zeta\omega_n s + \omega_n^2} \tag{10.7}$$

where ζ and ω_n are two system parameters known as damping ratio and undamped natural frequency, respectively.

Two commonly used time domain transient response specifications include overshoot (α) and settling time (t_{set}). Using relations developed in Chap. 6, the overshoot (α) can be used to obtain the desired damping ratio (ζ):

$$\alpha = e^{\frac{-\pi\zeta}{\sqrt{1-\zeta^2}}} \tag{10.8}$$

$$\zeta = -\frac{\frac{\ln(\alpha)}{\pi}}{\sqrt{1+\left(\frac{\ln(\alpha)}{\pi}\right)^2}} \tag{10.9}$$

Similarly, the relation developed in Chap. 6 can be used to obtain the desired undamped natural frequency (ω_n) for the specified settling time (t_{set}) as

$$t_{\text{set}} = \frac{4}{\zeta\omega_n} \tag{10.10}$$

$$\omega_n = \frac{4}{\zeta\, t_{\text{set}}} \tag{10.11}$$

The damping ratio and the undamped natural frequency can bc used to get the desired closed-loop poles in the s-plane as

$$s_{1,2} = -\zeta\omega_n \pm j\omega_n\sqrt{1-\zeta^2} \tag{10.12}$$

The corresponding desired closed-loop poles in the z-plane can be obtained, assuming T as the sampling time, as

$$z_{1,2} = e^{sT}\Big|_{s=s_{1,2}} = e^{-\zeta\omega_n T} e^{\pm j\omega_{nT}\sqrt{1-\zeta^2}} = r\angle \pm\theta \tag{10.13}$$

where

$$r = e^{-\zeta\omega_n T} \tag{10.14}$$

$$\theta = T\omega_n\sqrt{1-\zeta^2} \tag{10.15}$$

10.4 Frequency Domain Specifications

Desired frequency domain characteristics for a discrete-time system are specified in terms of gain margin (GM), phase margin (ϕ_m), and often gain cross-over frequency (ω_{gc}) as

$$GM \geq GM_{\text{desired}} \tag{10.16}$$

$$\phi_m \geq \phi_{m,\text{ desired}} \tag{10.17}$$

10.5 Relationship Between Specifications in Time and Frequency Domains

The relationship between phase margin and damping ratio of a closed-loop system can be derived using the open-loop continuous-time system, modeled as a dominantly second-order system, $G(s)$, as

$$G(s) = \frac{{\omega_n}^2}{s^2 + 2\zeta\omega_n s} \tag{10.18}$$

The frequency response of the open-loop system can be analyzed by substituting $s = j\omega$ as

$$G(j\omega) = \frac{{\omega_n}^2}{-\omega^2 + j2\zeta\omega_n\omega} \tag{10.19}$$

Equation (10.19) can be expressed in terms of frequency ratio:

$$r = \frac{\omega}{\omega_n} \tag{10.20}$$

$$G(j\omega) = \frac{1}{-r^2 + j2\zeta r} \tag{10.21}$$

Magnitude and angle relations can be expressed as

$$|G(j\omega)| = \frac{1}{\sqrt{r^4 + (2\zeta r)^2}} \tag{10.22}$$

$$\angle G(j\omega) = \tan^{-1}\left(\frac{-2\zeta r}{-r^2}\right) = 180^\circ + \tan^{-1}\left(\frac{2\zeta}{r}\right) \tag{10.23}$$

At the gain cross-over frequency (ω_{gc}), the magnitude can be expressed as

$$|G(j\omega_{gc})| = \frac{1}{\sqrt{r^4 + (2\zeta r)^2}} = 1 \tag{10.24}$$

It can be expressed as

$$r^4 + (2\zeta r)^2 - 1 = 0 \tag{10.25}$$

The solution for a real positive frequency ratio can be obtained as

$$r = \sqrt{\left(\sqrt{4\zeta^4 + 1}\right) - 2\zeta^2} \tag{10.26}$$

Phase margin (in degrees) can be expressed as

$$\phi_m = 180^\circ + \tan^{-1}\left(\frac{2\zeta}{r}\right) - 180^\circ = \tan^{-1}\left(\frac{2\zeta}{\sqrt{\left(\sqrt{4\zeta^4 + 1}\right) - 2\zeta^2}}\right) \tag{10.27}$$

The exact relation (10.27) and approximate linear relation (10.28) are plotted in Fig. 10.1:

$$\phi_m \approx 100\zeta \tag{10.28}$$

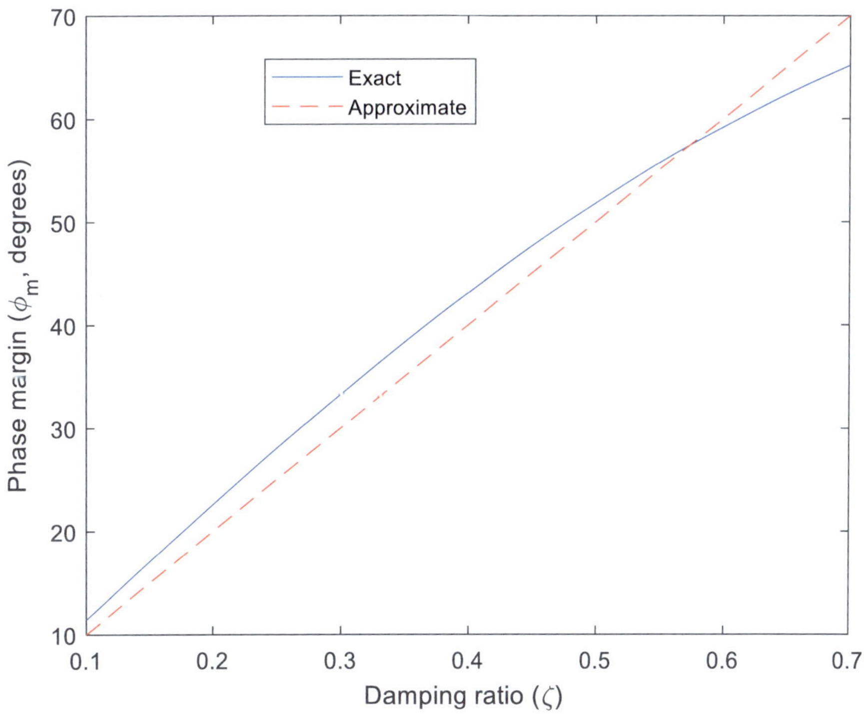

Fig. 10.1 Exact and approximate relations between phase margin and damping ratio

The phase margin (ϕ_m) can be related to settling time (t_{set}) combining (10.29) and (10.30):

$$\tan(\phi_m) = \frac{2\zeta}{r} = \frac{2\zeta\omega_n}{\omega_{gc}} \tag{10.29}$$

$$t_{set} = \frac{4}{\zeta\omega_n} \tag{10.30}$$

$$t_{set} = \frac{4}{\zeta\omega_n} = \frac{8}{\omega_{gc}\tan(\phi_m)} \tag{10.31}$$

Example 10.1 It is required to design a controller such that the closed-loop system step response has a percentage overshoot of 5% and a settling time of 2 s. Find the damping ratio and undamped natural frequency for the closed-loop system. Obtain closed-loop poles in the s- and z-domains. Use sampling time $T = 0.10$ s. Compare the values of percentage overshoot and settling time from the closed-loop step response of the system, assumed to be predominantly of second-order, in continuous time and discrete time.

Solution Given overshoot $\alpha = 5\ \% \ = 0.05$, settling time $t_{set} = 2$ s, and sampling time $T = 0.10$ s.

Using Eq. (10.9), the damping ratio (ζ) can be obtained as

$$\zeta = -\frac{\frac{\ln(\alpha)}{\pi}}{\sqrt{1 + \left(\frac{\ln(\alpha)}{\pi}\right)^2}} = -\frac{\frac{\ln(0.05)}{\pi}}{\sqrt{1 + \left(\frac{\ln(05)}{\pi}\right)^2}} = 0.6901$$

Using Eq. (10.11), undamped natural frequency (ω_n) can be obtained as

$$\omega_n = \frac{4}{\zeta\, t_{set}} = 2.8244 \text{ rad/s}$$

Using Eq. (10.12), the closed-loop poles are obtained as

$$s_{1,2} = -\zeta\omega_n \pm j\omega_n\sqrt{1-\zeta^2} = -1.9560 \pm 2.0512\text{i}$$

The corresponding desired closed-loop poles in the z-plane can be obtained using Eq. (10.13), as

$$z_{1,2} = e^{sT}\Big|_{s=s_{1,2}} = 0.8051 \pm 0.1675\text{i} = 0.8223\angle \pm 0.2051 \text{ rad}$$

Phase margin (in degrees) can be expressed, using Eq. (10.27), as

$$\phi_m = \tan^{-1}\left(\frac{2\zeta}{\sqrt{\left(\sqrt{4\zeta^4+1}\right)-2\zeta^2}}\right) = 72.7512^\circ$$

Approximate phase margin using the linear relation (10.28) is obtained as

$$\phi_m \approx 100\zeta = 69.01^\circ$$

Closed-loop step response of the system, obtained using Matlab code, in continuous time and discrete time is presented in Figs. 10.2 and 10.3, respectively. The percentage overshoot and settling are close to the corresponding specified values for the continuous-time system as shown in Fig. 10.2. However, the effect of sampling is evident as the corresponding values are relatively higher for the discrete-time system response as shown in Fig. 10.3.

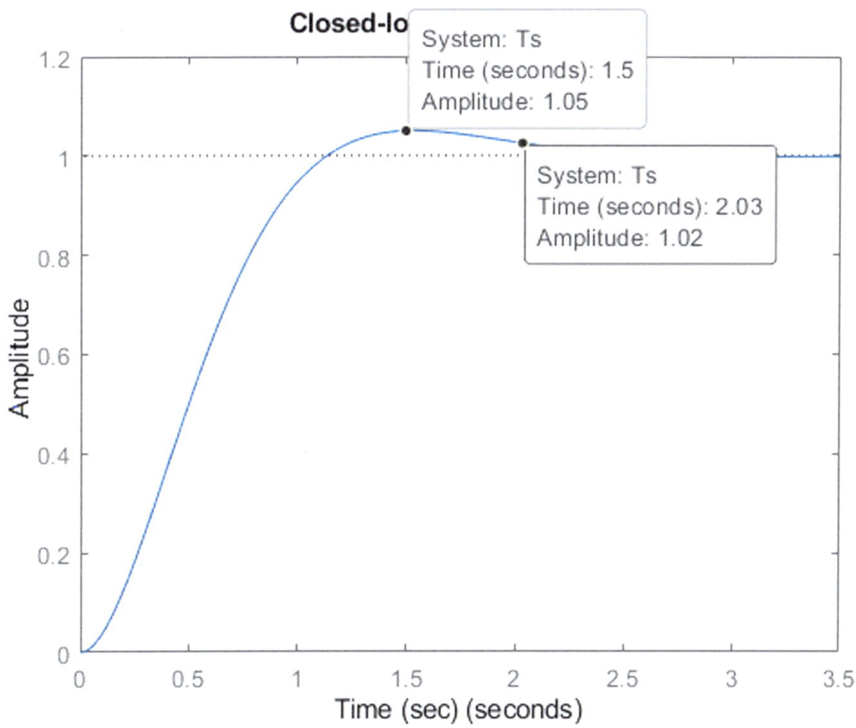

Fig. 10.2 Closed-loop system step response (continuous time)

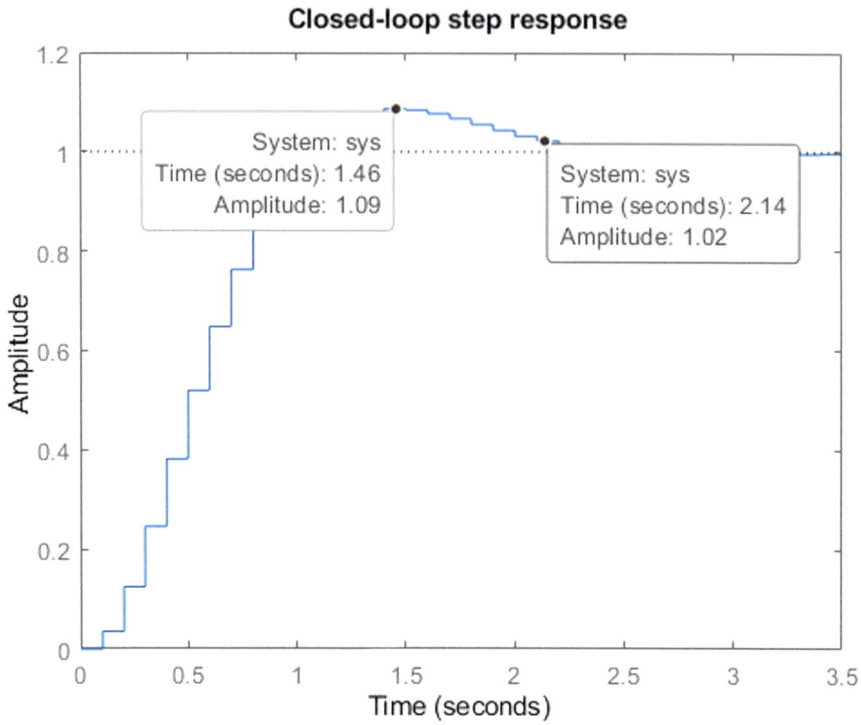

Fig. 10.3 Closed-loop system step response (discrete time)

Matlab code and results are presented here.

```
% Ch 10- ex10_1 desired pole locations s1, s2
%input %overshoot(alpha) & settling time(tset)
clear all; clc; close all
T = 0.10 % sampling time
alpha = 0.05; tset=2; beta=0.02;
% damping ratio
zeta = abs(log(alpha))/sqrt(pi^2+log(alpha)^2) % damping ratio
% natural frequency
wn = -log(beta)/(zeta*tset)  % natural frequency
% pole locations s1, s2
s1 = -zeta*wn + j*wn*sqrt(1-zeta^2)  % desired pole in s-plane
s2 = -zeta*wn - j*wn*sqrt(1-zeta^2)  % desired pole in s-plane
% desired pole locations z1, z2
z1 = exp(s1*T)  % desired pole in z-plane
z2 = exp(s2*T)  % desired pole in z-plane
% magnitude and angles of poles in z-plane
r = abs(z1) % magnitude of closed-loop poles
th1 = angle(z1) % angle of the closed-loop pole
th2 = angle(z2) % angle of the closed-loop pole
th1d=th1*180/pi
th2d=th2*180/pi
% estimated phase margin
pm = atand(2*zeta/(sqrt(4*zeta^4+1)-2*zeta^2)) % phase margin in degrees
pm_approx = 100*zeta % appromiate phase margin in degrees
% open-loop system Transfer function
Gs =tf(wn^2,[1 2*zeta*wn 0]) % G(s)
% closed-loop system transfer function
```

```
Ts = feedback(Gs,1) % T(s)
% closed-loop system step response
figure
step(Ts)
title('Closed-loop step response')
% G(z)
Gz = c2d(Gs,T)
% closed-loop system transfer function
Tz = feedback(Gz,1) % T(z)
% closed-loop system step response
figure
dstep(Tz,1)
title('Closed-loop step response')

Results:
T =     0.1000, zeta =     0.6901, wn =     2.8344
s1 =   -1.9560 + 2.0512i, s2 =   -1.9560 - 2.0512i
z1 =    0.8051 + 0.1675i, z2 =    0.8051 - 0.1675i
r =     0.8223, th1 =    0.2051, th2 =    -0.2051, th1d =    11.7528,
th2d =   -11.7528
pm =    72.7512, pm_approx =    69.0107
G(s) =
      8.034
  -----------------
  s^2 + 3.912 s
T(s) =
           8.034
  --------------------------
  s^2 + 3.912 s + 8.034
G(z) =
    0.0354 z + 0.03108
  -----------------------------
  z^2 - 1.676 z + 0.6762
T(z)=
    0.0354 z + 0.03108
  -----------------------------
  z^2 - 1.641 z + 0.7073
```

Example 10.2 It is required to design a controller such that the closed-loop system has a phase margin (ϕ_m) of 60° at gain cross-over frequency (ω_{gc}) of 2.5 rad/s. Find the damping ratio, percentage overshoot, and settling time for the closed-loop step response of the system. Obtain closed-loop poles in the s- and z-domains. Use sampling time $T = 0.10$ s. Compare the values of percentage overshoot and settling time from the closed-loop step response of the system, assumed to be predominantly of second-order, in continuous time and discrete time.

Solution Given phase margin $\phi_m = 60°$ at gain cross-over frequency $\omega_{gc} = 2.5\,\frac{\text{rad}}{\text{s}}$, $T = 0.1$ s.

The damping ratio can be obtained, using the approximate relation of Eq. (10.28), as

$$\zeta \approx \frac{\phi_m}{100} = \frac{60}{100} = 0.60$$

Percentage overshoot can be obtained, using Eq. (10.8), as

$$\alpha = e^{\frac{-\pi\zeta}{\sqrt{1-\zeta^2}}} = 0.0948$$

Settling time can be estimated, using Eq. (10.31), as

$$t_{set} = \frac{8}{\omega_{gc}\tan(\phi_m)} = \frac{8}{2.5\ \tan(60^\circ)} = 1.8475\ \text{s}$$

Undamped natural frequency can be obtained, using Eq. (10.11), as

$$\omega_n = \frac{4}{\zeta\, t_{\text{set}}} = 3.6084\ \text{rad/s}$$

Using Eq. (10.12), the closed-loop poles are obtained as

$$s_{1,2} = -\zeta\omega_n \pm j\omega_n\sqrt{1-\zeta^2} = -2.1651 \pm 2.8868\text{i}$$

The corresponding desired closed-loop poles in the z-plane can be obtained using Eq. (10.13), as

$$z_{1,2} = e^{sT}\Big|_{s=s_{1,2}} = 0.7720 \pm 0.2291\text{i} = 0.8053\angle \pm 0.2887\ \text{rad}$$

Closed-loop step response of the system, obtained using Matlab code, in continuous time and discrete time is presented in Figs. 10.4 and 10.5, respectively. The percentage overshoot and settling are close to the corresponding estimated values for the continuous-time system as shown in Fig. 10.4. However, the effect of sampling is evident as the corresponding values are relatively higher for the discrete-time system response as shown in Fig. 10.5.

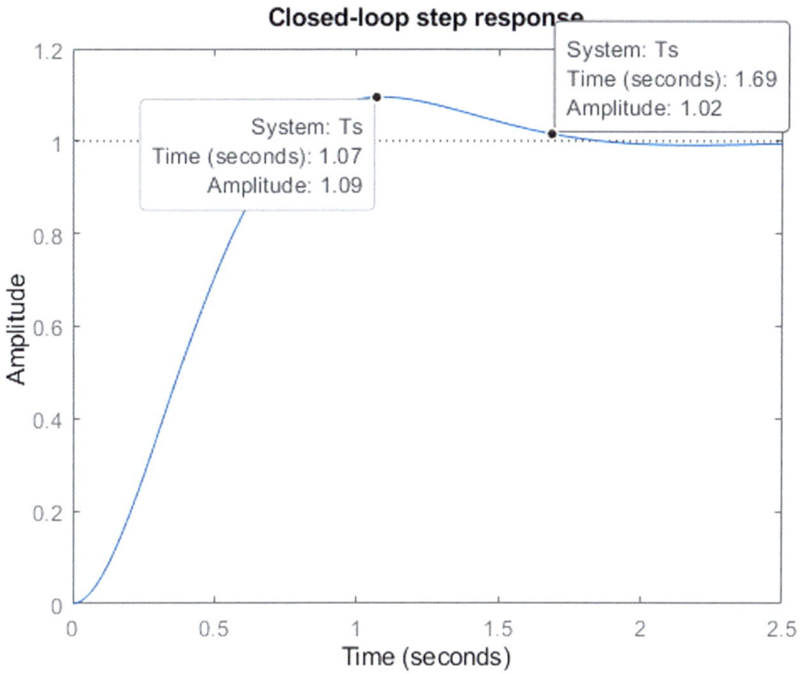

Fig. 10.4 Closed-loop system step response (continuous time)

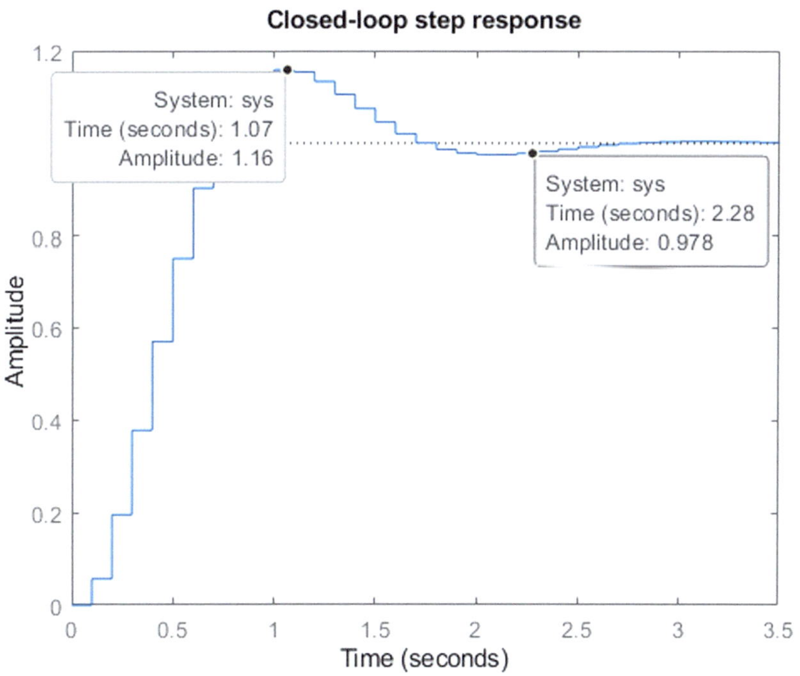

Fig. 10.5 Closed-loop system step response (discrete time)

Matlab code and results are presented here.

```
% Ch 10- ex10_2 desired pole locations s1, s2
clear all; clc; close all
T = 0.10 % sampling time
% Given phase amrgin, wgc find damping ratio, settling time
% then estimate wn, s1,s2, z1, z2, alpha, tset
beta=0.02;
pm = 60   % phase margin in degrees
wgc = 2.5 % gain cross-over frequency in rad/s

% damping ratio
zeta =pm/100 % damping rato using approximate lienar relation
% settling time
tset = 8/(wgc*tand(pm))
% natural frequency
wn = 4/(tset*zeta)
% overshoot
alpha = exp(-pi*zeta/sqrt(1-zeta^2))
% pole locations s1, s2
s1 = -zeta*wn + j*wn*sqrt(1-zeta^2)  % desired pole in s-plane
s2 = -zeta*wn - j*wn*sqrt(1-zeta^2)  % desired pole in s-plane
% desired pole locations z1, z2
z1 = exp(s1*T)  % desired pole in z-plane
z2 = exp(s2*T)  % desired pole in z-plane
% magnitude and angles of poles in z-plane
r = abs(z1) % magnitude of closed-loop poles
th1 = angle(z1) % angle of the closed-loop pole
th2 = angle(z2) % angle of the closed-loop pole
th1d=th1*180/pi
th2d=th2*180/pi
% open-loop system Transfer function
Gs =tf(wn^2,[1 2*zeta*wn 0]) % G(s)
% closed-loop system transfer function
Ts = feedback(Gs,1) % T(s)
% closed-loop system step response
figure
step(Ts)
title('Closed-loop step response')

% G(z)
Gz = c2d(Gs,T)
% closed-loop system transfer function
Tz = feedback(Gz,1) % T(z)
% closed-loop system step response
figure
dstep(Tz,1)
title('Closed-loop step response')

Results:
T =     0.1000, pm =     60,  wgc =     2.5000
zeta =     0.6000,  tset =     1.8475,  wn =     3.6084,  alpha =
0.0948
s1 =  -2.1651 + 2.8868i,  s2 =  -2.1651 - 2.8868i
z1 =    0.7720 + 0.2293i,  z2 =    0.7720 - 0.2293i
r =    0.8053, th1 =     0.2887,  th2 =   -0.2887,  th1d =
16.5399, th2d =  -16.5399
```

```
G(s) =
      13.02
  ---------------
  s^2 + 4.33 s

T(s) =
         13.02
  --------------------------
  s^2 + 4.33 s + 13.02

G(z) =
   0.05664 z + 0.04904
  ----------------------
  z^2 - 1.649 z + 0.6486

T(z) =
    0.05664 z + 0.04904
  ----------------------
  z^2 - 1.592 z + 0.6976
```

Exercises

10.1 It is required to design a controller such that the closed-loop system step response has a percentage overshoot of 7.5% and a settling time of 2.5 s. Find the damping ratio and undamped natural frequency for the closed-loop system. Obtain closed-loop poles in the s- and z-domains. Use sampling time $T = 0.10$ s. Compare the values of percentage overshoot and settling time from the closed-loop step response of the system, assumed to be predominantly of second-order, in continuous time and discrete time.

10.2 It is required to design a controller such that the closed-loop system step response has a percentage overshoot of 10% and a settling time of 3.0 s. Find the damping ratio and undamped natural frequency for the closed-loop system. Obtain closed-loop poles in the s- and z-domains. Use sampling time $T = 0.15$ s. Compare the values of percentage overshoot and settling time from the closed-loop step response of the system, assumed to be predominantly of second-order, in continuous time and discrete time.

10.3 It is required to design a controller such that the closed-loop system step response has a percentage overshoot of 8% and a settling time of 1.5 s. Find the damping ratio and undamped natural frequency for the closed-loop system. Obtain closed-loop poles in the s- and z-domains. Use sampling time $T = 0.20$ s. Compare the values of percentage overshoot and settling time from the closed-loop step response of the system, assumed to be predominantly of second-order, in continuous time and discrete time.

10.4 It is required to design a controller such that the closed-loop system step response has a percentage overshoot of 15% and a settling time of 4.5 s. Find the damping ratio and undamped natural frequency for the closed-loop system. Obtain closed-loop poles in the s- and z-domains. Use sampling time $T = 0.05$ s. Compare the values of percentage overshoot and settling time from the closed-loop step response of the system, assumed to be predominantly of second-order, in continuous time and discrete time.

10.5 It is required to design a controller such that the closed-loop system has a phase margin (ϕ_m) of 55° at gain cross-over frequency (ω_{gc}) of 4.5 rad/s. Find the damping ratio, percentage overshoot, and settling time for the closed-loop step response of the system. Obtain closed-loop poles in the s- and z-domains. Use sampling time $T = 0.10$ s. Compare the values of percentage overshoot and settling time from the closed-loop step response of the system, assumed to be predominantly of second-order, in continuous time and discrete time.

10.6 It is required to design a controller such that the closed-loop system has a phase margin (ϕ_m) of 45° at gain cross-over frequency (ω_{gc}) of 5.5 rad/s. Find the damping ratio, percentage overshoot, and settling time for the closed-loop step response of the system. Obtain closed-loop poles in the s- and z-domains. Use sampling time $T = 0.15$ s. Compare the values of percentage overshoot and settling time from the closed-loop step response of the system, assumed to be predominantly of second-order, in continuous time and discrete time.

10.7 It is required to design a controller such that the closed-loop system has a phase margin (ϕ_m) of 60° at gain cross-over frequency (ω_{gc}) of 3.5 rad/s. Find the damping ratio, percentage overshoot, and settling time for the closed-loop step response of the system. Obtain closed-loop poles in the s- and z-domains. Use sampling time $T = 0.20$ s. Compare the values of percentage overshoot and settling time from the closed-loop step response of the system, assumed to be predominantly of second-order, in continuous time and discrete time.

10.8 It is required to design a controller such that the closed-loop system has a phase margin (ϕ_m) of 65° at gain cross-over frequency (ω_{gc}) of 6.5 rad/s. Find the damping ratio, percentage overshoot, and settling time for the closed-loop step response of the system. Obtain closed-loop poles in the s- and z-domains. Use sampling time $T = 0.05$ s. Compare the values of percentage overshoot and settling time from the closed-loop step response of the system, assumed to be predominantly of second-order, in continuous time and discrete time.

Further Reading

1. Asadi F (2023) Signals and systems with MATLAB® and Simulink®. Springer, Cham
2. Astrom KJ, Murray RM (2021) Feedback systems: an introduction for scientists and engineers, 2nd edn. Princeton University Press, Princeton
3. Chen C-T (2006) Analog and digital control system design: transfer-function, state-space, and algebraic methods. Oxford University Press, New York
4. Dorf R, Bishop R (2016) Modern control systems, 13th edn. Pearson, Essex
5. Fadali MS, Visioli A (2020) Digital control engineering analysis and design, 3rd edn. Academic, London
6. Franklin GF, Powell JD, Emami-Naeini A (2019) Feedback control of dynamic systems, 8th edn. Pearson, Upper Saddle River

7. Franklin GF, Powell JD, Workman M (1998) Digital control of dynamic systems, 3rd edn. Addison-Wesley, Boston
8. Isermann R (1981) Digital control systems. Springer, Berlin
9. Jalili N, Candelino NW (2023) Dynamic systems and control engineering. Cambridge University Press, Cambridge
10. Keviczky L, Bars R, Hetthéssy J, Bányász (2019) Control engineering. Springer, Singapore
11. Koeing DM (2009) Practical control engineering: A guide for engineers, managers, and practitioners. McGraw-Hill, New York
12. Kuo BC (1995) Digital control systems, 2nd edn. Oxford University Press, Oxford
13. Landau ID, Gito G (2006) Digital control systems: design, identification and implementation. Springer-Verlag, London
14. Leigh JR (2006) Applied digital control: theory, design and implementation, 2nd edn. Dover, Mineola
15. Mathworks (2023a) Control System Toolbox. https://www.mathworks.com/products/control.html
16. Mathworks (2023b) Matlab. https://www.mathworks.com/products/matlab.html
17. Nise NS (2019) Control systems engineering, 8th edn. Wiley, Hoboken
18. Nof SY (ed) (2009) Springer handbook of automation. Springer, Berlin
19. Ogata K (1995) Discrete-time control systems, 2nd edn. Pearson, Hoboken
20. Phillips CL, Nagle HT, Chakrabortty A (2015) Digital control system analysis and design, 4th edn. Pearson, Essex
21. Rabbath CA, Lechevin N (2014) Discrete-time control system design with applications. Springer, New York
22. Samanta B (2023) Introduction to mechatronics – an integrated approach. Springer, Cham
23. Shertukde HM (2015) Digital control applications illustrated with MATLAB®. CRC Press, Boca Raton
24. Starr G (2020) Introduction to applied digital controls. Springer, Cham
25. Veloni A, Miridakis NI (2018) Digital control systems-theoretical problems and simulation tools. CRC Press, Boca Raton
26. Vilanova R, Visioli A (2012) PID control in the third millennium – lessons learned and new approaches. Springer, London
27. Visioli A (2006) Practical PID control. Springer, London

Chapter 11
Controller Design for Discrete-Time Systems Using Time Domain Specifications

11.1 Introduction

Methods of designing controllers for discrete-time systems using time domain specifications are presented in this chapter. Types of controllers covering proportional-integral-derivative (PID), phase-lag-lead, and their variations along with corresponding transfer functions for discrete-time systems are covered first. Next details of design steps for discrete-time controllers are presented. Examples along with Matlab codes are presented for the controllers to illustrate the procedure. Laboratory experiments for analysis, design, simulation, and real-time implementation of the controllers on a laboratory-scale DC motor module are presented. Analysis, design, and simulation are presented using Matlab, and implementation of the discrete-time controllers is presented using LabVIEW. End-of-chapter exercises are provided to enhance understanding of the topics covered and consolidate problem-solving skills.

11.2 Types of Controllers for Discrete-Time Systems

11.2.1 PI, PD, and PID Controllers

The transfer function of a PID controller consists of three components: proportional (P), integral (I), and derivative (D). In the continuous-time domain, the output, $m(t)$, of a PID controller can be expressed in terms of the input, $e(t)$ as

$$m(t) = K_P e(t) + K_I \int e(t)dt + K_D \frac{de(t)}{dt} \tag{11.1}$$

B. Samanta, *Introduction to Digital Control*,
https://doi.org/10.1007/978-3-031-66830-2_11

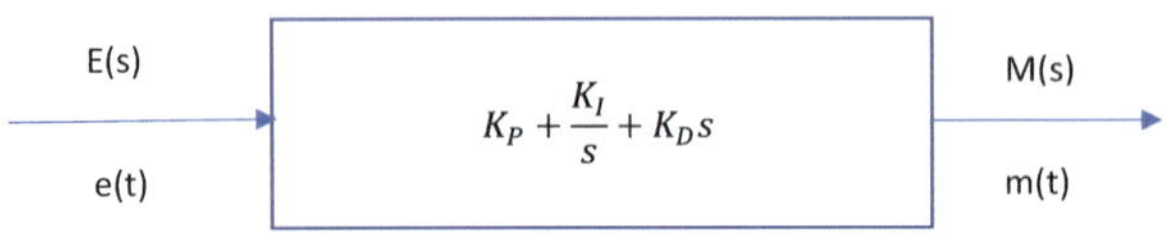

Fig. 11.1 Block diagram of a PID controller in continuous-time domain

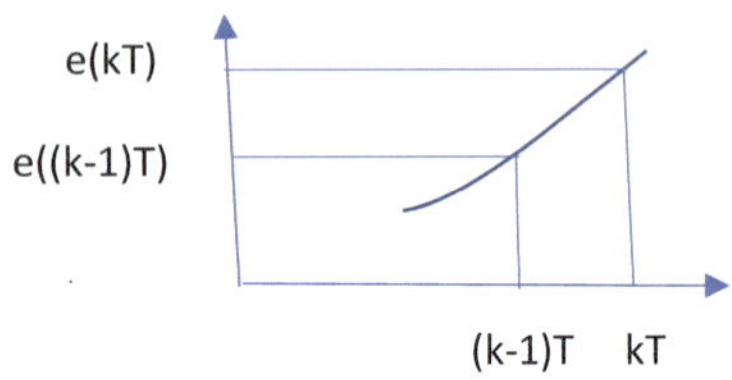

Fig. 11.2 Trapezoidal rule for numerical integration

In s-domain, the transfer function of a PID controller can be expressed as

$$G_{PID}(s) = \frac{M(s)}{E(s)} = K_P + \frac{K_I}{s} + K_D s \tag{11.2}$$

The block diagram for a PID controller in continuous-time domain is shown in Fig. 11.1.

It is necessary to derive the discrete-time version of the integral and derivative components to represent a PID controller transfer function in the discrete-time domain.

PI Controllers

The numerical integration in a discrete-time integral filter is represented in Fig. 11.2.

The output $m(kT)$ is obtained using the trapezoidal rule for numerical integration of the input $e(kT)$ as

$$m(kT) = m((k-1)T) + \frac{T}{2}(e((k-1)T) + e(kT)) \tag{11.3}$$

Taking z-transform, Eq. (11.3) can be expressed as

$$M(z) = z^{-1}M(z) + \frac{T}{2}\left(z^{-1}E(z) + E(z)\right) \tag{11.4}$$

After some rearrangement, the transfer function for the discrete-time integral filter can be expressed as

$$D_I(z) = \frac{M(z)}{E(z)} = \frac{T}{2}\frac{1+z^{-1}}{1-z^{-1}} = \frac{T}{2}\left(\frac{z+1}{z-1}\right) \tag{11.5}$$

The transfer function of a discrete-time PD controller can be expressed as

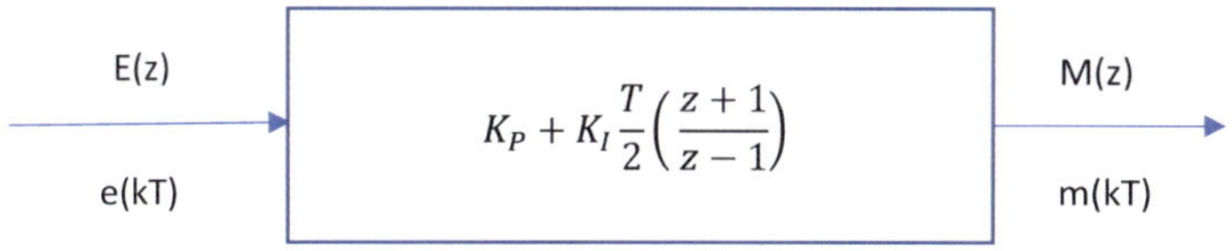

Fig. 11.3 Block diagram of a discrete-time PI controller

Fig. 11.4 Discrete-time derivative filter

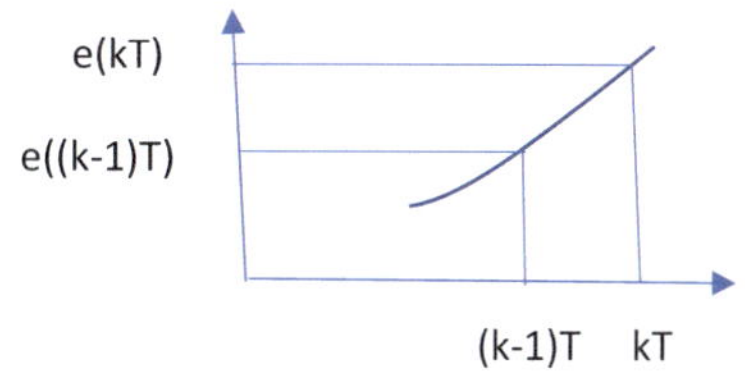

$$D_{PI}(z) = \frac{M(z)}{E(z)} = K_P + K_I \frac{T}{2}\left(\frac{z+1}{z-1}\right) \tag{11.6}$$

The block diagram of a discrete-time PI controller is shown in Fig. 11.3.

PD Controllers

A discrete-time derivative filter (differentiator) is presented in Fig. 11.4.

The output $m(kT)$ is obtained using numerical differentiation of the input $e(kT)$ as

$$m(kT) = \frac{e(kT) - e((k-1)T)}{T} \tag{11.7}$$

Taking z-transform, Eq. (11.6) can be expressed as

$$M(z) = \frac{E(z) - z^{-1}E(z)}{T} \tag{11.8}$$

After some rearrangement, the transfer function for the discrete-time derivative filter can be expressed as

$$D_D(z) = \frac{M(z)}{E(z)} = \frac{1 - z^{-1}}{T} = \frac{z-1}{Tz} \tag{11.9}$$

The transfer function of a discrete-time PD controller can be expressed as

$$D_{PD}(z) = \frac{M(z)}{E(z)} = K_P + K_D \frac{z-1}{Tz} \tag{11.10}$$

The block diagram of a discrete-time PD controller is shown in Fig. 11.5.

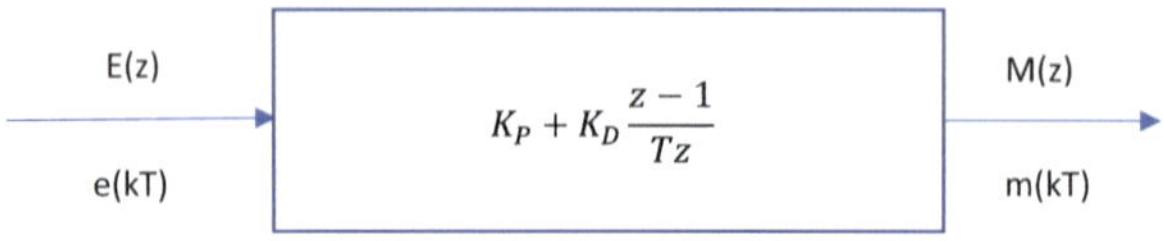

Fig. 11.5 Block diagram of a discrete-time PD controller

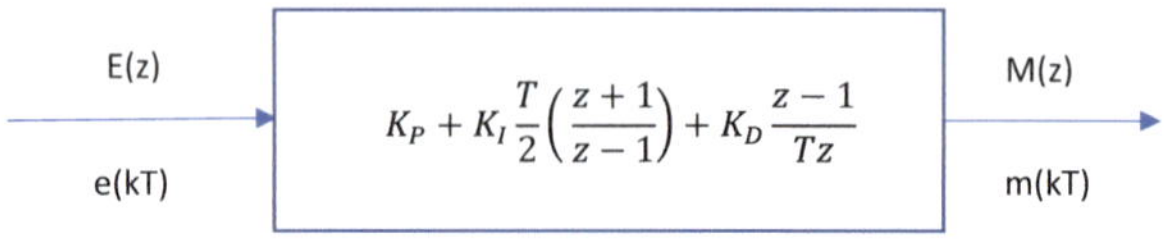

Fig. 11.6 Block diagram of a discrete-time PID controller

PID Controllers

The transfer function of a discrete-time PID controller can be expressed as

$$D_{PID}(z) = \frac{M(z)}{E(z)} = K_P + K_I \frac{T}{2}\left(\frac{z+1}{z-1}\right) + K_D \frac{z-1}{Tz} \tag{11.11}$$

The block diagram of a discrete-time PID controller is shown in Fig. 11.6.

11.2.2 Phase-Lag, Phase-Lead, and Phase-Lag-Lead Controllers

Phase-Lag Controllers

The transfer function of a phase-lag controller for continuous-time systems can be represented in s-domain as

$$G_{\text{lag}}(s) = a_0 \frac{1 + s/\omega_0}{1 + s/\omega_p} \tag{11.12}$$

where the controller zero is at $s = -\omega_0$ and the pole is at $s = -\omega_p$ and $\omega_p < \omega_0$. The phase-lag angle is expressed as the difference of angles contributed by the zero and the pole, shown in Fig. 11.7.

$$\phi_{\text{lag}} = \theta_p - \theta_z \tag{11.13}$$

In general, the phase-lag angle is in the range of 5–10° to keep the destabilizing effect of the phase-lag at a minimum and to improve the steady-state accuracy.

The transfer function of a discrete-time phase-lag controller in the z-domain as

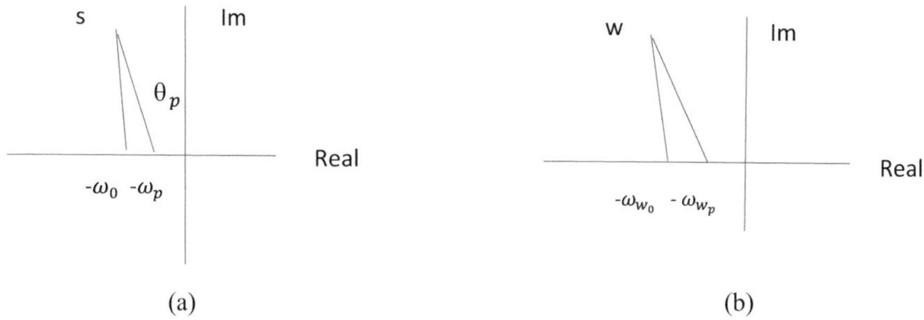

Fig. 11.7 Phase-lag controller zero and pole (**a**) s-plane, (**b**) w-plane

$$D_{\text{lag}}(z) = b_0 \frac{z - z_0}{z - z_p} \tag{11.14}$$

where the controller zero is at $z = z_0$ and the pole is at $z = z_p$ in the z-plane. For comparison with the phase-lag controller in the s-plane for continuous-time systems, the transfer function of Eq. (11.14) can be transformed to the w-domain using the bilinear transformation as

$$D_{\text{lag}}(\mathrm{w}) = b_0 \frac{z - z_0}{z - z_p}\bigg|_{z=\frac{1+\frac{Tw}{2}}{1-\frac{Tw}{2}}} \tag{11.15}$$

The phase-lag controller transfer function in the w-domain can be expressed, similar to Eq. (11.12) in the s-domain, as

$$D_{\text{lag}}(w) = c_0 \frac{1 + w/\omega_{w_0}}{1 + w/\omega_{w_p}} \tag{11.16}$$

where the controller zero is at $w = -\omega_{w_0}$ and the pole is at $w = -\omega_{w_p}$ and $\omega_{w_p} < \omega_{w_0}$.

The w-domain transfer function of the phase-lag controller can be converted to a z-domain in the form of Eq. (11.14):

$$D_{\text{lag}}(z) = c_0 \frac{1 + w/\omega_{w_0}}{1 + w/\omega_{w_p}}\bigg|_{w=\frac{2}{T}\frac{z-1}{z+1}} \tag{11.17}$$

After some rearrangement, the parameters of Eq. (11.14) can be expressed as

$$b_0 = c_0 \frac{\omega_{w_p}\left(\omega_{w_0} + \frac{2}{T}\right)}{\omega_{w_0}\left(\omega_{w_p} + \frac{2}{T}\right)} \tag{11.18}$$

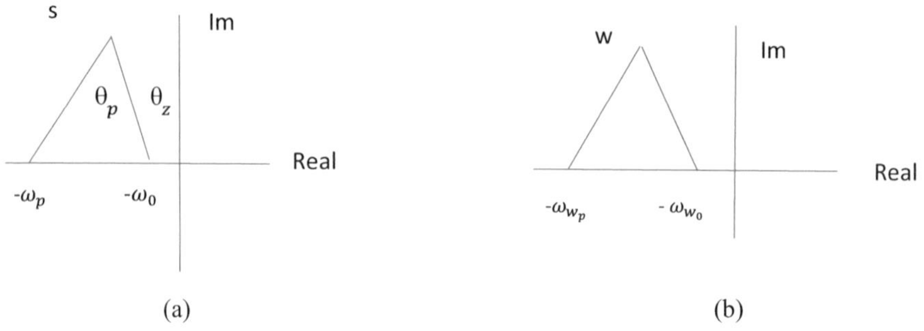

Fig. 11.8 Phase-lead controller zero and pole (**a**) s-plane, (**b**) w-plane

$$z_0 = \frac{\frac{2}{T} - \omega_{w_0}}{\frac{2}{T} + \omega_{w_0}} \tag{11.19}$$

$$z_p = \frac{\frac{2}{T} - \omega_{w_p}}{\frac{2}{T} + \omega_{w_p}} \tag{11.20}$$

Phase-Lead Controllers

The transfer function of a phase-lead controller for continuous-time systems can be represented in s-domain as

$$G_{\text{lead}}(s) = a'_0 \frac{1 + s/\omega'_0}{1 + s/\omega'_p} \tag{11.21}$$

where the controller zero is at $s = -\omega'_0$ and the pole is at $s = -\omega'_p$ and $\omega'_0 < \omega'_p$. The phase-lead angle is expressed as the difference of angles contributed by the zero and the pole, shown in Fig. 11.8:

$$\phi_{\text{lead}} = \theta_z - \theta_p \tag{11.22}$$

In general, the phase-lead angle is in the range of 25–35° to keep the stabilizing effect of the phase-lead at a high level and to improve the transient response.

The transfer function of a discrete-time phase-lead controller in the z-domain as

$$D_{\text{lead}}(z) = b'_0 \frac{z - z'_0}{z - z'_p} \tag{11.23}$$

where the controller zero is at $z = z'_0$ and the pole is at $z = z'_p$ in the z-plane. For comparison with the phase-lead controller in the s-plane for continuous-time systems, the transfer function of Eq. (11.23) can be transformed to the w-domain using the bilinear transformation as

$$D_{\text{lead}}(\mathrm{w}) = c'_0 \frac{z - z'_0}{z - z'_p}\bigg|_{z=\frac{1+\frac{Tw}{2}}{1-\frac{Tw}{2}}} \tag{11.24}$$

The phase-lead controller transfer function in the w-domain can be expressed, similar to Eq. (11.21) in the s-domain, as

$$D_{\text{lead}}(w) = c'_0 \frac{1 + w/\omega'_{w0}}{1 + w/\omega'_{wp}} \tag{11.25}$$

where the controller zero is at $w = -\omega'_{w0}$ and the pole is at $w = -\omega'_{wp}$ and $\omega'_{w0} < \omega'_{wp}$.

The w-domain transfer function of the phase-lead controller can be converted to a z-domain in the form of Eq. (11.23):

$$D_{\text{lead}}(z) = c'_0 \frac{1 + w/\omega'_{w0}}{1 + w/\omega'_{wp}}\bigg|_{w=\frac{2}{T}\frac{z-1}{z+1}} \tag{11.26}$$

After some rearrangement, the parameters of Eq. (11.23) can be expressed as

$$b'_0 = c'_0 \frac{\omega'_{wp}\left(\omega'_{w0} + \frac{2}{T}\right)}{\omega'_{w0}\left(\omega'_{wp} + \frac{2}{T}\right)} \tag{11.27}$$

$$z'_0 = \frac{\frac{2}{T} - \omega'_{w0}}{\frac{2}{T} + \omega'_{w0}} \tag{11.28}$$

$$z'_p = \frac{\frac{2}{T} - \omega'_{wp}}{\frac{2}{T} + \omega'_{wp}} \tag{11.29}$$

Phase-Lag-Lead Controllers

Phase-lag-lead controllers consist of both phase-lag and phase-lead components in cascade, with the transfer function in the w-domain expressed as

$$D_{\text{lag-lead}}(w) = \left(c_0 \frac{1 + w/\omega_{w0}}{1 + w/\omega_{wp}}\right)\left(c'_0 \frac{1 + w/\omega'_{w0}}{1 + w/\omega'_{wp}}\right) \tag{11.30}$$

where, for the phase-lag part, the zero is at $w = -\omega_{w_0}$ and the pole is at $w = -\omega_{w_p}$ and $\omega_{w_p} < \omega_{w_0}$ and, for the phase-lead part, zero is at $w = -\omega'_{w_0}$ and the pole is at $w = -\omega'_{w_p}$ and $\omega'_{w_0} < \omega'_{w_p}$. The phase-lag part is effective in the lower frequency range to improve the steady-state accuracy. The phase-lead part is effective in the higher frequency range to improve the transient response characteristics of the control system.

The transfer function can be expressed in the z-domain as

$$D_{\text{lag-lead}}(z) = \left(b_0 \frac{z - z_0}{z - z_p}\right) \left(b'_0 \frac{z - z'_0}{z - z'_p}\right) \tag{11.31}$$

In general, the phase-lag, phase-lead, and phase-lag-lead controller constants b_0, b'_0, c_0, c'_0 are determined using steady-state accuracy specifications. In the design stage, the controller zeros and poles are determined using the time domain specifications.

11.3 Design of PI, PD, and PID Controllers for Discrete-Time Systems

In this section, the procedure for designing PID controllers for discrete-time systems is presented. Other variants of PID controllers, namely, PI and PD controllers, are obtained by setting one of the controller gains to zero, i.e., $K_D = 0$ for PI and $K_I = 0$ for PD.

11.3.1 PID Controllers

The desired time domain specifications of overshoot (α) and settling time (t_{set}) can be used to get equivalent second-order system desired pole locations s_1, s_2 in the s-plane that can be converted to z_1, z_2 in the z-plane. It should be noted that these pole pairs, if complex consisting of real and imaginary parts, appear in conjugates. The characteristic equation for the closed-loop discrete-time system must be satisfied by these closed-loop system poles z_1, z_2 in the z-plane. It is also worth noting that if one of these poles (z_1) satisfies the characteristic equation, its conjugate will also satisfy the system characteristic equation.

$$1 + D(z_1)\overline{GH}(z_1) = 0 \tag{11.32}$$

In the context of the root locus, it should be noted that the root locus of the discrete-time system has to pass through the desired closed-loop poles.

From the characteristic equation, Eq. (11.32), the controller transfer function can be expressed as

$$D(z_1) = -\frac{1}{\overline{GH}(z_1)} \tag{11.33}$$

For a PID controller, the transfer function of Eq. (11.11) can be used in Eq. (11.32) as

$$\left(K_P + K_I \frac{T}{2}\left(\frac{z_1+1}{z_1-1}\right) + K_D \frac{z_1-1}{Tz_1}\right) = -\frac{1}{\overline{GH}(z_1)} \tag{11.34}$$

The known components in Eq. (11.33) can be expressed in terms of the desired pole z_1 as

$$\overline{GH}(z_1) = me^{j\Psi} \tag{11.35}$$

$$\frac{z_1+1}{z_1-1} = a_R + ja_I \tag{11.36}$$

$$\frac{z_1-1}{z_1} = b_R + jb_I \tag{11.37}$$

Equation (11.34) can be expressed as

$$K_P + K_I'(a_R + ja_I) + K_D'(b_R + jb_I) = -\frac{1}{m}(\cos\Psi - j\sin\Psi) \tag{11.38}$$

where

$$K_I' = K_I \frac{T}{2} \tag{11.39}$$

$$K_D' = \frac{K_D}{T} \tag{11.40}$$

Real and imaginary parts of both sides of Eq. (11.38) can be expressed as

$$K_P + K_I' a_R + K_D' b_R = -\frac{1}{m}\cos\Psi \tag{11.41}$$

$$K_I' a_I + K_D' b_I = \frac{1}{m}\sin\Psi \tag{11.42}$$

In Eqs. (11.41) and (11.42), there are 3 unknown parameters $\left(K_P,\ K'_I, K'_D\right)$ and 2 equations; thus, there is no unique solution. Instead, one of the unknowns is assumed, and the other 2 are expressed in terms of the assumed one. In general, the integral gain K_I is assumed, and the other two gains are obtained from Eqs. (11.41) and (11.42) as

$$K'_D = \frac{1}{b_I}\left(\frac{1}{m}\sin\Psi - K'_I a_I\right) \tag{11.43}$$

$$K_D = K'_D\,T \tag{11.44}$$

$$K_P = -\frac{1}{m}\cos\Psi - K'_I a_R - K'_D b_R \tag{11.45}$$

As a guideline, initially, K_I is assumed to be 0, and $K'_D, K_D,\ K_P$ are obtained. Next, an estimate of K'_I is obtained from

$$0.01 \le \frac{K'_I}{K_P} \le 0.10 \tag{11.46}$$

The values of $K'_D, K_D,\ K_P$ are updated using the new estimate of K'_I. The process is repeated till the values converge. Final values are fine-tuned after checking with simulation using the calculated values of the controller gains K_P, K_I, K_D.

The PID controller transfer function in terms of numerator and denominator polynomials can be expressed as

$$D_{PID}(z) = \frac{\left(K_P + K'_I + K'_D\right)z^2 + \left(-K_P + K'_I - 2K'_D\right)z + K'_D}{z^2 - z} \tag{11.47}$$

11.3.2 PI Controllers

PI controller gains can be obtained from Eqs. (11.42) and (11.41), with $K'_D = \frac{K_D}{T} = 0$, as

$$K'_I = \frac{1}{m\,a_I}\sin\Psi \tag{11.48}$$

$$K_P = -\frac{1}{m}\cos\Psi - K'_I\,a_R \tag{11.49}$$

It should be noted that both controller gains K_P and K_I should be positive. If K'_I from Eq. (11.48) turns out to be negative, K'_I can be assumed to be a small positive value, and K_P can be accordingly calculated using Eq. (11.49). The value of K'_I should be so selected that results in a reasonable response of the closed-loop system.

The PI controller transfer function in terms of numerator and denominator polynomials can be expressed as

$$D_{PI}(z) = \frac{(K_P + K_I')z + (-K_P + K_I')}{z-1} \tag{11.50}$$

11.3.3 PD Controllers

PD controller gains can be obtained from Eqs. (11.42) and (11.41), with $K'_I = K_I \frac{T}{2} = 0$

$$K_D' = \frac{1}{m\, b_I} \sin \Psi \tag{11.51}$$

$$K_P = -\frac{1}{m} \cos \Psi - K_D' b_R \tag{11.52}$$

The PD controller transfer function in terms of numerator and denominator polynomials can be expressed as

$$D_{PD}(z) = \frac{(K_P + K_D')z + (-K_D')}{z} \tag{11.53}$$

Example 11.1 Design a PI controller for $G_p(s) = \frac{10}{s^2+2.5s+1}$, $H(s) = 1$ such that the closed-loop system has an overshoot of 5% and a settling time of 2 sec (2% error beta) for a step input. Use sampling time $T = 0.10$ sec. Show the effect of the PI controller on the closed-loop system stability and step response.

Solution Matlab code and results are presented.

```
% design of PI controller
close all; clear all; clc
% plant tf
Gp=tf(10, [1 2.5 1]) % G(s) =10/(s^2 +2.5s+1)
% Get G(z)
T=0.10 % sampling time
Gz = c2d(Gp, T);  % G(z)
[ngz, dgz]=tfdata(Gz,'v');
% desired pole location s1
%input %overshoot(alpha) & settling time(tset)
alpha=0.05; tset=2; beta=0.02;
zeta=abs(log(alpha))/sqrt(pi^2+log(alpha)^2) % damping ratio
wn=-log(beta)/(zeta*tset)  % natural frequency
s1=-zeta*wn+j*wn*sqrt(1-zeta^2)  % desired pole in s-plane
z1=exp(s1*T)  % desired pole in z-plane

% Design of PI controller
a1=((z1+1)/(z1-1));
b1=(z1-1)/(z1);
a1r=real(a1); a1i=imag(a1);
b1r=real(b1); b1i=imag(b1);

%evaluate  G(z1)
nv=polyval(ngz,z1);dv=polyval(dgz,z1);
g=nv/dv;      % G(z1)
m=abs(g)      % magnitude
psi=angle(g) % angle

% PI controller gains kp, ki

kip=.01 % kip   You need to adjust this to get suitable response
% PI
kdp = 0
kp=-(cos(psi)/m+a1r*kip+b1r*kdp)
ki=kip/(T/2)
Dz=tf([kp+kip -kp+kip], [1 -1], T)     %PI: D(z)
% end
DGz= Dz*Gz; % D(z)G(z)
figure
% root locus of uncompensated system (without controller)
rlocus(Gz)
axis('equal')
title('Root locus of uncompensated system')
% root locus of compensated system (with controller)
figure(3)
rlocus(DGz)
axis('equal')
title('Root locus of compensated system')
% bode plot of uncompensated system (without controller)
Gw = d2c(Gz,'tustin')
Dw = d2c(Dz,'tustin')
DGw = Dw*Gw
% bode plot of the uncompensated system (without controller)
figure
bode(Gw)
title('Bode plot of uncompensated system')
[gm,pm,wpc,wgc]=margin(Gw)
```

```
% bode plot of the compensator (controller)
figure
bode(Dw)
title('Bode plot of PI controller')
% bode plot of the compensated system (with controller)
figure
bode(DGw)
title('Bode plot of compensated system')
[gm,pm,wpc,wgc]=margin(DGw)
wgs=atan(wgc*T/2)*2/T % in s-domain
wps=atan(wpc*T/2)*2/T % in s-domain

% closed loop step response of compensated system
figure
Tz=feedback(DGz,1)
dstep(Tz,1)
title('Closed loop step response of compensated system')

Results:
z1 = 0.8051 + 0.1675i
m = 2.1363, psi = 2.4398,kip = 0.0100, kdp = 0, kp = 0.4065

D(z) =
  0.4165 z - 0.3965
  -----------------
        z - 1
```

The stability of the closed-loop system has been enhanced with the PI controller as shown in the root locus in Figs. 11.9 and 11.10. This is also confirmed in Bode plots of Figs. 11.11, 11.12 and 11.13. However, the desired specifications are not exactly achieved with the PI controller as shown in the closed-loop step response of the system in Fig. 11.14. The steady-state error due to a step input is zero due to the integral part.

Example 11.2 PD: Design a PD controller for $G_p(s) = \frac{10}{s^2+2.5s+1}$, $H(s) = 1$ such that the closed-loop system has an overshoot of 5% and a settling time of 2 sec (2% error beta) for a step input. Use sampling time $T = 0.10$ sec. Show the effect of the PD controller on the closed-loop system stability and step response.

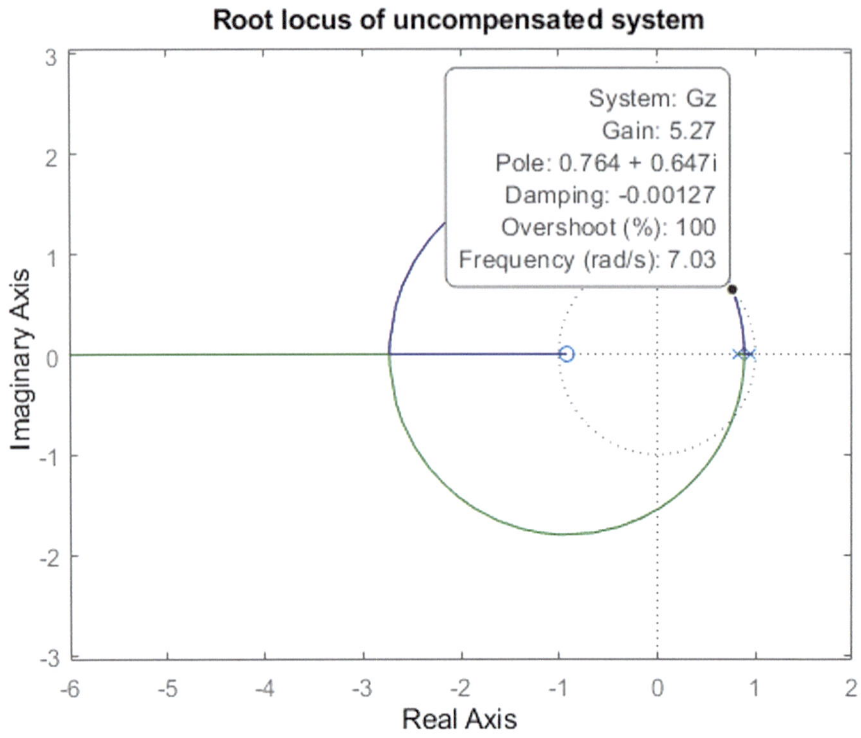

Fig. 11.9 Root locus of the uncompensated system

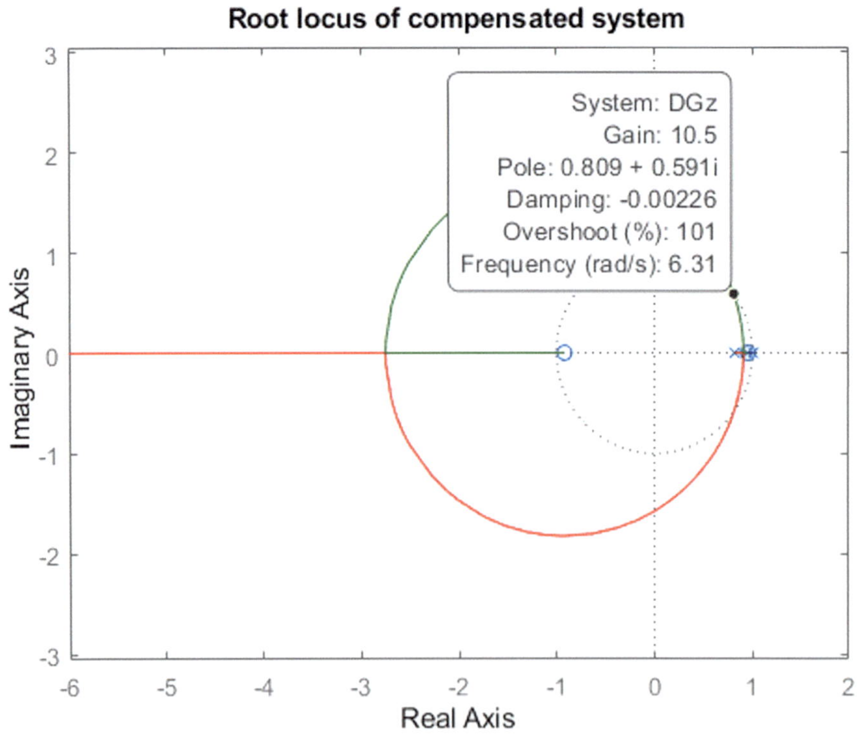

Fig. 11.10 Root locus of the compensated system with PI controller

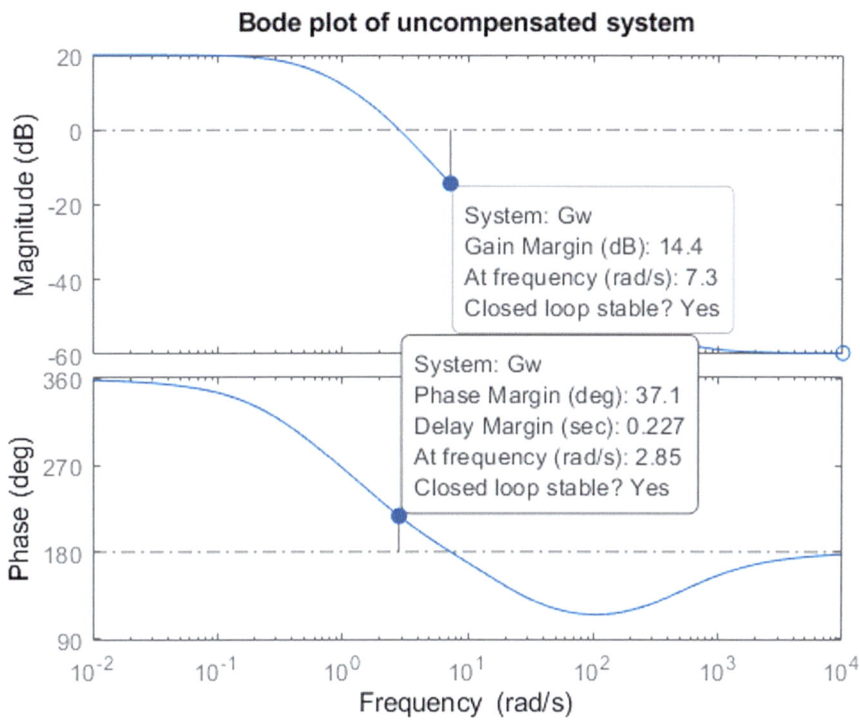

Fig. 11.11 Bode plots of the uncompensated system

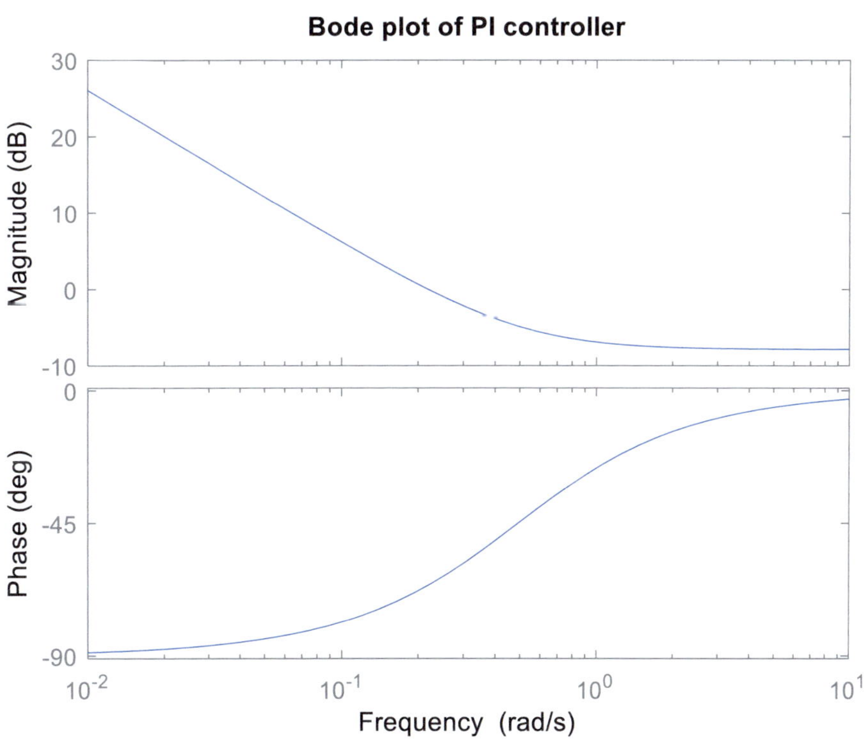

Fig. 11.12 Bode plots of the PI controller

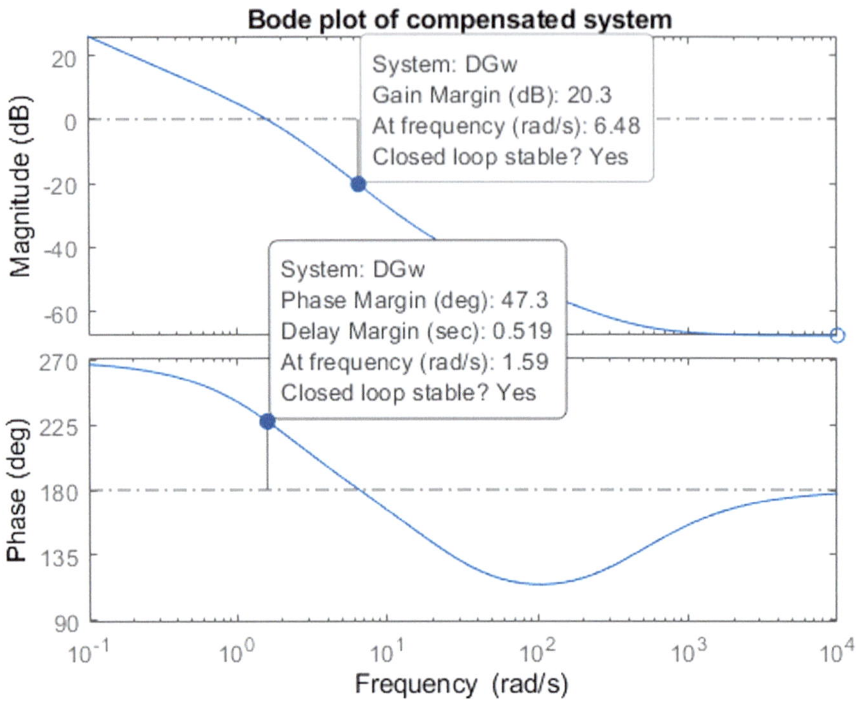

Fig. 11.13 Bode plots of the compensated system with PI controller

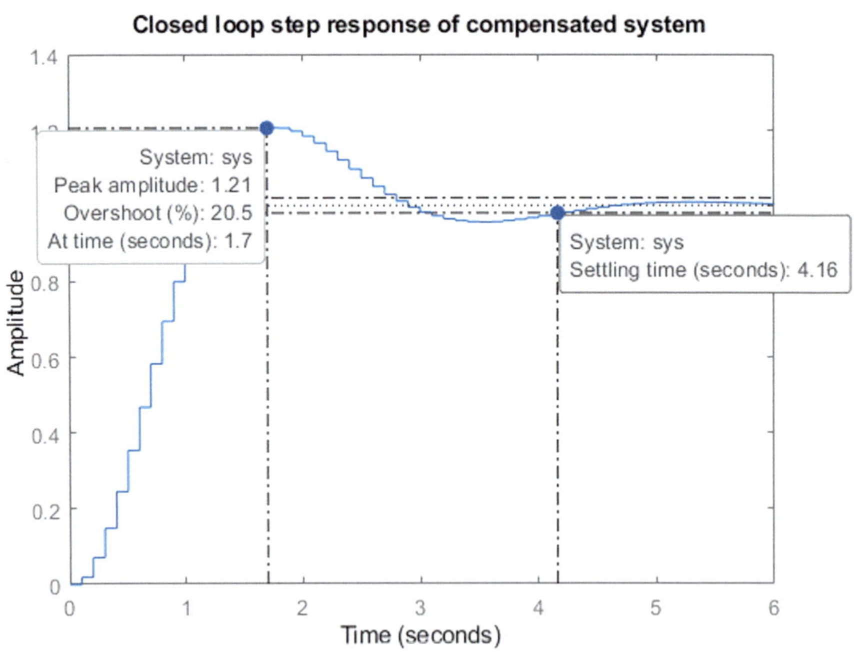

Fig. 11.14 Closed-loop step response of the compensated system with PI controller

Solution Matlab code and results are presented.

```
% design of PD controller
close all; clear all; clc
%G(s)
Gp=tf(10, [1 2.5 1]) % G(s) =10/(s^2 +2.5s+1)
% Get G(z)
T=0.10 % sampling time
Gz = c2d(Gp, T);  % G(z)
[ngz, dgz]=tfdata(Gz,'v');
% desired pole location s1
%input %overshoot(alpha) & settling time(tset)
alpha=0.05; tset=2; beta=0.02;
zeta=abs(log(alpha))/sqrt(pi^2+log(alpha)^2) % damping ratio
wn=-log(beta)/(zeta*tset)  % natural frequency
s1=-zeta*wn+j*wn*sqrt(1-zeta^2)  % desired pole in s-plane
z1=exp(s1*T)  % desired pole in z-plane

% Design of PID controller
a1=((z1+1)/(z1-1));
b1=(z1-1)/(z1);
a1r=real(a1); a1i=imag(a1);
b1r=real(b1); b1i=imag(b1);

%evaluate  G(z1)
nv=polyval(ngz,z1);dv=polyval(dgz,z1);
g=nv/dv;      % G(z1)
m=abs(g)      % magnitude
psi=angle(g) % angle

%PD controller gains kp, kd
kip=0 % kip   You need to adjust this to get suitable response
kdp=(sin(psi)/m-kip*a1i)/b1i
kp=-(cos(psi)/m+a1r*kip+b1r*kdp)
ki=kip/(T/2)
kd=kdp*T
Dz =tf([kp+kdp  -kdp]  ,[1 0],T)
%D(z)G(z)
DGz= Dz*Gz; % D(z)G(z)

figure
% root locus of uncompensated system (without controller)
rlocus(Gz)
axis('equal')
title('Root locus of uncompensated system')
% root locus of compensated system (with controller)
figure
rlocus(DGz)
axis('equal')
title('Root locus of compensated system')
% bode plot of uncompensated system (without controller)
Gw = d2c(Gz,'tustin')
Dw = d2c(Dz,'tustin')
DGw = Dw*Gw
% bode plot of the uncompensated system (without controller)
figure
bode(Gw)
title('Bode plot of uncompensated system')
[gm,pm,wpc,wgc]=margin(Gw)
```

```
% bode plot of the compensator (controller)
figure
bode(Dw)
title('Bode plot of PD controller')
% bode plot of the compensated system (with controller)
figure
bode(DGw)
title('Bode plot of compensated system')
[gm,pm,wpc,wgc]=margin(DGw)
wgs=atan(wgc*T/2)*2/T % in s-domain
wps=atan(wpc*T/2)*2/T % in s-domain

% closed loop step response of compensated system
figure
Tz=feedback(DGz,1)
dstep(Tz,1)
title('Closed loop step response of compensated system')

Results:
kip = 0, kdp = 1.2200, kp = 0.5899
ki =  0, kd = 0.1220,
D(z) =
  1.81 z - 1.22
  -------------
        z
gm = 12.6724, pm = 71.9699, wgc = 2.2141, wpc = 19.4187, wgs = 2.2052, wps
= 15.4130
```

The desired pole location is on the root locus as shown in Fig. 11.15. The stability of the closed-loop system has been enhanced with the PD controller as shown in Bode plots in Figs. 11.16 and 11.17. The desired specification of transient response is closely achieved with the PD controller as shown in the closed-loop step response of the system in Fig. 11.18. However, there is a non-zero steady-state error due to a step input (1/(1 + 0.5899*10) = 0.145) as shown in Fig. 11.18.

Example 11.3 PID: Design a PID controller for $G_p(s) = \frac{10}{s^2+2.5s+1}$, $H(s) = 1$ such that the closed-loop system has an overshoot of 5% and a settling time of 2 sec (2% error beta) for a step input. Use sampling time $T = 0.10$ sec. Show the effect of the PID controller on the closed-loop system stability and step response.

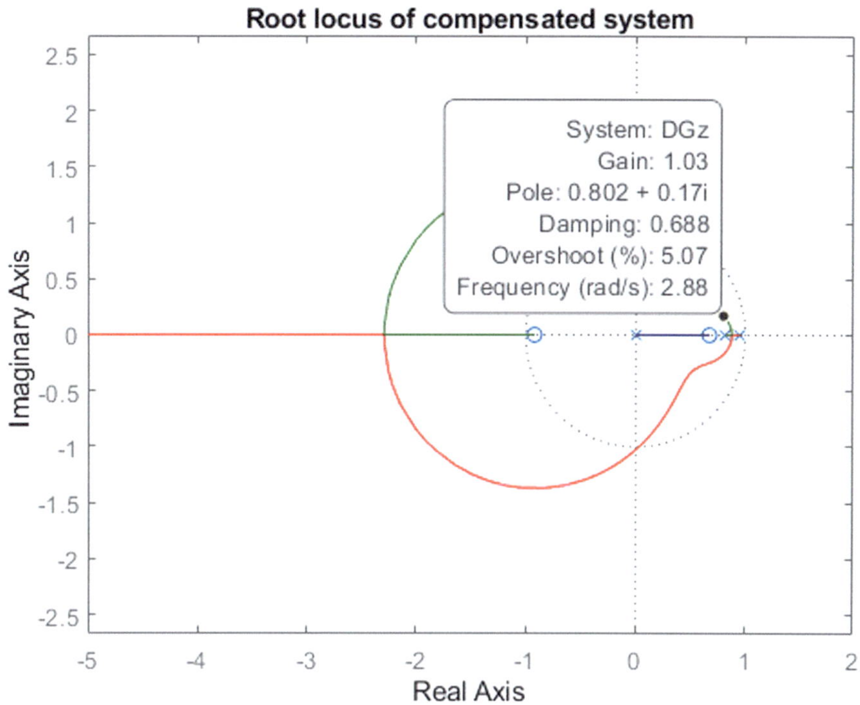

Fig. 11.15 Root locus of the compensated system with PD controller

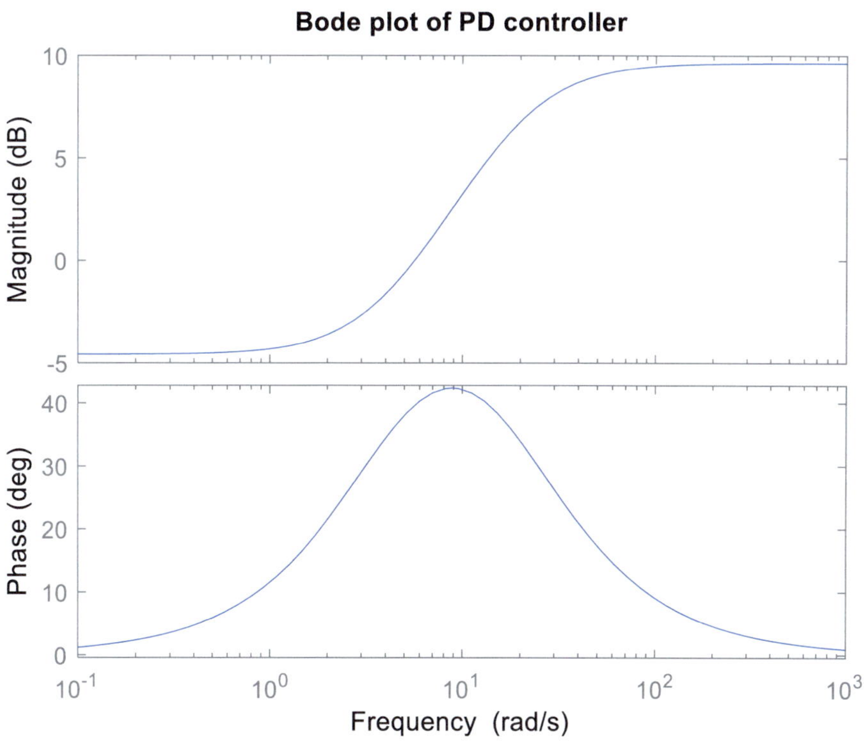

Fig. 11.16 Bode plots of the PD controller

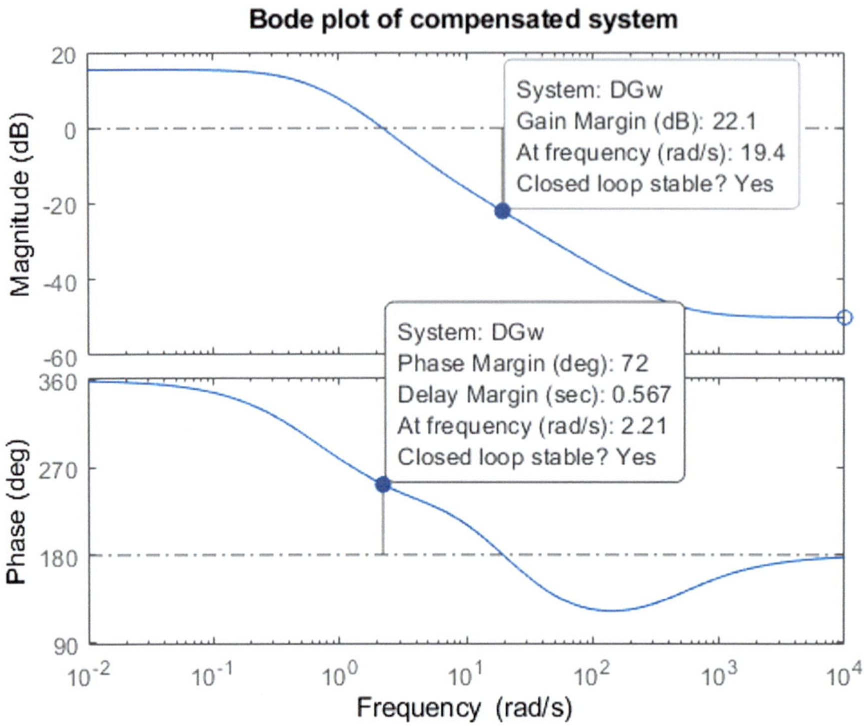

Fig. 11.17 Bode plots of the compensated system with PD controller

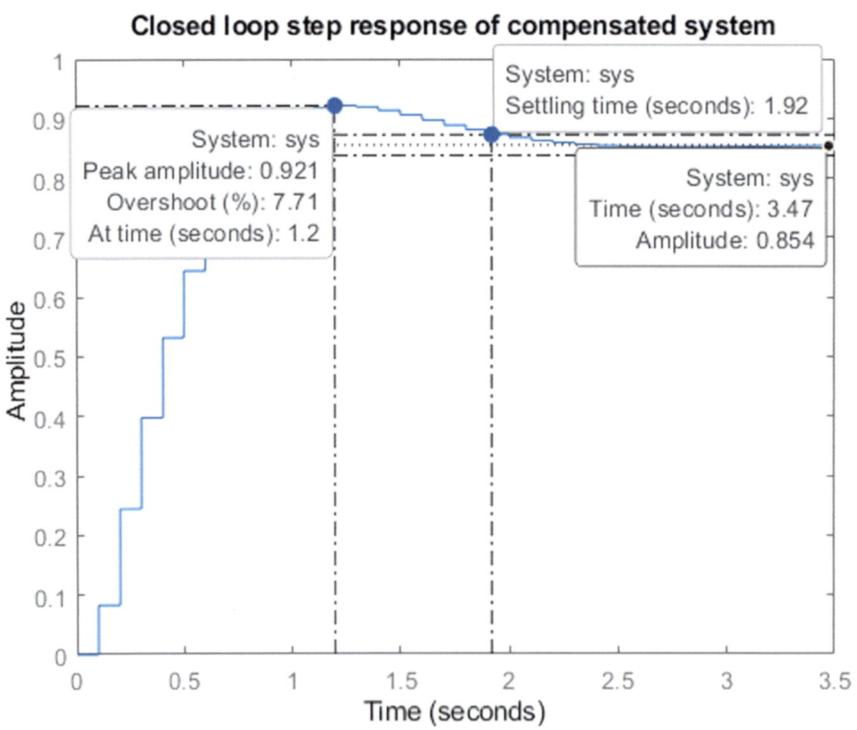

Fig. 11.18 Closed-loop step response of the compensated system with PD controller

Solution Matlab code and results are presented.

```
% design of PID controller[DIGITAL]
close all; clear all; clc
Gp=tf(10, [1 2.5 1]) % G(s) =10/(s^2 +2.5s+1)
% Get G(z)
T=0.10 % sampling time
Gz = c2d(Gp, T);  % G(z)
[ngz, dgz]=tfdata(Gz,'v');
% desired pole location s1
%input %overshoot(alpha) & settling time(tset)
alpha=0.05; tset=2; beta=0.02;
zeta=abs(log(alpha))/sqrt(pi^2+log(alpha)^2) % damping ratio
wn=-log(beta)/(zeta*tset)  % natural frequency
s1=-zeta*wn+j*wn*sqrt(1-zeta^2)  % desired pole in s-plane
z1=exp(s1*T)  % desired pole in z-plane

% Design of PID controller
a1=((z1+1)/(z1-1));
b1=(z1-1)/(z1);
a1r=real(a1); a1i=imag(a1);
b1r=real(b1); b1i=imag(b1);

%evaluate  G(z1)
nv=polyval(ngz,z1);dv=polyval(dgz,z1);
g=nv/dv;      % G(z1)
m=abs(g)      % magnitude
psi=angle(g) % angle

% pid controller gains kp, kd, ki

kip=.01 % kip   You need to adjust this to get suitable response
kdp=(sin(psi)/m-kip*a1i)/b1i
kp=-(cos(psi)/m+a1r*kip+b1r*kdp)
ki=kip/(T/2)
kd=kdp*T
Dz=tf([kp+kip+kdp -kp+kip-2*kdp kdp], [1 -1 0],T) %PID: D(z)
DGz= Dz*Gz; % D(z)G(z)

figure
% root locus of uncompensated system (without controller)
rlocus(Gz)
axis('equal')
title('Root locus of uncompensated system')
% root locus of comppensated system (with controller)
figure
rlocus(DGz)
axis('equal')
title('Root locus of compensated system')
% bode plot of uncompensated system (without controller)
Gw = d2c(Gz,'tustin')
Dw = d2c(Dz,'tustin')
DGw = Dw*Gw
% bode plot of the uncompensated system (without controller)
figure
bode(Gw)
title('Bode plot of uncompensated system')
[gm,pm,wpc,wgc]=margin(Gw)
% bode plot of the compensator (controller)
```

```
figure
bode(Dw)
title('Bode plot of PID controller')
% bode plot of the compensated system (with controller)
figure
bode(DGw)
title('Bode plot of compensated system')
[gm,pm,wpc,wgc]=margin(DGw)
wgs=atan(wgc*T/2)*2/T % in s-domain
wps=atan(wpc*T/2)*2/T % in s-domain

% closed loop step response of compensated system
figure(7)
Tz=feedback(DGz,1)
dstep(Tz,1)
title('Closed loop step response of compensated system')

Results:
zeta =    0.6901   wn =    2.8344
s1 =  -1.9560 + 2.0512i; z1 =    0.8051 + 0.1675i
m =    2.1363;   psi =    2.4398
kip =    0.0100; kdp =    1.4248; kp =    0.6780
ki =    0.2000; kd =    0.1425
D(z) =  2.113 z^2 - 3.518 z + 1.425
  ---------------------------
           z^2 - z

         -0.001037 w^2 - 0.4775 w + 9.965
 G(w) =  --------------------------------
           w^2 + 2.493 w + 0.9965

          3.528 w^2 + 13.76 w + 4
D(w) =   -----------------------
             w^2 + 20 w
gm =   10.9040, pm =   64.3781, wpc =   19.3984, wgc =    2.3996
wgs =    2.3882, wps =   15.4026
```

The desired pole location is on the root locus as shown in Fig. 11.19. The stability of the closed-loop system has slightly deteriorated with the PID controller compared to PD due to the integral part as shown in Bode plots in Figs. 11.20 and 11.21. The desired specification of percentage overshoot is achieved, but settling time increased with the PID controller as shown in the closed-loop step response of the system in Fig. 11.22. However, there is no steady-state error due to a step input, because of the integral part, as shown in Fig. 11.22.

11.4 Design of Phase-Lag, Phase-Lead, and Phase-Lag-Lead Controllers for Discrete-Time Systems

11.4.1 Design of Phase-Lag Controllers

The desired time domain specifications of overshoot and settling time can be used to get equivalent second-order system desired pole locations s_1, s_2 in the s-plane that can be converted to z_1, z_2 in the z-plane that can finally be converted to w_1, w_2 in the

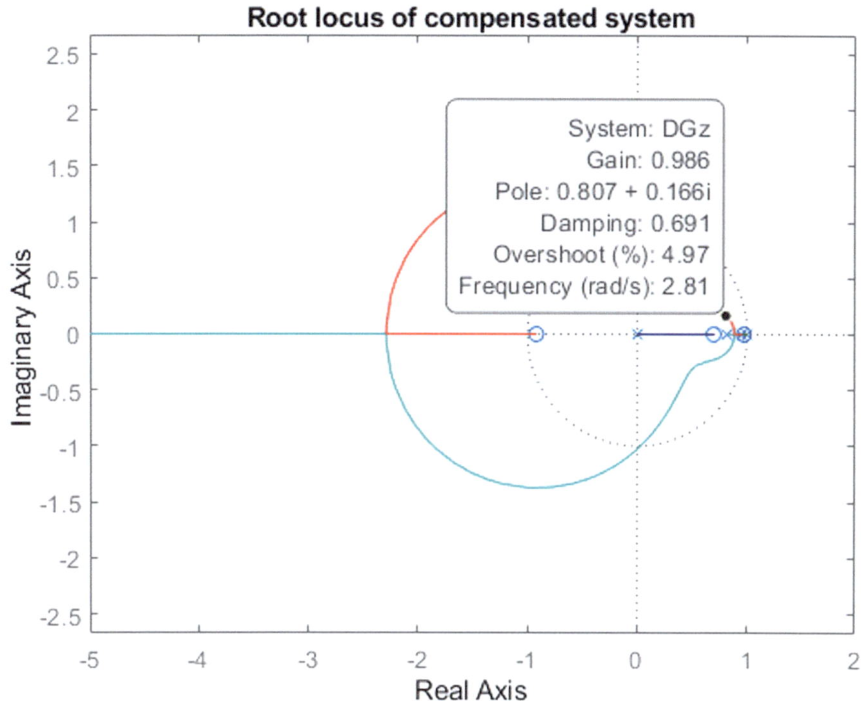

Fig. 11.19 Root locus of the compensated system with PID controller

w-plane. It should be noted that these pole pairs are complex conjugates. The characteristic equation of the discrete-time closed-loop system must be satisfied at the desired pole w_1 in the w-plane:

$$1 + D(w_1)\overline{GH}(w_1) = 0 \tag{11.54}$$

From the characteristic equation, Eq. (11.54), the controller transfer function can be expressed as

$$D(w_1) = -\frac{1}{\overline{GH}(w_1)} \tag{11.55}$$

For a phase-lag controller, the transfer function of Eq. (11.16) can be used in Eq. (11.55) as

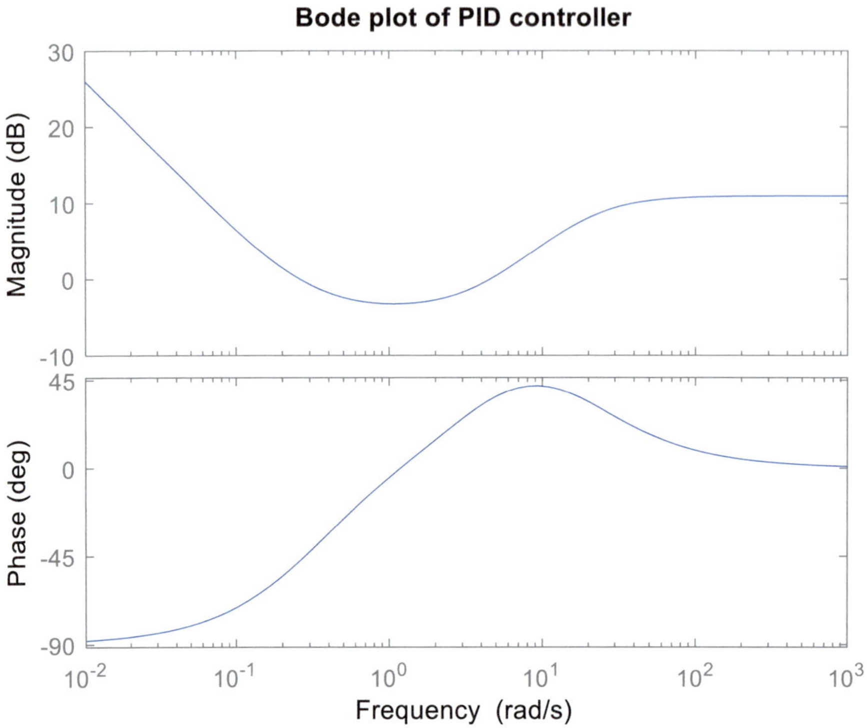

Fig. 11.20 Bode plots of the PID controller

$$\left(\frac{1 + w_1/\omega_{w_0}}{1 + w_1/\omega_{w_p}}\right) = -\frac{1}{c_0\overline{GH}(w_1)} \tag{11.56}$$

In general, the steady-state accuracy specification can be used to get c_0. c_0 is included with the open-loop transfer function as $c_0\overline{GH}(w_1)$. In subsequent calculations, c_0 is taken as 1.

Equation (11.56) can be expressed, after some rearrangements, as

$$1 + a_1 w_1 = -\frac{1}{c_0\overline{GH}(w_1)}\,(1 + b_1 w_1) \tag{11.57}$$

where

$$a_1 = 1/\omega_{w_0} \tag{11.58}$$

$$b_1 = 1/\omega_{w_p} \tag{11.59}$$

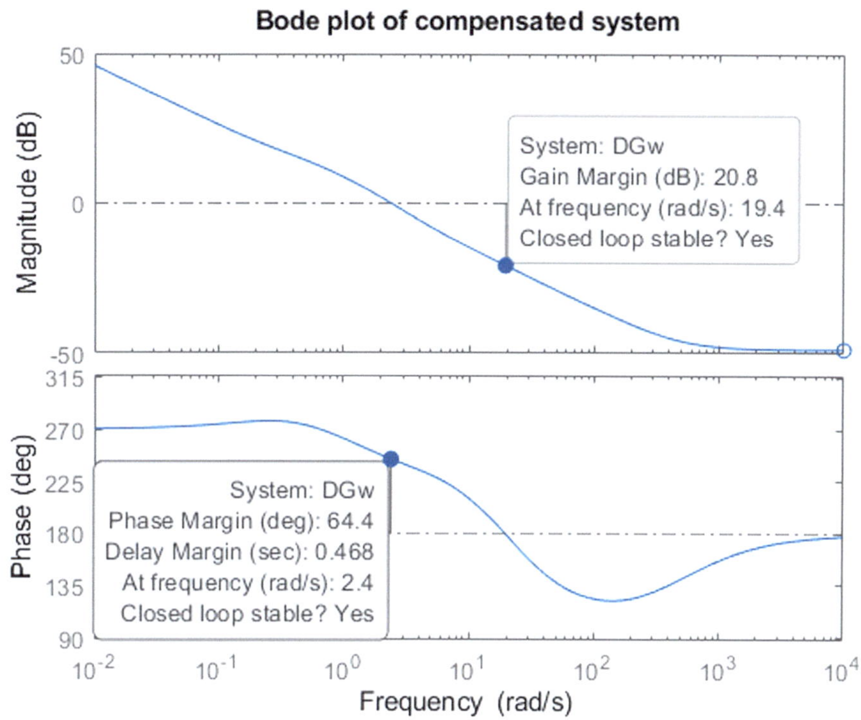

Fig. 11.21 Bode plots of the compensated system with PID controller

The known components of Eq. (11.57) can be expressed in terms of the desired pole $w_1$in the w-plane as

$$c_0\,\overline{GH}(w_1) = me^{j\Psi} \tag{11.60}$$

$$w_1 = re^{j\theta} \tag{11.61}$$

Equation (11.57) can be expressed as

$$1 + a_1 re^{j\theta} = -\frac{1}{me^{j\Psi}}\left(1 + b_1 re^{j\theta}\right) \tag{11.62}$$

Equation (11.62) can be further expanded, expressing each complex term with real and imaginary components, as

$$\begin{aligned} &1 + a_1 r(\cos\,\theta + j\,\sin\theta) \\ &\quad = q\,(cos\,\Psi - j\,\sin\Psi) + qb_1 r\,(\cos(\theta - \Psi) + j\,\sin(\theta - \Psi)) \end{aligned} \tag{11.63}$$

where

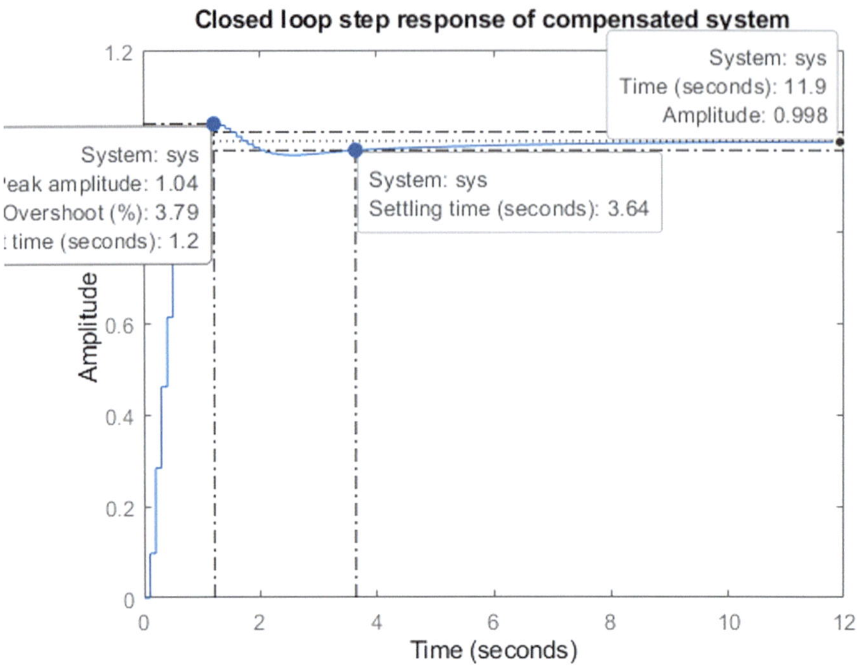

Fig. 11.22 Closed-loop step response of the compensated system with PID controller

$$q = -\frac{1}{m} \tag{11.64}$$

Real and imaginary parts of both sides of Eq. (11.63) can be expressed as

$$1 + a_1 r \cos\theta = q\cos\Psi + qb_1 r \cos(\theta - \Psi) \tag{11.65}$$

$$a_1 r \sin\theta = -q\sin\Psi + qb_1 r \sin(\theta - \Psi) \tag{11.66}$$

Equations (11.65) and (11.66) can be rearranged in matrix form to solve for the unknown parameters a_1, b_1 as

$$\begin{bmatrix} r\cos\theta & -qr\cos(\theta - \Psi) \\ r\sin\theta & -qr\sin(\theta - \Psi) \end{bmatrix} \begin{bmatrix} a_1 \\ b_1 \end{bmatrix} = \begin{bmatrix} q\cos\Psi - 1 \\ -q\sin\Psi \end{bmatrix} \tag{11.67}$$

The solutions of Eq. (11.67) can be obtained as

$$a_1 = \frac{\sin(\theta - \Psi) - q\sin\theta}{r\sin\Psi} \tag{11.68}$$

$$b_1 = \frac{\sin\theta \ \ - q\ \sin(\theta + \Psi)}{qr\ \sin\Psi} \tag{11.69}$$

The phase-lag controller transfer function, with $c_0 = 1$, in w-domain

$$D_{\text{lag}}(w) = \frac{1 + w/\omega_{w_0}}{1 + w/\omega_{w_p}} \tag{11.70}$$

The phase-lag controller zero is at $w = -\omega_{w_0} = -\frac{1}{a_1}$.

The phase-lag controller pole is at $w = -\omega_{w_p} = -\frac{1}{b_1}$.

The phase-lag controller transfer function can be expressed in the z-domain, similar to Eq. (11.14) with parameters defined in Eqs. (11.18)–(11.20), with $c_0 = 1$.

11.4.2 Design of Phase-Lead Controllers

The procedure for designing phase-lead controllers using time domain specifications is identical to that for designing phase-lag controllers discussed in Sect. 11.4.1. In the case of phase-lag controllers, the greater emphasis is on steady-state accuracy, whereas for phase-lead controllers, the emphasis is on the transient response characteristics. The time domain specifications for phase-lead controllers with greater emphasis on transient response characteristics are expected to result in switching the positions of the pole and the zero in the w-plane with the net phase-lead. The values of phase-lead controllers in the w-domain $\left(c_0', \omega_{w_0}', \omega_{w_p}', \ \ \omega_{w_0}' < \omega_{w_p}'\right)$ are calculated and next converted to z-domain (b'_0, z'_0, z'_p) using the same procedure as presented for phase-lead controllers.

11.4.3 Design of Phase-Lag-Lead Controllers

Phase-lag-lead controllers are designed in two stages. First, the phase-lag part is designed to satisfy the steady-state accuracy and relaxed (less restrictive) transient specifications using the procedure of Sect. 11.4.1. The phase-lag part is combined with the open-loop system transfer function and used as the new open-loop system for designing the phase-lead part satisfying the desired transient time response specifications. The transfer function of the phase-lag-lead controller is obtained by combining the transfer functions of the phase-lag and phase-lead parts.

Examples are presented to illustrate the procedure of designing phase-lag, phase-lead, and phase-lag-lead controllers using time domain specifications.

Example 11.4 Design a phase-lag controller for $G_p(s) = \frac{1}{s^2+3s+2}$, $H(s) = 1$ such that the closed-loop system has a steady-state error of 2%, an overshoot of 25%, and a settling time of 4 sec (2% error) for a step input. Use sampling time $T = 0.10$ sec. Show the effect of the phase-lag controller on the closed-loop system stability and step response.

Solution Matlab code and results are presented.

```
%Ex- 9.4  design of phase Lag controller
clear all; clc; close all
essStep=0.02
Kp=1/essStep-1
a0=2*Kp
ngs=a0; dgs=[1 3 2];  % K for ess,step =0.02  G(S)= K/(s+1)(s+2)
Gp =tf(ngs,dgs)  % G(s)
T=0.1
% G(z)
Gz=c2d(Gp, T)
%G(w)
Gw=d2c(Gz,'tustin')
[ngw,dgw]=tfdata(Gw,'v')
printsys(ngw,dgw,'w')

% desired pole location s1 for lag controller
%input %overshoot(alpha) & settling time(ts)
alpha=0.25; ts=4.; beta=0.02;
zeta=abs(log(alpha))/sqrt(pi^2+log(alpha)^2)
wn=abs(log(beta))/(zeta*ts)
s1=-zeta*wn+j*wn*sqrt(1-zeta^2)
z1=exp(s1*T)
w1=(2/T)*(z1-1)/(z1+1)
r=abs(w1)
theta=angle(w1)

%evaluate  G(s1) to get m and psi
nv=polyval(ngw,w1); dv=polyval(dgw,w1);
gv=nv/dv;     % G(s1)
%gv=polyval(sysg,s1)
m=abs(gv)
q=-1/m
psi=angle(gv)

% D(w)= (1+w/w0)/(1+w/wp)
a0=1
A=[ r*cos(theta)   -q*r*cos(theta-psi);
    r*sin(theta)   -q*r*sin(theta-psi)]
B=[q*cos(psi)-1;
   -q *sin(psi)]
a = A\B

%ngc=[a1 1]; dgc=[b1 1];
Dw = tf([a(1) 1], [a(2) 1])
%phase lag zero and pole
w0 = 1/a(1)
wp = 1/a(2)

% D(z) D
b0=wp*(2/T+w0)/(w0*(2/T+wp))
z0=(2/T-w0)/(2/T+w0)
zp=(2/T-wp)/(2/T+wp)
Dz=tf(b0*[1 -z0], [1 -zp],T)

% RL of uncompensated system G(s)
figure
axis equal
```

```
hold on
DGz = Dz*Gz
rlocus(Gz)
title('Root locus of uncompensated system')
% RL of compenasted system D(z)G(z)
figure
axis equal
hold on
rlocus(DGz)
title('Root locus of system with phase-lag controller')
%Closed loop system response without controller
figure
Tz = feedback(Gz, 1)
dstep(Tz,1)
title('Closed-loop step response of uncompensated system')
%CLTF: System Step Response
figure
Tz = feedback(DGz, 1)
dstep(Tz,1)
title('Closed-loop step response of system with phase-lag controller')
% bode plot of the uncompensated plant
figure
bode(Gw)
[gm,pm,wpc,wgc]=margin(Gw)
title('Bode plots of uncompensated system')
% bode plot of the controller only
figure
bode(Dw)
title('Bode plots of phase-lag controller')
% bode plot of the compensated plant
figure
DGw=Dw*Gw
bode(DGw)
[gm,pm,wpc,wgc]=margin(DGw)
title('Bode plots of system with phase-lag controller')
wps=(2/T)*atan(wpc*T/2)
wgs=(2/T)*atan(wgc*T/2)

Results:
Desired pole location for the phase-lag controller
z1 = 0.8846 + 0.1993i
w1 =-0.9893 + 2.2201i
r = 2.4305, theta = 1.9900, m = 18.9926, q = -0.0527 psi = -2.8122
w0 = 0.8294, wp = 0.0473

              1.206 w + 1
D(w)=    -----------
              21.14 w + 1

              0.05927 z - 0.05455
D(z)=    -------------------
                  z - 0.9953
```

The desired pole location has been achieved as shown in the root locus in Figs. 11.23 and 11.24. The system without the dynamic part of the phase-lag controller is unstable (with negative gain margin dB and negative phase margin) as shown in the Bode plots in Fig. 11.25. Bode plots for the phase-lag controller are shown in Fig. 11.26. The compensated system with the phase-lag controller is stable with a positive gain margin and phase margin as shown in the Bode plots in Fig. 11.27. The steady-state accuracy and the transient performance specifications are met with the phase-lag controller as shown in the closed-loop system step response in Fig. 11.28.

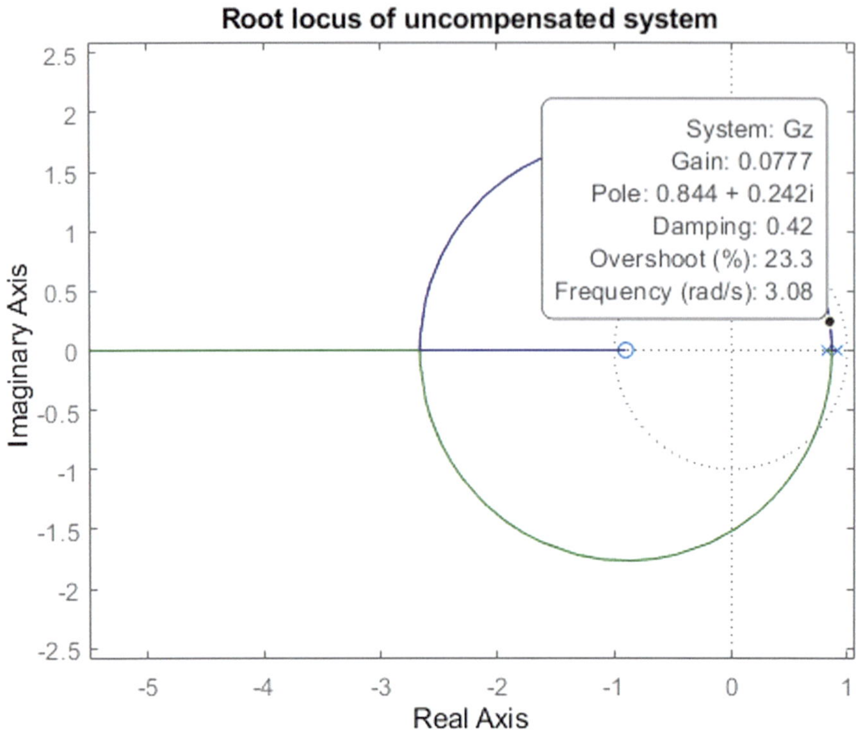

Fig. 11.23 Root locus of the uncompensated system

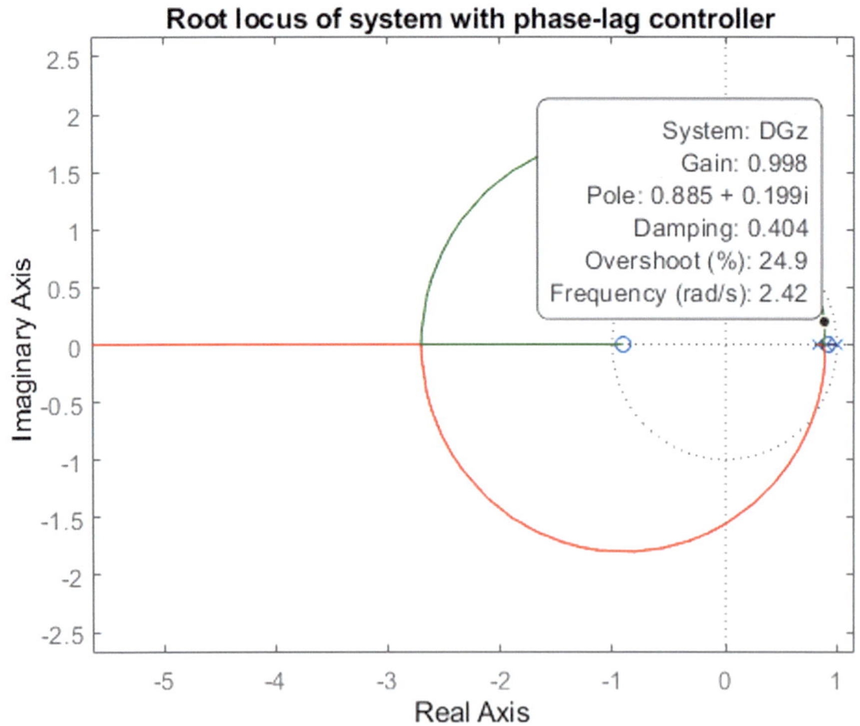

Fig. 11.24 Root locus of the compensated system with the phase-lag controller

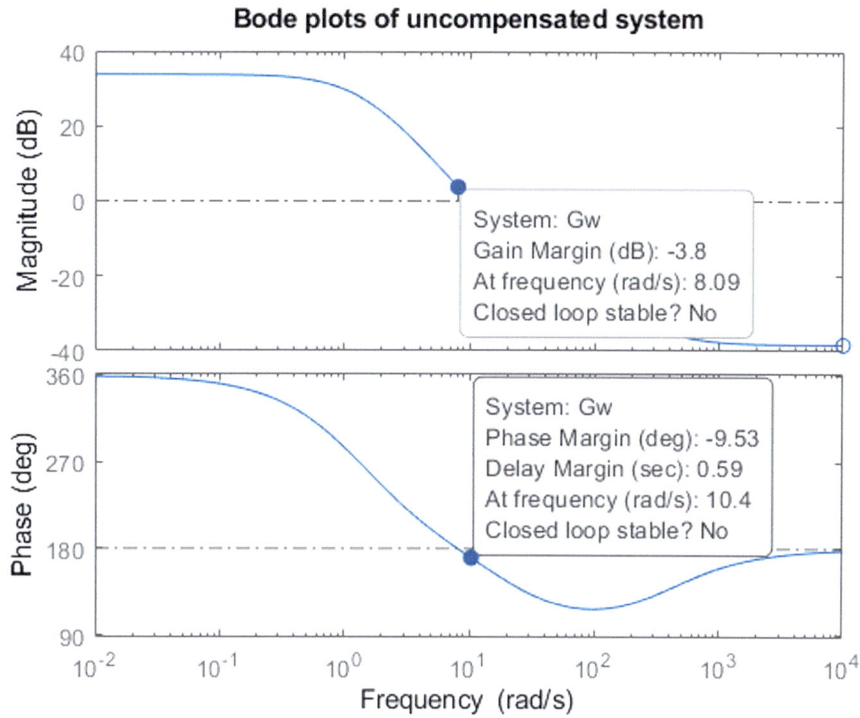

Fig. 11.25 Bode plots of the uncompensated system

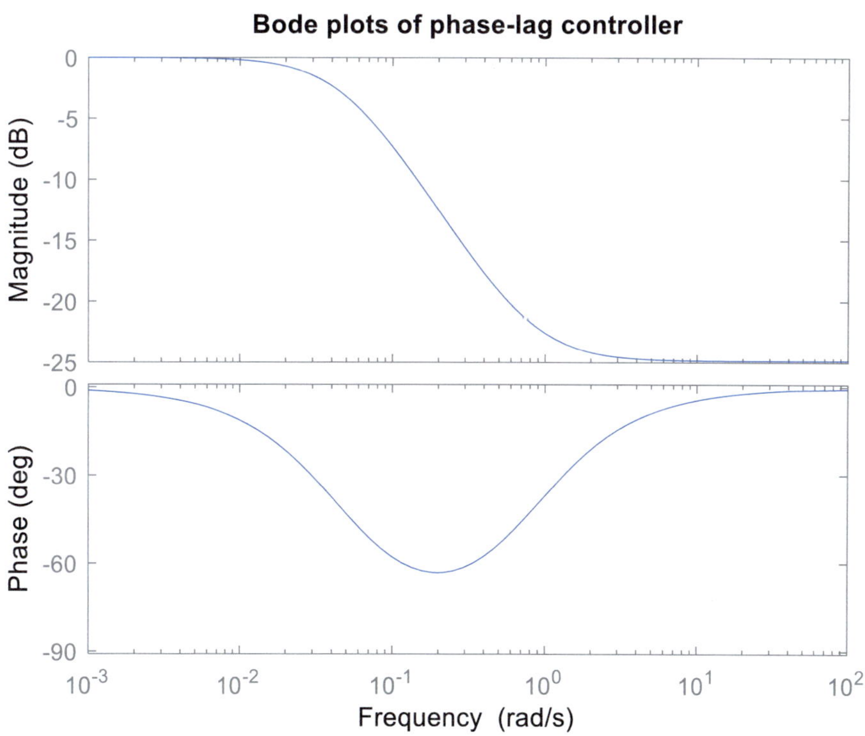

Fig. 11.26 Bode plots of the phase-lag controller

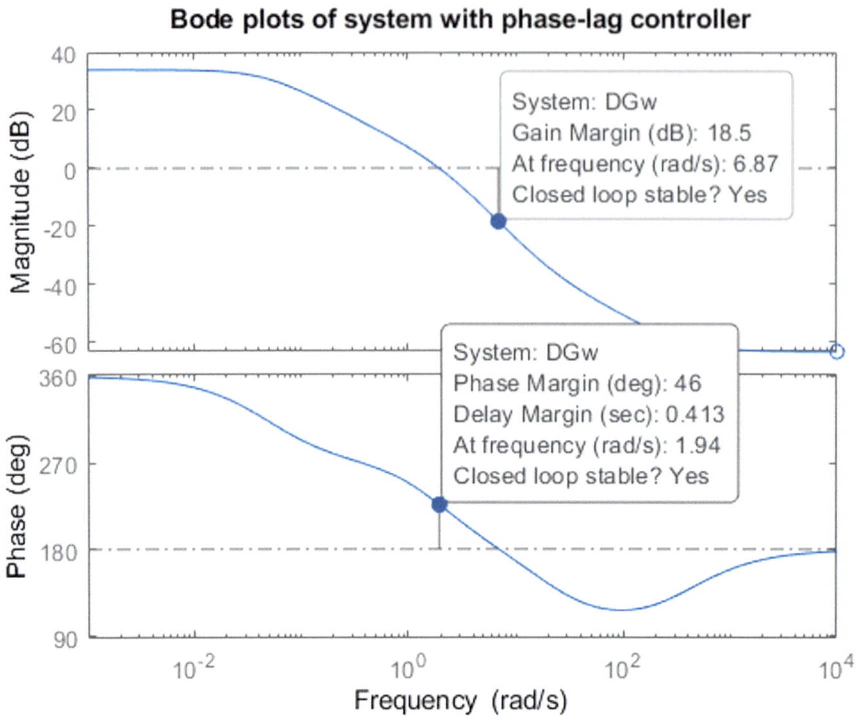

Fig. 11.27 Bode plots of the compensated system with the phase-lag controller

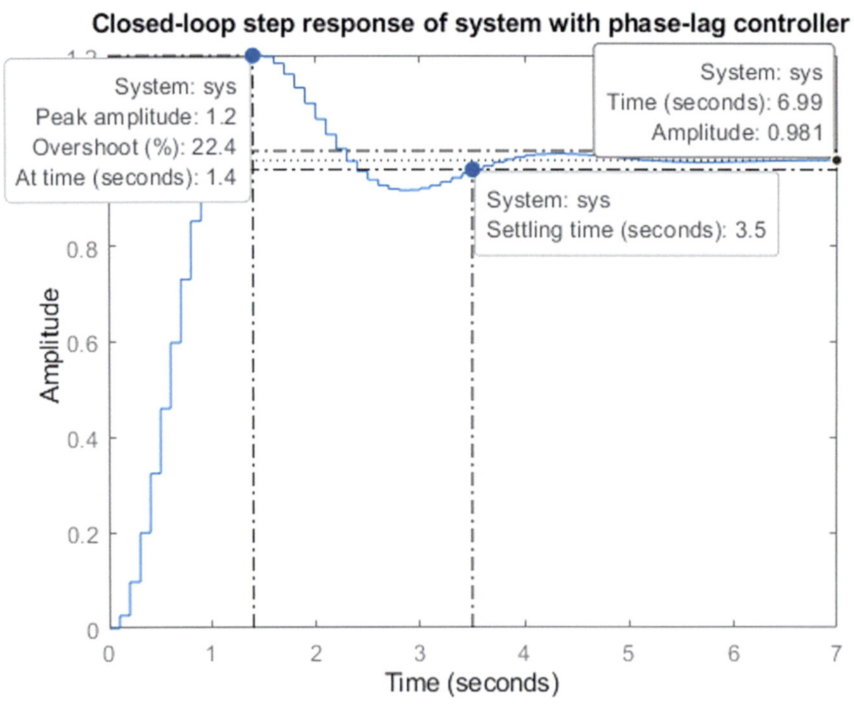

Fig. 11.28 Closed-loop step response of the compensated system with the phase-lag controller

Example 11.5 Design a phase-lead controller for $G_p(s) = \frac{1}{s^2+3s+2}$, $H(s) = 1$ such that the closed-loop system has a steady-state error of 10%, an overshoot of 5%, and a settling time of 1 sec (2% error for a step input). Use sampling time $T = 0.10$ sec. Show the effect of the phase-lead controller on the closed-loop system stability and step response.

Solution Matlab code is the same as that for the phase-lag controller design with the only change being in the desired time domain specification. Matlab code and results are presented next.

```
%Ex- 9.5  design of phase Lead controller
% Design a phase lead controller such that (1) ess,step =10%,
%  (2) overshoot =5%, settling time =1.0 s
% plant tf
clear all; clc
close all
essStep=0.10
Kp=1/essStep-1
a0=2*Kp
ngs=a0; dgs=[1 3 2];  % K for ess,step =0.02  G(S)= K/(s+1)(s+2)
Gp =tf(ngs,dgs)  % G(s)
T=0.1
% G(z)
Gz=c2d(Gp, T)
%G(w)
Gw=d2c(Gz,'tustin')
[ngw,dgw]=tfdata(Gw,'v')
printsys(ngw,dgw,'w')

% desired pole location s1 for lead controller
%input %overshoot(alpha) & settling time(ts)
alpha=0.05; ts=1.; beta=0.02;
zeta=abs(log(alpha))/sqrt(pi^2+log(alpha)^2)
wn=abs(log(beta))/(zeta*ts)
s1=-zeta*wn+j*wn*sqrt(1-zeta^2)
z1=exp(s1*T)
w1=(2/T)*(z1-1)/(z1+1)
r=abs(w1)
theta=angle(w1)

%evaluate  G(s1) to get m and psi
nv=polyval(ngw,w1); dv=polyval(dgw,w1);
gv=nv/dv;     % G(s1)
%gv=polyval(sysg,s1)
m=abs(gv)
q=-1/m
psi=angle(gv)

% D(w)= (1+w/w0)/(1+w/wp)
a0=1
A=[r*cos(theta)   -q*r*cos(theta-psi);
   r*sin(theta)   -q*r*sin(theta-psi)]
B=[ q*cos(psi)-1;
    -q*sin(psi)]
% a1=(a0*sin(theta-psi)-q*sin(theta))/(r*sin(psi))
% b1=(a0*sin(theta)-q*sin(theta+psi))/(q*r*sin(psi))
a = A\B

%ngc=[a1 1]; dgc=[b1 1];
Dw = tf([a(1) 1], [a(2) 1])
%phase lag zero and pole
w0 = 1/a(1)
wp = 1/a(2)

% D(z) D
b0=wp*(2/T+w0)/(w0*(2/T+wp))
z0=(2/T-w0)/(2/T+w0)
```

```
zp=(2/T-wp)/(2/T+wp)
Dz=tf(b0*[1 -z0], [1 -zp],T)

% RL of uncompensated system G(s)
figure
axis equal
DGz = Dz*Gz
rlocus(Gz)
% RL of compenasted system D(z)G(z)
figure
axis equal
rlocus(DGz)
title('Root locus of system with phase-lead controller')
%Closed loop system response without controller
figure
Tz = feedback(Gz, 1)
dstep(Tz,1)
title('Closed-loop step response of the uncompensated system')
%CLTF: System Step Response
figure
Tz = feedback(DGz, 1)
dstep(Tz,1)
title('Closed-loop step response of the system with the phase-lead
controller')
% bode plot of the uncompensated plant
figure
bode(Gw)
[gm,pm,wpc,wgc]=margin(Gw)
title('Bode plots of the uncompensated system')
% bode plot of the controller only
figure
bode(Dw)
title('Bode plots of the phase-lead controller')
% bode plot of the compensated plant
figure
DGw=Dw*Gw
bode(DGw)
title('Bode plots of the compensated system with the phase-lead
controller')
[gm,pm,wpc,wgc]=margin(DGw)

Results:
Desired pole location for the phase-lag controller
z1 = 0.6201 + 0.2697i
w1 = -4.0236 + 3.9993i
r = 5.6731, theta = 2.3592, m = 0.9609, q = -1.0407, psi = 1.8695
w0 = 4.5031, wp = 26.0787

       0.2221 w + 1
D(w)=  -------------
       0.03835 w + 1

       3.08 z - 1.948
D(z)=  -------------------
         z + 0.1319
```

The root locus of the uncompensated system is shown in Fig. 11.29. The desired pole location has been achieved as shown in the root locus in Fig. 11.30. The system without the dynamic part of the phase-lead controller is stable (with a positive gain margin of 10.9 dB and a phase margin of 30°) as shown in the Bode plots in

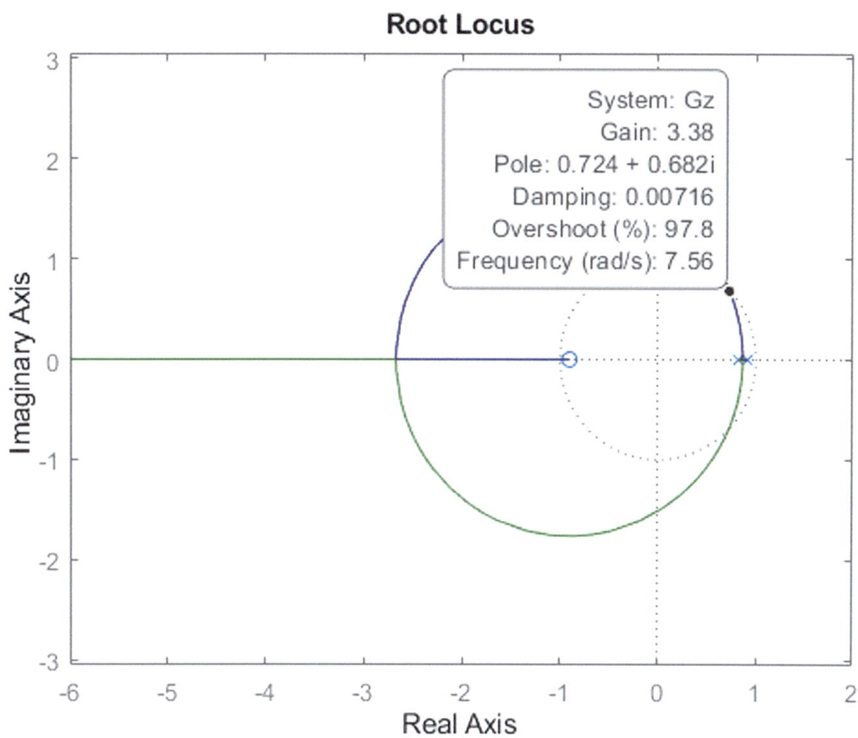

Fig. 11.29 Root locus of the uncompensated system

Fig. 11.31. Bode plots for the phase-lead controller are shown in Fig. 11.32. The compensated system with the phase-lead controller is stable with a positive gain margin of 13.8 dB and a phase margin of 57.1° as shown in the Bode plots in Fig. 11.33. The closed-loop step response of the uncompensated system is shown in Fig. 11.34. The steady-state accuracy and the transient performance specifications are closely met with the phase-lead controller as shown in the closed-loop system step response in Fig. 11.35 except the percentage overshoot calculated based on the actual steady-state value of 0.90 (overshoot 16.7%) instead of the ideal steady-state value of 1 (overshoot 5%).

Example 11.6 Design a phase-lag-lead controller for $G_p(s) = \frac{1}{s^2+3s+2}$, $H(s) = 1$ such that the closed-loop system has a steady-state error of 2%, an overshoot of 5%, and a settling time of 1 sec (2% error) for a step input. Use sampling time $T = 0.10$ sec. Show the effect of the phase-lag-lead controller on the closed-loop system stability and step response.

Solution Matlab code and results are presented next. The code has two parts—the first part is for the design of the phase-lag controller satisfying the steady-state accuracy specification and relaxed transient specifications of the overshoot of 25% and settling time of 4 sec. This part of the code is the same as that presented in

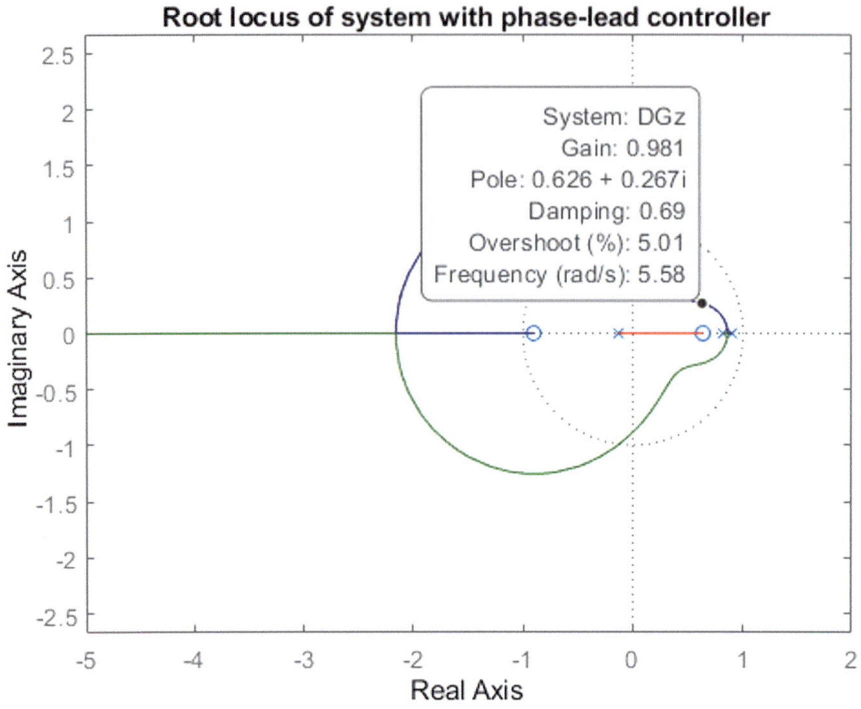

Fig. 11.30 Root locus of the compensated system with the phase-lead controller

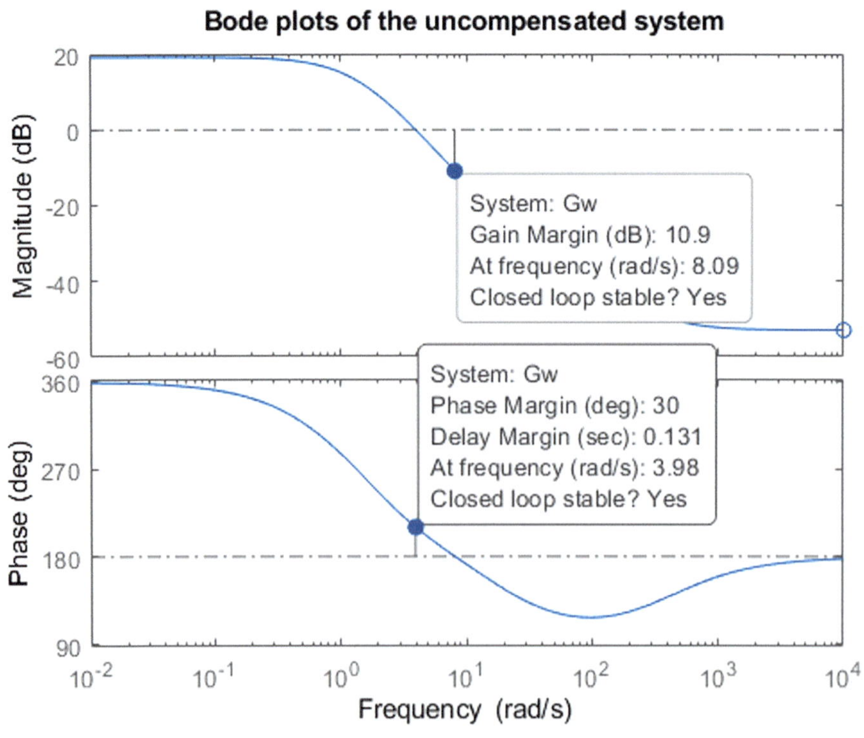

Fig. 11.31 Bode plots of the uncompensated system

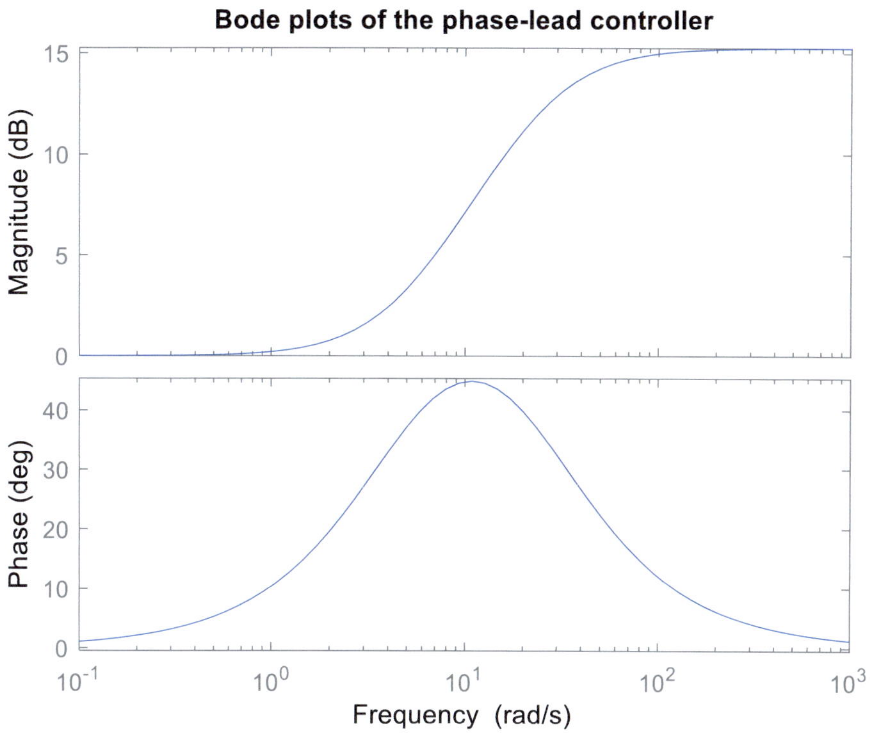

Fig. 11.32 Bode plots of the phase-lead controller

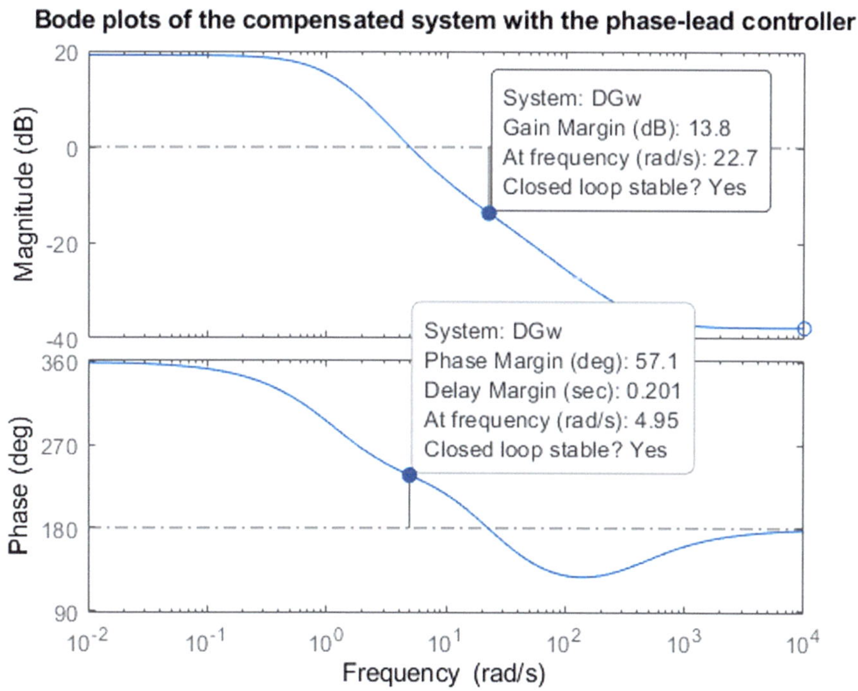

Fig. 11.33 Bode plots of the compensated system with the phase-lead controller

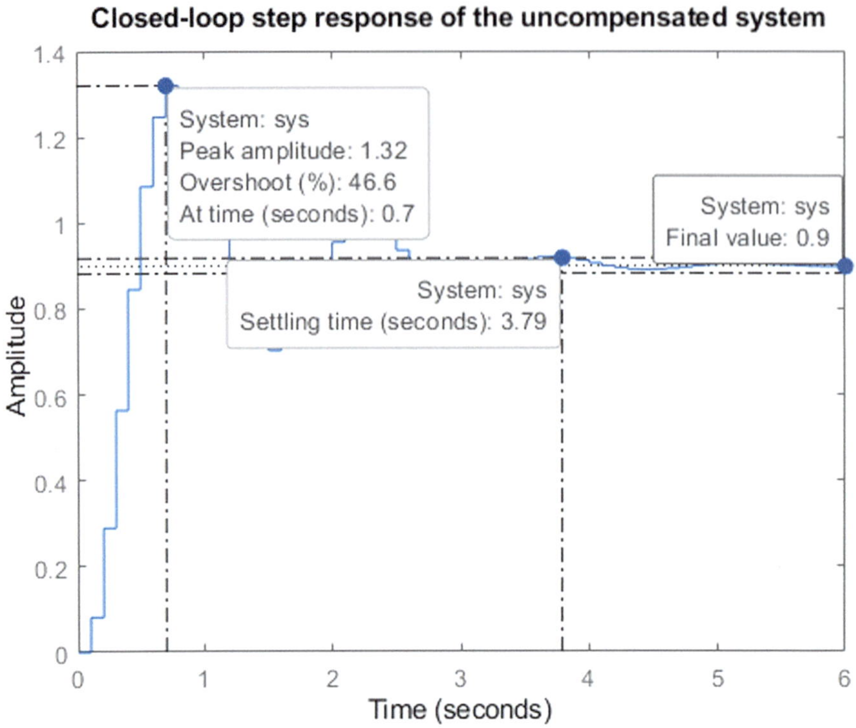

Fig. 11.34 Closed-loop step response of the uncompensated system

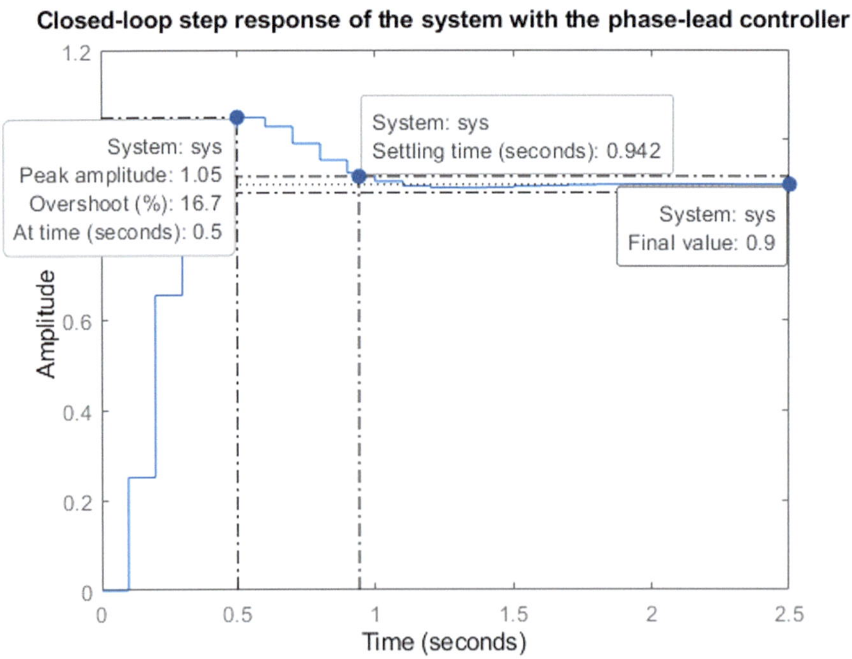

Fig. 11.35 Closed-loop step response of the compensated system with the phase-lead controller

Example 11.4 for the design of the phase-lag controller. The second part is for the design of the phase-lead controller with the desired transient specifications of 5% overshoot and settling of 1 sec. The phase-lead part code is similar to that presented in Example 11.5. However, for completeness, the entire code is presented here along with the corresponding results.

```
%Ex- 9.6  design of phase Lag Lead controller
% Design a phase-lag-lead controller such that (1) ess,step =2%,
%  (2) overshoot =5%, settling time =1.0 s
% plant tf
clear all; clc; close all
essStep=0.02
Kp=1/essStep-1
a0=2*Kp
ngs=a0; dgs=[1 3 2];  % K for ess,step =0.02  G(S)= K/(s+1)(s+2)
Gp =tf(ngs,dgs)  % G(s)
T=0.1
% G(z)
Gz=c2d(Gp, T)
%G(w)
Gw=d2c(Gz,'tustin')
[ngw,dgw]=tfdata(Gw,'v')
printsys(ngw,dgw,'w')

% desired pole location s1 for lag controller -relaxed spec
%input %overshoot(alpha) & settling time(ts)
alpha=0.25; ts=4.; beta=0.02;
zeta=abs(log(alpha))/sqrt(pi^2+log(alpha)^2)
wn=abs(log(beta))/(zeta*ts)
s1=-zeta*wn+j*wn*sqrt(1-zeta^2)
z1=exp(s1*T)
w1=(2/T)*(z1-1)/(z1+1)
r=abs(w1)
theta=angle(w1)

%evaluate  G(s1) to get m and psi
nv=polyval(ngw,w1); dv=polyval(dgw,w1);
gv=nv/dv;     % G(s1)
%gv=polyval(sysg,s1)
m=abs(gv)
q=-1/m
psi=angle(gv)

% D(w)= (1+w/w0)/(1+w/wp)
A=[r*cos(theta)    -q*r*cos(theta-psi);
   r*sin(theta)    -q*r*sin(theta-psi)]
B=[ q*cos(psi)-1;
    -q*sin(psi)]
% a1=(sin(theta-psi)-q*sin(theta))/(r*sin(psi))
% b1=(sin(theta)-q*sin(theta+psi))/(q*r*sin(psi))
a = A\B

%ngc=[a1 1]; dgc=[b1 1];
Dw = tf([a(1) 1], [a(2) 1])
%phase lag zero and pole
w0 = 1/a(1)
wp = 1/a(2)

% D(z)
b0=wp*(2/T+w0)/(w0*(2/T+wp))
z0=(2/T-w0)/(2/T+w0)
zp=(2/T-wp)/(2/T+wp)
Dz=tf(b0*[1 -z0], [1 -zp],T)
```

```
% RL of uncompensated system G(s)
figure

DGz = Dz*Gz
rlocus(Gz)
axis equal
title('Root locus of uncompensated system')
% RL of compenasted system D(z)G(z)
figure
rlocus(DGz)
axis equal
title('Root locus of system with phase-lag controller')
%Closed loop system response without controller
figure
Tz = feedback(Gz, 1)
dstep(Tz,1)
title('Closed-loop step response of uncompensated system')
%CLTF: System Step Response
figure
Tz = feedback(DGz, 1)
dstep(Tz,1)
title('Closed-loop step response of system with phase-lag controller')

% bode plot of the uncompensated plant
figure
bode(Gw)
[gm,pm,wpc,wgc]=margin(Gw)
title('Bode plots of uncompensated system')
% bode plot of the controller only
figure
bode(Dw)
title('Bode plots of phase-lag controller')
% bode plot of the compensated plant
figure
DGw=Dw*Gw
[ngw1, dgw1]=tfdata(DGw,'v') % D(w)G(w)
bode(DGw)
[gm,pm,wpc,wgc]=margin(DGw)
title('Bode plots of system with phase-lag controller')
%%%%%%%%%%%%%%% phase lead part %%%%%%%%%%%%%%%%%%%%%%%%
% desired pole location s1 for lead part - full transient spec
%input %overshoot(alpha) & settling time(ts)
alpha=0.05; ts=1.; beta=0.02;
zeta=abs(log(alpha))/sqrt(pi^2+log(alpha)^2)
wn=abs(log(beta))/(zeta*ts)
s1=-zeta*wn+j*wn*sqrt(1-zeta^2)
z1=exp(s1*T)
w1=(2/T)*(z1-1)/(z1+1)
r=abs(w1)
theta=angle(w1)

%evaluate  G(s1) to get m and psi
nv=polyval(ngw1,w1); dv=polyval(dgw1,w1);
gv=nv/dv;     % G(s1)
%gv=polyval(sysg,s1)
m=abs(gv)
q=-1/m
```

```
psi=angle(gv)

% D(w)= (1+w/w0)/(1+w/wp)
A=[r*cos(theta)    -q*r*cos(theta-psi);
   r*sin(theta)    -q*r*sin(theta-psi)]
B=[ q*cos(psi)-1;
    -q*sin(psi)]
% a1=(sin(theta-psi)-q*sin(theta))/(r*sin(psi))
% b1=(sin(theta)-q*sin(theta+psi))/(q*r*sin(psi))
a = A\B

%ngc=[a1 1]; dgc=[b1 1];
Dwp = tf([a(1) 1], [a(2) 1])
%phase lag zero and pole
w0p = 1/a(1)
wpp = 1/a(2)

% D(z) D
b0p=wpp*(2/T+w0p)/(w0p*(2/T+wpp))
z0p=(2/T-w0p)/(2/T+w0p)
zpp=(2/T-wpp)/(2/T+wp)
Dzp=tf(b0p*[1 -z0p], [1 -zpp],T)

% RL of compenasted system D(z)G(z)

figure
DGzp = Dzp*DGz
rlocus(DGzp)
axis equal
title('Root locus of system with phase-lag-lead controller')
%CLTF: System Step Response
figure
Tz = feedback(DGzp, 1)
dstep(Tz,1)
title('Closed-loop step response of system with phase-lag-lead controller')
% bode plot of the controller only
Dwf=Dw*Dwp
figure
bode(Dwf)
% bode plot of the compensated plant
figure
DGw=Dw*Gw*Dwp
bode(DGw)
title('Bode plots of system with phase-lag-lead controller')
[gm,pm,wpc,wgc]=margin(DGw)
wps=(2/T)*atan(wpc*T/2)
wgs=(2/T)*atan(wgc*T/2)

Results:
Desired pole location for phase-lag part
z1 = 0.8846 + 0.1993i
w1 =-0.9893 + 2.2201i
w0 = 0.8294, wp = 0.0473

           1.206 w + 1
Dlag(w) =  -----------
           21.14 w + 1
```

```
          0.05927 z - 0.05455
Dlag(z) = -------------------
              z - 0.9953

Desired pole location for the phase-lead part
z1 = 0.6201 + 0.2697i
w1 =-4.0236 + 3.9993i
w0p = 1.7596, wpp =    8.7215

           0.5683 w + 1
Dlead(w) = ------------
           0.1147 w + 1

            3.755 z - 3.148
Dlead(z) =  ---------------
              z - 0.5626

Phae lag-lead controller TF

           1.206 w + 1     0.5683 w + 1
D(w)=     ------------   --------------
           21.14 w + 1     0.1147 w + 1

           0.05927 z - 0.05455    3.755 z - 3.148
D(z)=      -------------------   ---------------
                z - 0.9953          z - 0.5626


gm = 7.0109, pm = 71.2663, wpc = 14.2040, wgc =  2.8753, wps = 12.3508, wgs
= 2.8557
```

The desired pole location cannot be achieved for the phase-lag-lead controller as shown in the root locus in Fig. 11.36 due to the arbitrary transient specifications for the phase-lag part. The system with the phase-lag controller is similar to that presented in Example 11.4. Bode plots for the phase-lag-lead controller are shown in Fig. 11.37. The compensated system with the phase-lag-lead controller is stable with a positive gain margin of 16.9 dB and a phase margin of 71.3° as shown in the Bode plots in Fig. 11.38. The closed-loop step response of the compensated system is shown in Fig. 11.39. The steady-state accuracy and the transient performance specifications are approximately met with the phase-lag-lead controller as shown in the closed-loop system step response in Fig. 11.39 with the percentage overshoot calculated based on the actual steady-state value of 0.986 (overshoot 13.6%), settling of 1.76 sec.

11.5 Implementation of Controllers

11.5.1 Laboratory Experiment for Analysis, Design, Simulation, and Implementation of PI, PD, and PID Controllers for Discrete-Time Systems Using Time Domain Specifications

Objectives The objectives are to design, simulate, implement, and validate PI, PD, and PID controllers in the z-domain to achieve the desired transient response of a DC motor position control system.

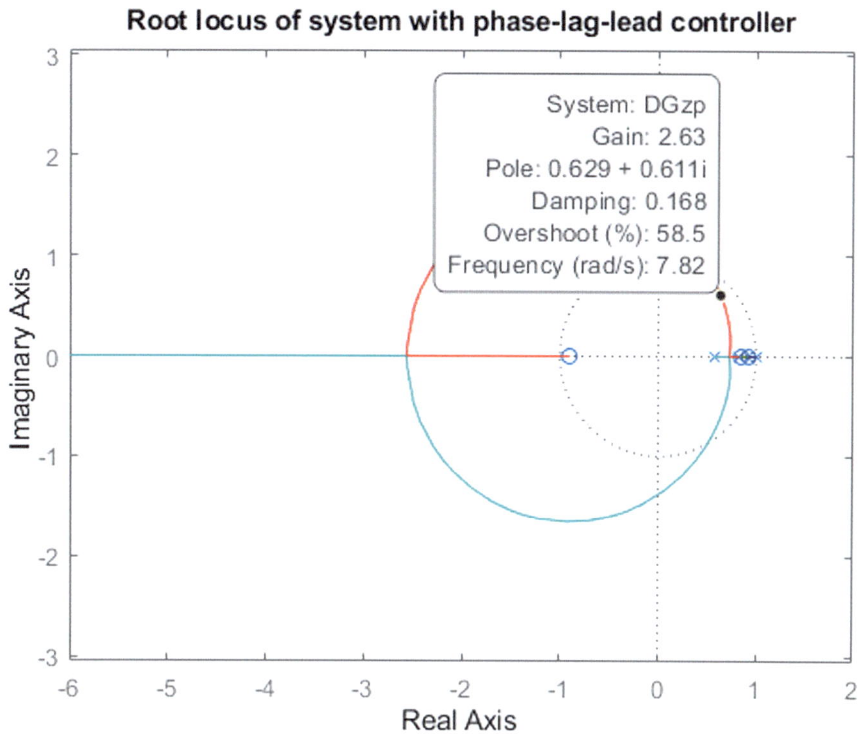

Fig. 11.36 Root locus of the compensated system with the phase-lag-lead controller

Introduction The transfer function for a PID controller in z-domain is

$$D_{PID}(z) = K_P + K_I \frac{T}{2}\left(\frac{z+1}{z-1}\right) + K_D\left(\frac{z-1}{Tz}\right)$$

In Part I, it is required to design the controller and simulate its performance in Matlab as presented in Sect. 11.3. In Part II, it is required to implement the controller in LabVIEW and validate its performance with reference to the design specifications.

I. *Controller Design and Simulation*

1. Use the DC motor position control model with previously estimated values of K and τ to get the discrete-time open-loop transfer function in z-domain $G(z)$ selecting a suitable sampling time T. For the DC motor control trainer (DCMCT) module, the sampling time (T) is taken as 0.01 s.
2. Determine the desired pole location in both s- and z-planes such that closed-loop DC motor position response has an overshoot (α) within 15% and a settling time (t_{set}) of 0.25 s for an error limit (β) of $\pm 2\%$ for a step input. Refer to Chap. 10, Sect. 10.3 for details.

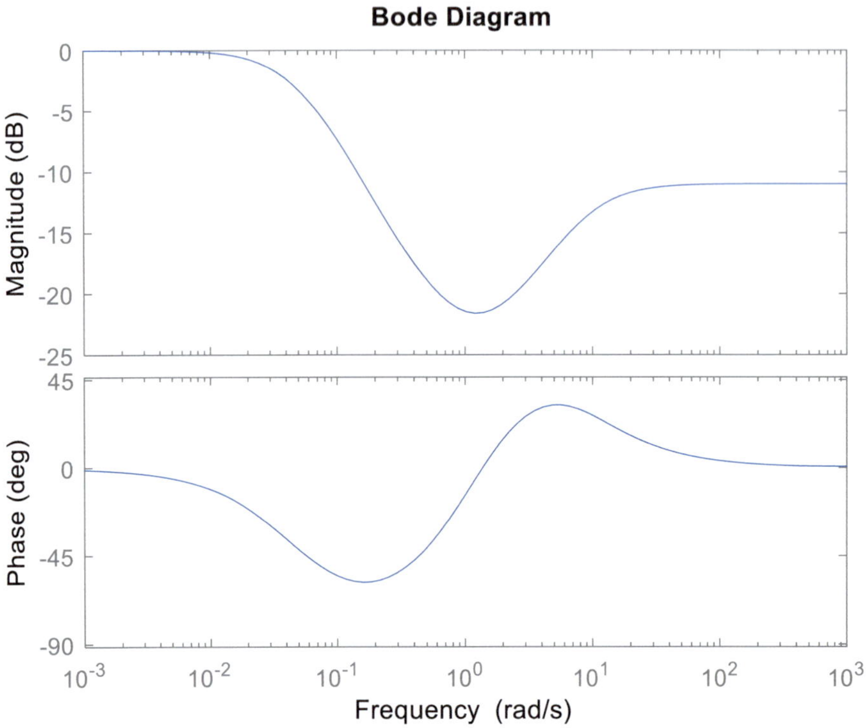

Fig. 11.37 Bode plots of the phase-lag-lead controller

3. Design a PID controller for the position control system in MATLAB for the desired transient specifications. Select a suitable integral gain K_I. Refer to Sect. 11.3 for details.
4. Note down the controller gains K_P, K_I, and K_D and corresponding controller transfer function $D(z)$ for each variation: PI ($K_D = 0$), PD ($K_I = 0$), and PID in Table 11.1.
5. Simulate closed-loop step response for each controller. Locate response characteristics by right-clicking the figure in Matlab, selecting peak response, settling time, and steady state (right-click → response characteristics → peak response → settling time → steady state → cursor on identified point). Note down in Table 11.1 overshoot (α) and settling time (t_{set}) from the simulated response for each type of controller. Copy closed-loop step response with each controller for presentation in the lab report.
6. Draw the root locus of the system without the controller and with each type of controller: PI, PD, and PID. Locate the closed-loop poles of the system with each controller using the cursor on the root locus near the desired pole, and present these in Table 11.2. Copy root locus with each controller for presentation in the lab report.

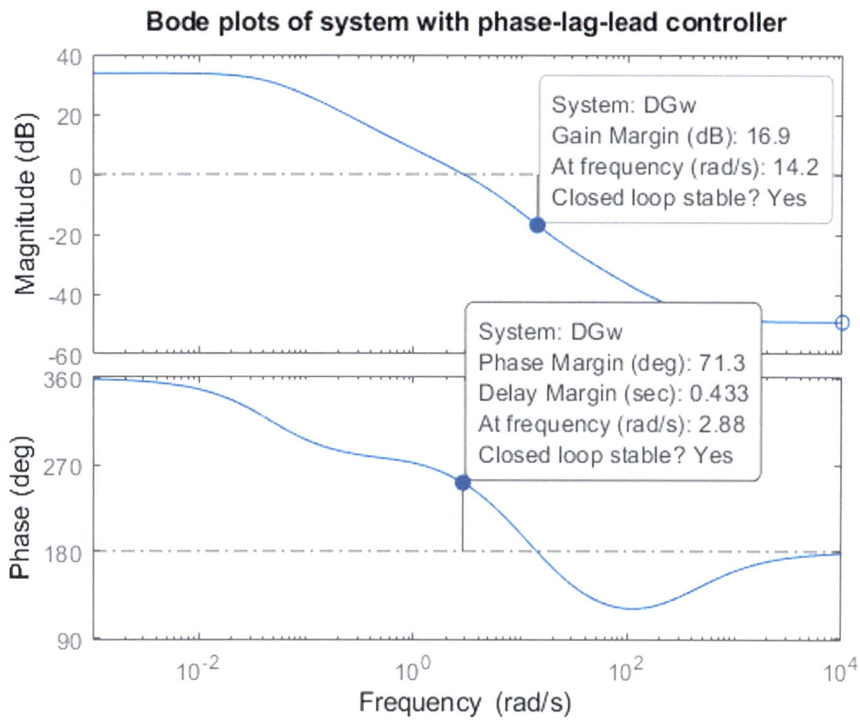

Fig. 11.38 Bode plots of the compensated system with the phase-lag-lead controller

7. Convert DG(z) to DG(w) using the bilinear transformation. Draw Bode plots of the system without the controller and with each type of controller: PI, PD, and PID. Obtain gain margin (GM), phase margin (ϕ_M), gain cross-over frequency (ω_{gc}), and phase cross-over frequency (ω_{pc}) for each type of controller by right-clicking on any of the Bode plots, selecting all stability margins, and placing cursors on the identified points (right-click on Bode plots → all stability margins → cursor on the stability points). Present these in Table 11.2. Copy Bode plots with each controller for presentation in the lab report.

II. *Implementation and Validation*

1. Set up NI ELVIS II board and DC motor module. Select the device, and set the sampling rate. For the DC motor control trainer (DCMCT) module, a sampling rate of 100 Hz is used.
2. Study the block diagram for the position control experiment (*01-QNET_DCMCT_Position_Control.vi*).
3. Open the block diagram VI, and identify the different components of the control system, including the input, error, controller, plant, sensor, and output in Fig. 11.40.
4. Modify the block diagram VI of Fig. 11.40 by removing the control parameters block and adding a subtraction block to calculate the error, as well as a

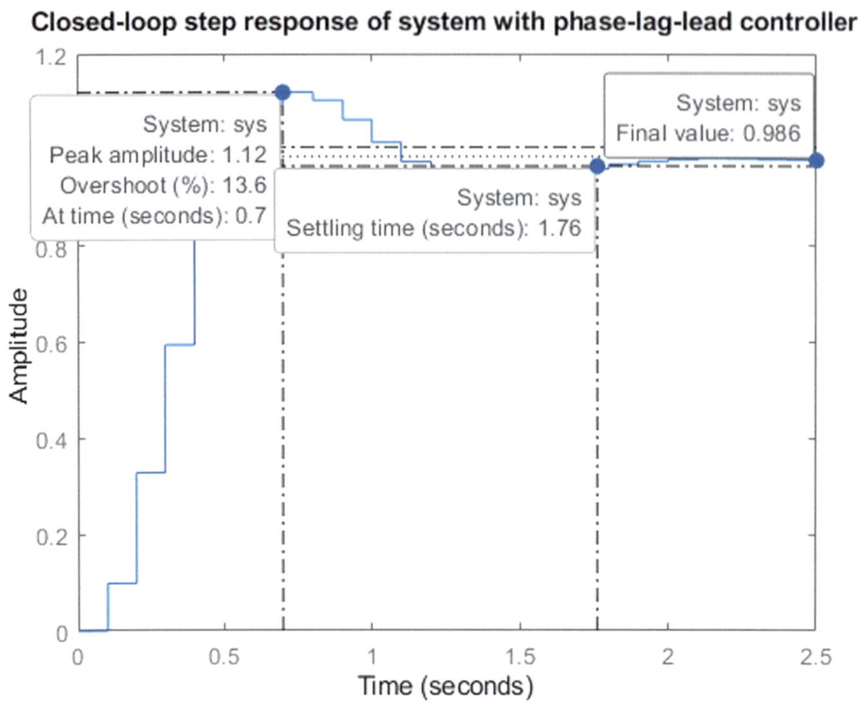

Fig. 11.39 Closed-loop step response of the compensated system with the phase-lag-lead controller

discrete transfer function block for implementation of the designed controller, as shown in the VI of Fig. 11.41.

5. Put the discrete-time z-domain transfer function $D(z)$, and follow the format of setting the controller transfer function coefficients. Save the VI.
6. Select your input parameters. For the DCMCT module, typical values for the input square wave voltage are as follows: amplitude of 1.0 V, frequency of 0.40 Hz, and offset of 3.0 V.
7. Run the VI, and observe the closed-loop system response. Copy graphs of closed-loop system response from the front panel VI for presentation in the lab report.
8. Make sure to set the number of significant figures to at least 5 for the display format of each graph on the front panel VI (right-click on the graph → display format → decimal → 6). Export the time response graphs, both motor rotation $\theta(t)$ and motor input voltage $V(t)$, to Excel by right-clicking on each graph and selecting the Export to Excel option (right-click → Export to Excel). Estimate overshoot and settling time from closed-loop system response graph.
9. Repeat steps (5)–(8) for each type of controller: PI, PD, and PID. Complete the last two columns of Table 11.1.

Table 11.1 Summary of controller parameters and performance

Controller	K_P	K_I	K_D	Controller transfer function $D(z)$	Closed-loop system step response			
					Simulation		Implementation	
					Overshoot α (%)	Settling time $t_{\text{set}}(s)$	Overshoot α (%)	Settling time $t_{\text{set}}(s)$
PI			0					
PD		0						
PID								

Table 11.2 Summary of controller performance using root locus and Bode plots

Controller	Root locus closed-loop poles	Bode plots			
		GM	ϕ_M	ω_{gc} (rad/s)	ω_{pc} (rad/s)
Without controller					
PI					
PD					
PID					

III. *Deliverables*

1. Complete Tables 11.1 and 11.2.
2. Present graphs of simulated closed-loop step response for each controller with response characteristics (peak response, settling time, and steady-state value) clearly annotated.
3. Present graphs of root locus analysis for each controller with closed-loop poles near the desired ones clearly annotated.
4. Include Bode plots for each controller with all stability margins clearly annotated.
5. Include graphs of closed-loop system step response with each implemented controller.
6. Discuss results comparing the performance of controllers in simulation and experimental validation.
7. Discuss the stability of the closed-loop system with controllers from the results of root locus and Bode plots.

Lab Report
Present all deliverables in the lab report and discuss your results.

11.5.2 *Lab Experiment for Analysis, Design, Simulation, and Implementation of Phase-Lag, Phase-Lead, and Phase-Lag-Lead Controllers for Discrete-Time Systems Using Time Domain Specifications*

Objectives The objectives are to design, simulate, implement, and validate phase-lag, phase-lead, and phase-lag-lead controllers in the z-domain to achieve the desired transient response of a DC motor position control system.

Introduction The transfer function for a phase-lag-lead controller in z-domain

$$D_{\text{lag}-\text{lead}}(z) = \left(b_0 \frac{z - z_0}{z - z_p}\right)\left(b_0' \frac{z - z_0'}{z - z'_p}\right)$$

In Part I, it is required to design the phase-lag, phase-lead, and phase-lag-lead controllers and simulate the performance of each controller in Matlab as presented in Sect. 11.4. In Part II, it is required to implement each controller in LabVIEW and validate its performance with reference to the design specifications.

I. *Controller Design and Simulation*

1. Use the DC motor position control model with previously estimated values of K and τ to get the discrete-time open-loop transfer function in z-domain $G(z)$ selecting a suitable sampling time T. For the DC motor control trainer (DCMCT) module, the sampling time (T) is taken as 0.01 s. Convert the open-loop transfer function $G(z)$ to $G(w)$ using bilinear transformation.
2. Determine the desired pole location in both s- and z-planes such that closed-loop DC motor position response has an overshoot (α) within 25% and a settling time (t_{set}) of 1 s for an error limit (β) of $\pm$2% for a step input. Refer to Chap. 10, Sect. 10.3 for details. Convert the desired pole location in the z-domain (z_1) to w-domain (w_1).
3. Design a phase-lag controller for the position control system in MATLAB for the desired pole location w_1. Refer to Sect. 11.4.1 for details. Convert the phase-lag controller transfer function $D(w)$ to $D(z)$. Note down the controller zero and pole and controller transfer function $D(z)$ for the phase-lag controller in Table 11.3.
4. Design a phase-lead controller for the position control system step response with an overshoot (α) within 15% and a settling time (t_{set}) of 0.25 s. Refer to Sect. 11.4.2 for details. Convert the phase-lead controller transfer function $D(w)$ to $D(z)$. Note down the controller zero and pole and controller transfer function $D(z)$ for the phase-lead controller in Table 11.3.
5. Design a phase-lag-lead controller for the position control system step response with an overshoot (α) within 15% and a settling time (t_{set}) of 0.25 s. Assume the specification of step (2) for designing the phase-lag part of the controller. Include the phase-lag controller with the original $G(w)$ for designing the phase-lead part of the phase-lag-lead controller. Refer to Sect. 11.4.3 for details. Convert the phase-lag-lead controller transfer function $D(w)$ to $D(z)$. Note down the controller zeros and poles and controller transfer function $D(z)$ for the phase-lag-lead controller in Table 11.3.
6. Simulate closed-loop step response for each controller. Locate response characteristics by right-clicking the figure in Matlab, selecting peak response, settling time, and steady state (right-click → response characteristics → peak response → settling time → steady state → cursor on

Table 11.3 Summary of controller parameters and performance

				Closed-loop system step response			
				Simulation		Implementation	
Controller	Zeros (z_i)	Poles (p_i)	Controller transfer function $D(z)$	Overshoot α (%)	Settling time $t_{set}(s)$	Overshoot α (%)	Settling time $t_{set}(s)$
Phase-lag							
Phase-lead							
Phase-lag-lead							

Table 11.4 Summary of controller performance using root locus and Bode plots

Controller	Root locus closed-loop poles	Bode plots			
		GM	ϕ_M	ω_{gc} (rad/s)	ω_{pc} (rad/s)
Without controller					
Phase-lag					
Phase-lead					
Phase-lag-lead					

identified point). Note down in Table 11.3 overshoot (α) and settling time (t_{set}) from the simulated response for each type of controller. Copy closed-loop step response with each controller for presentation in the lab report.

7. Draw the root locus of the system without the controller and with each type of controller. Locate the closed-loop poles of the system with each controller using the cursor on the root locus near the desired pole and present these in Table 11.4. Copy root locus with each controller for presentation in the lab report.
8. Draw Bode plots of the system without the controller and with each type of controller. Obtain gain margin (GM), phase margin (ϕ_M), gain cross-over frequency (ω_{gc}), and phase cross-over frequency (ω_{pc}) for each type of controller by right-clicking on any of the Bode plots, selecting all stability margins, and placing cursors on the identified points (right-click on Bode plots → all stability margins → cursor on the stability points). Present these in Table 11.4. Copy Bode plots with each controller for presentation in the lab report.

II. *Implementation and Validation*

1. Set up NI ELVIS II board and DC motor module. Select the device, and set the sampling rate. For the DC motor control trainer (DCMCT) module, a sampling rate of 100 Hz is used.
2. Study the block diagram for the position control experiment (*01-QNET_DCMCT_Position_Control.vi*).
3. Open the block diagram VI, and identify the different components of the control system, including the input, error, controller, plant, sensor, and output in Fig. 11.40.
4. Modify the block diagram VI of Fig. 11.40 by removing the control parameters block and adding a subtraction block to calculate the error, as well as a discrete transfer function block for implementation of the designed controller, as shown in the VI of Fig. 11.41.
5. Put the discrete-time z-domain transfer function $D(z)$, and follow the format of setting the controller transfer function coefficients. Save the VI.
6. Select your input parameters. For the DCMCT module, typical values for the input square wave voltage are as follows: amplitude of 1.0 V, frequency of 0.40 Hz, and offset of 3.0 V.

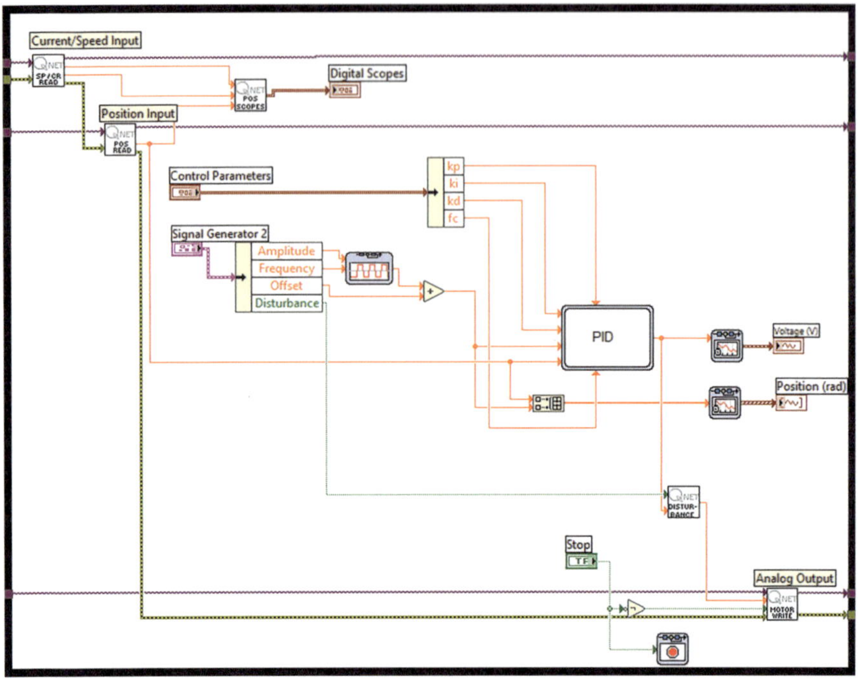

Fig. 11.40 Block diagram VI for DC motor control trainer position control VI before modification

7. Run the VI, and observe the closed-loop system response. Copy graphs of closed-loop system response from the front panel VI for presentation in the lab report.
8. Make sure to set the number of significant figures to at least 5 for the display format of each graph on the front panel VI (right-click on the graph → display format → decimal → 6). Export the time response graphs, both motor rotation $\theta(t)$ and motor input voltage $V(t)$, to Excel using right-click on each graph and selecting the Export to Excel option (right-click → Export to Excel). Estimate overshoot and settling time from closed-loop system response graph.
9. Repeat steps (5)–(8) for each type of controller: phase-lag, phase-lead, and phase-lag-lead. Complete the last two columns of Table 11.4.

III. *Deliverables*

1. Complete Tables 11.3 and 11.4.
2. Present graphs of simulated closed-loop step response for each controller with response characteristics (peak response, settling time, and steady-state value) clearly annotated.

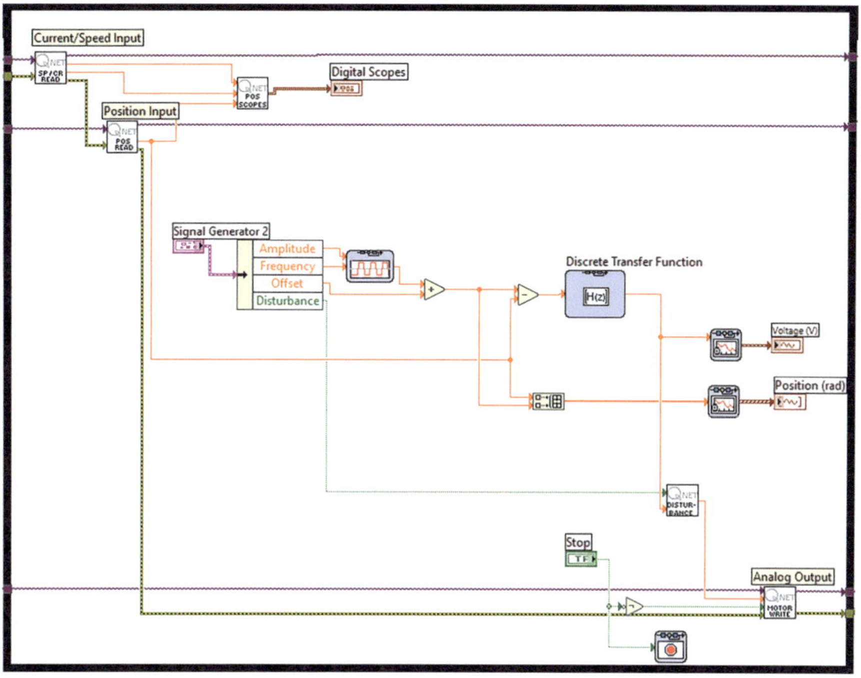

Fig. 11.41 Block diagram VI for DC motor control trainer position control VI after modification

3. Present graphs of root locus analysis for each controller with closed-loop poles near the desired ones clearly annotated.
4. Include Bode plots for each controller with all stability margins clearly annotated.
5. Include graphs of closed-loop system step response with each implemented controller.
6. Discuss results comparing the performance of controllers in simulation and experimental validation.
7. Discuss the stability of the closed-loop system with controllers from the results of root locus and Bode plots.

Lab Report

Present all deliverables in the lab report and discuss your results.

Exercises

11.1 It is required to design a PI controller for a system shown in Fig. P11.1 with $G_p(s) = \frac{5}{s^2+3.5s+2.5}$, $H(s) = 1$ such that the closed-loop system has an overshoot of 10% and a settling time of 3 sec for a step input. Use sampling time $T = 0.10$ sec. Use suitable value for the integral gain. Show the effect of the controller on the closed-loop system stability and step response using root locus, Bode plots, and closed-loop step response without and with the PI controller.

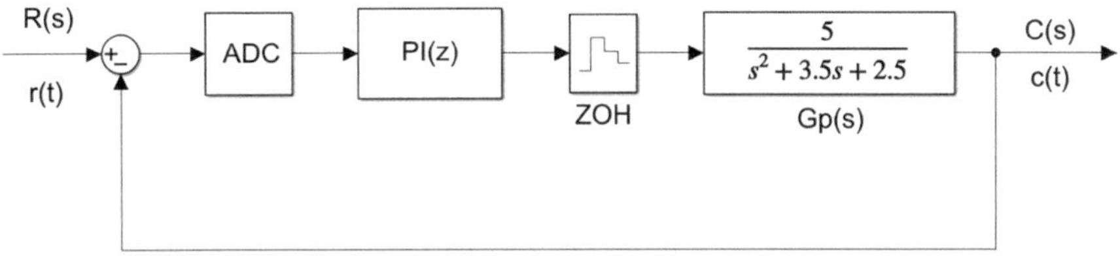

Fig. P11.1 Model of a discrete-time closed-loop system for Problem 11.1

11.2 It is required to design a PI controller for a system shown in Fig. P11.2 with $G_p(s) = \frac{5}{s^2+6.5s+1.5}$, $H(s) = 1$ such that the closed-loop system has an overshoot of 10% and a settling time of 3 sec for a step input. Use sampling time $T = 0.10$ sec. Use suitable value for the integral gain. Show the effect of the controller on the closed-loop system stability and step response using root locus, Bode plots, and closed-loop step response without and with the PI controller.

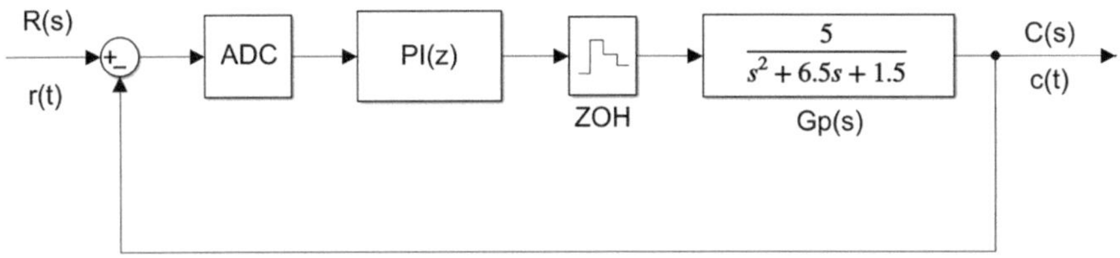

Fig. P11.2 Model of a discrete-time closed-loop system for Problem 11.2

11.3 It is required to design a PI controller for a system shown in Fig. P11.3 with $G_p(s) = \frac{K}{(s+1)(s^2+10s+4)}$, $H(s) = 1$. Obtain K such that the closed-loop system has a steady-state error of 20% for a step input without the controller. Design a PI controller such that the closed-loop system has an overshoot of 15% and a settling time of 5 sec for a step input. Use sampling time $T = 0.05$ sec. Use suitable value for the integral gain. Show the effect of the controller on the closed-loop system stability and step response using root locus, Bode plots, and closed-loop step response without and with the PI controller.

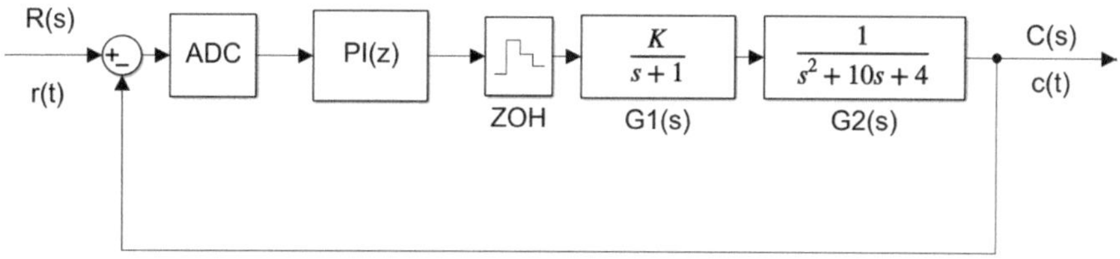

Fig. P11.3 Model of a discrete-time closed-loop system for Problem 11.3

11.4 It is required to design a PD controller for a system shown in Fig. P11.4 with $G_p(s) = \frac{15}{s^2+3.5s+1.5}$, $H(s) = 1$ such that the closed-loop system has an overshoot of 5% and a settling time of 2 sec for a step input. Use sampling time $T = 0.10$ sec. Show the effect of the controller on the closed-loop system stability and step response using root locus, Bode plots, and closed-loop step response without and with the PD controller.

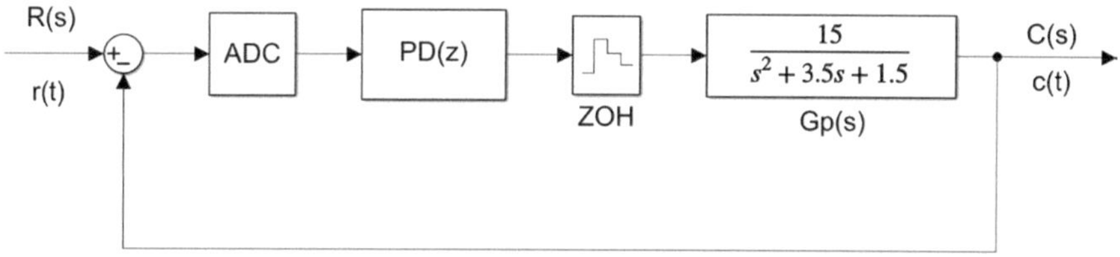

Fig. P11.4 Model of a discrete-time closed-loop system for Problem 11.4

11.5 It is required to design a PD controller for a system shown in Fig. P11.5 with $G_p(s) = \frac{K}{(s+1)(s^2+10s+4)}$, $H(s) = 1$ such that the closed-loop system has a steady-state error of 10%, an overshoot of 5%, and a settling time of 1.5 sec for a step input. Use sampling time $T = 0.05$ sec. Select K to satisfy the steady-state error specification without the controller. Show the effect of the controller on the closed-loop system stability and step response using root locus, Bode plots, and closed-loop step response without and with the PD controller.

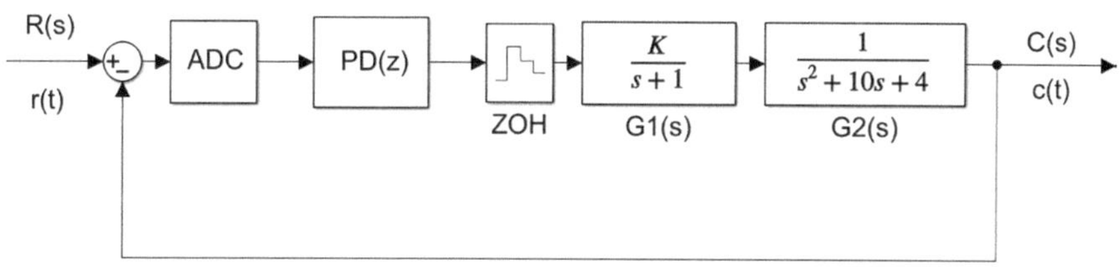

Fig. P11.5 Model of a discrete-time closed-loop system for Problem 11.5

11.6 It is required to design a PID controller for a system shown in Fig. P11.6 with $G_p(s) = \frac{15}{s^2+3.5s+1.5}$, $H(s) = 1$ such that the closed-loop system has an overshoot of 5% and a settling time of 2 sec for a step input. Use sampling time $T = 0.10$ sec. Use suitable value for the integral gain. Show the effect of the controller on the closed-loop system stability and step response using root locus, Bode plots, and closed-loop step response without and with the PID controller.

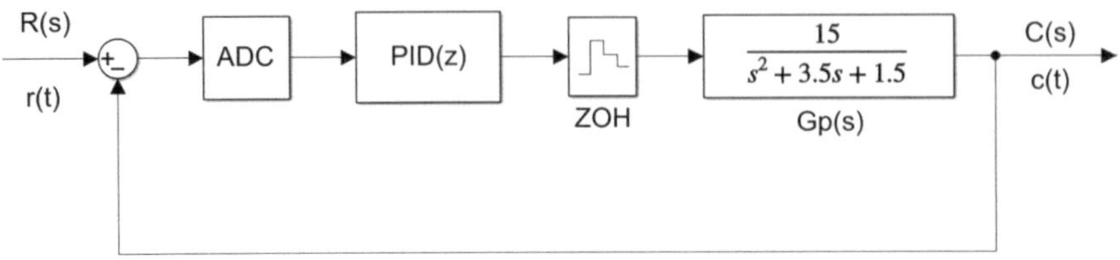

Fig. P11.6 Model of a discrete-time closed-loop system for Problem 11.6

11.7 It is required to design a PID controller for a system shown in Fig. P11.7 with $G_p(s) = \frac{K}{(s+1)(s^2+10s+4)}$, $H(s) = 1$ such that the closed-loop system has a steady-state error of 10%, an overshoot of 5%, and a settling time of 1 sec for a step input. Use sampling time $T = 0.05$ sec. Select K to satisfy the steady-state error specification without the controller. Use suitable value for the integral gain. Show the effect of the controller on the closed-loop system stability and step response using root locus, Bode plots, and closed-loop step response without and with the PID controller.

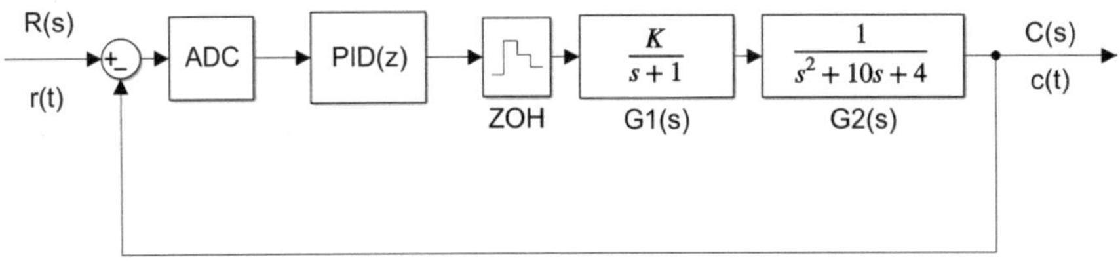

Fig. P11.7 Model of a discrete-time closed-loop system for Problem 11.7

11.8 It is required to design a PID controller for an automated drive speed control system shown in Fig. P11.8 with $G(s) = \frac{K}{(s+2.5)(s+5)}$, $H(s) = 1$. Obtain K such that the steady-state error of the closed-loop system without any controller remains at 8% for a unit step input. Design a PID controller such that the closed-loop system has an overshoot of 4% and a settling time of 1 sec for a step input. Use sampling time $T = 0.10$ sec. Use suitable value for the integral gain. Show the effect of the controller on the closed-loop system stability and step response using root locus, Bode plots, and closed-loop step response without and with the PID controller.

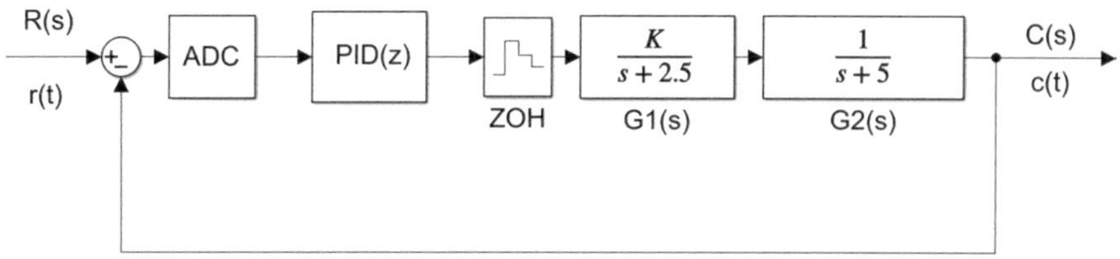

Fig. P11.8 Model of a discrete-time closed-loop system for Problem 11.8

11.9 The closed-loop vehicle distance control system for a traffic control system can be represented as shown in Fig. P11.9 with $G_p(s) = \frac{1.5}{s^2+12s+18}$. The sampling time is given as $T = 0.10$ sec . It is required to design a PI controller such that the closed-loop system has an overshoot of 10% and a settling time of 2.5 sec for a step input. Use suitable value for the integral gain. Show the effect of the controller on the closed-loop system stability and step response using root locus, Bode plots, and closed-loop step response without and with the PI controller.

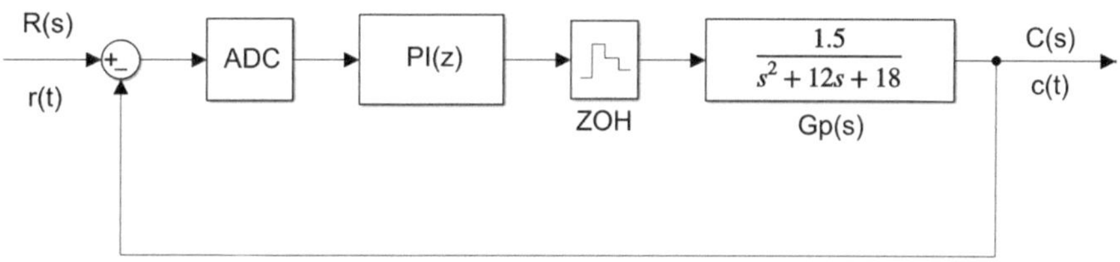

Fig. P11.9 Model of a discrete-time closed-loop system for Problem 11.9

11.10 It is required to design a PD controller such that the closed-loop system of Fig. P11.10 with $G_p(s) = \frac{1.5}{s^2+12s+18}$ has a steady-state error of 8%, an overshoot of 10%, and a settling time of 0.80 sec for a step input. Use sampling time $T = 0.10$ sec. Show the effect of the controller on the closed-loop system stability and step response using root locus, Bode plots, and closed-loop step response without and with the PD controller.

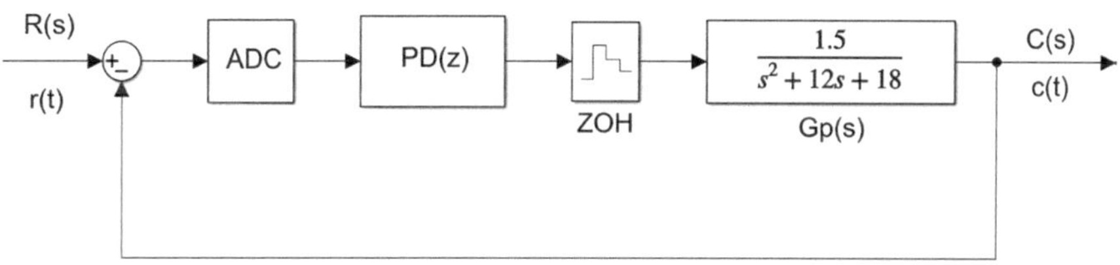

Fig. P11.10 Model of a discrete-time closed-loop system for Problem 11.10

11.11 It is required to design a PID controller such that the closed-loop system of Fig. P11.11 with $G_p(s) = \frac{1.5}{s^2+12s+18}$ has an overshoot of 5% and a settling time of 0.60 sec for a step input. Use sampling time $T = 0.10$ sec. Use suitable value for the integral gain. Show the effect of the controller on the closed-loop system stability and step response using root locus, Bode plots, and closed-loop step response without and with the PID controller.

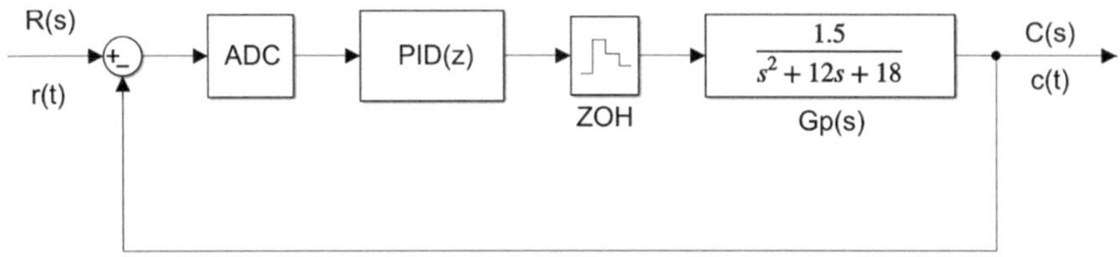

Fig. P11.11 Model of a discrete-time closed-loop system for Problem 11.11

11.12 The closed-loop roll angle control system for a jet airplane can be represented as shown in Fig. P11.12. The plant transfer function of the roll control system is represented in s-domain as $G_p(s) = \frac{10}{(s+10)(s^2+2.5s+5)}$. The sampling time is given as $T = 0.05$ sec. It is required to design a PID controller such that the closed-loop system has an overshoot of 5% and a settling time of 2 sec for a step input. Use suitable value for the integral gain. Show the effect of the controller on the closed-loop system stability and step response using root locus, Bode plots, and closed-loop step response without and with the PID controller.

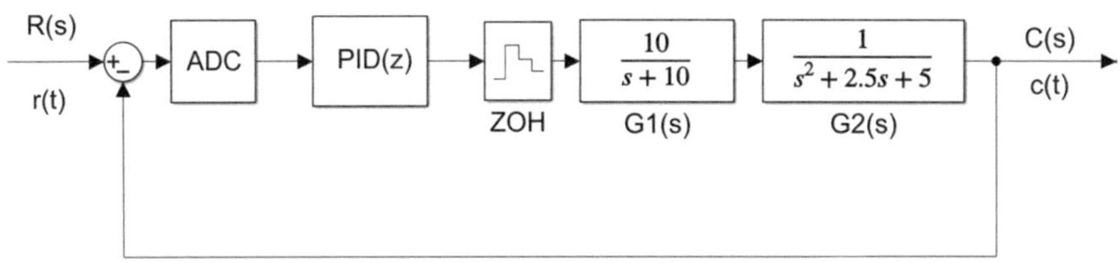

Fig. P11.12 Model of a discrete-time closed-loop system for Problem 11.12

11.13 It is required to design a phase-lag controller for a system shown in Fig. P11.13 with $G_p(s) = \frac{1}{s^2+4.5s+3}$, $H(s) = 1$ such that the closed-loop system has a steady-state error of 5%, an overshoot of 20%, and a settling time of 5 sec for a step input. Use sampling time $T = 0.10$ sec. Show the effect of the controller on the closed-loop system stability and step response using root locus, Bode plots, and closed-loop step response without and with the phase-lag controller.

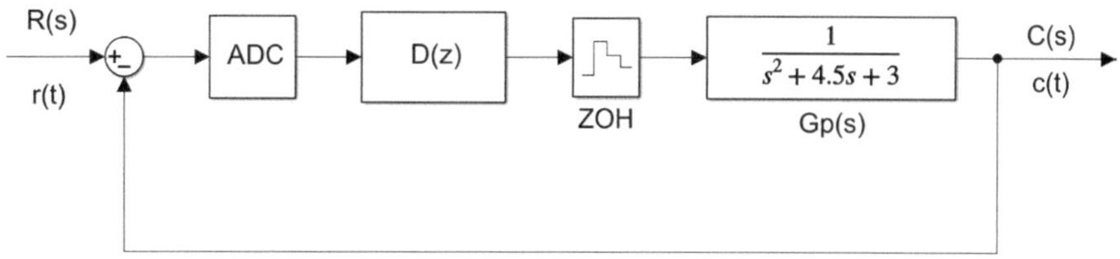

Fig. P11.13 Model of a discrete-time closed-loop system for Problem 11.13

11.14 It is required to design a phase-lag controller for a system shown in Fig. P11.14 with $G_p(s) = \frac{K}{(s+1)(s^2+10s+4)}$, $H(s) = 1$ such that the closed-loop system has a steady-state error of 8%, an overshoot of 20%, and a settling time of 10 sec for a step input. Use sampling time $T = 0.10$ sec. Select K to satisfy the steady-state error specification without the controller. Show the effect of the controller on the closed-loop system stability and step response using root locus, Bode plots, and closed-loop step response without and with the phase-lag controller.

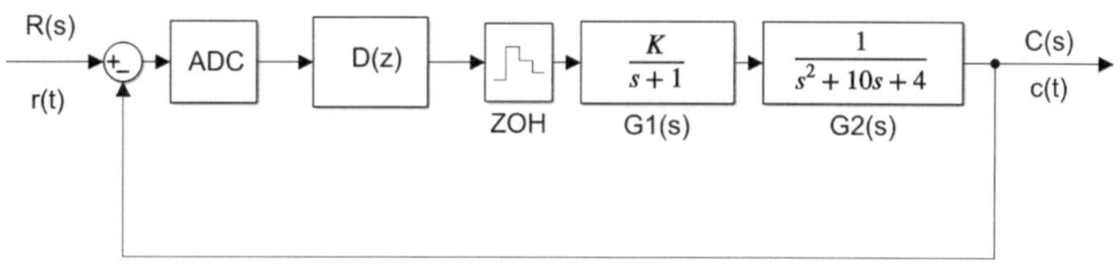

Fig. P11.14 Model of a discrete-time closed-loop system for Problem 11.14

11.15 It is required to design a phase-lead controller for a system shown in Fig. P11.15 with $G_p(s) = \frac{1}{s^2+4.5s+3}$, $H(s) = 1$ such that the closed-loop system has a steady-state error of 12%, an overshoot of 5%, and a settling time of 1 sec for a step input. Use sampling time $T = 0.10$ sec. Show the effect of the controller on the closed-loop system stability and step response using root locus, Bode plots, and closed-loop step response without and with the phase-lead controller.

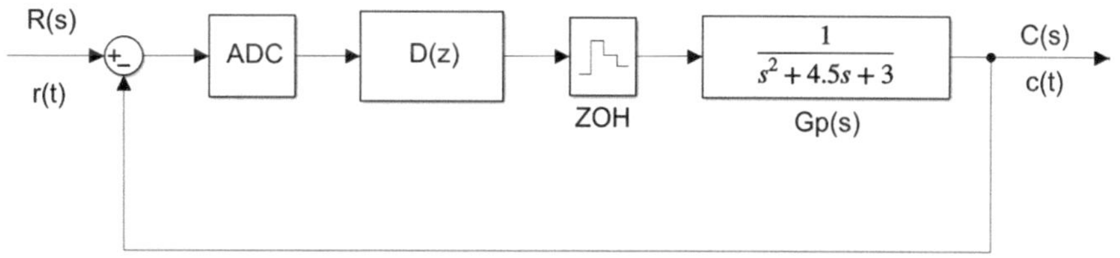

Fig. P11.15 Model of a discrete-time closed-loop system for Problem 11.15

11.16 It is required to design a phase-lead controller for a system shown in Fig. P11.16 with $G_p(s) = \frac{K}{(s+1)(s^2+10s+4)}$, $H(s) = 1$ such that the closed-loop system has a steady-state error of 12%, an overshoot of 8%, and a settling time of 1.5 sec (2% error) for a step input. Use sampling time $T = 0.10$ sec. Select K to satisfy the steady-state error specification without the controller. Use suitable design specifications for the phase-lag part. Show the effect of the controller on the closed-loop system stability and step response using root locus, Bode plots, and closed-loop step response without and with the phase-lead controller.

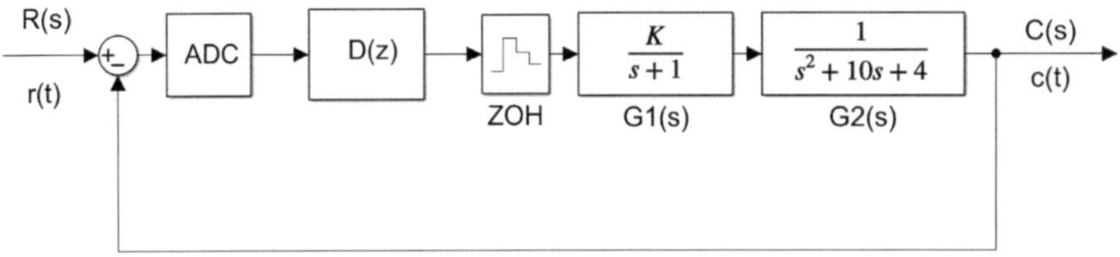

Fig. P11.16 Model of a discrete-time closed-loop system for Problem 11.16

11.17 It is required to design a phase-lag-lead controller for a system shown in Fig. P11.17 with $G_p(s) = \frac{1}{s^2+4.5s+3}$, $H(s) = 1$ such that the closed-loop system has a steady-state error of 5%, an overshoot of 5%, and a settling time of 1 sec for a step input. Use sampling time $T = 0.10$ sec. Use suitable design specifications for the phase-lag part. Show the effect of the controller on the closed-loop system stability and step response using root locus, Bode plots, and closed-loop step response without and with the phase-lead controller.

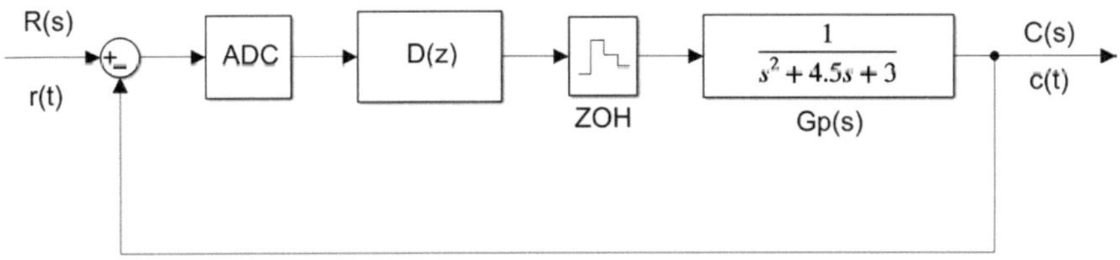

Fig. P11.17 Model of a discrete-time closed-loop system for Problem 11.17

11.18 It is required to design a phase-lag-lead controller for a system shown in Fig. P11.18 with $G_p(s) = \frac{K}{(s+1)(s^2+10s+4)}$, $H(s) = 1$ such that the closed-loop system has a steady-state error of 5%, an overshoot of 5%, and a settling time of 1.5 sec for a step input. Use sampling time $T = 0.10$ sec. Select K to satisfy the steady-state error specification without the controller. Use suitable design specifications for the phase-lag part. Show the effect of the controller on the closed-loop system stability and step response using root locus, Bode plots, and closed-loop step response without and with the phase-lead controller.

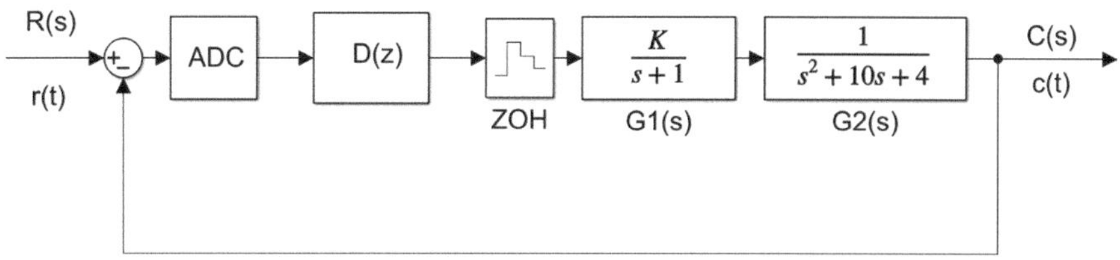

Fig. P11.18 Model of a discrete-time closed-loop system for Problem 11.18

11.19 It is required to design a phase-lag-lead controller for a system shown in Fig. P11.19 with $G_p(s) = \frac{1}{s(s+1)}$, $H(s) = 1$ such that the closed-loop system has a steady-state error of 20% for a ramp input, an overshoot of 7.5%, and a settling time of 1 sec for a step input. Use sampling time $T = 0.10$ sec. Use suitable design specifications for the phase-lag part. Show the effect of the controller on the closed-loop system stability and step response using root locus, Bode plots, and closed-loop step response without and with the phase-lead controller.

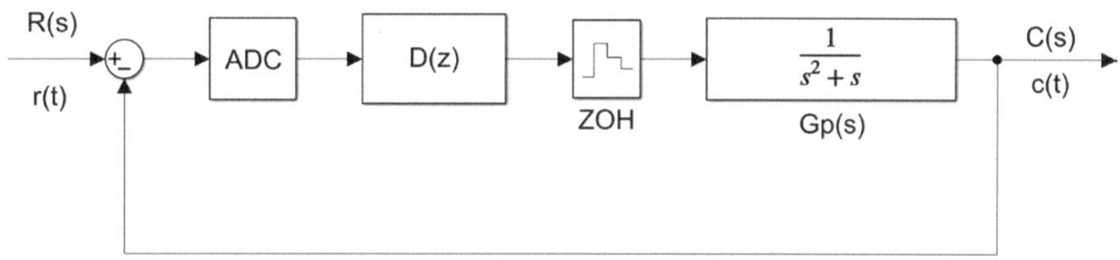

Fig. P11.19 Model of a discrete-time closed-loop system for Problem 11.19

11.20 It is required to design a phase-lag-lead controller for a system shown in Fig. P11.20 with $G_p(s) = \frac{1}{(s+3)(s^2+7s+2)}$, $H(s) = 1$ such that the closed-loop system has a steady-state error of 8%, an overshoot of 7.5%, and a settling time of 2 sec for a step input. Use sampling time $T = 0.10$ sec. Use suitable design specifications for the phase-lag part. Show the effect of the controller on the closed-loop system stability and step response using root locus, Bode plots, and closed-loop step response without and with the phase-lead controller.

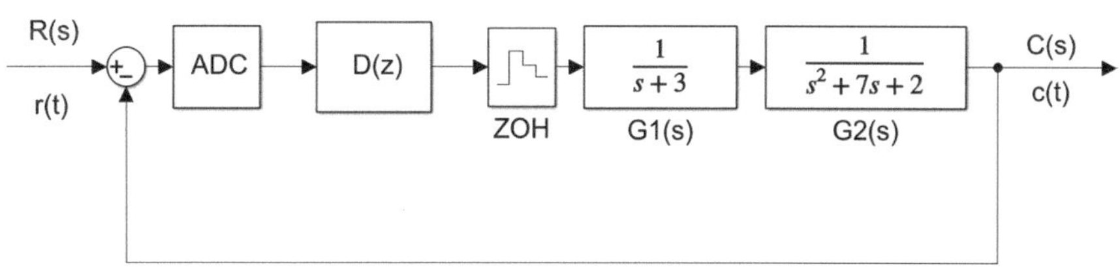

Fig. P11.20 Model of a discrete-time closed-loop system for Problem 11.20

Further Reading

1. Asadi F (2023) Signals and systems with MATLAB® and Simulink®. Springer, Cham
2. Astrom KJ, Murray RM (2021) Feedback systems: an introduction for scientists and engineers, 2nd edn. Princeton University Press, Princeton
3. Chen C-T (2006) Analog and digital control system design: transfer-function, state-space, and algebraic methods. Oxford University Press, New York
4. Dorf R, Bishop R (2016) Modern control systems, 13th edn. Pearson, Essex
5. Fadali MS, Visioli A (2020) Digital control engineering analysis and design, 3rd edn. Academic Press, London
6. Franklin GF, Powell JD, Emami-Naeini A (2019) Feedback control of dynamic systems, 8th edn. Pearson, New Jersey
7. Franklin GF, Powell JD, Workman M (1998) Digital control of dynamic systems, 3rd edn. Addison-Wesley, Boston
8. Isermann R (1981) Digital control systems. Springer-Verlag, Berlin
9. Jalili N, Candelino NW (2023) Dynamic systems and control engineering. Cambridge University Press, Cambridge
10. Keviczky L, Bars R, Hetthéssy J, Bányász (2019) Control engineering. Springer, Singapore
11. Koeing DM (2009) Practical control engineering: a guide for engineers, managers, and practitioners. McGraw-Hill, New York
12. Kuo BC (1995) Digital control systems, 2nd edn. Oxford University Press, Oxford
13. Landau ID, Gito G (2006) Digital control systems: design, identification and implementation. Springer-Verlag, London
14. Leigh JR (2006) Applied digital control: theory, design and implementation, 2nd edn. Dover, Mineola
15. Mathworks (2024a) Control System Toolbox. https://www.mathworks.com/products/control.html
16. Mathworks (2024b) Matlab. https://www.mathworks.com/products/matlab.html
17. National Instruments (2024a) LabVIEW. https://www.ni.com/en/support/downloads/software-products/download.labview.html#487445
18. National Instruments (2024b) NI-ELVIS-II. https://www.ni.com/en-us/support/model.ni-elvis-ii.html
19. National Instruments (2024c) Labs for Quanser QNET DC Motor Add-On Board. https://education.ni.com/teach/resources/180/labs-for-quanser-qnet-dc-motor-add-on-board
20. Nise NS (2019) Control systems engineering, 8th edn. Wiley, Hoboken/New Jersey
21. Nof SY (ed) (2009) Springer handbook of automation. Springer-Verlag, Berlin
22. Ogata K (1995) Discrete-time control systems, 2nd edn. Pearson, New Jersey
23. Phillips CL, Nagle HT, Chakrabortty A (2015) Digital control system analysis and design, 4th edn. Pearson, Essex
24. Quanser (2024) QUBE Servo 2. https://www.quanser.com/products/qube-servo-2/
25. Rabbath CA, Lechevin N (2014) Discrete-time control system design with applications. Springer, New York
26. Samanta B (2023) Introduction to mechatronics- an integrated approach. Springer, Cham
27. Shertukde HM (2015) Digital control applications illustrated with MATLAB®, CRC Press, Boca Raton
28. Starr G (2020) Introduction to applied digital controls. Springer, Cham
29. Veloni A, Miridakis NI (2018) Digital control systems-theoretical problems and simulation tools. CRC Press, Boca Raton
30. Vilanova R, Visioli A (2012) PID control in the third millennium - lessons learned and new approaches. Springer-Verlag, London
31. Visioli A (2006) Practical PID control. Springer, London

Chapter 12
Controller Design for Discrete-Time Systems Using Frequency-Domain Specifications

12.1 Introduction

Methods of designing controllers for discrete-time systems using frequency-domain specifications are presented in this chapter. Types of controllers covering proportional-integral-derivative (PID), phase-lag-lead, and their variations along with corresponding transfer functions in the w-domain for discrete-time systems are presented. Details of design steps for discrete-time controllers are presented. Examples along with Matlab codes are presented for the controllers to illustrate the procedure. Laboratory experiments for analysis, design, simulation, and real-time implementation of the controllers on a DC motor module are presented. Analysis, design, and simulation are presented using Matlab, and implementation of the discrete-time controllers is presented using LabVIEW. End-of-chapter exercise problems are provided to enhance understanding of the topics covered and consolidate problem-solving skills.

12.2 Design of PI, PD, and PID Controllers for Discrete-Time Systems

The frequency-domain specifications include the gain cross-over frequency (ω_{gc}), gain margin (GM), and phase margin (ϕ_m). The w-domain gain cross-over frequency can be expressed as

$$\omega_{w_{gc}} = \left(\frac{2}{T}\right) \tan\left(\frac{\omega_{gc} T}{2}\right) \tag{12.1}$$

B. Samanta, *Introduction to Digital Control*,
https://doi.org/10.1007/978-3-031-66830-2_12

12.2.1 PID Controller Transfer Function in w-Domain

The PID controller transfer function in z-domain is expressed as

$$D_{PID}(z) = K_P + K_I \frac{T}{2}\left(\frac{z+1}{z-1}\right) + K_D \frac{z-1}{Tz} \tag{12.2}$$

To design a controller using frequency domain specifications, it is necessary to convert the transfer function from z-domain to w-domain using bilinear transformation as

$$D_{PID}(w) = \left(K_P + K_I \frac{T}{2}\left(\frac{z+1}{z-1}\right) + K_D \frac{z-1}{Tz}\right)\Bigg|_{z=\frac{1+\left(\frac{T}{2}\right)w}{1-\left(\frac{T}{2}\right)w}} \tag{12.3}$$

The PID controller transfer function in the w-domain can be expressed as

$$D_{PID}(w) = K_P + \frac{K_I}{w} + K_D \frac{w}{1+\left(\frac{T}{2}\right)w} \tag{12.4}$$

12.2.2 Design of PID Controllers

At the gain cross-over frequency $\omega_{w_{gc}}$, the combined controller and the system transfer function must have a magnitude of 1 and a phase angle of $(180^\circ + \phi_m)$ as

$$\overline{GH}\left(j\omega_{w_{gc}}\right)D\left(j\omega_{w_{gc}}\right) = 1e^{j\left(180^\circ + \phi_m\right)} \tag{12.5}$$

Using the controller as the PID, the controller transfer function can be expressed as

$$\left(K_P + \frac{K_I}{w} + K_D \frac{w}{1+\left(\frac{T}{2}\right)w}\right)\Bigg|_{w=j\omega_{w_{gc}}} = \frac{1}{\overline{GH}\left(j\omega_{w_{gc}}\right)} e^{j\left(180^\circ + \phi_m\right)} \tag{12.6}$$

After some rearrangements

$$\left(K_P - j\frac{K_I}{\omega_{w_{gc}}} + K_D \frac{j\omega_{w_{gc}}\left(1-\left(\frac{T}{2}\right)j\omega_{w_{gc}}\right)}{\left(1+\left(\frac{T}{2}\right)j\omega_{w_{gc}}\right)\left(1-\left(\frac{T}{2}\right)j\omega_{w_{gc}}\right)}\right) = \frac{1}{m} e^{j\left(180^\circ + \phi_m - \Psi\right)} \tag{12.7}$$

$$\left(K_P - j\frac{K_I}{\omega_{w_{gc}}} + K_D\frac{\left(j\omega_{w_{gc}} + \left(\frac{T}{2}\right)\left(\omega_{w_{gc}}\right)^2\right)}{\left(1 + \left(\left(\frac{T}{2}\right)\omega_{w_{gc}}\right)^2\right)}\right) = \frac{1}{m}(\cos\,\theta + j\sin\theta) \tag{12.8}$$

where

$$\overline{GH}\left(j\omega_{w_{gc}}\right) = me^{j\Psi} \tag{12.9}$$

$$\theta = 180^{\circ} + \phi_m - \Psi \tag{12.10}$$

Real and imaginary parts of both sides of Eq. (12.8) can be expressed as

$$K_P + K_D\frac{\left(\frac{T}{2}\right)\left(\omega_{w_{gc}}\right)^2}{\left(1 + \left(\left(\frac{T}{2}\right)\omega_{w_{gc}}\right)^2\right)} = \frac{1}{m}\cos\,\theta \tag{12.11}$$

$$-\frac{K_I}{\omega_{w_{gc}}} + K_D\frac{\omega_{w_{gc}}}{\left(1 + \left(\left(\frac{T}{2}\right)\omega_{w_{gc}}\right)^2\right)} = \frac{1}{m}\sin\theta \tag{12.12}$$

There are two equations and three unknowns (K_P, K_I, K_D). In general, the gains K_D and K_P are expressed in terms of K_I as

$$K_D = \left(\frac{1}{m}\sin\theta + \frac{K_I}{\omega_{w_{gc}}}\right)\left(1 + \left(\left(\frac{T}{2}\right)\omega_{w_{gc}}\right)^2\right)/\omega_{w_{gc}} \tag{12.13}$$

$$K_P = \frac{1}{m}\cos\,\theta - K_D\frac{\left(\frac{T}{2}\right)\left(\omega_{w_{gc}}\right)^2}{\left(1 + \left(\left(\frac{T}{2}\right)\omega_{w_{gc}}\right)^2\right)} \tag{12.14}$$

As a guideline, initially, K_I is assumed to be 0, and K_D, K_P are obtained. Next, an estimate of K_I is obtained from

$$0.01 \leq \frac{K_I}{K_P} \leq 0.10 \tag{12.15}$$

The values of K_D, K_P are updated using the new estimate of K_I. The process is repeated till the values converge. Final values are fine-tuned after checking with simulation with the calculated values of the controller gains K_P, K_I, K_D.

The PID controller transfer function in terms of numerator and denominator polynomials can be expressed as

$$D_{PID}(w) = \frac{\left(\frac{K_P T}{2} + K_D\right)w^2 + \left(K_P + \frac{K_I T}{2}\right)w + K_I}{w\left(1 + \left(\frac{T}{2}\right)w\right)} \tag{12.16}$$

12.2.3 PI Controllers

PI controller gains can be obtained from Eqs. (12.11) and (12.12), with $K_D = 0$,as

$$K_P = \frac{1}{m} \cos\,\theta \tag{12.17}$$

$$K_I = -\frac{\omega_{W_{gc}}}{m} \sin\,\theta \tag{12.18}$$

The PI controller transfer function in terms of numerator and denominator polynomials can be expressed as

$$D_{PI}(w) = \frac{K_P w + K_I}{w} \tag{12.19}$$

12.2.4 PD Controllers

PD controller gains can be obtained from Eqs. (12.11) and (12.12), with $K_I = 0$,

$$K_D = \left(\frac{1}{m} \sin\theta\right)\left(1 + \left(\left(\frac{T}{2}\right)\omega_{W_{gc}}\right)^2\right) \Big/ \omega_{W_{gc}} \tag{12.20}$$

$$K_P = \frac{1}{m} \cos\,\theta - K_D \frac{\left(\frac{T}{2}\right)\left(\omega_{W_{gc}}\right)^2}{\left(1 + \left(\left(\frac{T}{2}\right)\omega_{W_{gc}}\right)^2\right)} \tag{12.11.21}$$

The PD controller transfer function in terms of numerator and denominator polynomials can be expressed as

$$D_{PD}(w) = \frac{\left(\frac{K_P T}{2} + K_D\right)w + K_P}{1 + \left(\frac{T}{2}\right)w} \tag{12.22}$$

It should be noted that even if the controllers are designed in the w-domain, discrete-time controllers are converted to the z-domain using bilinear transformation for implementation.

Example 12.1 PI Controller: Design a PI controller for $G_p(s) = \frac{5}{s^2+2.5s+1}$, $H(s) = 1$ such that the closed-loop system has a gain cross-over frequency of 2.5 rad/s, a phase margin of 55°, and a gain margin of 10 dB. Use sampling time $T = 0.10$ sec. Show the effect of the PI controller on the closed-loop system stability and step response.

Solution Matlab code and results are presented.

```
% ch12_ex3 _design of pid controller-w-domain (DIGITAL)%
close all;clear all;clc
T=0.1;% sampling time
Gp=tf(5,[1 2.5 1]) % Gp(s)=10/(s^2+2.5s+1)
% desired design spec
wgs=2.50; % gain cross-over frequency (s-plane) wgcs
dpm=55; dgm=10; %Desired phase margin, gain Margin
wgw=tan(wgs*T/2)*2/T; % w-domain wgc
Gz=c2d(Gp, T)  % G(z)
% G(w)
Gw=d2c(Gz,'tustin')
[ngw,dgw]=tfdata(Gw,'v') % G(w)
printsys(ngw,dgw,'w')

% bode plots of uncompensated system
figure
bode(Gw)
title('Bode plots of uncompensated system')
[gm,pm,wgc,wpc]=margin(Gw)

%Design of PID controller in w-domain
nv=polyval(ngw,j*wgw);dv=polyval(dgw,j*wgw);
g=nv/dv; %G(wgc)
m=abs(g) %mag(G(wgc)) at wgc
psi=angle(g) % angle(G(wgc))

%PI controller gains
ki=0.4 % to be selected
theta=(180+dpm)*pi/180-psi % angle in radians
kd=(sin(theta)/m+ki/wgw)*(1+(T*wgw/2)^2)/wgw
kd=0
kp=cos(theta)/m-kd*(T/2)*wgw^2/(1+(T*wgw/2)^2)

%Dw = tf([kp*T/2+kd kp+ki*T/2 ki],[T/2 1 0]) %D(w)
Dw = tf([kp ki], [1 0])
DGw = Dw*Gw

% bode plots of the controller
figure
bode(Dw)
title('Bode plots of the PI controller')

% bode plots of system with the controller
figurc
bode(DGw)
title('Bode plots of the compensated system with the PI controller')
[gm,pm,wpc,wgc]=margin(DGw)
wgs=atan(wgc*T/2)*2/T % in s-domain
wps=atan(wpc*T/2)*2/T % in s-domain

% Root Locus
Dz=c2d(Dw,T,'tustin'); % D(z)

%G(z)d(z)
DGz=Dz*Gz; %G(z)D(z)

% root locus of uncompensated system (without controller)
```

```
figure
rlocus(Gz)
axis('equal')
title('Root locus of the uncompensated system')
% root locus of comppensated system (with controller)
figure
rlocus(DGz)% Root locus
axis('equal')
title('Root locus of the system with the PI controller')

%Closed loop system step response of the uncompensated system
Tz=feedback(Gz,1)
figure
dstep(Tz,1)
title('Closed-loop step response of the uncompensated system')
%Closed loop system step response of the system with PID controller
Tz=feedback(DGz,1)
figure
dstep(Tz,1)
title('Closed-loop step response of the system with the PI controller')

Results:
G(z) =
   0.02302 z + 0.02118
  ---------------------
  z^2 - 1.77 z + 0.7788
G(w) =
   -0.0005186 s^2 - 0.2387 s + 4.982
  ---------------------------------
        s^2 + 2.493 s + 0.9965
gm = 10.4432, pm = 58.6155, wgc = 1.7961, wpc = 7.3018

m = 0.6110, psi = -2.3944, theta = 6.4960

PI controller
ki = 0.4000, kd = 0, kp = 1.5998

         1.6 w + 0.4
D(w) = -------------
               W

        1.62 z - 1.58
D(z)= --------------
            z - 1
For the uncompensated system (without the PI controller)
gm = 10.4432, pm = 58.6155, wgc = 1.7961, wpc = 7.3018, wgs = 1.7913, wps =
7.0011

For the compensated system with the PI controller
gm = 5.8790, pm = 37.5843, wgc = 2.4846, wpc = 6.8941, wgs = 2.4719, wps =
6.6390
```

It should be noted that the design point was chosen somewhat arbitrarily to illustrate the controller design process. The integral gain was not optimized but was chosen to get a reasonable response from the closed-loop system with the PI controller. The gain margin and the phase margin reduced with the PI controller as shown in Bode plots in Figs. 12.1, 12.2, and 12.3. Root locus plots are shown in

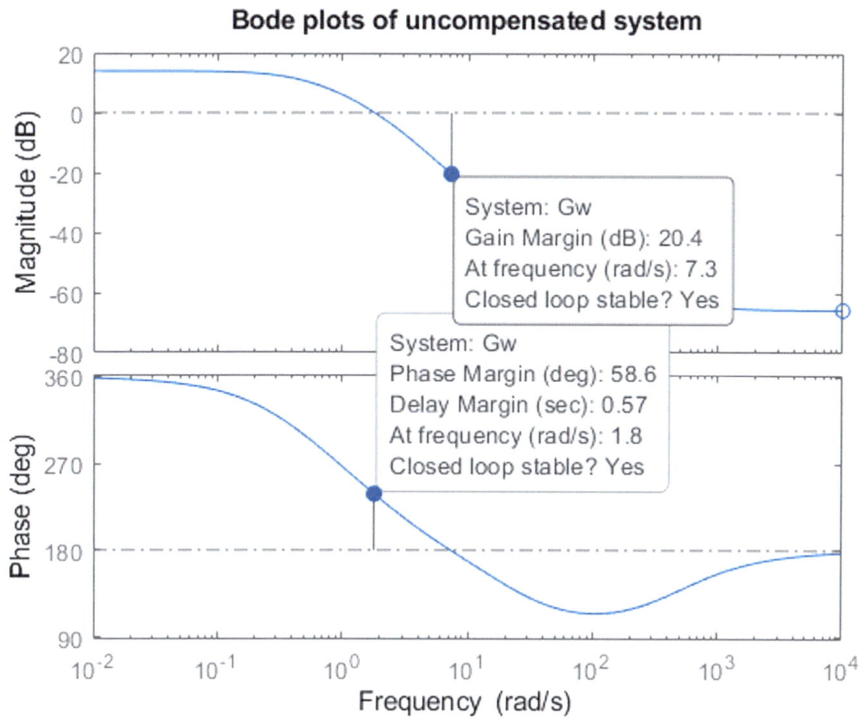

Fig. 12.1 Bode plots of the uncompensated system

Figs. 12.4 and 12.5. The percentage overshoot and settling time increased with the selected integral gain of the PI controller as shown in closed-loop system step response in Figs. 12.6 and 12.7. The beneficial effect of the PI controller to eliminate the steady-state error due to step input is evident in the closed-loop step response of the compensated system in Fig. 12.7. The value of the integral gain and the design point specifications can be adjusted to suit the requirements of the closed-loop compensated system with the PI controller.

Example 12.2 PD Controller: Design a PD controller for $G_p(s) = \frac{5}{s^2+2.5s+1}$, $H(s) = 1$ such that the closed-loop system has a gain cross-over frequency of 2.5 rad/s, a phase margin of 55°, and a gain margin of 10 dB. Use sampling time $T = 0.10$ sec. Show the effect of the PD controller on the closed-loop system stability and step response.

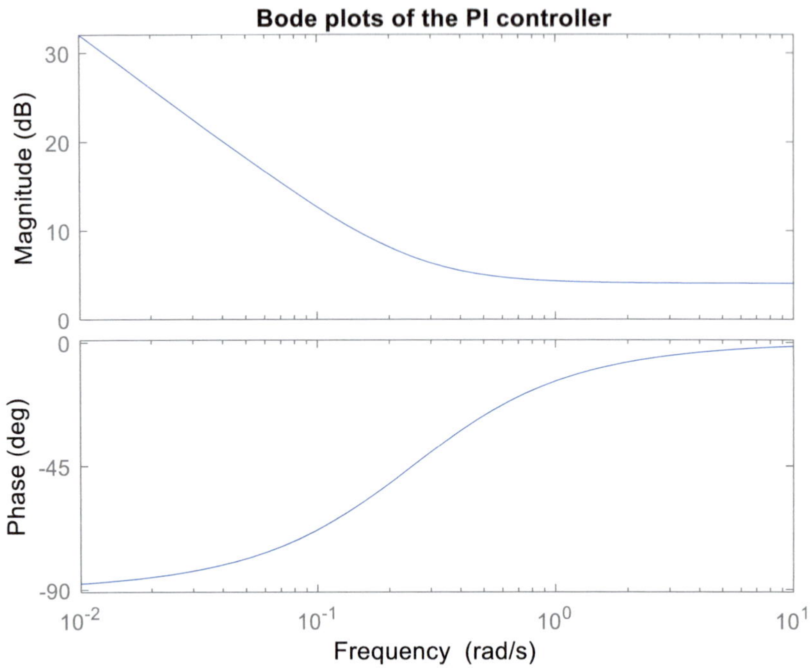

Fig. 12.2 Bode plots of the PI controller

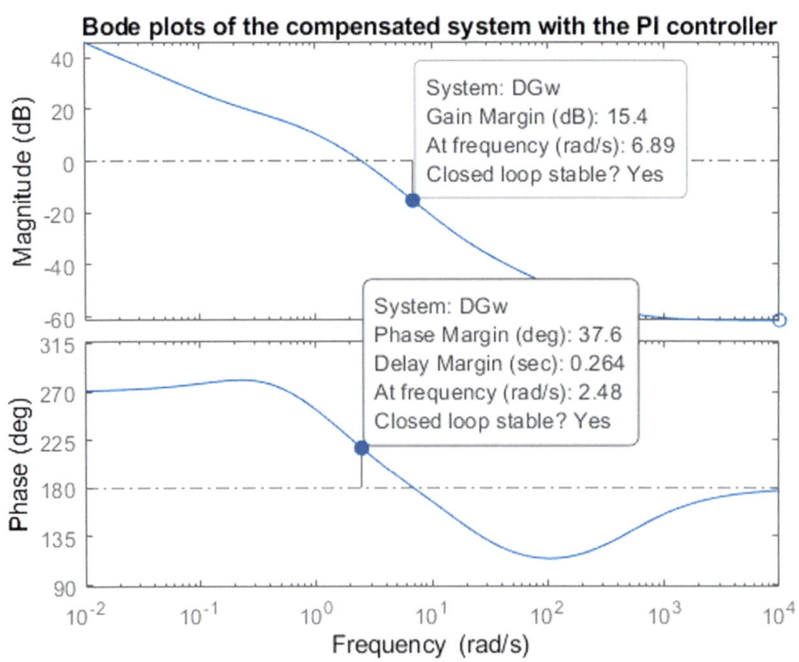

Fig. 12.3 Bode plots of the compensated system with the PI controller

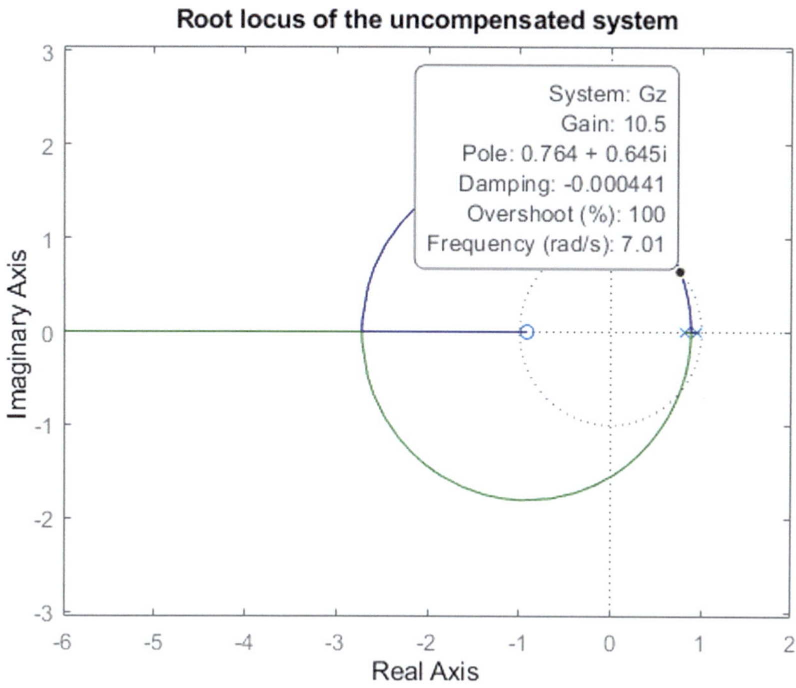

Fig. 12.4 Root locus of the uncompensated system

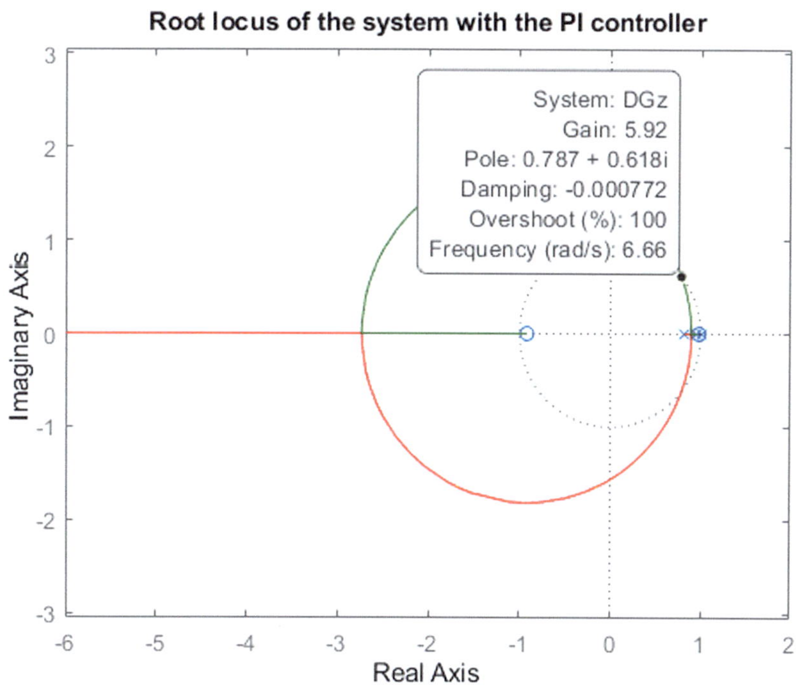

Fig. 12.5 Root locus of the compensated system with the PI controller

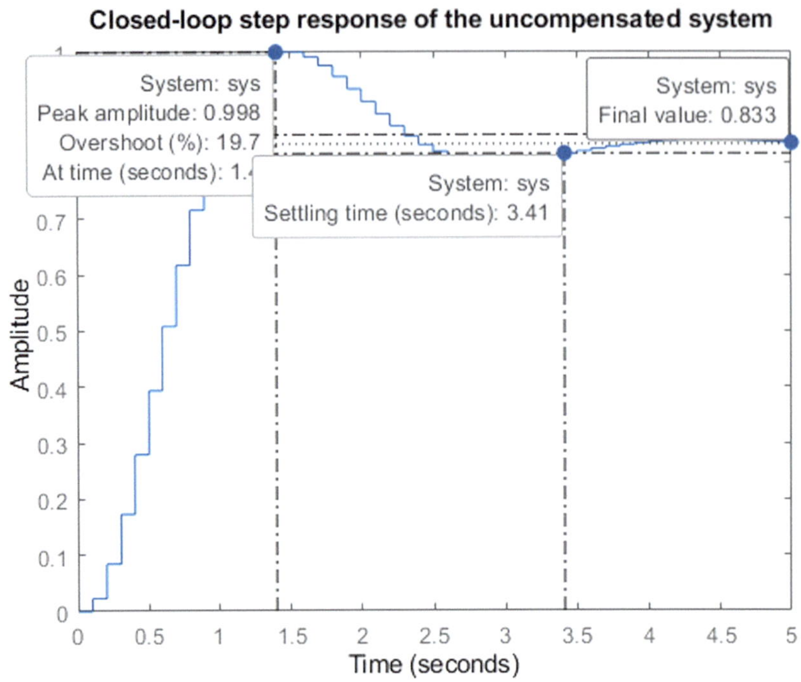

Fig. 12.6 Closed-loop step response of the uncompensated system

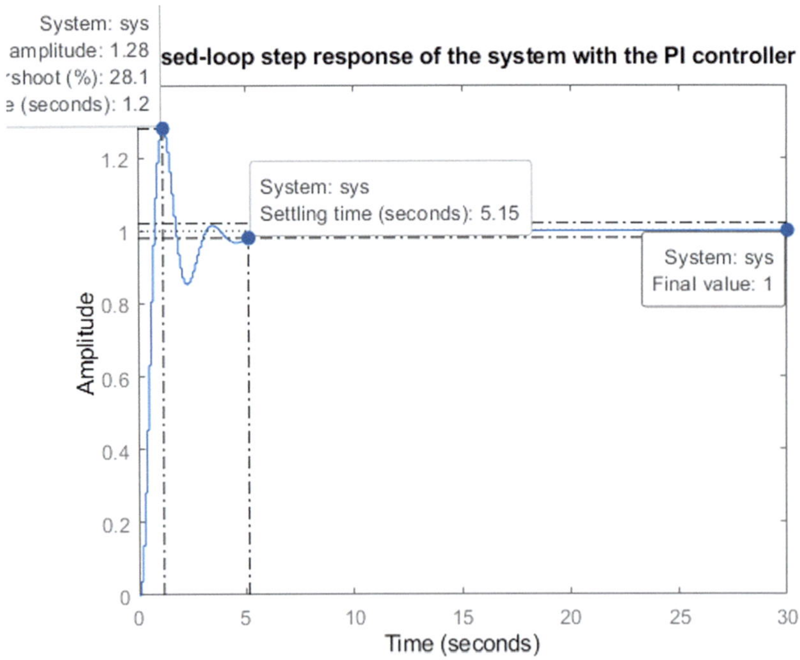

Fig. 12.7 Closed-loop step response of the system with the PI controller

Solution Matlab code and results are presented.

```
% ch12_ex2 _design of PD controller-w-domain
close all;clear all;clc
T=0.1;% sampling time
Gp=tf(5,[1 2.5 1]) % Gp(s)=10/(s^2+2.5s+1)
% desired design spec
wgs=2.50; % gain cross-over frequency (s-plane) wgcs
dpm=55; dgm=10; %Desired phase margin, gain Margin
wgw=tan(wgs*T/2)*2/T; % w-domain wgc
Gz=c2d(Gp, T)  % G(z)
% G(w)
Gw=d2c(Gz,'tustin')
[ngw,dgw]=tfdata(Gw,'v') % G(w)
printsys(ngw,dgw,'w')

% bode plots of uncompensated system
figure
bode(Gw)
title('Bode plots of uncompensated system')
[gm,pm,wgc,wpc]=margin(Gw)

%Design of PID controller in w-domain
nv=polyval(ngw,j*wgw);dv=polyval(dgw,j*wgw);
g=nv/dv; %G(wgc)
m=abs(g) %mag(G(wgc)) at wgc
psi=angle(g) % angle(G(wgc))

%PD controller gains
ki=0 % to be selected
theta=(180+dpm)*pi/180-psi % angle in radians
kd=(sin(theta)/m+ki/wgw)*(1+(T*wgw/2)^2)/wgw
kp=cos(theta)/m-kd*(T/2)*wgw^2/(1+(T*wgw/2)^2)

%Dw = tf([kp*T/2+kd kp+ki*T/2 ki],[T/2 1 0]) %PID(w)
%Dw = tf([kp ki], [1 0]) % PI D(w)
Dw = tf([kp*T/2+kd kp], [T/2 1]) % PD D(w)
DGw = Dw*Gw

% bode plots of the controller
figure
bode(Dw)
title('Bode plots of the PD controller')

% bode plots of system with the controller
figure
bode(DGw)
title('Bode plots of the compensated system with the PD controller')
[gm,pm,wpc,wgc]=margin(DGw)
wgs=atan(wgc*T/2)*2/T % in s-domain
wps=atan(wpc*T/2)*2/T % in s-domain

% Root Locus
Dz=c2d(Dw,T,'tustin') % D(z)

%G(z)d(z)
DGz=Dz*Gz; %G(z)D(z)

% root locus of uncompensated system (without controller)
```

```
figure
rlocus(Gz)
axis('equal')
title('Root locus of the uncompensated system')
% root locus of comppensated system (with controller)
figure
rlocus(DGz)% Root locus
axis('equal')
title('Root locus of the system with the PD controller')

%Closed loop system step response of the uncompensated system
Tz=feedback(Gz,1)
figure
dstep(Tz,1)
title('Closed-loop step response of the uncompensated system')
%Closed loop system step response of the system with PID controller
Tz=feedback(DGz,1)
figure
dstep(Tz,1)
title('Closed-loop step response of the system with the PD controller')

Results:
PD Controller
ki = 0, kd = 0.1397, kp = 1.5564

        0.2175 w + 1.556
D(w) = ----------------
          0.05 s + 1

        2.953 z - 1.397
D(z) = ----------------
              z

For the compensated system with the PD controller
gm = 13.3491, pm = 55.0000, wpc = 15.7633, wgc = 2.5131, wgs = 2.5000, wps
= 13.3497
```

It should be noted that the same design point for the PI controller was used for illustrating the PD controller design process. Bode plots for the PD controller are shown in Fig. 12.8. The gain margin increased, and the phase margin reduced slightly with the PD controller as shown in Bode plots in Fig. 12.9 and root locus in Fig. 12.10. The percentage overshoot and settling time reduced slightly with the PD controller as shown in closed-loop system step response in Fig. 12.11. The steady-state error reduced slightly with the PD controller.

Example 12.3 PID Controller: Design a PID controller for $G_p(s) = \frac{5}{s^2+2.5s+1}$, $H(s) = 1$ such that the closed-loop system has a gain cross-over frequency of 2.5 rad/s, a phase margin of 55°, and a gain margin of 10 dB. Use sampling time $T = 0.10$ sec. Show the effect of the PID controller on the closed-loop system stability and step response.

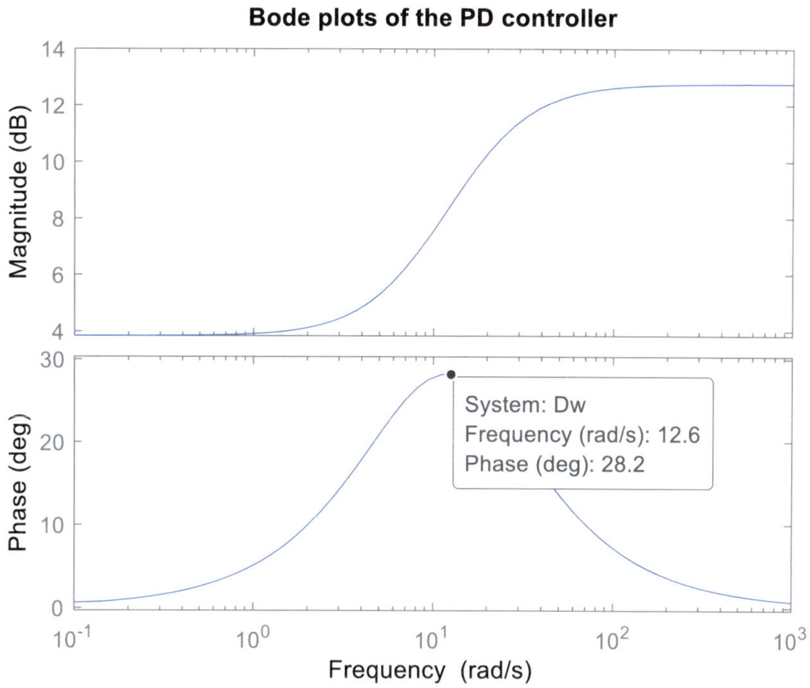

Fig. 12.8 Bode plots of the PD controller

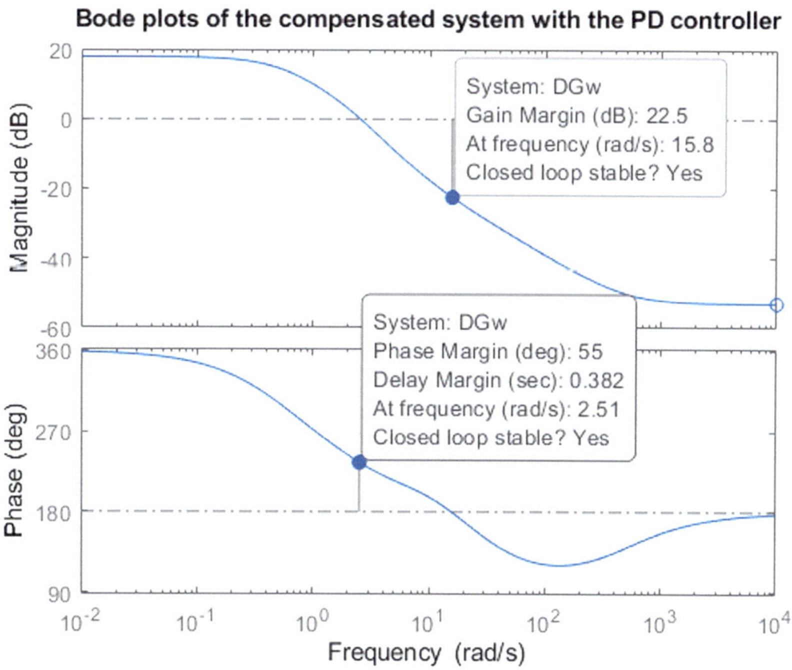

Fig. 12.9 Bode plots of the compensated system with the PD controller

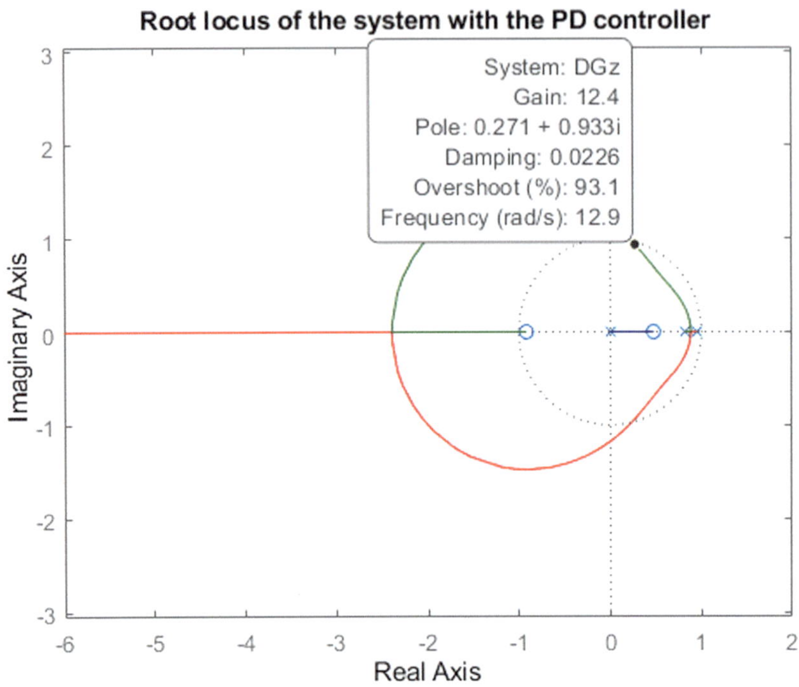

Fig. 12.10 Root locus of the system with the PD controller

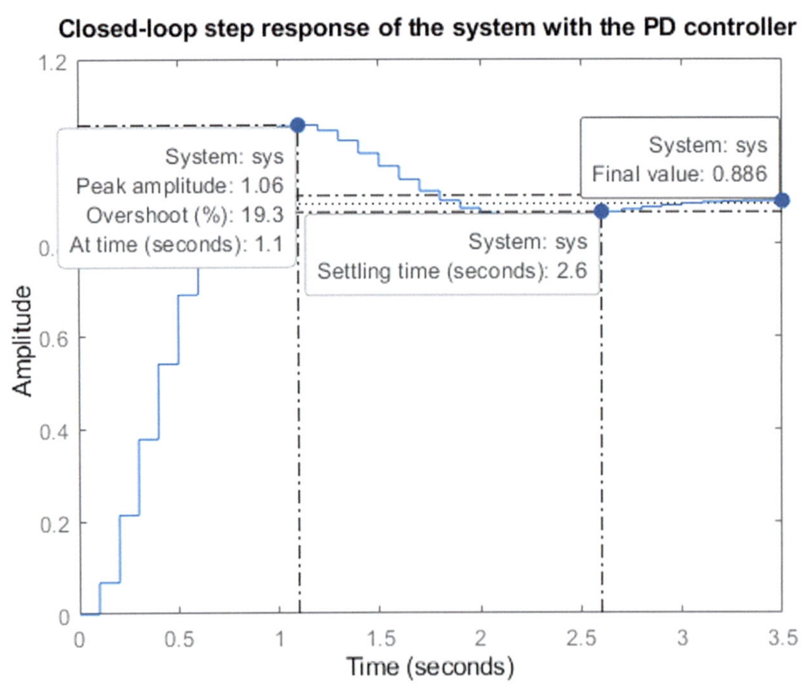

Fig. 12.11 Closed-loop step response of the system with the PD controller

Solution Matlab code and results are presented.

```
% ch12_ex3 _design of pid controller-w-domain (DIGITAL)%
close all;clear all;clc
T=0.1;% sampling time
Gp=tf(5,[1 2.5 1]) % Gp(s)=5/(s^2+2.5s+1)
% desired design spec
wgs=2.50; % gain cross-over frequency (s-plane) wgcs
dpm=55; dgm=10; %Desired phase margin, gain Margin
wgw=tan(wgs*T/2)*2/T; % w-domain wgc
Gz=c2d(Gp, T)  % G(z)
% G(w)
Gw=d2c(Gz,'tustin')
[ngw,dgw]=tfdata(Gw,'v') % G(w)
printsys(ngw,dgw,'w')

% bode plots of uncompensated system
figure
bode(Gw)
title('Bode plots of uncompensated system')
[gm,pm,wgc,wpc]=margin(Gw)

%Design of PID controller in w-domain
nv=polyval(ngw,j*wgw);dv=polyval(dgw,j*wgw);
g=nv/dv; %G(wgc)
m=abs(g) %mag(G(wgc)) at wgc
psi=angle(g) % angle(G(wgc))

%PID controller gains
ki=0.4 % to be selected
theta=(180+dpm)*pi/180-psi % angle in radians
kd=(sin(theta)/m+ki/wgw)*(1+(T*wgw/2)^2)/wgw
kp=cos(theta)/m-kd*(T/2)*wgw^2/(1+(T*wgw/2)^2)

Dw = tf([kp*T/2+kd kp+ki*T/2 ki],[T/2 1 0]) %D(w)
DGw = Dw*Gw

% bode plots of the controller
figure
bode(Dw)
title('Bode plots of the PID controller')

% bode plots of system with the controller
figure
bode(DGw)
title('Bode plots of the compensated system with the PID controller')
[gm,pm,wpc,wgc]=margin(DGw)
wgs=atan(wgc*T/2)*2/T % in s-domain
wps=atan(wpc*T/2)*2/T % in s-domain

% Root Locus
Dz=c2d(Dw,T,'tustin') % D(z)

%G(z)d(z)
DGz=Dz*Gz; %G(z)D(z)

% root locus of uncompensated system (without controller)
figure
rlocus(Gz)
```

```
axis('equal')
title('Root locus of the uncompensated system')
% root locus of comppensated system (with controller)
figure
rlocus(DGz)% Root locus
axis('equal')
title('Root locus of the system with the PID controller')

%Closed loop system step response of the uncompensated system
Tz=feedback(Gz,1)
figure
dstep(Tz,1)
title('Closed-loop step response of the uncompensated system')
%Closed loop system step response of the system with PID controller
Tz=feedback(DGz,1)
figure
dstep(Tz,1)
title('Closed-loop step response of the system with the PID controller')
% Estimation of damping ratio from phase margin and overshoot
z= dpm/100
alpha=exp(-pi*z/sqrt(1-z^2))
tset = 8/(wgs*tan(dpm*pi/180))
aa=0.107
delta=-log(aa)/pi;
zeta=abs(delta)/sqrt(1+delta^2)
delta=-log(aa)/pi;
zeta=abs(delta)/sqrt(1+delta^2)
```

```
Results:
PID controller
ki = 0.4000, kd = 0.2040, kp = 1.5364

        0.2809 w^2 + 1.556 w + 0.4
D(w) = --------------------------
              0.05 w^2 + w

        3.597 z^2 - 5.597 z + 2.04
D(z) = --------------------------
               z^2 - z

For the compensated system with the PID controller
gm = 12.1022, pm = 55.0000, wgc = 2.5131, wpc = 17.5746, wgs = 2.5000, wps
= 14.4188

Estimation of damping ratio and settling time from the specifications and
comparing with the step response
From desired specifications:
ζ = 0.5500, α = 0.1263, tset = 2.2407 s
From closed-loop system step response:
α = 0.1070, ζ = 0.5797, tset = 3.82 s
```

Bode plots for the PID controller are shown in Fig. 12.12. The gain margin slightly decreased compared to the PD controller, and the phase margin specification was met with the PID controller as shown in Bode plots in Fig. 12.13 and root locus in Fig. 12.14. The percentage overshoot reduced considerably with the PID controller as shown in the closed-loop system step response in Fig. 12.15. The steady-state error was eliminated with the PID controller. In fact, the PID controller combined the beneficial effects of PI and PD controllers as expected.

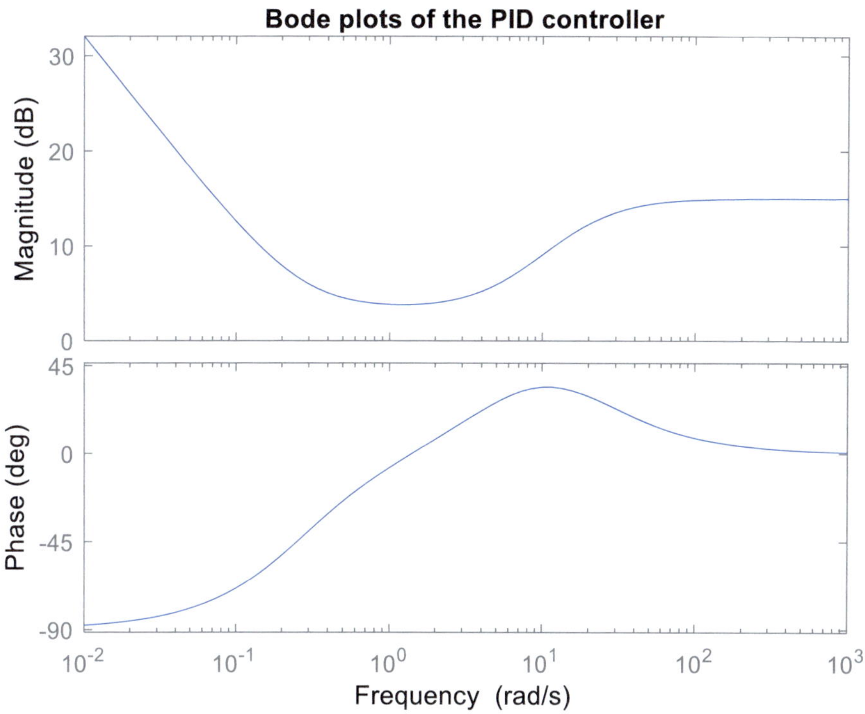

Fig. 12.12 Bode plots of the PID controller

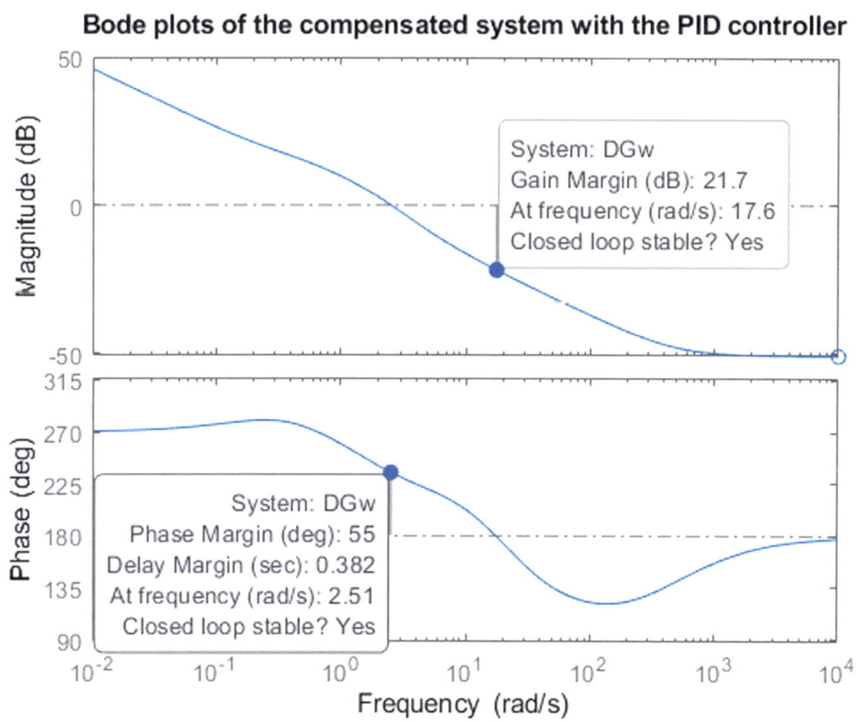

Fig. 12.13 Bode plots of the compensated system with the PID controller

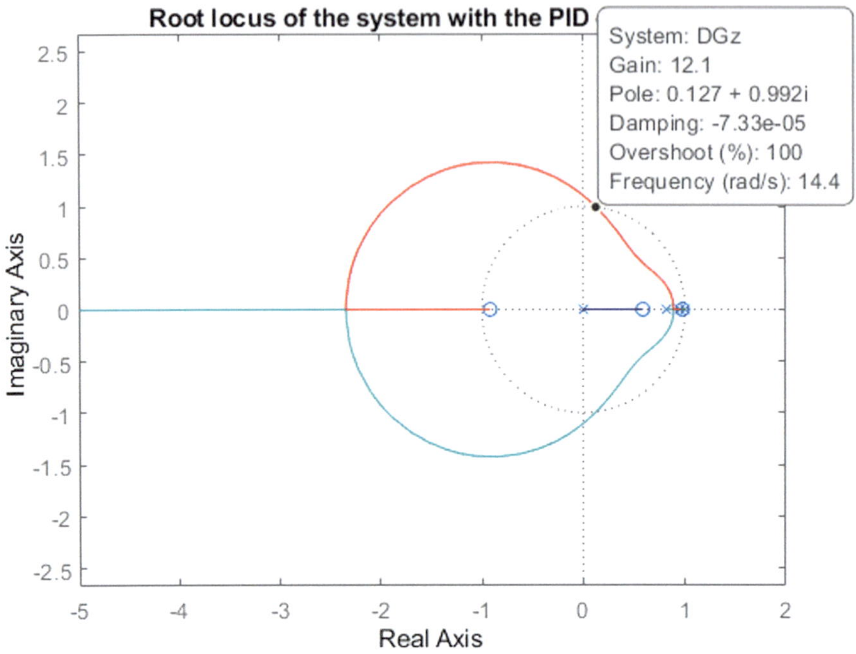

Fig. 12.14 Root locus of the system with the PID controller

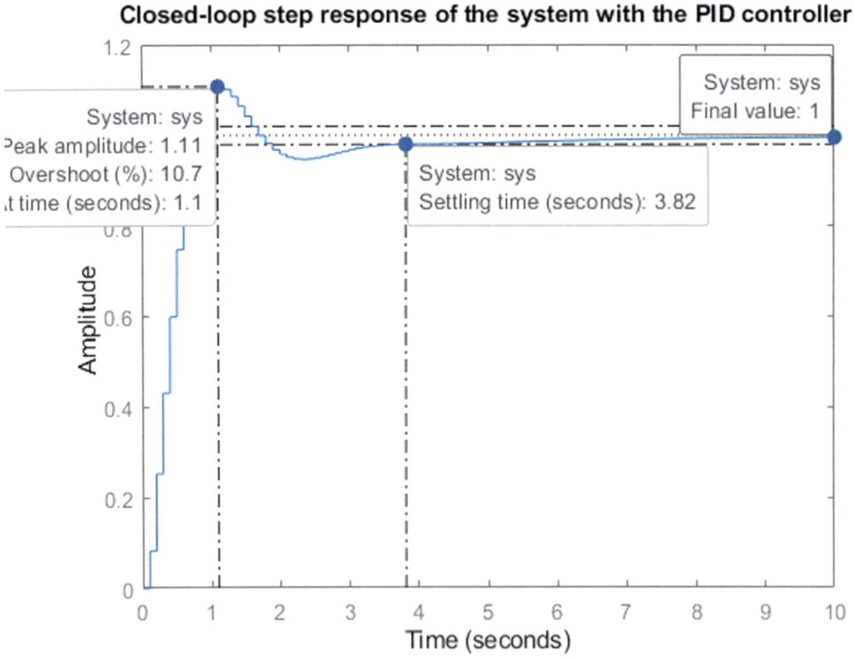

Fig. 12.15 Closed-loop step response of the system with the PID controller

12.3 Design of Phase-Lag, Phase-Lead, and Phase-Lag-Lead Controllers for Discrete-Time Systems Using Analytical Approach

12.3.1 Design of Phase-Lag Controllers

At the gain cross-over frequency $\omega_{w_{gc}}$, the combined controller and the system transfer function must have a magnitude of 1 and a phase angle of $(180^\circ + \phi_m)$ as

$$\overline{GH}\left(j\omega_{w_{gc}}\right)D\left(j\omega_{w_{gc}}\right) = 1e^{j\left(180^\circ + \phi_m\right)} \tag{12.23}$$

Using the controller as the phase-lag, the controller transfer function can be expressed as

$$\left.\left(\frac{1 + w/\omega_{w_0}}{1 + w/\omega_{w_p}}\right)\right|_{w = j\omega_{w_{gc}}} = \frac{1}{c_0\overline{GH}\left(j\omega_{w_{gc}}\right)}e^{j\left(180^\circ + \phi_m\right)} \tag{12.24}$$

Equation (12.24) can be expressed, after some rearrangements, as

$$1 + ja_1\omega_{w_{gc}} = \frac{1}{m}e^{j\left(180^\circ + \phi_m - \Psi\right)}\left(1 + jb_1\omega_{w_{gc}}\right) \tag{12.25}$$

where

$$c_0\overline{GH}\left(j\omega_{w_{gc}}\right) = me^{j\Psi} \tag{12.26}$$

$$a_1 = \frac{1}{\omega_{w_0}} \tag{12.27}$$

$$b_1 = \frac{1}{\omega_{w_p}} \tag{12.28}$$

Equation (2.25) can be rewritten, after some rearrangements, as

$$1 + ja_1\omega_{w_{gc}} = \frac{1}{m}(\cos\theta + j\sin\theta)\left(1 + jb_1\omega_{w_{gc}}\right) \tag{12.29}$$

where

$$\theta = 180^\circ + \phi_m - \Psi \tag{12.30}$$

Real and imaginary parts of both sides of Eq. (12.29) can be expressed as

$$1 = \frac{1}{m}\left(\cos\theta - \sin\theta\, b_1 \omega_{w_{gc}}\right) \tag{12.31}$$

$$a_1 \omega_{w_{gc}} = \frac{1}{m}\left(\cos\theta b_1 \omega_{w_{gc}} + \sin\theta\right) \tag{12.32}$$

Equations (12.31) and (12.32) can be analyzed to express the unknown parameters a_1, b_1

$$a_1 = \frac{1 - \mathrm{m}\,\cos\theta}{m\,\omega_{w_{gc}}\,\sin\theta} \tag{12.33}$$

$$b_1 = \frac{\cos\theta - m}{\omega_{w_{gc}}\,\sin\theta} \tag{12.34}$$

The phase-lag controller transfer function in the w-domain as

$$D_{\text{lag}}(w) = \frac{1 + a_1 w}{1 + b_1 w} \tag{12.35}$$

It should be noted that the constant c_0 can be found from steady-state accuracy specification and included in the transfer function, as shown in Eq. (12.26).

12.3.2 Design of Phase-Lead Controllers

The process is identical to that presented for phase-lag controllers. The desired specifications can determine the type of controllers: phase-lag (for improvement in steady-state performance) or phase-lead (for improvement in transient response).

12.3.3 Design of Phase-Lag-Lead Controllers

The phase-lag-lead controllers are designed in a two-step process: first, the phase-lag part is designed to meet the steady-state specification and a relaxed (less restrictive) transient response specification in terms of a phase margin of 20–30° (for a full specification of phase margin of 50°) and, finally, the phase-lag is included in the updated system transfer for designing the phase-lead part to meet the full phase margin specification.

Examples of designing phase-lag, phase-lead, and phase-lag-lead controllers are presented to illustrate the procedure.

Example 12.4 Phase-Lag Controller: Design a phase-lag controller for $G_p(s) = \frac{5}{s^2+2.5s+1}$, $H(s) = 1$ such that the closed-loop system has a maximum steady-state error of 5% due to a step input, a gain cross-over frequency of 2.5 rad/s, a phase margin of 35°, and a gain margin of 10 dB. Use sampling time $T = 0.10$ sec. Show the effect of the phase-lag controller on the closed-loop system stability and step response.

Solution Matlab code and results are presented.

```
% ch12_ex4_Analytical Design of phase-lead controller

clear all; close all; clc
T=0.1
dpm=35;  % desired phase margin
Gp=tf(5,[1 2.5 1] ) % Gp(s)=5/(s^2+2.5s+1)
essStep=0.05
kp=1/essStep-1
c0=kp/dcgain(Gp)
%G(z)
% G(z)
Gz = c2d(c0*Gp, T)
%G(w)
Gw= d2c(Gz,'tustin')
[ngw, dgw]= tfdata(Gw, 'v')
figure
bode(Gw)
title('Bode plots of the uncompensated system')
[gm,pm,wpc,wgc]=margin(Gw)
wgs=atan(wgc*T/2)*2/T
wps=atan(wpc*T/2)*2/T

% get m and psi at w1
wgc=2.5
w1=(2/T)*tan(wgc*T/2) % wgc in w-domain

nv=polyval(ngw,j*w1)
dv=polyval(dgw,j*w1)
gv=nv/dv
m=abs(gv)
psi=angle(gv)
thetad= 180+dpm-psi*180/pi
thetar=thetad*pi/180
% get a1, b1
a1=(1-m*cos(thetar))/(m*w1*sin(thetar))
b1=(cos(thetar)-m)/(w1*sin(thetar))

%D(w)
Dw=tf([a1 1], [b1 1])
figure
bode(Dw)
title('Bode plots of the phase-lag controller')
% D(w)G(w)
DGw = Dw*Gw

%bode plot G(w)D(w)
figure
bode(DGw)
title('Bode plots of the system with the phase-lag controller')
[gmc,pmc,wpc,wgc]=margin(DGw)
wgs=atan(wgc*T/2)*2/T
wps=atan(wpc*T/2)*2/T

% D(z)
Dz = c2d(Dw,T,'tustin')
%D(z)G(z)
DGz = Dz*Gz
```

```
% root locus of uncompensated system (without controller)
figure
rlocus(Gz)
axis('equal')
title('Root locus of the uncompensated system')
% root locus of compensated system (with controller)
figure
rlocus(DGz)
axis('equal')
title('Root locus of the system with the phase-lag controller')

% calculation of the closed loop step response
Tz= feedback(Gz/c0,1)
figure
dstep(Tz,1)
title('Closed-loop step response of the uncompensated system')

% Closed-loop step response of the compensated system
Tz=feedback(DGz,1)
figure
dstep(Tz,1)
title('Closed-loop step response of the system with the phase-lag
controller')
Results:
essStep = 0.0500, kp = 19, c0 = 3.8000

G(z) =
   0.08748 z + 0.08049
  ---------------------
  z^2 - 1.77 z + 0.7788

G(w) =
   -0.001971 w^2 - 0.9072 w + 18.93
  --------------------------------
       w^2 + 2.493 w + 0.9965

uncompensated system
gm = 2.7482, pm = 21.1879, wpc = 7.3018, wgc = 4.1614, wgs = 4.1028, wps =
7.0011
Phase-lag controller design
m =  2.3217,psi = -2.3944,thetad = 352.1910, thetar = 6.1469
a1 = 1.6400, b1 = 3.8978
Phase-lag controller
D(w) =
   1.64 w + 1
  -----------
  3.898 w + 1

D(z) =
   0.4281 z - 0.4028
  -----------------
      z - 0.9747
system with phase-lag controller
gmc = 5.6032, pmc = 35.0002, wpc = 6.7215, wgc = 2.5131, wgs = 2.5000, wps
= 6.4843
```

Bode plots of the uncompensated system are shown in Fig. 12.16. Bode plots of the phase-lag controller and that of the compensated system are shown in Figs. 12.17 and 12.18, respectively. Root locus plots of the system without and with the phase-lag controller are shown in Figs. 12.19 and 12.20. The closed-loop step response of

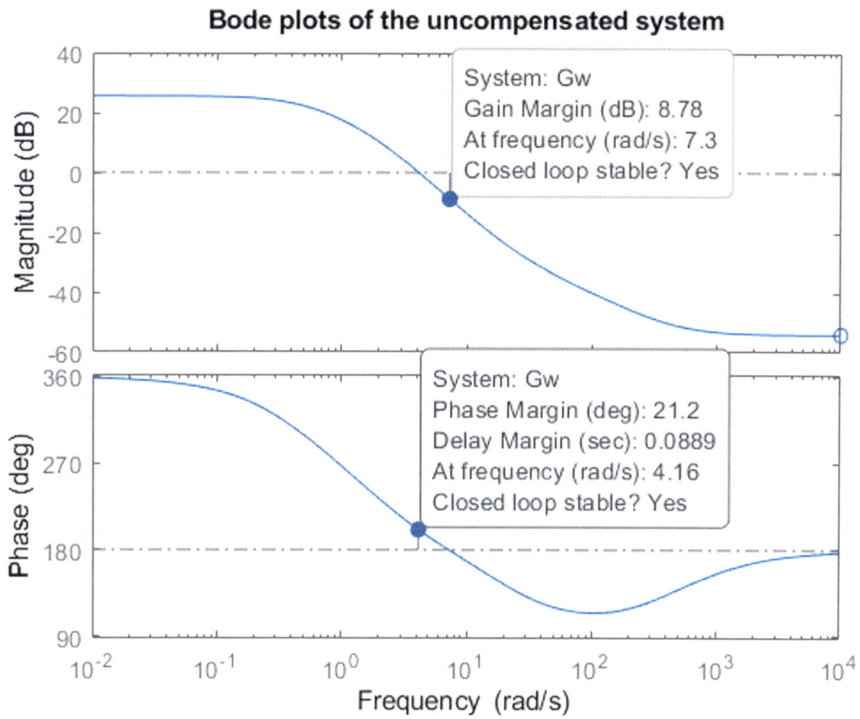

Fig. 12.16 Bode plots of the uncompensated system

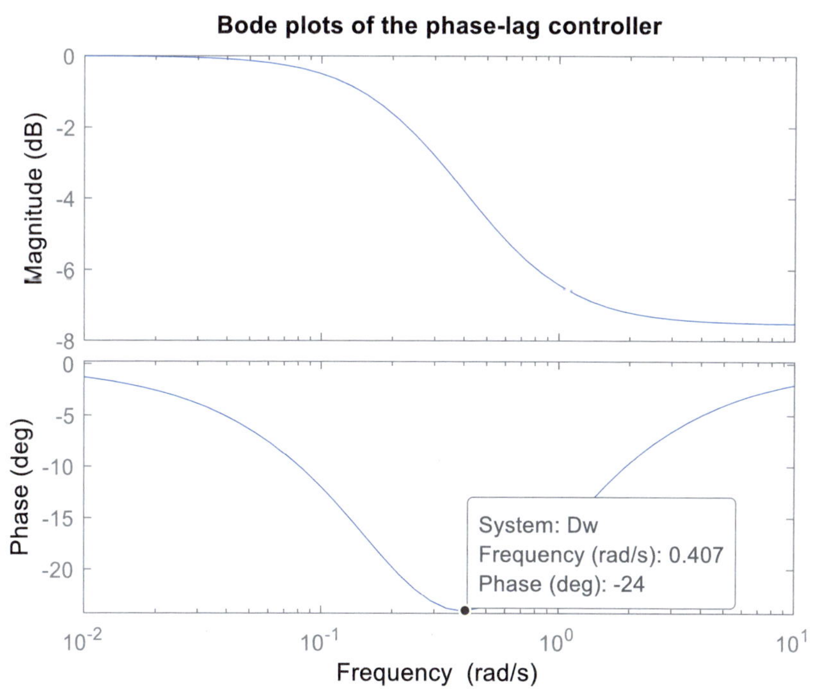

Fig. 12.17 Bode plots of the phase-lag controller

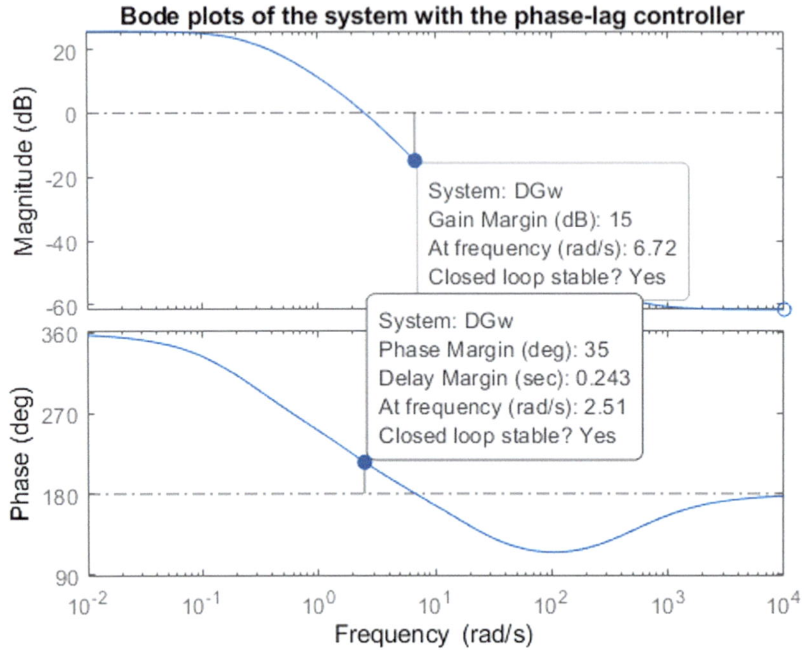

Fig. 12.18 Bode plots of the system with the phase-lag controller

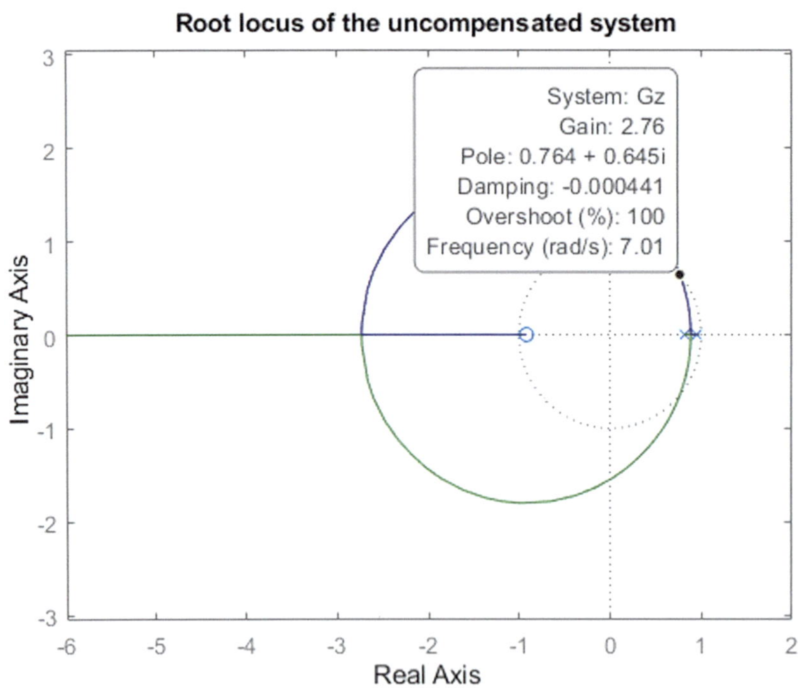

Fig. 12.19 Root locus of the uncompensated system

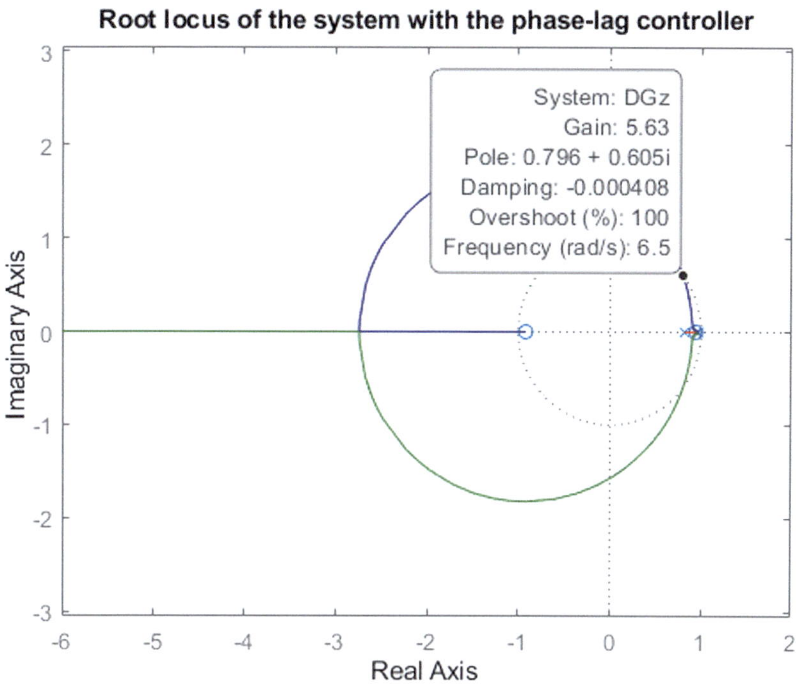

Fig. 12.20 Root locus of the system with the phase-lag controller

the system without and with the phase-lag controller is shown in Figs. 12.21 and 12.22, respectively.

Example 12.5 Phase-Lead Controller: Design a phase-lead controller for $G_p(s) = \frac{5}{s^2+2.5s+1}$, $H(s) = 1$ such that the closed-loop system has a maximum steady-state error of 15% due to a step input, a gain cross-over frequency of 2.5 rad/s, a phase margin of 55°, and a gain margin of 10 dB. Use sampling time $T = 0.10$ sec. Show the effect of the phase-lead controller on the closed-loop system stability and step response.

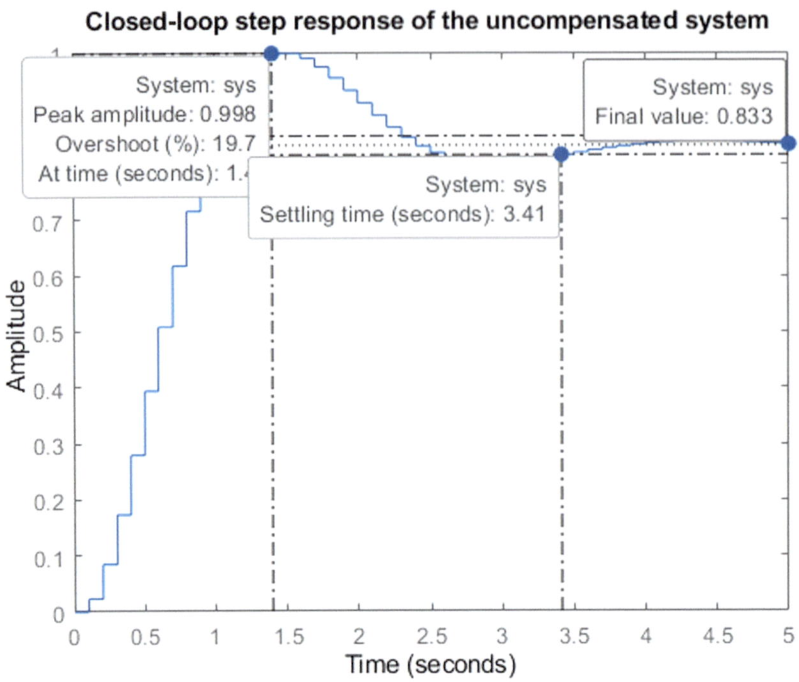

Fig. 12.21 Closed-loop step response of the uncompensated system

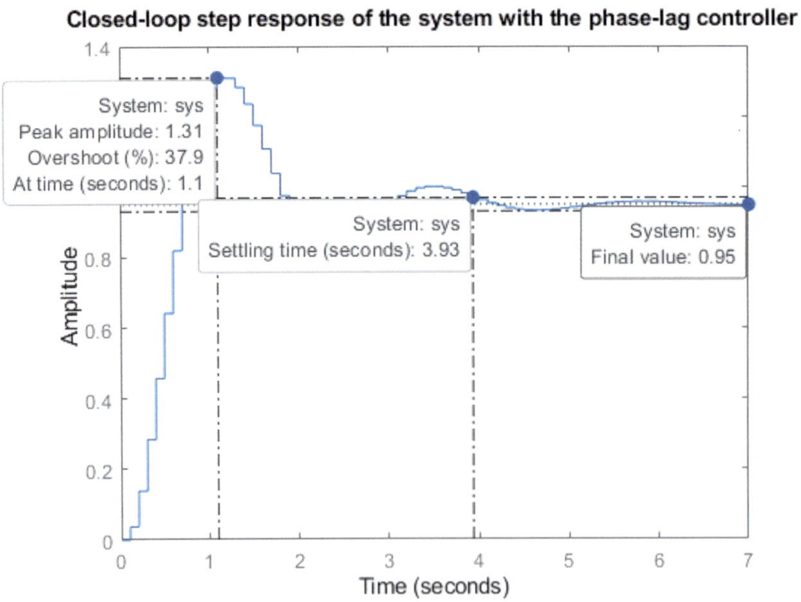

Fig. 12.22 Closed-loop step response of the system with the phase-lag controller

Solution Matlab code and results are presented.

```
% ch12_ex5_Analytical Design of phase-lead ontroller
clear all; close all; clc
T=0.1
dpm=55;  % desired phase margin
Gp=tf(5,[1 2.5 1] ) % Gp(s)=10/(s^2+2.5s+1)
essStep=0.15
kp=1/essStep-1
c0=kp/dcgain(Gp)
%G(z)
% G(z)
Gz = c2d(c0*Gp, T)
%G(w)
Gw= d2c(Gz,'tustin')
[ngw, dgw]= tfdata(Gw, 'v')

figure
bode(Gw)
title('Bode plots of the uncompensated system')
[gm,pm,wpc,wgc]=margin(Gw)
wgs=atan(wgc*T/2)*2/T
wps=atan(wpc*T/2)*2/T

% get m and psi at w1
wgc=2.5
w1=(2/T)*tan(wgc*T/2) % wgc in w-domain

nv=polyval(ngw,j*w1)
dv=polyval(dgw,j*w1)
gv=nv/dv
m=abs(gv)
psi=angle(gv)
thetad= 180+dpm-psi*180/pi
thetar=thetad*pi/180
% get a1, b1
a1=(1-m*cos(thetar))/(m*w1*sin(thetar))
b1=(cos(thetar)-m)/(w1*sin(thetar))

%D(w)
Dw=tf([a1 1], [b1 1])
figure
bode(Dw)
title('Bode plots of the phase-lead controller')
% D(w)G(w)
DGw = Dw*Gw

%bode plot G(w)D(w)
figure
bode(DGw)
title('Bode plots of the system with the phase-lead controller')
[gmc,pmc,wpc,wgc]=margin(DGw)
wgs=atan(wgc*T/2)*2/T
wps=atan(wpc*T/2)*2/T

% D(z)
Dz = c2d(Dw,T,'tustin')
%D(z)G(z)
DGz = Dz*Gz
```

```
% root locus of uncompensated system (without controller)
figure
rlocus(Gz)
axis('equal')
title('Root locus of the uncompensated system')
% root locus of compensated system (with controller)
figure
rlocus(DGz)
axis('equal')
title('Root locus of the system with the phase-lead controller')

% calculation of the closed loop step response
Tz= feedback(Gz,1)
figure
dstep(Tz,1)
title('Closed-loop step response of the uncompensated system')

% Closed-loop step response of the compensated system
Tz=feedback(DGz,1)
figure
dstep(Tz,1)
title('Closed-loop step response of the system with the phase-lead
controller')

Results:
essStep = 0.1500, kp = 5.6667, c0 = 1.1333

G(z) =
   0.02609 z + 0.02401
  ---------------------
  z^2 - 1.77 z + 0.7788

G(w) =
   -0.0005878 w^2 - 0.2706 w + 5.647
  --------------------------------
       w^2 + 2.493 w + 0.9965
uncompensated system
gm = 9.2146, pm = 54.3153, wpc = 7.3018, wgc = 1.9647, wgs = 1.9584, wps =
7.0011
phase-lead controller design
m = 0.6924, psi = -2.3944, thetad = 372.1910, thetar = 6.4960, a1 = 0.8795,
b1 = 0.5371
     0.8795 w + 1
Dw = ------------
     0.5371 w + 1
      1.583 z - 1.413
D(z)= ---------------
       z - 0.8297
System with phase-lead controller
gmc = 7.2756, pmc =  55.0071, wpc = 8.3510, wgc = 2.5128, wgs = 2.4997, wps
=  7.9108
```

Bode plots of the phase-lead controller and that of the compensated system are shown in Figs. 12.23 and 12.24, respectively. The root locus plot of the system with the phase-lead controller is shown in Fig. 12.25. The closed-loop step response of the system without and with the phase-lead controller is shown in Figs. 12.26 and 12.27, respectively.

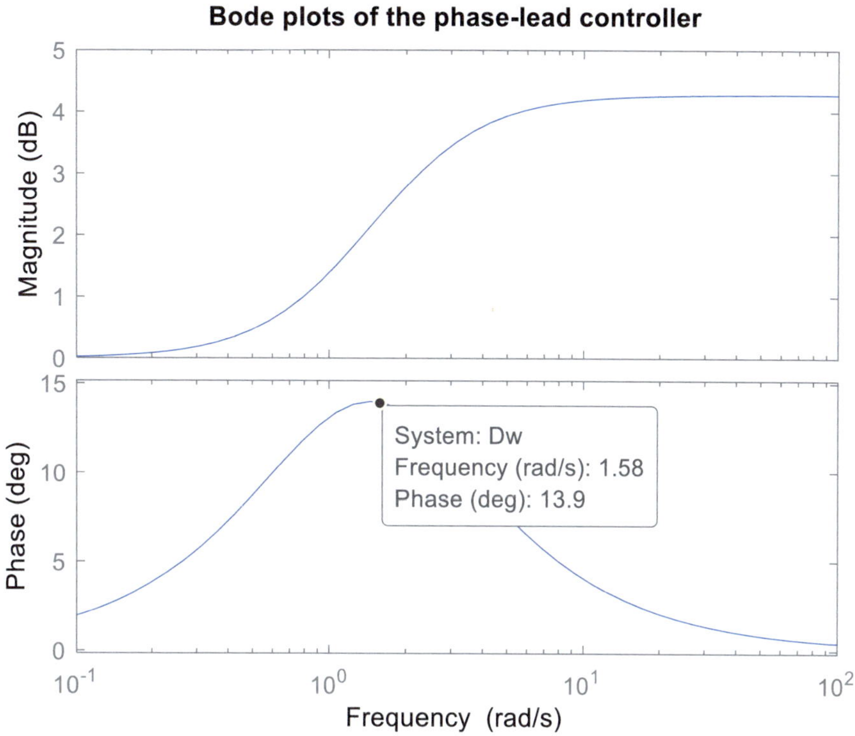

Fig. 12.23 Bode plots of the phase-lead controller

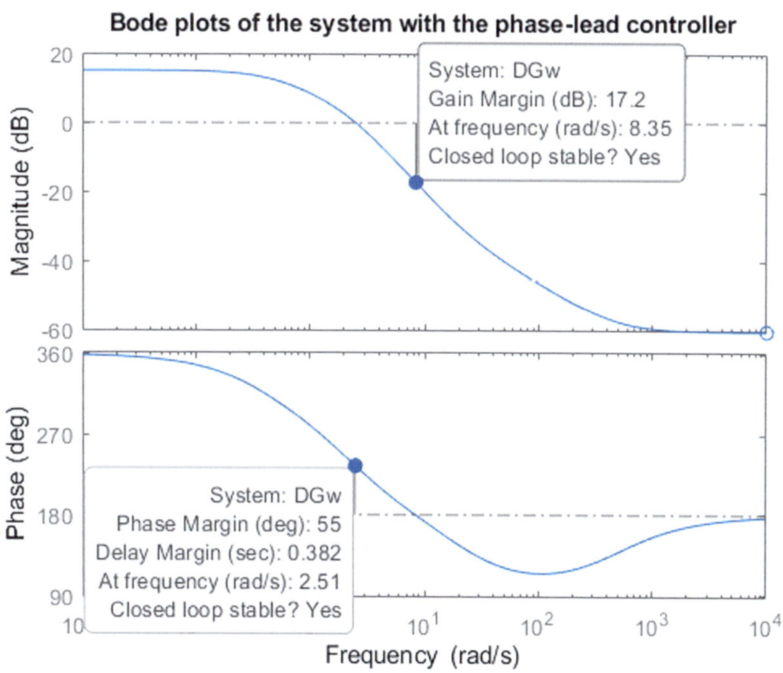

Fig. 12.24 Bode plots of the system with the phase-lead controller

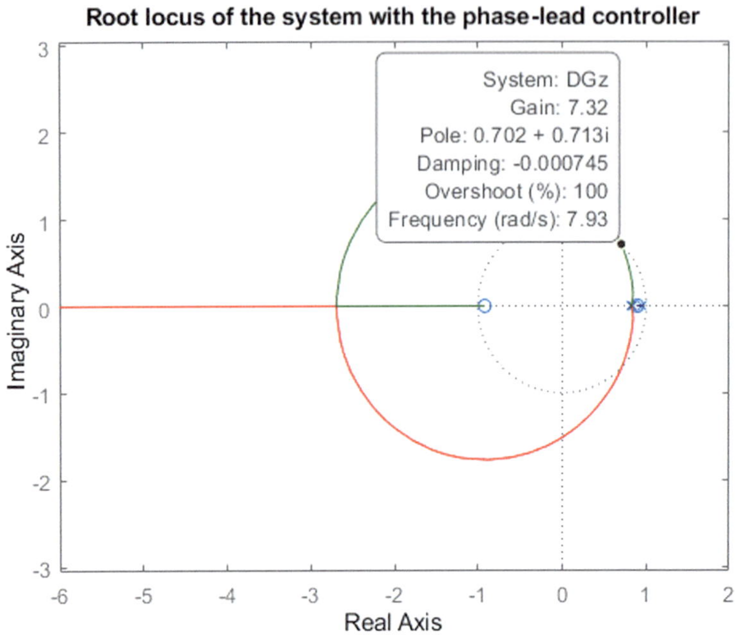

Fig. 12.25 Root locus of the uncompensated system

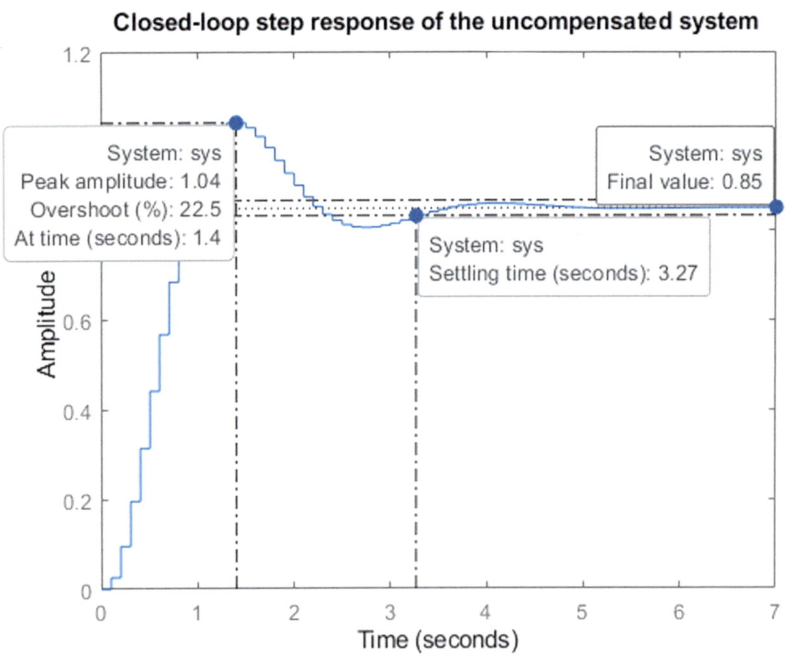

Fig. 12.26 Closed-loop step response of the uncompensated system

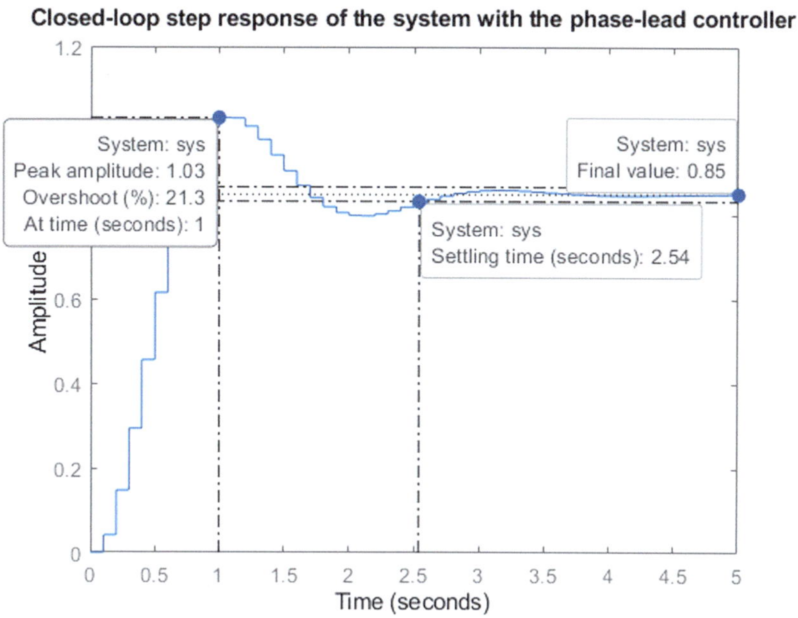

Fig. 12.27 Closed-loop step response of the system with the phase-lead controller

Example 12.6 Phase-Lag-Lead Controller: Design a phase-lag-lead controller for $G_p(s) = \frac{5}{s^2+2.5s+1}$, $H(s) = 1$ such that the closed-loop system has a maximum steady-state error of 5% due to a step input, a gain cross-over frequency of 2.5 rad/s, a phase margin of 55°, and a gain margin of 10 dB. Use sampling time $T = 0.10$ sec. Show the effect of the phase-lag-lead controller on the closed-loop system stability and step response.

Solution Matlab code and results are presented.

```
% ch12_ex6_Analytical Design of phase-lag-lead controller
clear all; close all; clc
T=0.1
essStep=0.05 % steady-state error
dpm=55;  % desired phase margin
Gp=tf(5,[1 2.5 1] ) % Gp(s)=5/(s^2+2.5s+1)

kp=1/essStep-1
c0=kp/dcgain(Gp)
%G(z)
% G(z)
Gz = c2d(c0*Gp, T)
%G(w)
Gw= d2c(Gz,'tustin')
[ngw, dgw]= tfdata(Gw, 'v')

figure
bode(Gw)
title('Bode plots of the uncompensated system')
[gm,pm,wpc,wgc]=margin(Gw)
wgs=atan(wgc*T/2)*2/T
wps=atan(wpc*T/2)*2/T

% Phase-lag controller part
% get m and psi at w1
dpm=30 % partial phase margin specification for phase-lag part
wgc=2.0
w1=(2/T)*tan(wgc*T/2) % wgc in w-domain

nv=polyval(ngw,j*w1)
dv=polyval(dgw,j*w1)
gv=nv/dv
m=abs(gv)
psi=angle(gv)
thetad= 180+dpm-psi*180/pi
thetar=thetad*pi/180
% get a1, b1
a1=(1-m*cos(thetar))/(m*w1*sin(thetar))
b1=(cos(thetar)-m)/(w1*sin(thetar))

%D(w)
Dw=tf([a1 1], [b1 1])
figure
bode(Dw)
title('Bode plots of the phase-lag controller')
% D(w)G(w)
DGw = Dw*Gw
[ndgw, ddgw]=tfdata(DGw,'v')
%bode plot G(w)D(w)
figure
bode(DGw)
title('Bode plots of the system with the phase-lag controller')
[gmc,pmc,wpc,wgc]=margin(DGw)
wgs=atan(wgc*T/2)*2/T
wps=atan(wpc*T/2)*2/T

% D(z)
```

```
Dz = c2d(Dw,T,'tustin')
%D(z)G(z)
DGz = Dz*Gz

% root locus of uncompensated system (without controller)
figure
rlocus(Gz)
axis('equal')
title('Root locus of the uncompensated system')
% root locus of compensated system (with controller)
figure
rlocus(DGz)
axis('equal')
title('Root locus of the system with the phase-lag controller')

% calculation of the closed loop step response
Tz= feedback(Gz,1)
figure
dstep(Tz,1)
title('Closed-loop step response of the uncompensated system')

% Closed-loop step response of the compensated system
Tz=feedback(DGz,1)
figure
dstep(Tz,1)
title('Closed-loop step response of the system with the phase-lag
controller')

%%%%%%%%%%%%%%%%%%%%%%%% phase-lead controller %%%%%%%%%%%%%%%%%%%%%%%%
dpm = 55 % actual phase margin
wgc=2.5
w1=(2/T)*tan(wgc*T/2) % wgc in w-domain

nv=polyval(ndgw,j*w1)
dv=polyval(ddgw,j*w1)
gv=nv/dv
m=abs(gv)
psi=angle(gv)
thetad= 180+dpm-psi*180/pi
thetar=thetad*pi/180
% get a1, b1
a1=(1-m*cos(thetar))/(m*w1*sin(thetar))
b1=(cos(thetar)-m)/(w1*sin(thetar))

%D(w)
Dwp=tf([a1 1], [b1 1])
Dwf=Dw*Dwp % Phase-lag-lead controller
figure
bode(Dwf)
title('Bode plots of the phase-lag-lead controller')
% D(w)G(w)
DGwf = Dwp*DGw

%bode plot G(w)D(w)
figure
bode(DGwf)
title('Bode plots of the system with the phase-lag-lead controller')
```

```
Dz = c2d(Dw,T,'tustin')
%D(z)G(z)
DGz = Dz*Gz

% root locus of uncompensated system (without controller)
figure
rlocus(Gz)
axis('equal')
title('Root locus of the uncompensated system')
% root locus of compensated system (with controller)
figure
rlocus(DGz)
axis('equal')
title('Root locus of the system with the phase-lag controller')

% calculation of the closed loop step response
Tz= feedback(Gz,1)
figure
dstep(Tz,1)
title('Closed-loop step response of the uncompensated system')

% Closed-loop step response of the compensated system
Tz=feedback(DGz,1)
figure
dstep(Tz,1)
title('Closed-loop step response of the system with the phase-lag
controller')

%%%%%%%%%%%%%%%%%%%%%%%% phase-lead controller %%%%%%%%%%%%%%%%%%%%%%%%
dpm = 55 % actual phase margin
wgc=2.5
w1=(2/T)*tan(wgc*T/2) % wgc in w-domain

nv=polyval(ndgw,j*w1)
dv=polyval(ddgw,j*w1)
gv=nv/dv
m=abs(gv)
psi=angle(gv)
thetad= 180+dpm-psi*180/pi
thetar=thetad*pi/180
% get a1, b1
a1=(1-m*cos(thetar))/(m*w1*sin(thetar))
b1=(cos(thetar)-m)/(w1*sin(thetar))

%D(w)
Dwp=tf([a1 1], [b1 1])
Dwf=Dw*Dwp % Phase-lag-lead controller
figure
bode(Dwf)
title('Bode plots of the phase-lag-lead controller')
% D(w)G(w)
DGwf = Dwp*DGw

%bode plot G(w)D(w)
figure
bode(DGwf)
title('Bode plots of the system with the phase-lag-lead controller')
```

```
[gmc,pmc,wpc,wgc]=margin(DGwf)
wgs=atan(wgc*T/2)*2/T
wps=atan(wpc*T/2)*2/T

% D(z)
Dzp = c2d(Dwp,T,'tustin')
%D(z)G(z)
DGzf = Dzp*DGz

% root locus of uncompensated system (without controller)
% figure
% rlocus(Gz)
% axis('equal')
% title('Root locus of the uncompensated system')
% root locus of compensated system (with controller)
figure
rlocus(DGzf)
axis('equal')
title('Root locus of the system with the phase-lag-lead controller')

% calculation of the closed loop step response
% Tz= feedback(Gz,1)
% figure
% dstep(Tz,1)
% title('Closed-loop step response of the uncompensated system')

% Closed-loop step response of the compensated system
Tz=feedback(DGzf,1)
figure
dstep(Tz,1)
title('Closed-loop step response of the system with the phase-lag-lead
controller')

Results:
Phase-lag controller design
dpm = 30, wgc = 2, m = 3.2530, psi = -2.2112, thetad = 336.6927, thetar =
5.8764
a1 = 0.7695,  b1 = 2.9405
       0.7695 w + 1
D(w) = ------------
       2.94 w + 1

      0.2741 z - 0.2406
D(z)= -----------------
         z - 0.9666

gmc = 6.3930, pmc = 30.0000, wpc = 5.6261, wgc = 2.0067, wgs = 2.0000, wps
= 5.4844

Phase-lead controller design
dpm = 55, wgc = 2.5000, m = 0.6778, psi = -2.7371, thetad = 391.8271,
thetar = 6.8387

a1 = 0.4721, b1 = 0.1296
        0.4721 w + 1
D(w) = --------------
        0.1296 w + 1
        2.906 z - 2.35
D(z) = --------------
        z - 0.4433
Lag-lead controller

        0.3633 w^2 + 1.242 w + 1
D(w) = ------------------------
        0.3812 w^2 + 3.07 w + 1
System with phase-lag-lead controller
gmc = 8.1947, pmc = 54.9996, wpc = 12.1149, wgc = 2.5132, wgs = 2.5000, wps
= 10.8926
```

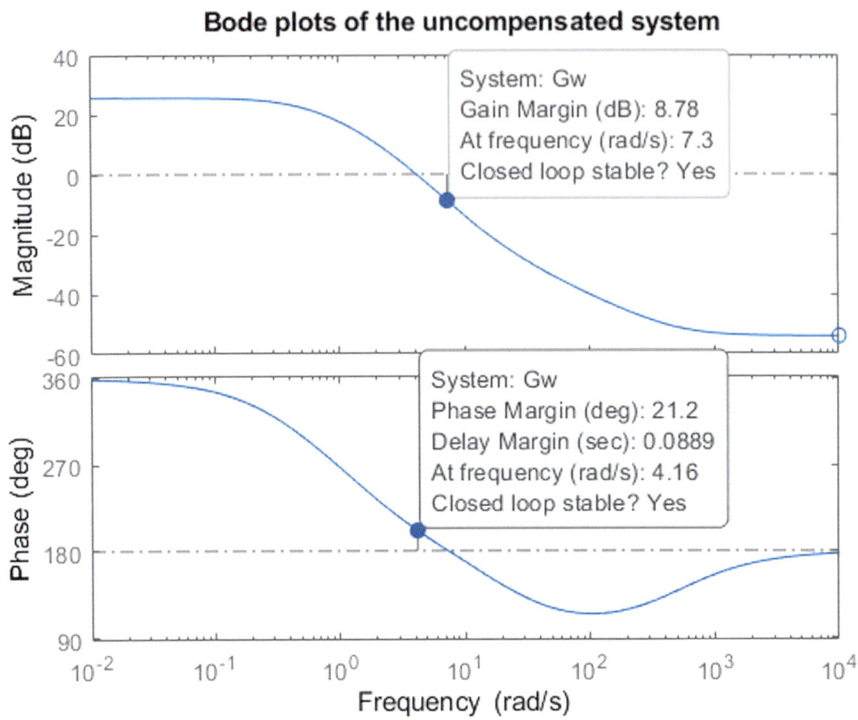

Fig. 12.28 Bode plots of the uncompensated system

Bode plots of the uncompensated system are shown in Fig. 12.28. Bode plots of the phase-lag part of the controller and that of the system with the phase-lag part are shown in Figs. 12.29 and 12.30, respectively. Root locus plots of the system without and with the phase-lag part of the controller are shown in Figs. 12.31 and 12.32. The closed-loop step response of the system without and with the phase-lag part of the controller is shown in Figs. 12.33 and 12.34, respectively.

Bode plots of the phase-lag-lead controller and that of the compensated system are shown in Figs. 12.35 and 12.36, respectively. The root locus plot of the system with the phase-lag-lead controller is shown in Fig. 12.37. The closed-loop step response of the system with the phase-lag-lead controller is shown in Fig. 12.38.

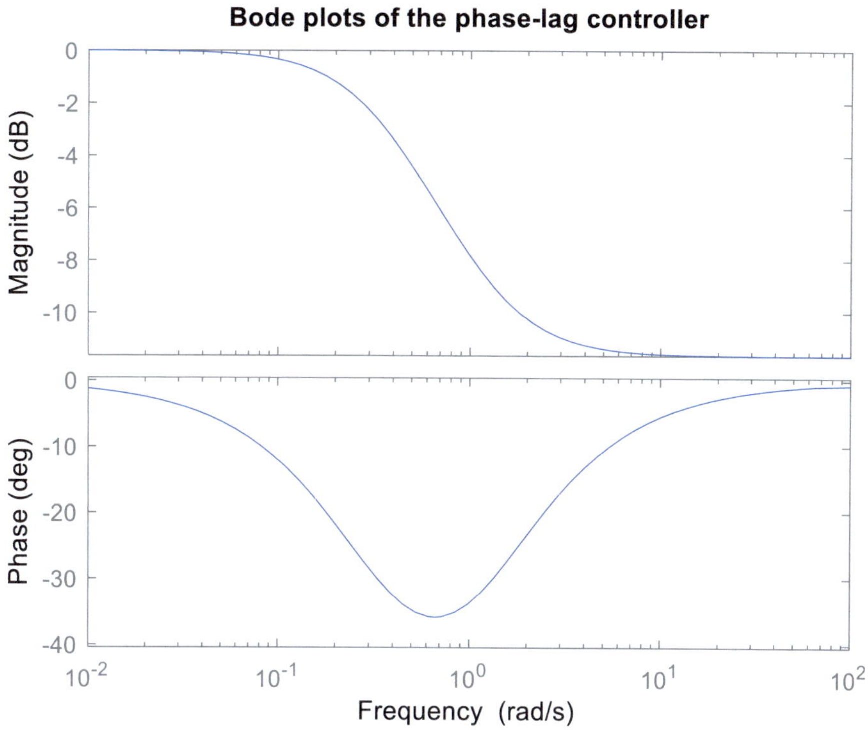

Fig. 12.29 Bode plots of the phase-lag controller

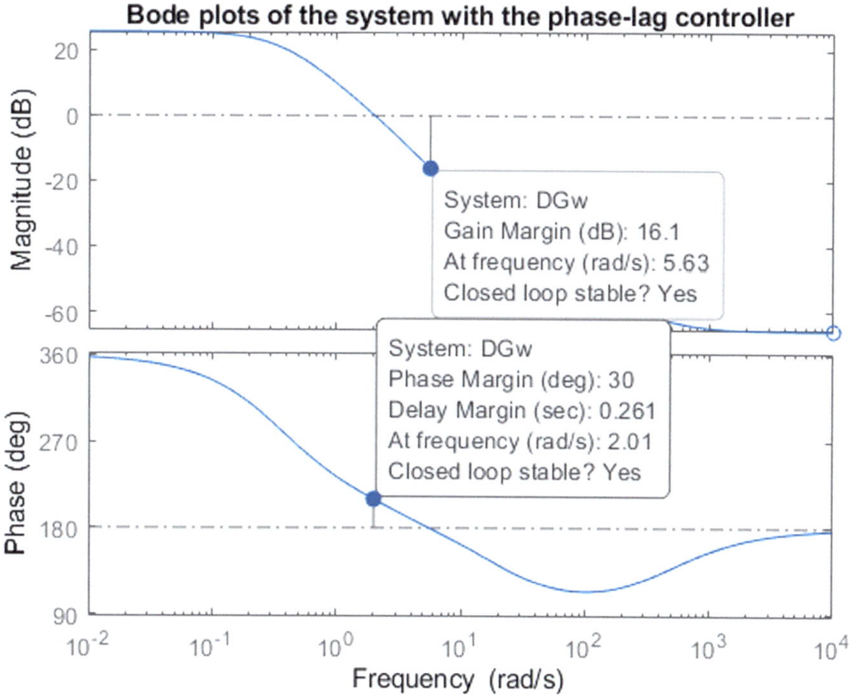

Fig. 12.30 Bode plots of the system with the phase-lag controller

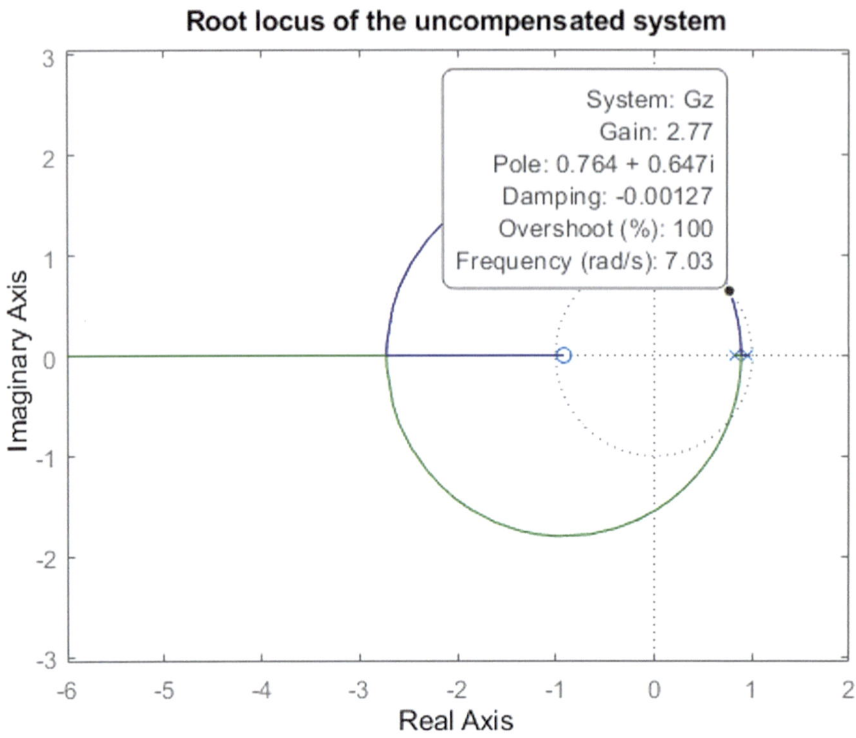

Fig. 12.31 Root locus of the uncompensated system

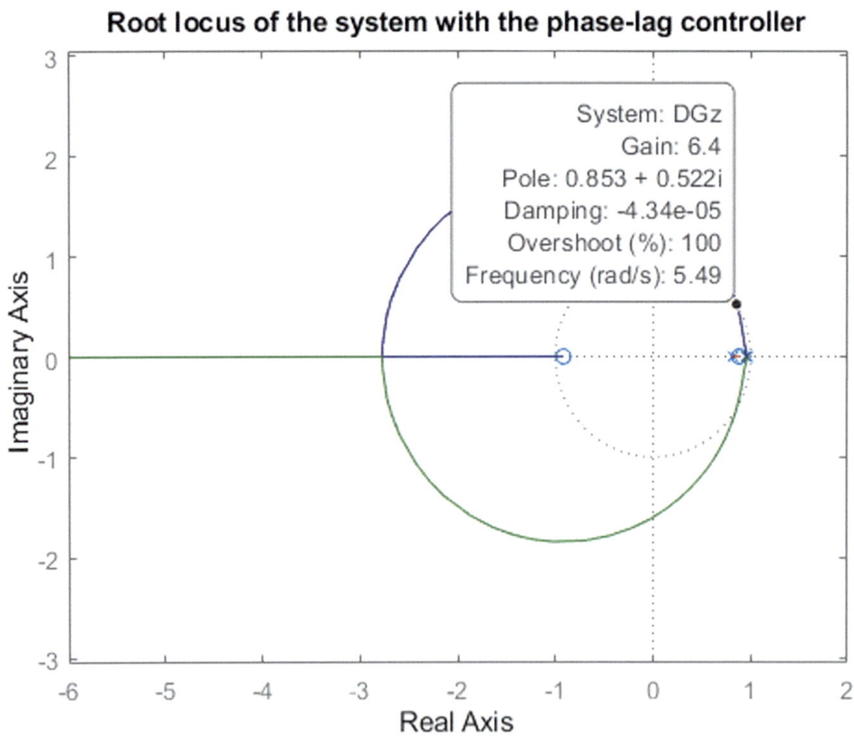

Fig. 12.32 Root locus of the system with the phase-lag controller

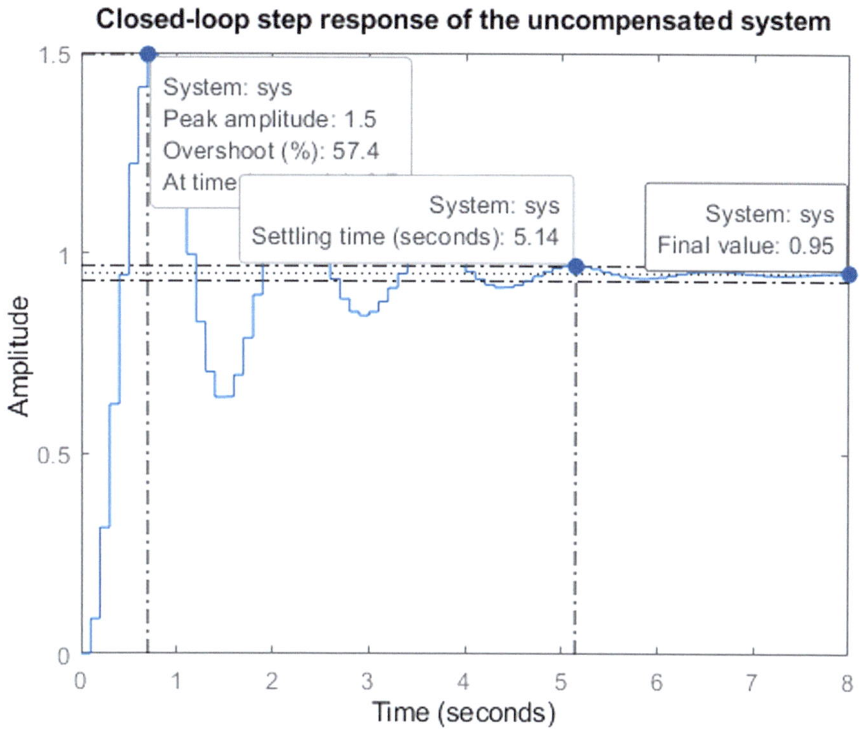

Fig. 12.33 Closed-loop step response of the uncompensated system

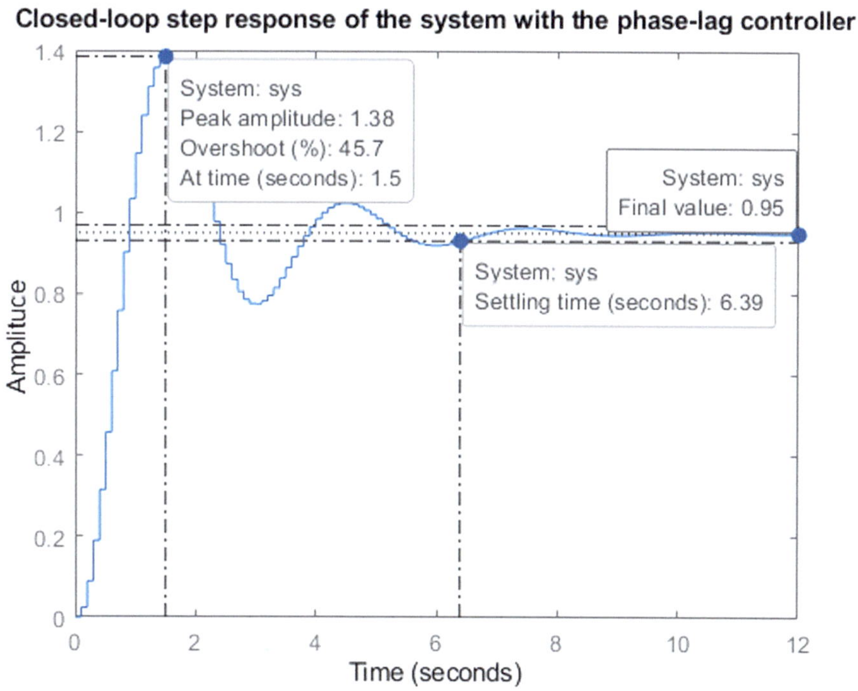

Fig. 12.34 Closed-loop step response of the system with the phase-lag controller

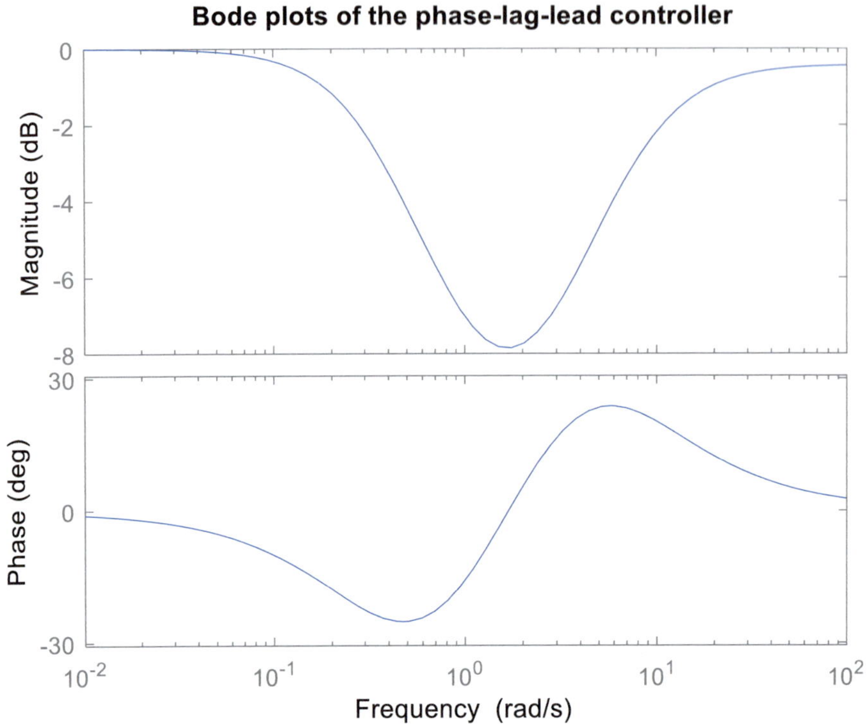

Fig. 12.35 Bode plots of the phase-lag-lead controller

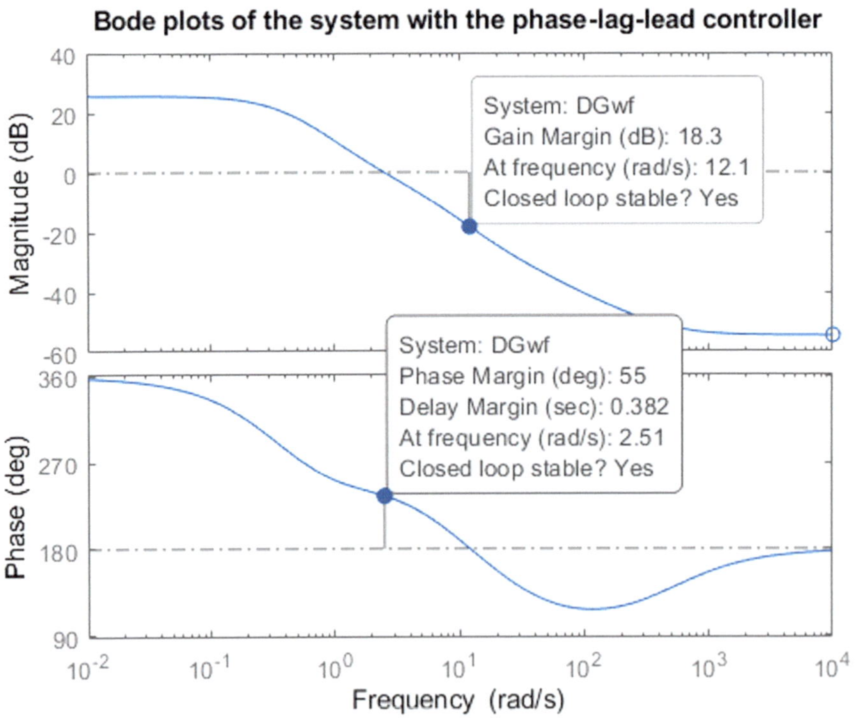

Fig. 12.36 Bode plots of the system with the phase-lag-lead controller

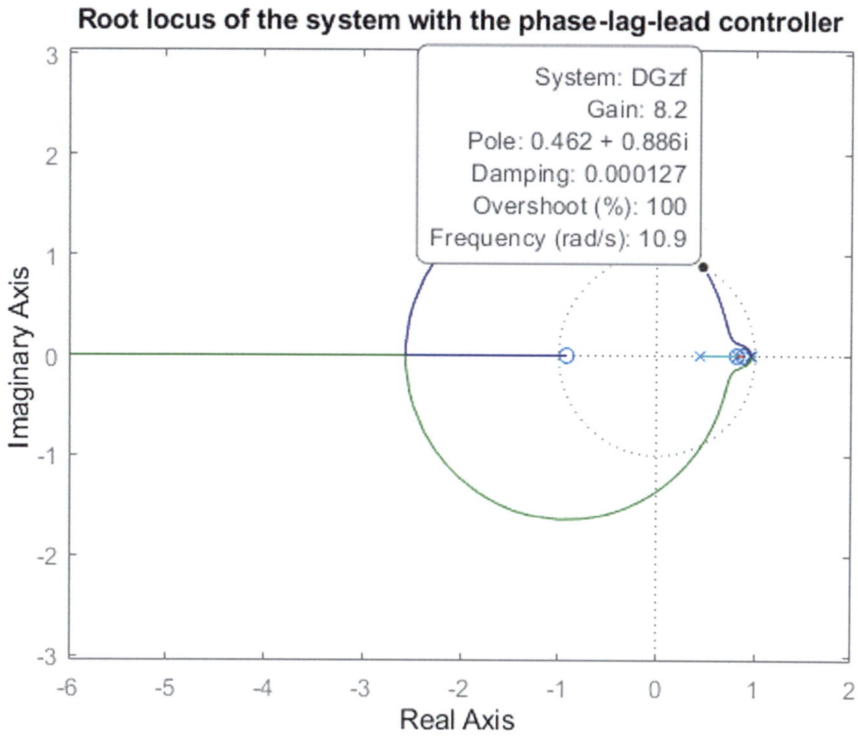

Fig. 12.37 Root locus of the system with the phase-lag-lead controller

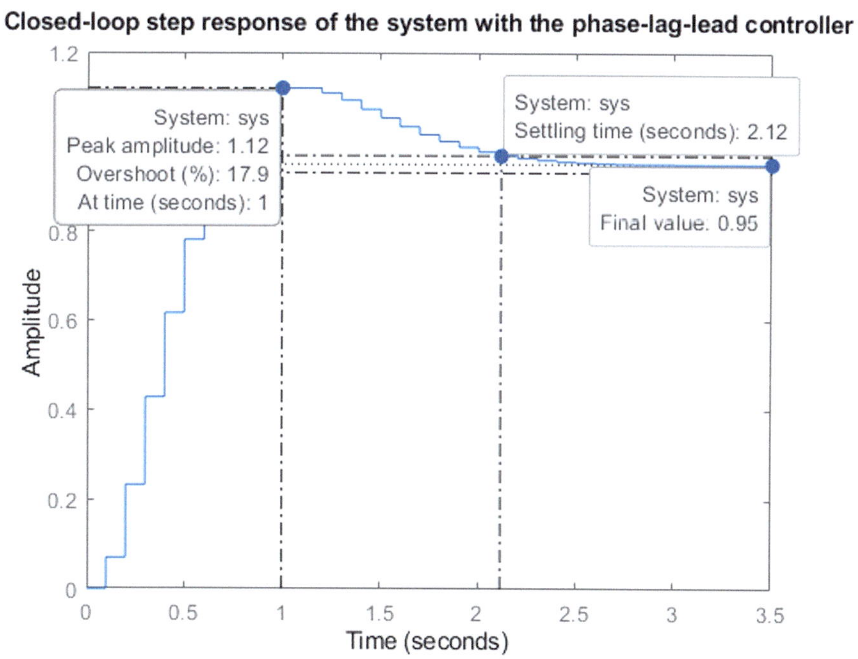

Fig. 12.38 Closed-loop step response of the system with the phase-lag-lead controller

12.4 Design of Phase-Lag, Phase-Lead, and Phase-Lag-Lead Controllers for Discrete-Time Systems Using Bode Plots

12.4.1 Magnitude and Phase Angle Relations for Phase-Lag and Phase-Lead Controllers

Phase-Lag Controllers

The w-domain transfer function of the phase-lag controller can be used to analyze frequency response with $w = j\omega_w$ as

$$D(j\omega_w) = \frac{1 + j\omega_w/\omega_{w0}}{1 + j\omega_w/\omega_{wp}} \tag{12.36}$$

where the controller pole magnitude is smaller than the controller zero magnitude, i.e., $\omega_{wp} < \omega_{w0}$.

Magnitude and phase angle can be expressed as

$$M(\omega_w) = \sqrt{\frac{1 + (\omega_w/\omega_{w0})^2}{1 + (\omega_w/\omega_{wp})^2}} \tag{12.37}$$

$$\phi(\omega_w) = \tan^{-1}\left(\frac{\omega_w}{\omega_{w0}}\right) - \tan^{-1}\left(\frac{\omega_w}{\omega_{wp}}\right) \tag{12.38}$$

Equation (12.38) can be rewritten as

$$\tan(\phi(\omega_w)) = \frac{\frac{\omega_w}{\omega_{w0}} - \frac{\omega_w}{\omega_{wp}}}{1 + \frac{\omega_w}{\omega_{w0}}\frac{\omega_w}{\omega_{wp}}} \tag{12.39}$$

To obtain the maximum or minimum value of phase angle, Eq. (12.39) can be differentiated with respect to ω_w and setting the derivative to 0, $\frac{d}{d\omega_w}\tan(\phi(\omega_w)) = 0$:

$$\frac{d}{d\omega_w}\tan(\phi(\omega_w)) = \frac{\left(\frac{1}{\omega_{w0}} - \frac{1}{\omega_{wp}}\right)\left(1 + \frac{\omega_w}{\omega_{w0}}\frac{\omega_w}{\omega_{wp}}\right) - \left(\frac{\omega_w}{\omega_{w0}} - \frac{\omega_w}{\omega_{wp}}\right)\left(\frac{2\omega_w}{\omega_{w0}\omega_{wp}}\right)}{\left(1 + \frac{\omega_w}{\omega_{w0}}\frac{\omega_w}{\omega_{wp}}\right)^2} = 0 \tag{12.40}$$

$$\frac{\left(\frac{1}{\omega_{w0}}-\frac{1}{\omega_{wp}}\right)\left[\left(1+\frac{\omega_w}{\omega_{w0}}\frac{\omega_w}{\omega_{wp}}\right)-\left(\frac{2\omega_w^{\ 2}}{\omega_{w0}\omega_{wp}}\right)\right]}{\left(1+\frac{\omega_w}{\omega_{w0}}\frac{\omega_w}{\omega_{wp}}\right)^2}=0 \tag{12.41}$$

$$\left(\frac{1}{\omega_{w0}}-\frac{1}{\omega_{wp}}\right)\left[1-\left(\frac{\omega_w^{\ 2}}{\omega_{w0}\omega_{wp}}\right)\right]=0 \tag{12.42}$$

The frequency at the extremum (maximum or minimum) can be expressed as

$$\omega_{wM}=\sqrt{\omega_{w0}\omega_{wp}} \tag{12.43}$$

The phase angle at the extremum can be expressed as

$$\tan(\phi_M(\omega_{wM}))=\frac{\frac{\omega_{wM}}{\omega_{w0}}-\frac{\omega_{wM}}{\omega_{wp}}}{1+\frac{\omega_{wM}}{\omega_{w0}}\frac{\omega_{wM}}{\omega_{wp}}}=\frac{\alpha-1}{2\sqrt{\alpha}} \tag{12.44}$$

where

$$\frac{\omega_{wp}}{\omega_{w0}}=\alpha<1 \tag{12.45}$$

Magnitude at frequency ω_{wM} can be expressed as

$$M(\omega_{wM})=\sqrt{\frac{1+(\omega_{wM}/\omega_{w0})^2}{1+(\omega_{wM}/\omega_{wp})^2}}=\sqrt{\frac{1+\left(\sqrt{\omega_{wp}/\omega_{w0}}\right)^2}{1+\left(\sqrt{\omega_{w0}/\omega_{wp}}\right)^2}}=\sqrt{\alpha} \tag{12.46}$$

It should be noted that the phase-lag controllers are designed not at the frequency (ω_{wM}) of minimum phase angle. Instead, the phase-lag controllers are designed far away from ω_{wM} to keep the phase-lag angle to a small value (5°–10°). Design steps are presented in Sect. 12.4.2.

Phase-Lead Controllers

The w-domain transfer function of the phase-lead controller can be used to analyze frequency response with $w=j\omega_w$ as

$$D(j\omega_w)=\frac{1+j\omega_w/\omega'_{w0}}{1+j\omega_w/\omega'_{wp}} \tag{12.47}$$

where the controller zero magnitude is smaller than the controller pole magnitude, i.e., $\omega'_{w0} < \omega'_{wp}$.

Magnitude and phase angle can be expressed as

$$M(\omega_w) = \sqrt{\frac{1 + (\omega_w/\omega\prime_{w0})^2}{1 + (\omega_w/\omega\prime_{wp})^2}} \tag{12.48}$$

$$\phi(\omega_w) = \tan^{-1}\left(\frac{\omega_w}{\omega'_{w0}}\right) - \tan^{-1}\left(\frac{\omega_w}{\omega'_{wp}}\right) \tag{12.49}$$

Equation (12.49) can be rewritten as

$$\tan(\phi(\omega_w)) = \frac{\frac{\omega_w}{\omega'_{w0}} - \frac{\omega_w}{\omega'_{wp}}}{1 + \frac{\omega_w}{\omega'_{w0}} \frac{\omega_w}{\omega'_{wp}}} \tag{12.50}$$

To obtain the maximum or minimum value of phase angle, Eq. (12.50) can be differentiated with respect to ω_w and setting the derivative to 0, $\frac{d}{d\omega_w} \tan(\phi(\omega_w)) = 0$:

$$\frac{d}{d\omega_w} \tan(\phi(\omega_w)) = \frac{\left(\frac{1}{\omega'_{w0}} - \frac{1}{\omega'_{wp}}\right)\left(1 + \frac{\omega_w}{\omega'_{w0}} \frac{\omega_w}{\omega'_{wp}}\right) - \left(\frac{\omega_w}{\omega'_{w0}} - \frac{\omega_w}{\omega'_{wp}}\right)\left(\frac{2\omega_w}{\omega'_{w0}\omega'_{wp}}\right)}{\left(1 + \frac{\omega_w}{\omega\prime_{w0}} \frac{\omega_w}{\omega\prime_{wp}}\right)^2} = 0 \tag{12.51}$$

$$\frac{\left(\frac{1}{\omega'_{w0}} - \frac{1}{\omega'_{wp}}\right)\left[\left(1 + \frac{\omega_w}{\omega'_{w0}} \frac{\omega_w}{\omega'_{wp}}\right) - \left(\frac{2\omega_w^2}{\omega'_{w0}\omega'_{wp}}\right)\right]}{\left(1 + \frac{\omega_w}{\omega\prime_{w0}} \frac{\omega_w}{\omega\prime_{wp}}\right)^2} = 0 \tag{12.52}$$

$$\left(\frac{1}{\omega'_{w0}} - \frac{1}{\omega'_{wp}}\right)\left[1 - \left(\frac{\omega_w^2}{\omega'_{w0}\omega'_{wp}}\right)\right] = 0 \tag{12.53}$$

The frequency at the extremum (maximum or minimum) can be expressed as

$$\omega'_{wM} = \sqrt{\omega'_{w0}\omega'_{wp}} \tag{12.54}$$

The phase angle at the extremum ($(\omega'_{wM}$) can be expressed as

$$\tan(\phi'_M) = \frac{\frac{\omega'_{wM}}{\omega'_{w0}} - \frac{\omega'_{wM}}{\omega'_{wp}}}{1 + \frac{\omega'_{wM}}{\omega'_{w0}} \frac{\omega'_{wM}}{\omega'_{wp}}} = \frac{1 - \alpha'}{2\sqrt{\alpha'}} \tag{12.55}$$

where

$$\frac{\omega'_{w0}}{\omega'_{wp}} = \alpha' < 1 \tag{12.56}$$

Equation (12.55) can be used to obtain the expression

$$\sin(\phi'_M) = \frac{1-\alpha'}{1+\alpha'} \tag{12.57}$$

Equation (12.57) can be used to express β in terms of ϕ'_M as

$$\alpha' = \frac{1-\sin(\phi'_M)}{1+\sin(\phi'_M)} \tag{12.58}$$

Magnitude at frequency ω'_{wM} can be expressed as

$$M(\omega'_{wM}) = \sqrt{\frac{1+(\omega\prime_{wM}/\omega\prime_{w0})^2}{1+(\omega\prime_{wM}/\omega\prime_{wp})^2}} = \sqrt{\frac{1+\left(\sqrt{\omega\prime_{wp}/\omega\prime_{w0}}\right)^2}{1+\left(\sqrt{\omega\prime_{w0}/\omega\prime_{wp}}\right)^2}} = 1/\sqrt{\alpha'} \tag{12.59}$$

It should be noted that the phase-lead controllers are designed at the frequency (ω'_{wM}) of maximum phase angle to maximize the phase-lead angle with a range of (25°–45°). Design steps are presented in Sect. 12.4.3.

12.4.2 Design of Phase-Lag Controllers

Design Steps for Phase-Lag Controllers

1. Obtain c_0 to satisfy the steady-state accuracy specification.
2. Include c_0 to obtain bode plots for $c_0\overline{GH}(w)$.
3. Locate from the Bode plots the frequency ω_{w1} where the phase angle is $\phi_1 = \phi_m + \lambda - 180°$ where ϕ_m is the desired phase margin and λ is the allowance (5°–15°) to compensate for the phase at the design point.

   Use spline command in Matlab as *ww1=spline(ph, w, ph1)* where ph and w are vectors of phase angle and frequency in Bode phase plot and ph1 is the phase angle ϕ_1.

The frequency ω_{w1} is the gain cross-over frequency ($\omega_{w1} = \omega_{wgc}$) of the control system with the phase-lag controller. Obtain from Bode plots the magnitude:

$$m = \left| c_0 \overline{GH}(j\omega_{w1}) \right|.$$

4. Obtain the magnitude of the controller zero as $\omega_{w0} = \frac{\omega_{w1}}{10}$.
5. Obtain the controller pole magnitude ω_{wp} satisfying the magnitude condition at the gain cross-over frequency as

$$\left| \frac{1 + j\omega_{w1}/\omega_{w0}}{1 + j\omega_{w1}/\omega_{wp}} \right| \left| c_0 \overline{GH}(j\omega_{w1}) \right| = 1$$

Recognizing $\frac{\omega_{w1}}{\omega_{w0}} = 10$ and $\frac{\omega_{w1}}{\omega_{wp}} > 10$, the magnitude of the controller pole can be approximated as

$$\frac{\omega_{w1}/\omega_{w0}}{\omega_{w1}/\omega_{wp}} m = 1$$

$$\omega_{wp} = \frac{\omega_{w0}}{m}$$

6. The controller transfer function in w-domain $D_{\text{lag}}(w) = \frac{1+w/\omega_{w0}}{1+w/\omega_{wp}}$.
7. Obtain the system bode plots with the controller $D_{\text{lag}}(j\omega_w) c_0 \overline{GH}(j\omega_w)$.
8. Verify if the desired specifications of phase margin and gain margin are achieved. Convert the controller transfer function to z-domain, get the system response, and verify if the steady-state accuracy specification is met.
9. If the specifications are met, accept the current values for the phase-lag controller. Otherwise, check on the specifications, adjust the phase-angle allowance (λ), and repeat the process till the desired specifications are closely met.

12.4.3 Design of Phase-Lead Controllers

Design Steps for Phase-Lead Controllers

1. Obtain c'_0 to satisfy the steady-state accuracy specification.
2. Include c'_0 to obtain bode plots for $c'_0 \overline{GH}(w)$.

3. Obtain from the Bode plots the available phase margin ϕ_a. Obtain the desired phase angle from the phase-lead controller as $\phi_M = \phi_m + \lambda - \phi_a$ where ϕ_m is the desired phase margin and λ is the allowance (5°–15°) to compensate for the phase at the design point. This phase angle is the maximum phase angle ($\phi'_M = \phi_M$) from the phase-lead controller at ω'_{wM}.This is the design point and also the gain cross-over frequency of the system with the controller.
4. Use Eq. (12.58) to obtain the magnitude ratio of controller zero and pole $\left(\frac{\omega'_{w0}}{\omega'_{wp}} = \alpha'\right)$ as

$$\alpha' = \frac{1 - \sin(\phi'_M)}{1 + \sin(\phi'_M)}$$

5. Obtain the system (without the controller) magnitude at the gain cross-over frequency satisfying the magnitude condition of the controller and the overall system, with the controller as

$$\left|\frac{1 + j\omega'_{wM}/\omega'_{w0}}{1 + j\omega'_{wM}/\omega'_{wp}}\right| \left|c_0\overline{GH}(j\omega'_{wM})\right| = 1$$

$$\frac{1}{\sqrt{\alpha'}}\left|c_0\overline{GH}(j\omega'_{wM})\right| = 1$$

$$\left|c_0\overline{GH}(j\omega'_{wM})\right| = \sqrt{\alpha'}$$

6. From Bode plots, obtain the frequency (ω'_{wM}) at which the magnitude is

$$\left|c_0\overline{GH}(j\omega'_{wM})\right| = \sqrt{\alpha'}$$

Use spline command in Matlab as *wwMp=spline(mag, w, mag_wM)* where mag and w are vectors of magnitude and frequency in Bode magnitude plot and mag_wM is the magnitude $\sqrt{\alpha'}$.
7. Obtain the magnitude of controller zero and pole from Eqs. (12.54) and (12.56) as

$$\omega'_{wM} = \sqrt{\omega'_{w0}\omega'_{wp}} = \frac{\omega'_{w0}}{\sqrt{\alpha'}}$$

$$\omega'_{w0} = \omega'_{wM}\sqrt{\alpha'}, \quad \omega'_{wp} = \omega'_{w0}/\alpha'$$

8. The controller transfer function in w-domain $D_{\text{lead}}(w) = \frac{1+w/\omega'_{w0}}{1+w/\omega'_{wp}}$.
9. Obtain the system bode plots with the controller $D_{\text{lead}}(j\omega_w)c'_0\overline{GH}(j\omega_w)$.
10. Verify if the desired specifications of phase margin and gain margin are achieved. Convert the controller transfer function to z-domain, get the system response, and verify if the steady-state accuracy specification is met.
11. If the specifications are met, accept the current values for the phase-lead controller. Otherwise, check on the specifications, adjust the phase-angle allowance (λ), and repeat the process till the desired specifications are closely met.

12.4.4 Design of Phase-Lag-Lead Controllers

The phase-lag-lead controllers are designed in a two-step process: first, the phase-lag part is designed to meet the steady-state specification and partially meet the phase margin specification and, finally, the phase-lag is included in the updated system transfer for designing the phase-lead part to meet the full phase margin specification, similar to Sect. 12.4.3.

Example 12.7 Phase-Lag Controller: Design a phase-lag controller for $G_p(s) = \frac{1}{s^2+3s+2}$, $H(s) = 1$ such that the closed-loop system has a maximum steady-state error of 5% due to a step input, a phase margin of 50°, and a gain margin of 10 dB. Use sampling time $T = 0.10$ sec. Show the effect of the phase-lag controller on the closed-loop system stability and step response.

Solution Matlab code and results are presented.

```
% Ch12-Example 7-Design of phase- lag controller
% Design a phase lag controller for plant Gp(s) so that
% ess,step=5%, phase margin=50 deg, Gain margin= 10 dB.
clear all; ;close all;clc
essStep =0.05
kp=1/essStep -1
a0=kp*2
T=0.10
dpm=50;  % desired phase margin
lambda=6.75; % allowance for shift in phase
Gp=tf(a0,[1 3 2]) % plant tf with a0/(s+1)(s+2)
% G(z)
Gz = c2d(Gp, T)
%G(w)
Gw = d2c(Gz, 'tustin')
[ngw, dgw]= tfdata(Gw, 'v')

figure
bode(Gw)
title('Bode plot of the uncompensated system')
[mag, ph, w]=bode(Gw);% check if angles are >0
ph=ph-360;% check if angles are >0 then subtract 360 to make angles  <0
[gm,pm,wpc,wgc]=margin(Gw)
ph1=(dpm+lambda-180)
% convert ph to phv
phv = permute(ph, [3 2 1]);

% inetrpolate to get wm at ph1
wm=spline(phv,w,ph1)
w0=wm/10

ngv=polyval(ngw,j*wm);dgv=polyval(dgw,j*wm)
g=abs(ngv/dgv)
wp=0.1*wm/g;

Dw = tf([1/w0 1], [1/wp 1]) % D(w)

%DwGw
DGw = Dw*Gw
figure
bode(Dw)
title('Bode plot of the phase-lag controller')
% bode plot of the compensated plant
figure
bode(DGw)
title('Bode plot of the system with the phase-lag controller')
[gmc,pmc,wpc,wgc]=margin(DGw)
wgs=atan(wgc*T/2)*2/T  % wg in s-plane
wps=atan(wpc*T/2)*2/T  % wp in s-plane


% calculation of the closed loop step response
% D(z)
Dz = c2d(Dw,T,'tustin')
figure
rlocus(Gz)
axis('equal')
```

```
title('Root locus of the uncompensated system')
%D(z)G(z)
DGz = Dz*Gz
figure
rlocus(DGz)
axis('equal')
title('Root locus of the system with the phase-lag controller')

% calculation of the closed loop step response
Tz= feedback(Gz,1)
figure
dstep(Tz,1)
title('Closed-loop step response of the uncompensated system')
% Closed-loop step response of the compensated system
Tz=feedback(DGz,1)
figure
dstep(Tz,1)
title('Closed-loop step response of the system with the phase-lag
controller')
```

```
Results:

             0.1721 z + 0.1557
G(z)  =   ----------------------
          z^2 - 1.724 z + 0.7408

         -0.004726 w^2 - 1.798 w + 37.84
G(w)  =  -------------------------------
             w^2 + 2.993 w + 1.992

ph1 = -123.2500, wm = 2.4827, w0 = 0.2483, g= 4.4752, wp= 0.0555

          4.028 w + 1
D(w)  =  -------------
          18.03 w + 1

          0.2256 z - 0.2201
D(z)  =   -----------------
            z - 0.9945
```

Bode plots of the uncompensated system are shown in Fig. 12.39. Bode plots of the phase-lag controller and that of the compensated system are shown in Figs. 12.40 and 12.41, respectively. Root locus plots of the system without and with the phase-lag controller are shown in Figs. 12.42 and 12.43. The closed-loop step response of the system without and with the phase-lag controller is shown in Figs. 12.44 and 12.45, respectively.

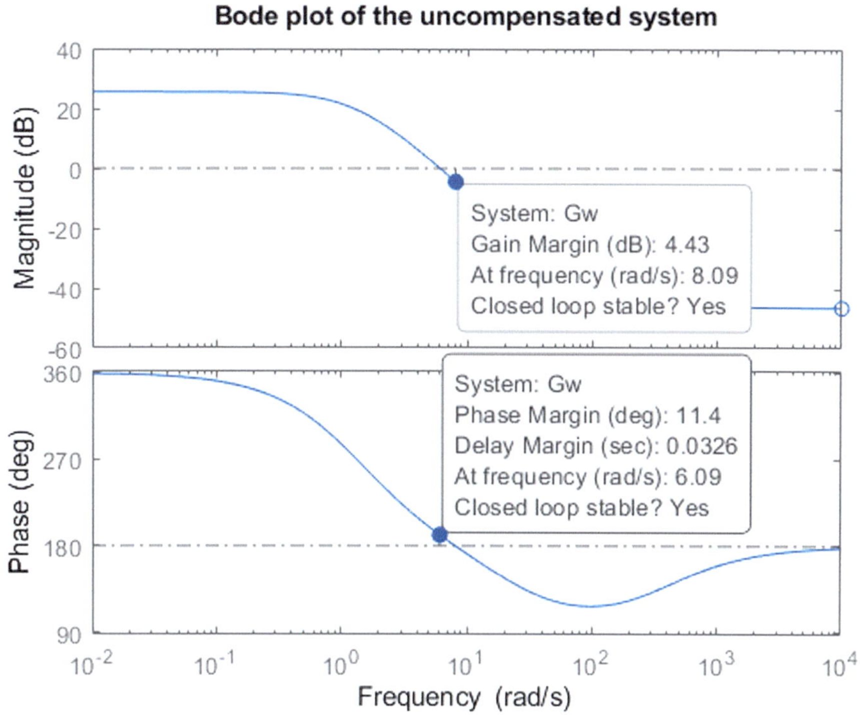

Fig. 12.39 Bode plots of the uncompensated system

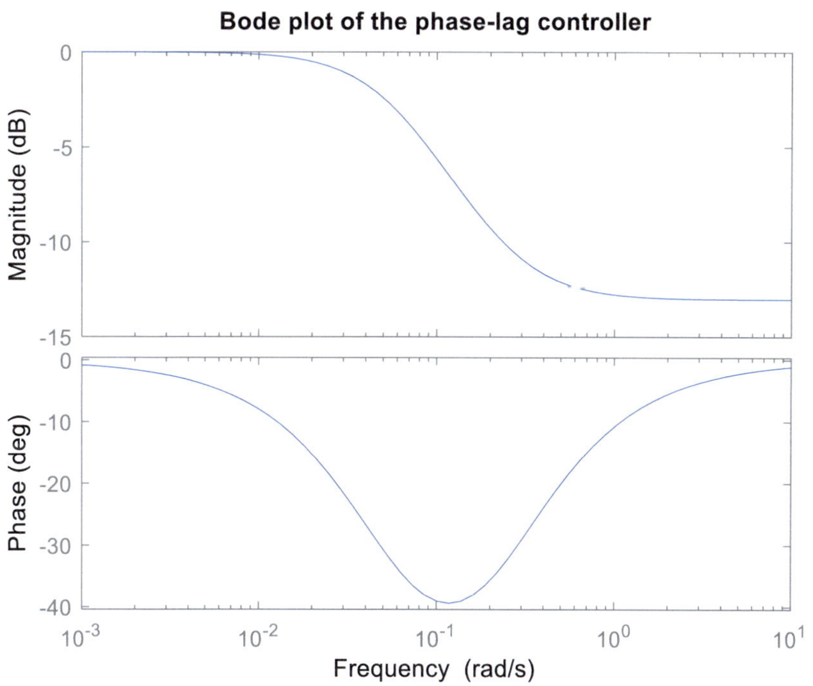

Fig. 12.40 Bode plots of the phase-lag controller

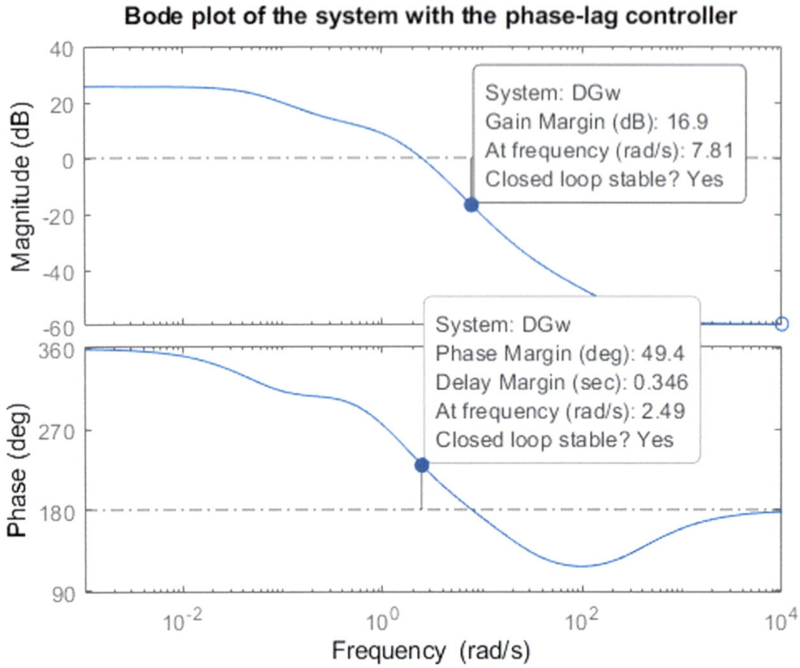

Fig. 12.41 Bode plots of the system with the phase-lag controller

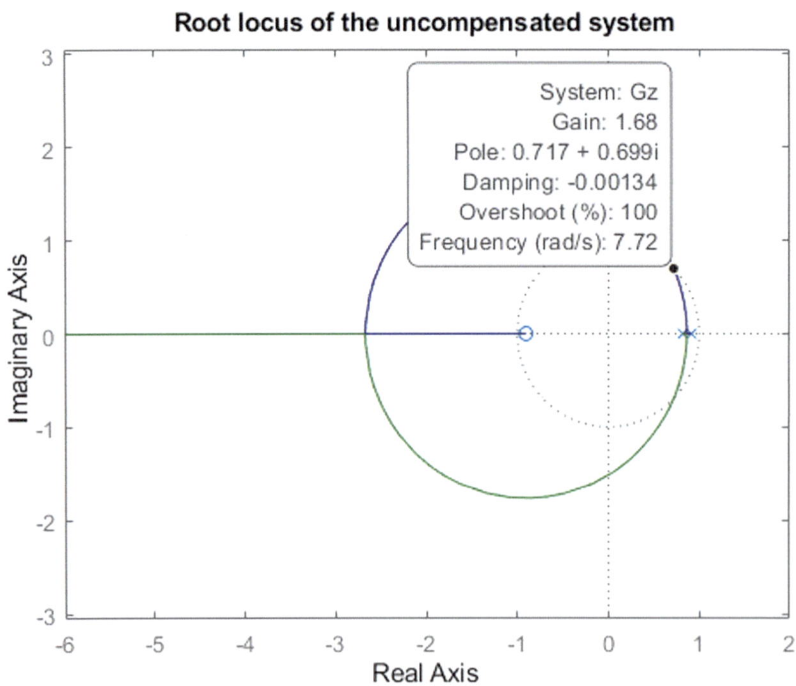

Fig. 12.42 Root locus of the uncompensated system

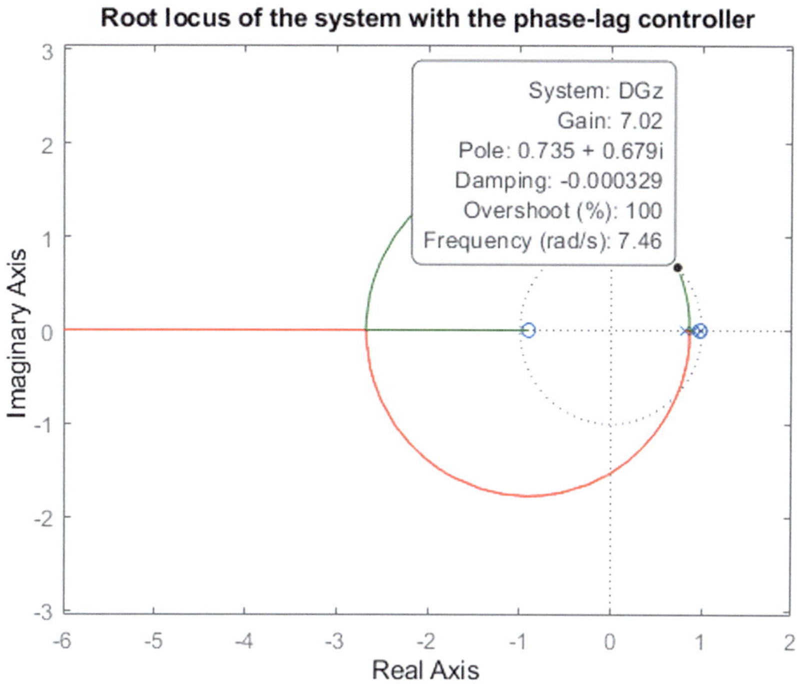

Fig. 12.43 Root locus of the system with the phase-lag controller

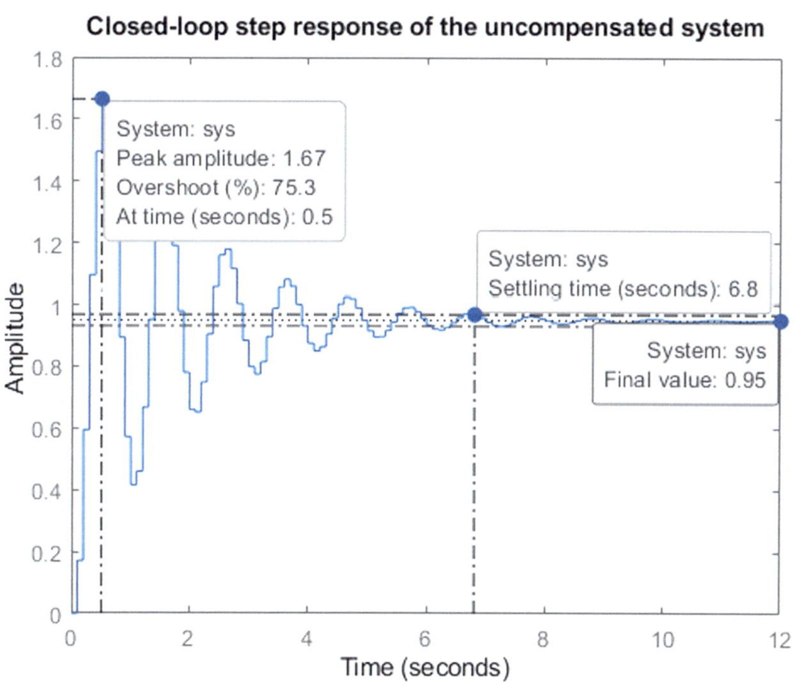

Fig. 12.44 Closed-loop step response of the uncompensated system

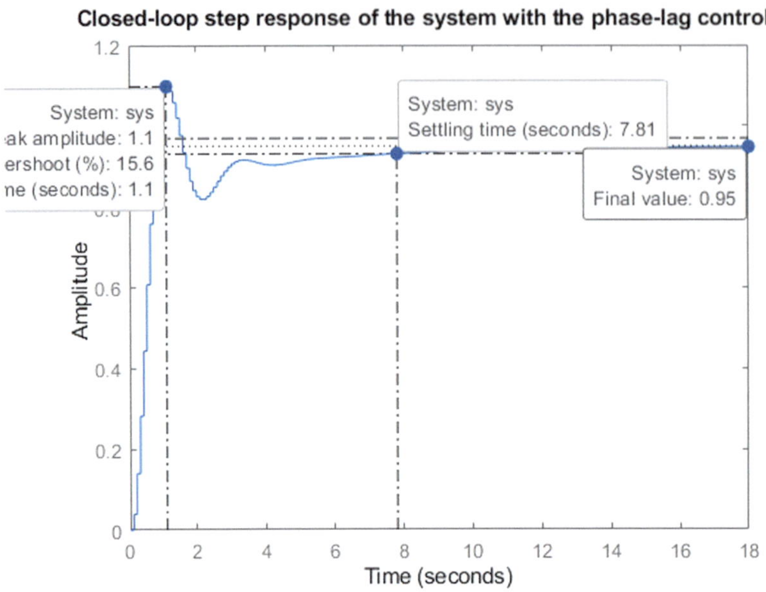

Fig. 12.45 Closed-loop step response of the system with the phase-lag controller

Example 12.8 Phase-Lead Controller: Design a phase-lead controller for $G_p(s) = \frac{1}{s^2+3s+2}$, $H(s) = 1$ such that the closed-loop system has a maximum steady-state error of 10% due to a step input, a phase margin of 50°, and a gain margin of 10 dB. Use sampling time $T = 0.10$ sec. Show the effect of the phase-lag controller on the closed-loop system stability and step response.

Solution Matlab code and results are presented.

```
% ch12_ex8_Design a phase lead controller for plant Gp(s) so that
% ess,step=10%, phase margin=55 deg, Gain margin= 10 dB.
clear all; ;close all;clc
essStep =0.10
kp=1/essStep -1
a0=kp*2
T=0.10
dpm=50;  % desired phase margin
lambda=15; % allowance for shift in phase
Gp=tf(a0,[1 3 2]) % plant tf with a0/(s+1)(s+2)
Gz = c2d(Gp, T)
% G(w)
Gw = d2c(Gz, 'tustin')
[ngw, dgw] = tfdata(Gw,'v')

figure
bode(Gw)
title('Bode plots of the uncompensated system')
% Design of phase-lead controller
[mag,ph,w]=bode(Gw);
[gm,pm,wpc,wgc]=margin(Gw)
phim=(dpm- pm +lambda)*pi/180;
alpha= (1-sin(phim))/(1+sin(phim))
magv = permute(mag,[3 2 1]);
m_db=20*log10(magv); % magnitude in dB scale
m_wg=-20*log10(1/sqrt(alpha));
wm=spline(m_db,w,m_wg)
w0=wm*sqrt(alpha)
wp=w0/alpha
Dw = tf([1/w0 1], [1/wp 1])
% Bode plots of phase-lead controller
figure
bode(Dw)
title('Bode plots of the phase-lead controller')
% D(w)G(w)
DGw = Dw*Gw
figure
bode(DGw)
title('Bode plots of the system with the phase-lead controller')
[gmc,pmc,wpc,wgc]=margin(DGw)
wgs=atan(wgc*T/2)*2/T
wps=atan(wpc*T/2)*2/T

% D(z)
Dz = c2d(Dw,T,'tustin')
figure
rlocus(Gz)
axis('equal')
title('Root locus of the uncompensated system')
%D(z)G(z)
DGz = Dz*Gz
figure
rlocus(DGz)
axis('equal')
title('Root locus of the system with the phase-lead controller')

% calculation of the closed loop step response
```

```
Tz= feedback(Gz,1)
figure
dstep(Tz,1)
title('Closed-loop step response of the uncompensated system')
% Closed-loop step response of the compensated system
Tz=feedback(DGz,1)
figure
dstep(Tz,1)
title('Closed-loop step response of the uncompensated system')

Results:

         0.0815 z + 0.07375
G(z) =   ----------------------
         z^2 - 1.724 z + 0.7408

           -0.002239 s^2 - 0.8515 s + 17.93
G(w) =     --------------------------------
               w^2 + 2.993 w + 1.992

alpha = 0.2709, wm = 5.7790, w0 = 3.0079, wp = 11.1029

        0.3325 s + 1
D(w) =  ------------
        0.09007 s + 1

        2.731 z - 2.017
D(z) =  ---------------
          z - 0.2861
```

Bode plots of the uncompensated system are shown in Fig. 12.46. Bode plots of the phase-lead controller and that of the compensated system are shown in Figs. 12.47 and 12.48, respectively. Root locus plots of the system without and with the phase-lead controller are shown in Figs. 12.49 and 12.50, respectively. The closed-loop step response of the system without and with the phase-lead controller is shown in Figs. 12.51 and 12.52, respectively.

Example 12.9 Phase-Lag-Lead Controller: Design a phase-lag-lead controller for $G_p(s) = \frac{1}{s^2+3s+2}$, $H(s) = 1$ such that the closed-loop system has a maximum steady-state error of 5% due to a step input, a phase margin of 50°, and a gain margin of 10 dB. Use sampling time $T = 0.10$ sec. Show the effect of the phase-lag-lead controller on the closed-loop system stability and step response.

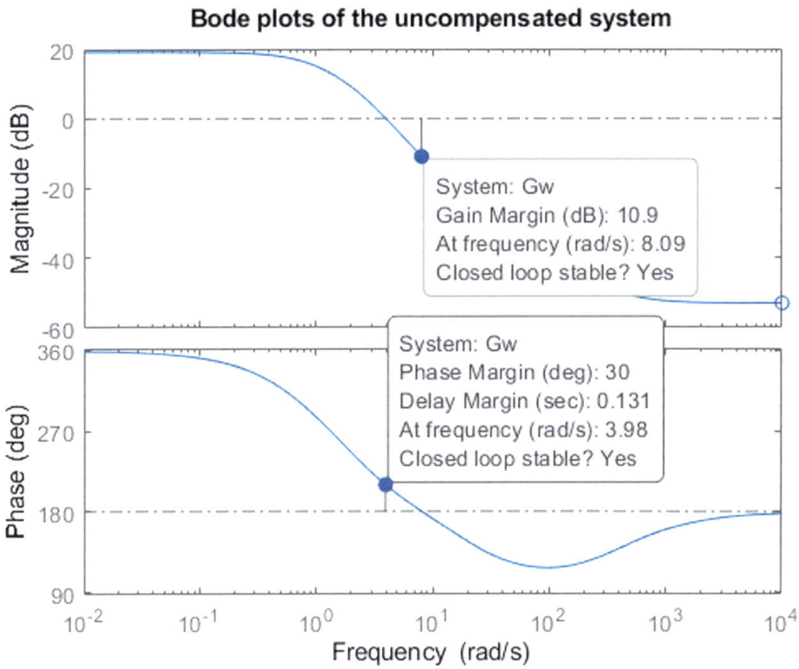

Fig. 12.46 Bode plots of the uncompensated system

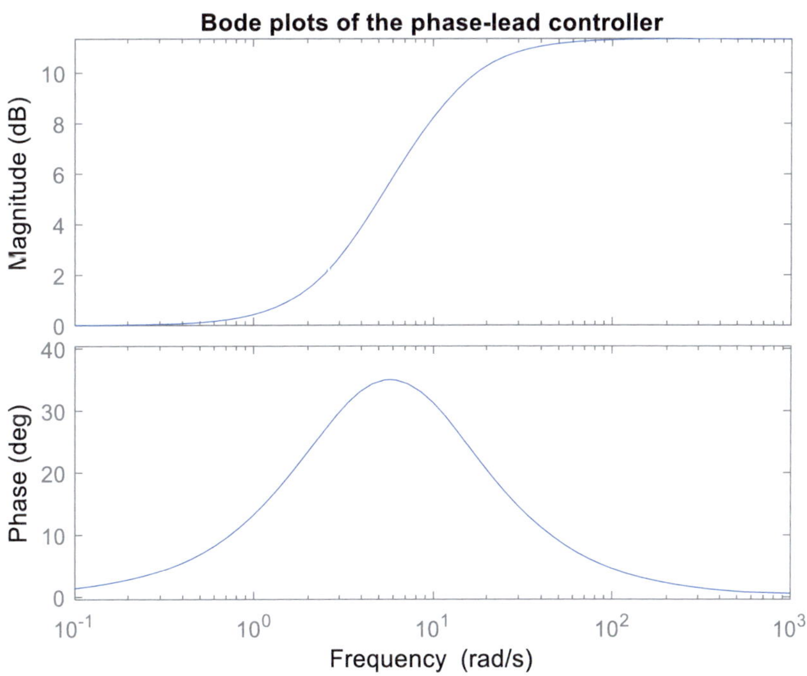

Fig. 12.47 Bode plots of the phase-lead controller

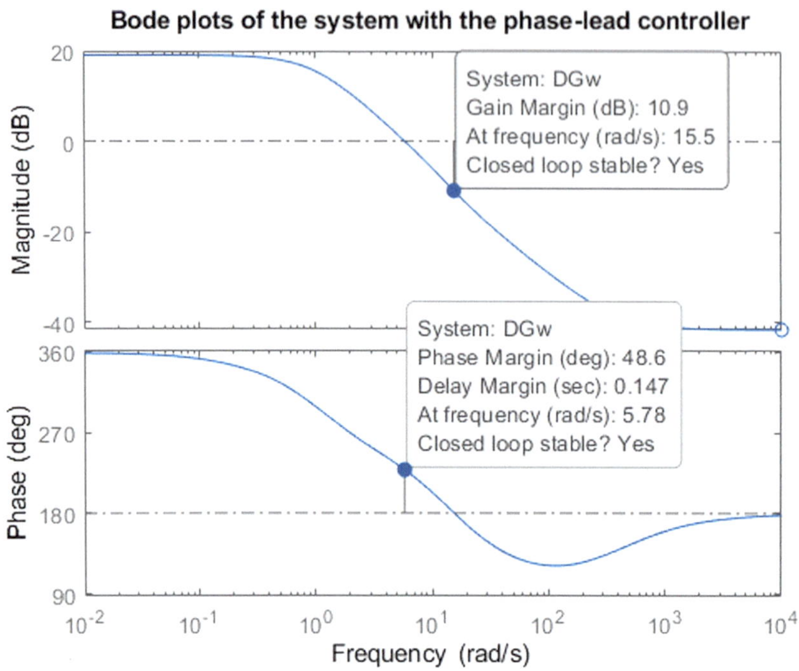

Fig. 12.48 Bode plots of the system with the phase-lead controller

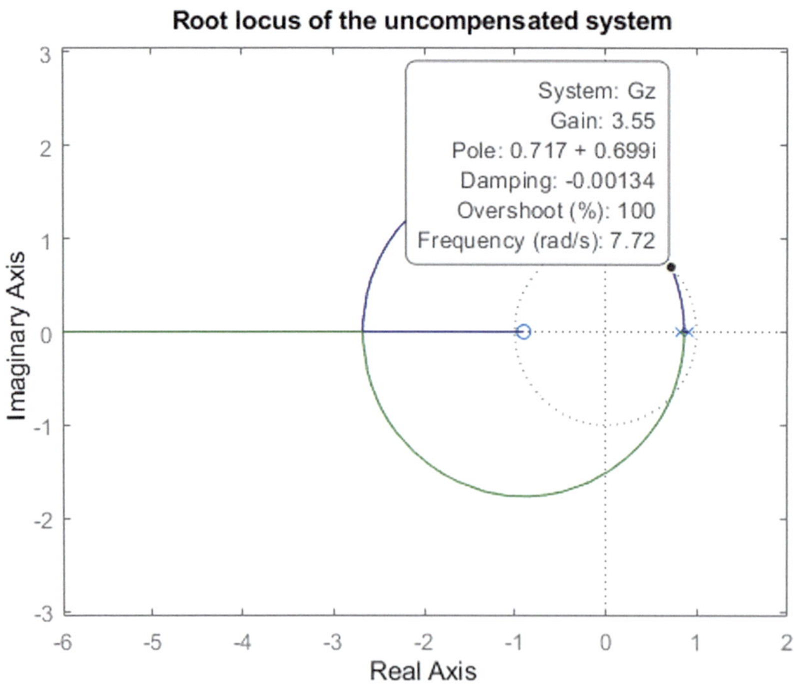

Fig. 12.49 Root locus of the uncompensated system

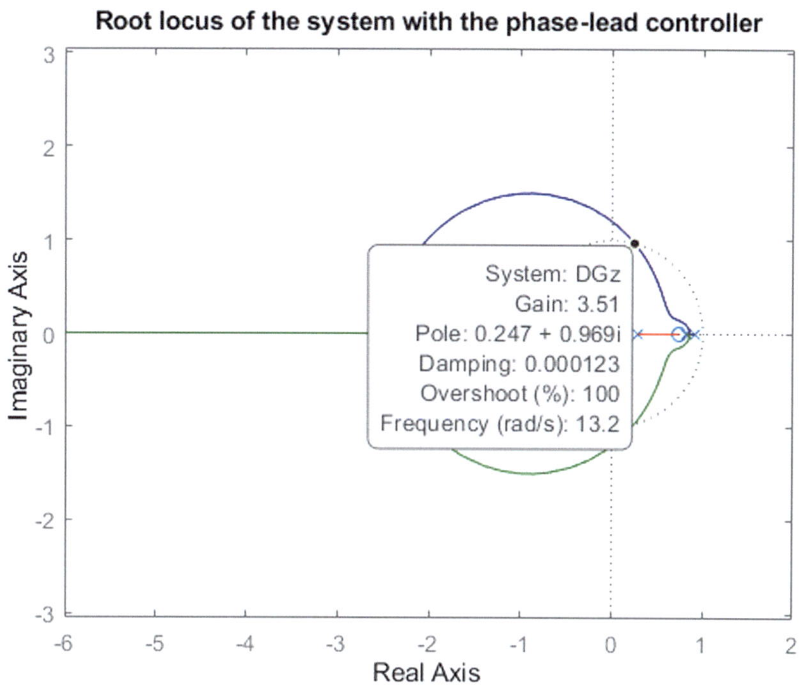

Fig. 12.50 Root locus of the system with the phase-lead controller

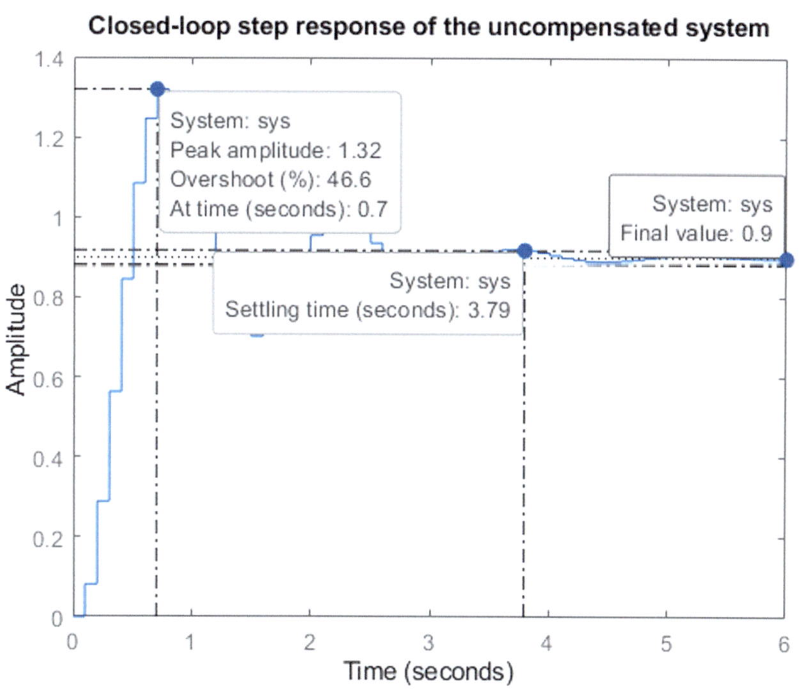

Fig. 12.51 Closed-loop step response of the uncompensated system

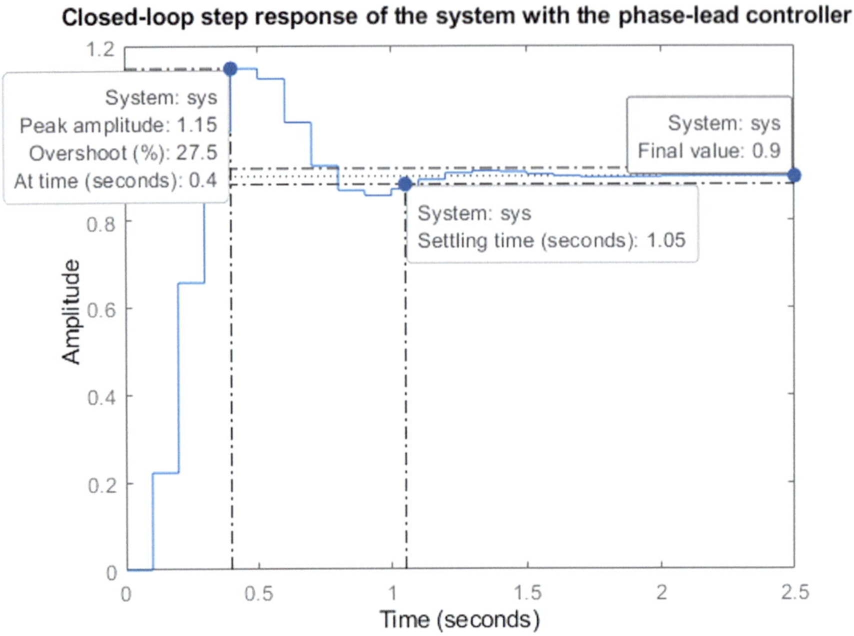

Fig. 12.52 Closed-loop step response of the system with the phase-lead controller

Solution Matlab code and results are presented.

```
% Ch12-Example 9-Design of phase-lag-lead controller
% Design a phase lag controller for plant Gp(s) so that
% ess,step=5%, phase margin=55 deg, Gain margin= 10 dB.
clear all; close all; clc
essStep =0.05
kp=1/essStep -1
a0=kp*2
T=0.10
dpm=55;  % desired phase margin
Gp=tf(a0,[1 3 2]) % plant tf with a0/(s+1)(s+2)
% G(z)
Gz = c2d(Gp, T)
%G(w)
Gw = d2c(Gz, 'tustin')
[ngw, dgw]= tfdata(Gw, 'v')

% phase- lag controller
figure
bode(Gw)  % bode plot of G(w)
title('Bode plots of the uncompensated system')
[mag,ph,w]=bode(Gw);
ph=ph-360;    % to make angle <0
[gm,pm,wpc,wgc]=margin(Gw)
dpmlg=40; % phase margin with the lag part only
lambda=10; % allowance for shift in phase
```

```
ph1=(dpmlg+lambda-180);
phv = permute(ph, [3 2 1]);
wm=spline(phv,w,ph1)
w0p=wm/10
ngv=polyval(ngw,j*wm);dgv=polyval(dgw,j*wm);
g=abs(ngv/dgv)
wpp=0.1*wm/g

Dw = tf([1/w0p 1], [1/wpp 1]) %lag part

figure
bode(Dw)  % bode of Dlag(w)
title('Bode plots of the phase-lag controller')
% D(w)G(w)
DGw = Dw*Gw

figure
bode(DGw) % bode plot of G(w)*Dlag(w)
title('Bode plots of the system with the phase-lag controller')
Dz = c2d(Dw,T,'tustin') % Dz _lag

% for phase-lead part

dpm=55  % desired phase margin
lambda=25; % allowance for shift in phase

[mc,pc,w]=bode(DGw);
[gm,pm,wgc,wpc]=margin(DGw)

phim=(dpm- pm +lambda)*pi/180;
alpha= (1-sin(phim))/(1+sin(phim))
mcv =  permute(mc, [3 2 1]);
m_db=20*log10(mcv); % magnitude in dB scale
m_wg=-20*log10(1/sqrt(alpha));
wm=spline(m_db,w,m_wg)
w0=wm*sqrt(alpha)
wp=w0/alpha
Dw2 = tf([1/w0 1], [1/wp 1])
% Dw*Dw2
Dwf = Dw*Dw2
figure
bode(Dwf)
title('Bode plots of the phase-lag-lead controller')

figure
DGw2 = Dw2*DGw % compensated tf
bode(DGw2) % bode plot of G(w)D(w)
title('Bode plots of the system with the phase-lag-lead controller')
[gmc_c,pmc_c,wpc_c,wgc_c]=margin(DGw2)
wgs=atan(wgc_c*T/2)*2/T  % wg in s-plane
wps=atan(wpc_c*T/2)*2/T  % wp in s-plane
% D(z)
Dz2 = c2d(Dw2,T,'tustin')
%D(z)G(z)
DGz2 = Dz2*Dz*Gz
figure
rlocus(Gz)
axis('equal')
title('Root locus of the uncompensated system')
figure
rlocus(DGz2)
axis('equal')
title('Root locus of the system with the phase-lag-lead controller')
```

```
% calculation of the closed loop step response
Tz= feedback(Gz,1)
figure
dstep(Tz,1)
title('Closed-loop step response of the uncompensated system')

% Closed-loop step response of the compensated system
Tz=feedback(DGz2,1)
figure
dstep(Tz,1)
title('Closed-loop step response of the system with the phase-lag-lead
controller')
```

```
Results:

G(z) =

    0.1721 z + 0.1557
  ----------------------
  z^2 - 1.724 z + 0.7408

G(w) =

  -0.004726 s^2 - 1.798 s + 37.84
  -------------------------------
       s^2 + 2.993 s + 1.992

Phase-lag controller:
gm = 1.6648, pm = 11.3680, wpc = 8.0935, wgc = 6.0914
wm = 3.1244, w0p = 0.3124, g = 3.1506, wpp = 0.0992

D(w) =

  3.201 w + 1
  -----------
  10.08 w + 1

D(z) =

  0.3208 z - 0.3109
  -----------------
     z - 0.9901

Phase-lead controller:

dpm = 55, gm = 4.8782, pm = 37.8158, wgc = 7.7753, wpc = 3.1332
alpha = 0.1965, wm = 5.0548, w0 = 2.2408, wp = 11.4025

D(w) =

  0.4463 w + 1
  ------------
  0.0877 w + 1

D(z) =

  3.604 z - 2.878
  ---------------
    z - 0.2738

gm= 4.1109, pm= 59.0140, wp= 16.3095, wg= 5.0548, wgs = 4.9511, wps =
13.6821
```

Bode plots of the uncompensated system are shown in Fig. 12.53. Bode plots of the phase-lag part of the controller and that of the system with the phase-lag part are shown in Figs. 12.54 and 12.55, respectively.

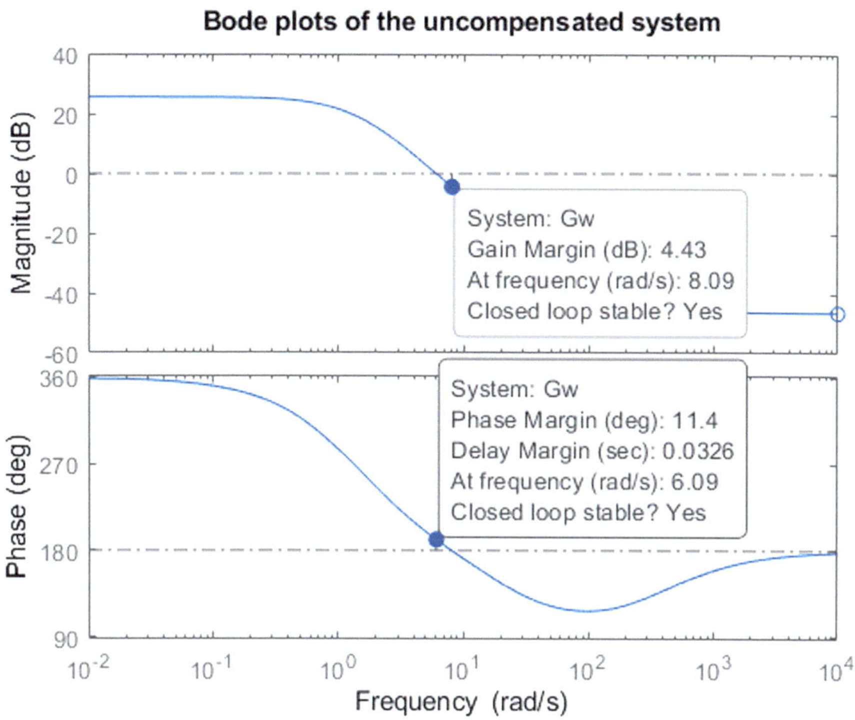

Fig. 12.53 Bode plots of the uncompensated system

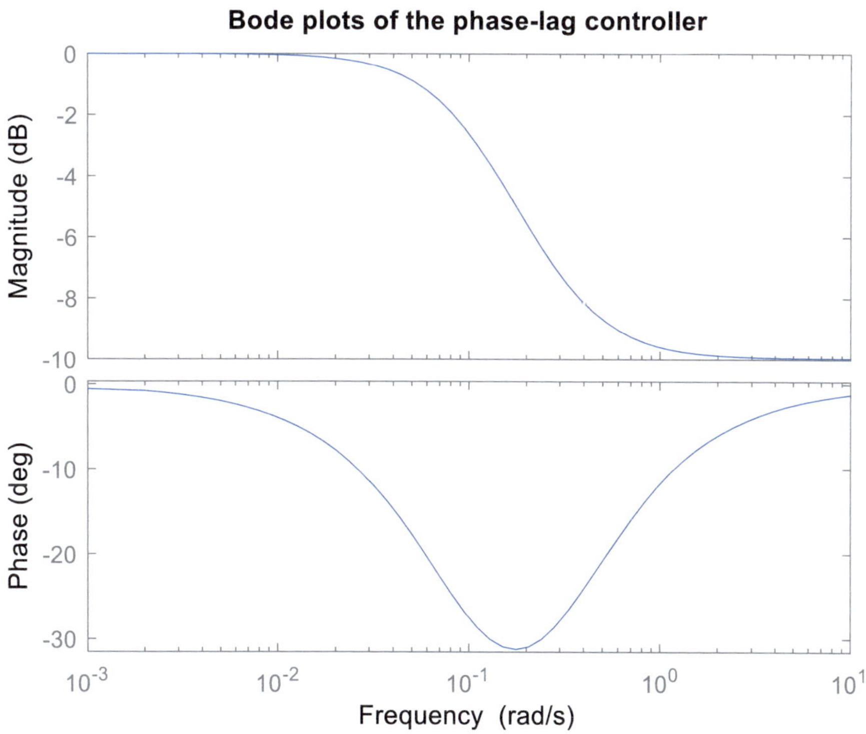

Fig. 12.54 Bode plots of the phase-lag controller

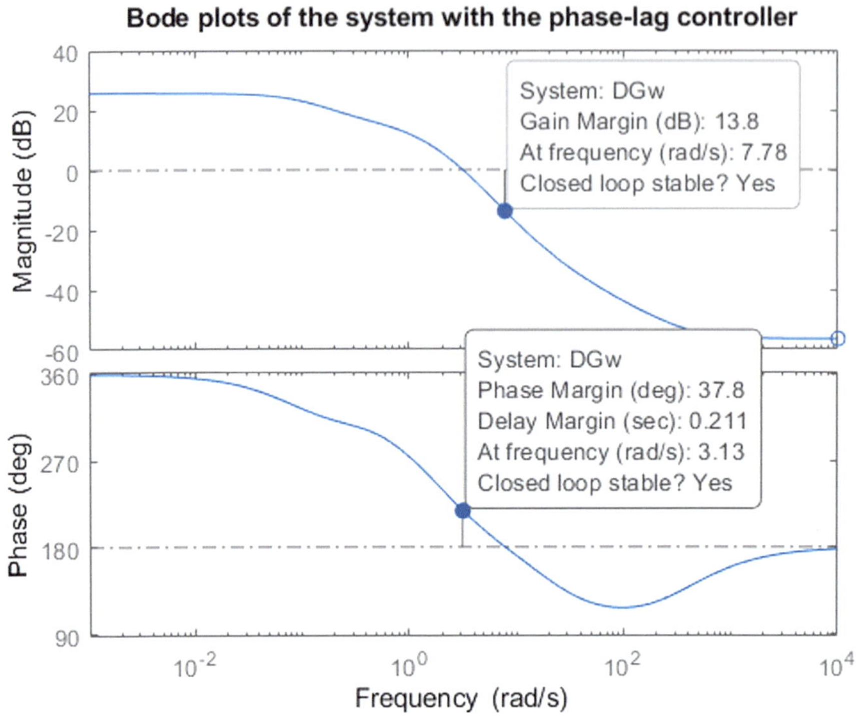

Fig. 12.55 Bode plots of the system with the phase-lag controller

Bode plots of the phase-lag-lead controller and that of the compensated system are shown in Figs. 12.56 and 12.57, respectively. Root locus plots of the system without and with the phase-lag-lead controller are shown in Figs. 12.58 and 12.59, respectively. The closed-loop step response of the system without and with the phase-lag-lead controller is shown in Figs. 12.60 and 12.61, respectively.

12.5 Implementation of Controllers

12.5.1 Lab Experiment for Analysis, Design, Simulation, and Implementation of PI, PD, and PID Controllers for Discrete-Time Systems Using Frequency-Domain Specifications

Objectives The objectives are to design, simulate, implement, and validate PI, PD, and PID controllers in the *w*-domain to achieve the desired frequency domain specification of a DC motor position control system.

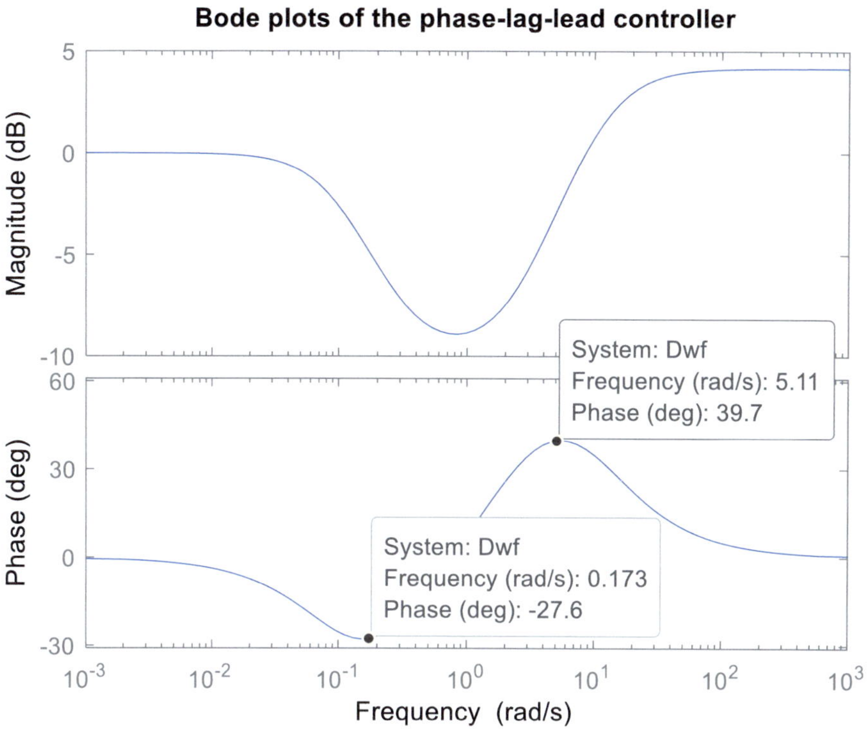

Fig. 12.56 Bode plots of the phase-lag-lead controller

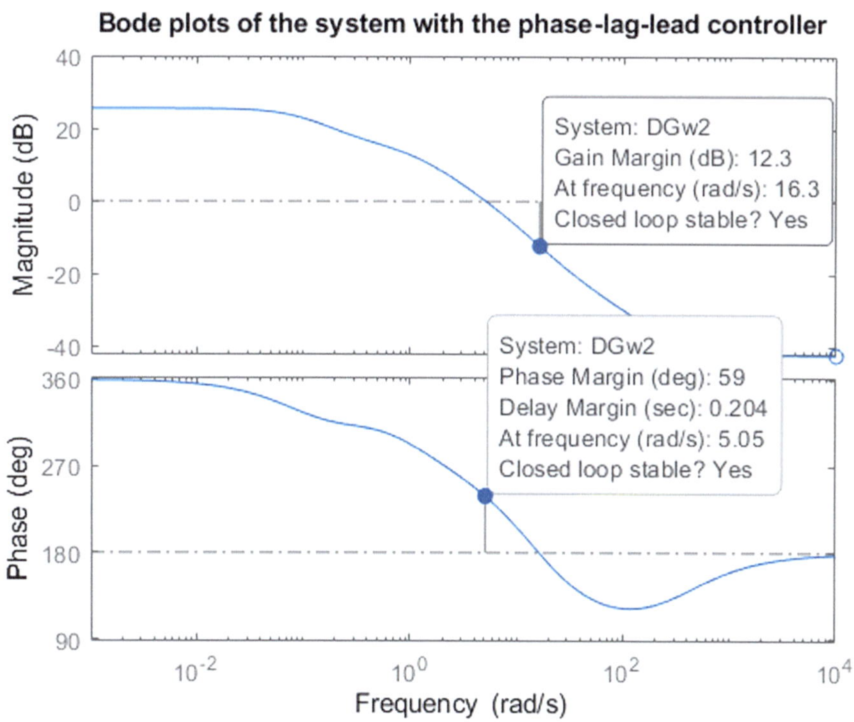

Fig. 12.57 Bode plots of the system with the phase-lag-lead controller

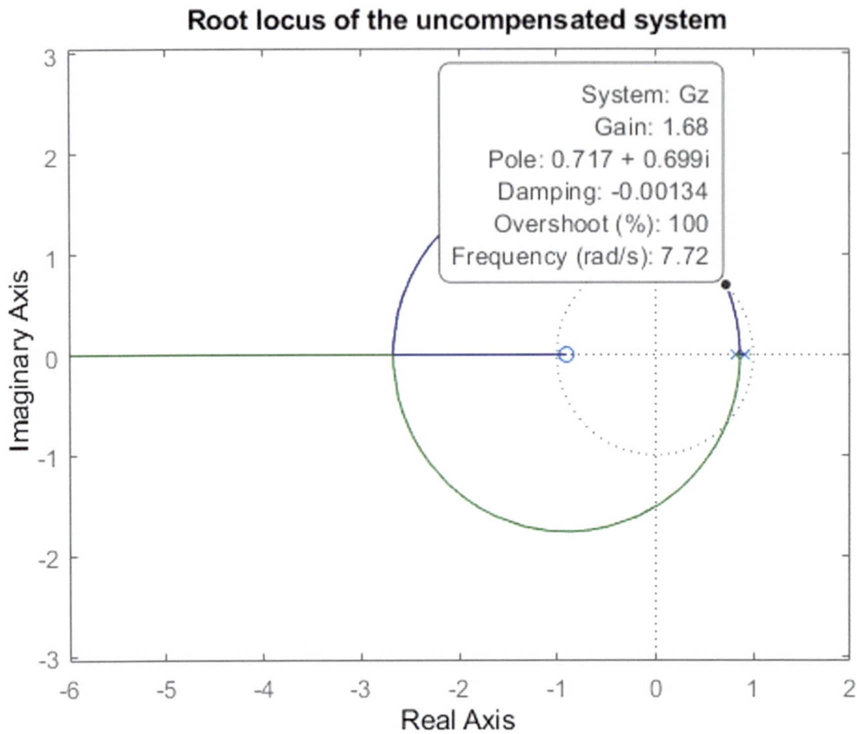

Fig. 12.58 Root locus of the uncompensated system

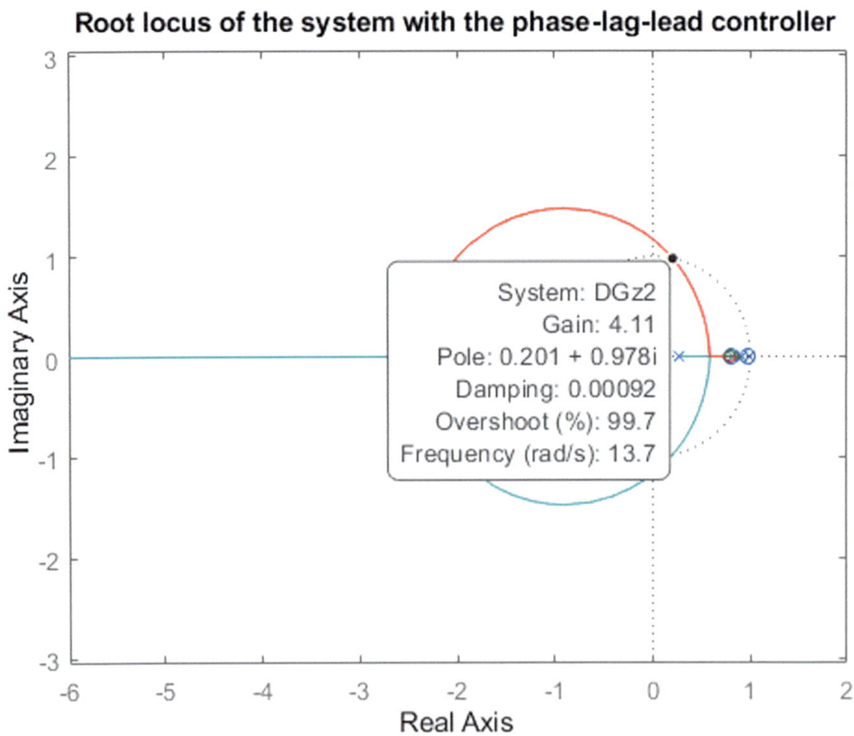

Fig. 12.59 Root locus of the system with the phase-lag-lead controller

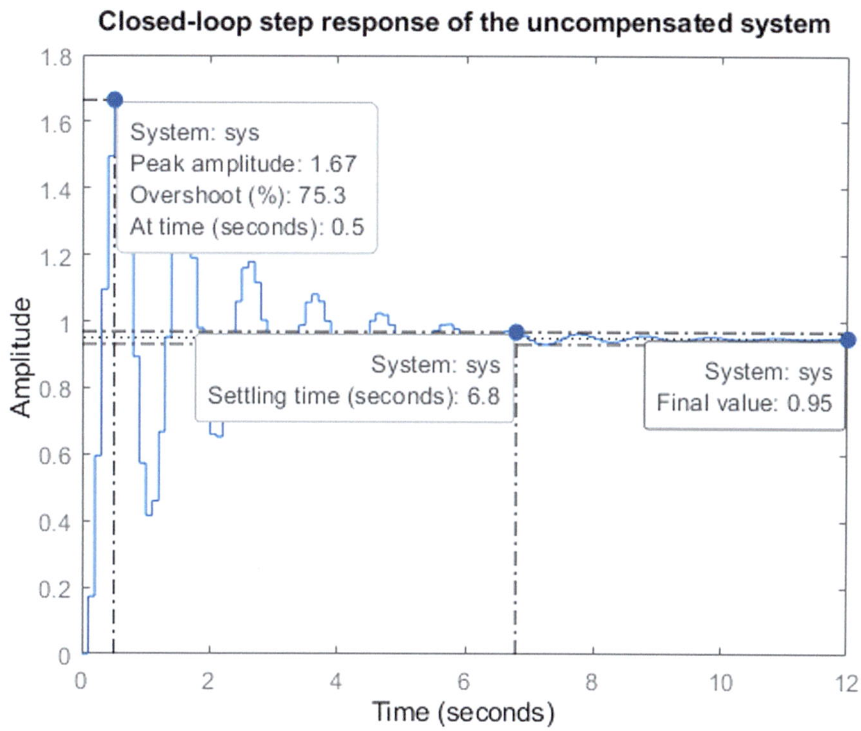

Fig. 12.60 Closed-loop step response of the uncompensated system

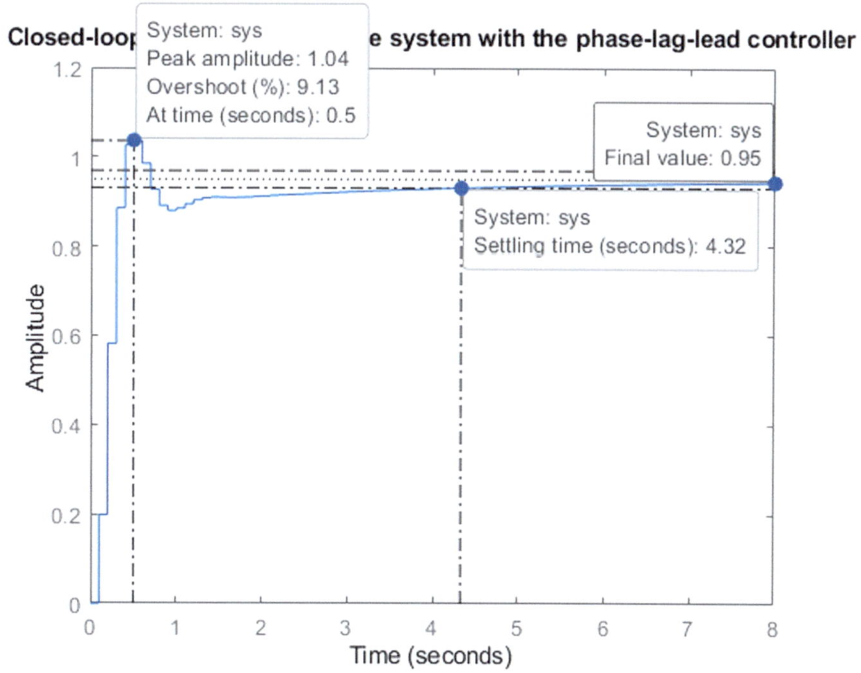

Fig. 12.61 Closed-loop step response of the system with the phase-lag-lead controller

Introduction The transfer function for a PID controller in w-domain is

$$D_{PID}(w) = K_P + \frac{K_I}{w} + K_D \frac{w}{1 + \left(\frac{T}{2}\right)w}$$

In Part I, it is required to design the controller using frequency domain specifications and simulate its performance in Matlab as presented in Sect. 12.2. In Part II, it is required to implement the controller in LabVIEW and validate its performance with reference to the design specifications.

I. *Controller Design and Simulation*

1. Use the DC motor position control model with previously estimated values of K and τ to get the discrete-time open-loop transfer function in z-domain $G(z)$ selecting a suitable sampling time T. For the DC motor control trainer (DCMCT) module, the sampling time (T) is taken as 0.01 s. Convert the open-loop transfer function $G(z)$ to $G(w)$ using bilinear transformation.
2. Note down the frequency domain specifications of gain cross-over frequency (ω_{gc}), gain margin (GM), and phase margin (ϕ_m). For the DC motor control trainer (DCMCT) module, a gain cross-over frequency of 29.5 rad/s, a gain margin of 10 dB, and a phase margin of 50° can be used as frequency domain specifications. Convert gain cross-over frequency to w-domain using Eq. (12.1).
3. Design a PID controller for the position control system in MATLAB for the desired frequency domain specifications. Select a suitable integral gain K_I. Refer to Sect. 12.2 for details.
4. Note down the controller gains K_P, K_I, and K_D and corresponding controller transfer function $D(w)$ for each variation: PI ($K_D = 0$), PD ($K_I = 0$), and PID. Convert the w-domain transfer function $D(w)$ of each controller to z-domain $D(z)$. Record the controller gains and z-domain transfer function $D(z)$ for each controller in Table 12.1.
5. Simulate closed-loop step response for each controller. Locate response characteristics by right-clicking the figure in Matlab, selecting peak response, settling time, and steady state (right-click → response characteristics → peak response → settling time → steady state → cursor on identified point). Note down in Table 12.1 overshoot (α) and settling time (t_{set}) from the simulated response for each type of controller. Copy closed-loop step response with each controller for presentation in the lab report.
6. Draw the root locus of the system without the controller and with each type of controller in the z-domain: PI, PD, and PID. Locate the closed-loop poles of the system with each controller using the cursor on the root locus near the observed step response, and present these in Table 12.2. (Hints: Observed step response overshoot and settling time can be converted to closed-loop system poles in s- and z-domains using relations presented in Sect. 10.3) Copy root locus with each controller for presentation in the lab report.

Table 12.1 Summary of controller parameters and performance

Controller	K_P	K_I	K_D	Controller transfer function $D(z)$	Closed-loop system step response			
					Simulation		Implementation	
					Overshoot α (%)	Settling time $t_{set}(s)$	Overshoot α (%)	Settling time $t_{set}(s)$
PI			0					
PD		0						
PID								

Table 12.2 Summary of controller performance using root locus and Bode plots

		Bode plots			
Controller	Root locus closed-loop poles	GM	ϕ_M	ω_{gc} (rad/s)	ω_{pc} (rad/s)
Without controller					
PI					
PD					
PID					

7. Draw Bode plots of the system without the controller and with each type of controller in w-domain: PI, PD, and PID. Obtain gain margin (GM), phase margin (ϕ_M), gain cross-over frequency (ω_{gc}), and phase cross-over frequency (ω_{pc}) for each type of controller by right-clicking on any of the Bode plots, selecting all stability margins, and placing cursors on the identified points (right-click on Bode plots → all stability margins → cursor on the stability points). Present these in Table 12.2. Copy Bode plots with each controller for presentation in the lab report.

II. *Implementation and Validation*

1. Set up NI ELVIS II board and DC motor module. Select the device, and set the sampling rate. For the DC motor control trainer (DCMCT) module, a sampling rate of 100 Hz is used.
2. Study the block diagram for the position control experiment (*01-QNET_DCMCT_Position_Control.vi*).
3. Open the block diagram VI, and identify the different components of the control system, including the input, error, controller, plant, sensor, and output in Fig. 12.62.
4. Modify the block diagram VI of Fig. 12.62 by removing the control parameters block and adding a subtraction block to calculate the error, as well as a discrete transfer function block for implementation of the designed controller, as shown in the VI of Fig. 12.63.
5. Put the discrete-time z-domain transfer function $D(z)$, and follow the format of setting the controller transfer function coefficients. Save the VI.
6. Select your input parameters. For the DCMCT module, typical values for the input square wave voltage are as follows: amplitude of 1.0 V, frequency of 0.40 Hz, and offset of 3.0 V.
7. Run the VI, and observe the closed-loop system response. Copy graphs of closed-loop system response from the front panel VI for presentation in the lab report.
8. Make sure to set the number of significant figures to at least 5 for the display format of each graph on the front panel VI (right-click on the graph → display format → decimal → 6). Export the time response graphs, both motor rotation $\theta(t)$ and motor input voltage $V(t)$, to Excel by right-clicking on each graph and

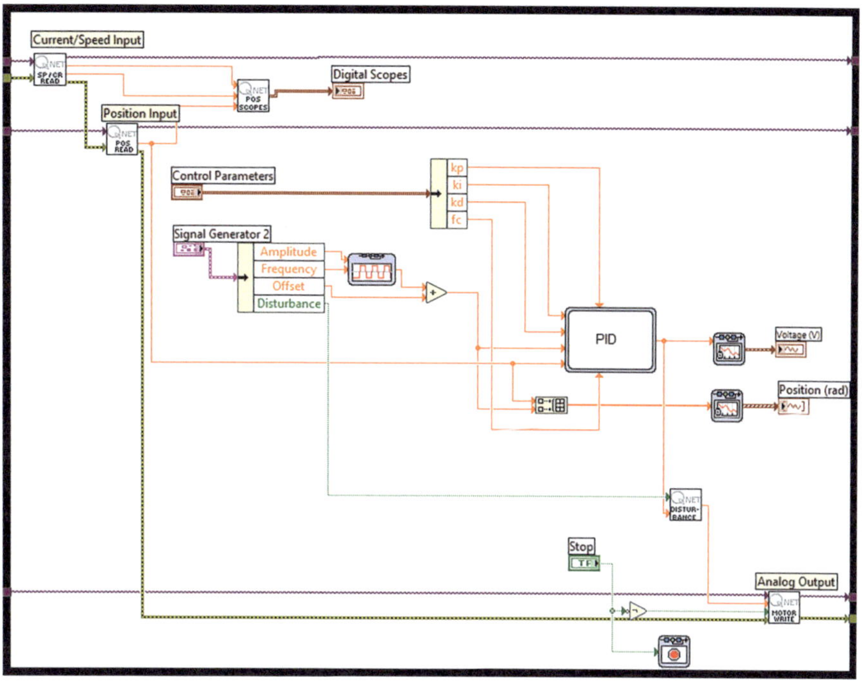

Fig. 12.62 Block diagram VI for DC motor control trainer position control VI before modification

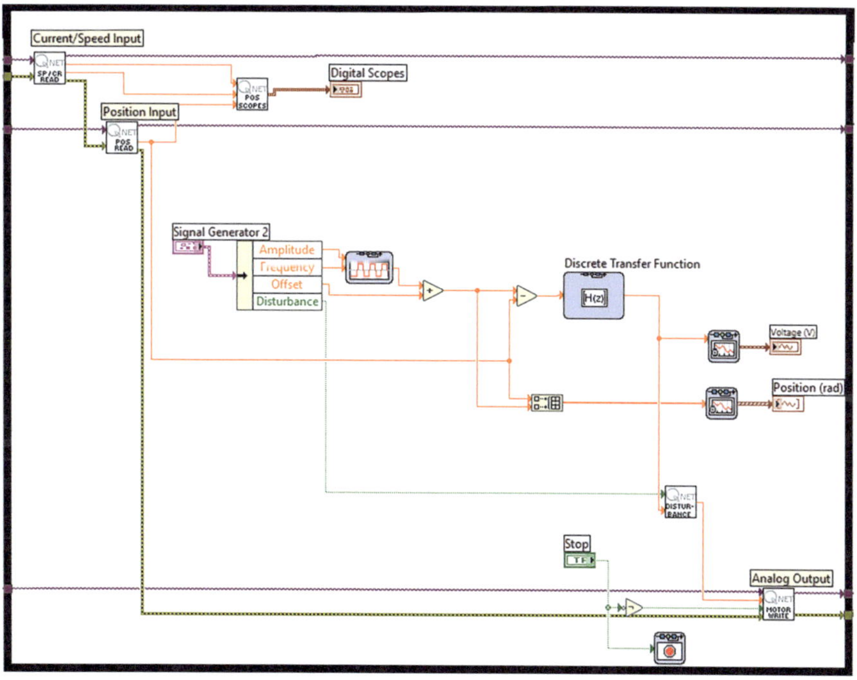

Fig. 12.63 Block diagram VI for DC motor control trainer position control VI after modification

selecting the Export to Excel option (right-click → Export to Excel). Estimate overshoot and settling time from closed-loop system response graph.

9. Repeat steps (5)–(8) for each type of controller: PI, PD, and PID. Complete the last two columns of Table 12.1.

III. *Deliverables*

1. Complete Tables 12.1 and 12.2.
2. Present graphs of simulated closed-loop step response for each controller with response characteristics (peak response, settling time, and steady-state value) clearly annotated.
3. Present graphs of root locus analysis for each controller with closed-loop poles near the desired ones clearly annotated.
4. Include Bode plots for each controller with all stability margins clearly annotated.
5. Include graphs of closed-loop system step response with each implemented controller.
6. Discuss results comparing the performance of controllers in simulation and experimental validation.
7. Discuss the stability of the closed-loop system with controllers from the results of root locus and Bode plots.

Lab Report
Present all deliverables in the lab report and discuss your results.

12.5.2 Lab Experiment for Analysis, Design, Simulation, and Implementation of Phase-Lag, Phase-Lead, and Phase-Lag-Lead Controllers for Discrete-Time Systems Using Frequency-Domain Specifications

Objectives The objectives are to design, simulate, implement, and validate PI, PD, and PID controllers in the z-domain to achieve the desired transient response of a DC motor position control system.

Introduction The transfer function for a phase-lag-lead controller in the w-domain is

$$D_{\text{lag-lead}}(w) = \left(c_0 \frac{1 + \frac{w}{\omega_{w0}}}{1 + \frac{w}{\omega_{wp}}}\right)\left(c_0' \frac{1 + \frac{w}{\omega_{w0}'}}{1 + \frac{w}{\omega_{wp}'}}\right)$$

In Part I, it is required to design the phase-lag, phase-lead, and phase-lag-lead controllers using frequency domain specifications and simulate the performance of each controller as presented in Sect. 12.4. In Part II, it is required to implement each controller in LabVIEW and validate its performance with reference to the design specifications.

I. *Controller Design and Simulation*

1. Use the DC motor position control model with previously estimated values of K and τ to get the discrete-time open-loop transfer function in z-domain $G(z)$ selecting a suitable sampling time T. For the DC motor control trainer (DCMCT) module, the sampling time (T) is taken as 0.01 s. Convert the open-loop transfer function $G(z)$ to $G(w)$ using bilinear transformation.
2. Note down the frequency domain specifications of gain margin (GM) and phase margin (ϕ_m). For the DC motor control trainer (DCMCT) module, a gain margin of 10 dB, and a phase margin of 50° can be used as frequency domain specifications.
3. Design a phase-lag controller for the position control system in MATLAB with a gain margin of 10 dB and a phase margin of 30°. Refer to Sect. 12.4.1 for details. Convert the phase-lag controller transfer function $D(w)$ to $D(z)$. Note down the controller zero and pole and controller transfer function $D(z)$ for the phase-lag controller in Table 12.3.
4. Design a phase-lead controller for the position control system with the desired frequency domain specifications of a gain margin of 10 dB and a phase margin of 50°. Refer to Sect. 12.4.2 for details. Convert the phase-lead controller transfer function $D(w)$ to $D(z)$. Note down the controller zero and pole and controller transfer function $D(z)$ for the phase-lead controller in Table 12.3.
5. Design a phase-lag-lead controller for the position control system step response with the desired frequency domain specifications of a gain margin of 10 dB and a phase margin of 50°. Assume the specification of step (3) for designing the phase-lag part of the controller. Include the phase-lag controller with the original $G(w)$ for designing the phase-lead part of the phase-lag-lead controller. Refer to Sect. 12.4.3 for details. Convert the phase-lag-lead controller transfer function $D(w)$ to $D(z)$. Note down the controller zeros and poles and controller transfer function $D(z)$ for the phase-lag-lead controller in Table 12.3.
6. Simulate closed-loop step response for each controller. Locate response characteristics by right-clicking the figure in Matlab, selecting peak response, settling time, and steady state (right-click → response characteristics → peak response → settling time → steady state → cursor on identified point). Note down in Table 12.3 overshoot (α) and settling time (t_{set}) from the simulated

Table 12.3 Summary of controller parameters and performance

				Closed-loop system step response			
				Simulation		Implementation	
Controller	Zeros (z_i)	Poles (p_i)	Controller transfer function $D(z)$	Overshoot α (%)	Settling time $t_{set}(s)$	Overshoot α (%)	Settling time $t_{set}(s)$
Phase-lag							
Phase-lead							
Phase-lag-lead							

Table 12.4 Summary of controller performance using root locus and Bode plots

Controller	Root locus closed-loop poles	Bode plots			
		GM	ϕ_M	ω_{gc} (rad/s)	ω_{pc} (rad/s)
Without controller					
Phase-lag					
Phase-lead					
Phase-lag-lead					

response for each type of controller. Copy closed-loop step response with each controller for presentation in the lab report.

7. Draw the root locus of the system without the controller and with each type of controller. Locate the closed-loop poles of the system with each controller using the cursor on the root locus near the desired pole and present these in Table 12.4. Copy root locus with each controller for presentation in the lab report.
8. Draw Bode plots of the system without the controller and with each type of controller. Obtain gain margin (GM), phase margin (ϕ_M), gain cross-over frequency (ω_{gc}), and phase cross-over frequency (ω_{pc}) for each type of controller by right-clicking on any of the Bode plots, selecting all stability margins, and placing cursors on the identified points (right-click on Bode plots → all stability margins → cursor on the stability points). Present these in Table 12.4. Copy Bode plots with each controller for presentation in the lab report.

II. *Implementation and Validation*

1. Set up NI ELVIS II board and DC motor module. Select the device, and set the sampling rate. For the DC motor control trainer (DCMCT) module, a sampling rate of 100 Hz is used.
2. Study the block diagram for the position control experiment (*01-QNET_DCMCT_Position_Control.vi*).
3. Open block diagram VI, and identify the different components of the control system, including the input, error, controller, plant, sensor, and output in 12.62.
4. Modify the block diagram VI of Fig. 12.62 by removing the control parameters block and adding a subtraction block to calculate the error, as well as a discrete transfer function block for implementation of the designed controller, as shown in the VI of Fig. 12.63.
5. Put the discrete-time z-domain transfer function $D(z)$, and follow the format of setting the controller transfer function coefficients. Save the VI.
6. Select your input parameters. For the DCMCT module, typical values for the input square wave voltage are as follows: amplitude of 1.0 V, frequency of 0.40 Hz, and offset of 3.0 V.

7. Run the VI, and observe the closed-loop system response. Copy graphs of closed-loop system response from the front panel VI for presentation in the lab report.
8. Make sure to set the number of significant figures to at least 5 for the display format of each graph on the front panel VI (right-click on the graph → display format → decimal → 6). Export the time response graphs, both motor rotation $\theta(t)$ and motor input voltage $V(t)$, to Excel by right-clicking on each graph and selecting the Export to Excel option (right-click → Export to Excel). Estimate overshoot and settling time from closed-loop system response graph.
9. Repeat steps (5)–(8) for each type of controller: phase-lag, phase-lead, and phase-lag-lead. Complete the last two columns of Table 12.4.

III. *Deliverables*

1. Complete Tables 12.3 and 12.4.
2. Present graphs of simulated closed-loop step response for each controller with response characteristics (peak response, settling time, and steady-state value) clearly annotated.
3. Present graphs of root locus analysis for each controller with closed-loop poles near the desired ones clearly annotated.
4. Include Bode plots for each controller with all stability margins clearly annotated.
5. Include graphs of closed-loop system step response with each implemented controller.
6. Discuss results comparing the performance of controllers in simulation and experimental validation.
7. Discuss the stability of the closed-loop system with controllers from the results of root locus and Bode plots.

Lab Report

Present all deliverables in the lab report and discuss your results.

Exercises

12.1 Design a PI controller for the system shown in Fig. P12.1 with $G_p(s) = \frac{10}{s^2+6.5s+1.5}$, $H(s) = 1$ such that the closed-loop system has a gain cross-over frequency of 3.5 rad/s, a phase margin of 55°, and a gain margin of 10 dB. Use sampling time $T = 0.10$ sec. Use suitable value for the integral gain. Show the effect of the PI controller on the closed-loop system stability and step response.

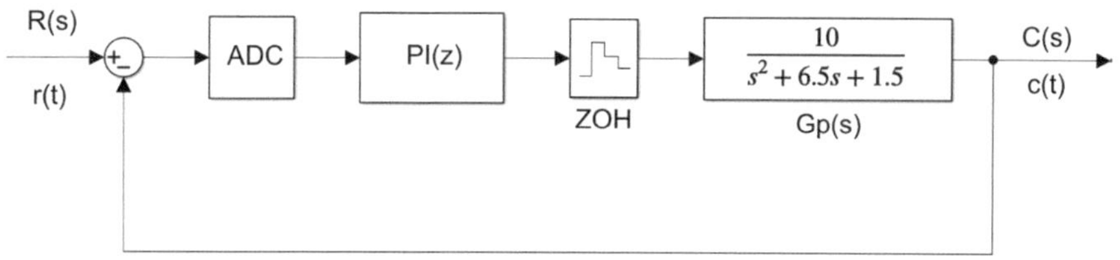

Fig. P12.1 Model of a discrete-time closed-loop system for Problem 12.1

12.2 Design a PD controller for the system shown in Fig. P12.2 with $G_p(s) = \frac{10}{s^2+6.5s+1.5}$, $H(s) = 1$ such that the closed-loop system has a gain cross-over frequency of 4.5 rad/s, a phase margin of 55°, and a gain margin of 10 dB. Use sampling time $T = 0.10$ sec. Show the effect of the PD controller on the closed-loop system stability and step response.

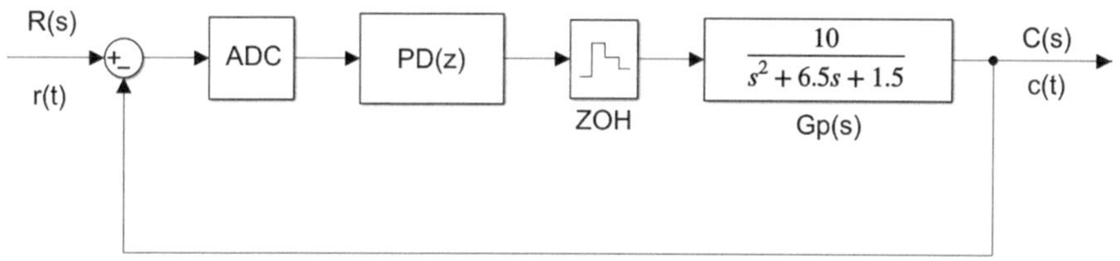

Fig. P12.2 Model of a discrete-time closed-loop system for Problem 12.2

12.3 Design a PID controller for the system shown in Fig. P12.3 with $G_p(s) = \frac{10}{s^2+6.5s+1.5}$, $H(s) = 1$ such that the closed-loop system has a gain cross-over frequency of 4.5 rad/s, a phase margin of 55°, and a gain margin of 10 dB. Use sampling time $T = 0.10$ sec. Show the effect of the PID controller on the closed-loop system stability and step response.

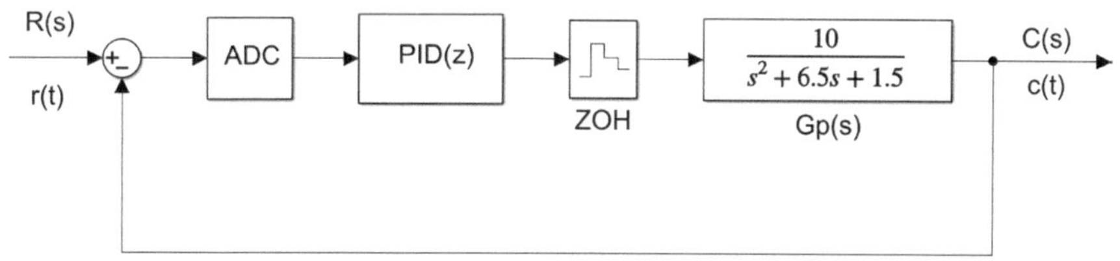

Fig. P12.3 Model of a discrete-time closed-loop system for Problem 12.3

12.4 It is required to design a PID controller for a system shown in Fig. P12.4 with $G_p(s) = \frac{K}{(s+1)(s^2+10s+4)}$, $H(s) = 1$. Obtain K such that the closed-loop system has a steady-state error of 10% for a unit ramp input with the controller. Design a PID controller such that the closed-loop system has a gain cross-over frequency of 2 rad/s, a phase margin of 60°, and a gain margin of 10 dB. Use sampling time $T = 0.10$ sec. Use suitable value for the integral gain. Show the effect of the PID controller on the closed-loop system stability and step response.

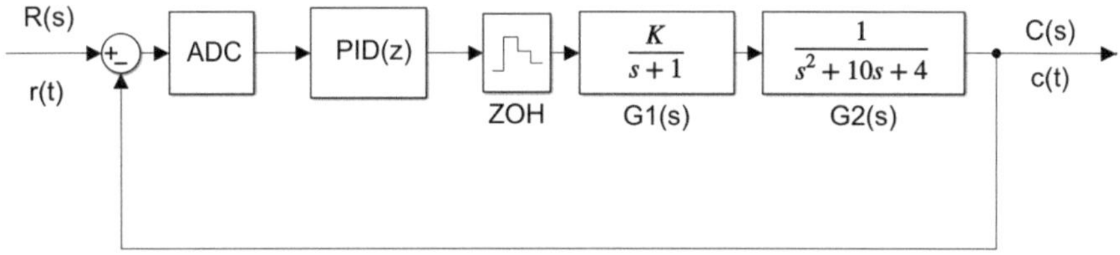

Fig. P12.4 Model of a discrete-time closed-loop system for Problem 12.4

12.5 Design a phase-lag controller for the system shown in Fig. P12.5 with $G_p(s) = \frac{10}{s^2+6.5s+1.5}$, $H(s) = 1$ such that the closed-loop system has a maximum steady-state error of 5% due to a step input, a gain cross-over frequency of 3.5 rad/s, a phase margin of 40°, and a gain margin of 10 dB. Use sampling time $T = 0.10$ sec. Show the effect of the phase-lag controller on the closed-loop system stability and step response.

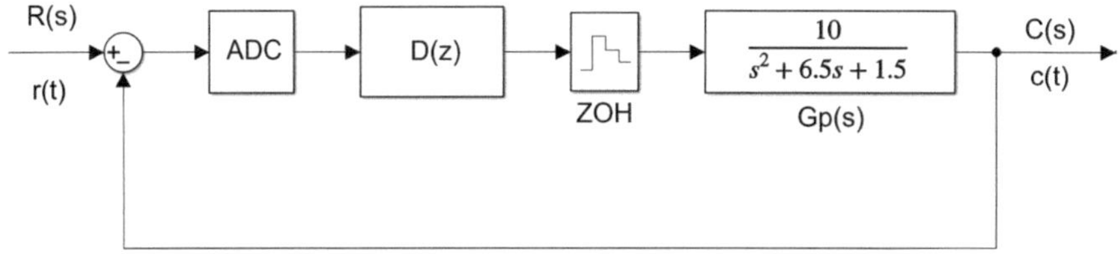

Fig. P12.5 Model of a discrete-time closed-loop system for Problem 12.5

12.6 Design a phase-lead controller for the system shown in Fig. P12.6 with $G_p(s) = \frac{10}{s^2+6.5s+1.5}$, $H(s) = 1$ such that the closed-loop system has a maximum steady-state error of 8% due to a step input, a gain cross-over frequency of 3.5 rad/s, a phase margin of 60°, and a gain margin of 10 dB. Use sampling time $T = 0.10$ sec. Show the effect of the phase-lead controller on the closed-loop system stability and step response.

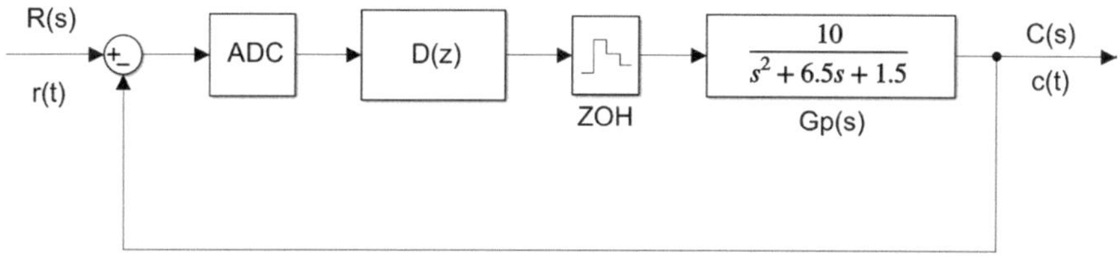

Fig. P12.6 Model of a discrete-time closed-loop system for Problem 12.6

12.7 Design a phase-lag-lead controller for the system shown in Fig. P12.7 with $G_p(s) = \frac{10}{s^2+6.5s+1.5}$, $H(s) = 1$ such that the closed-loop system has a maximum steady-state error of 2% due to a step input, a gain cross-over frequency of 4.5 rad/s, a phase margin of 60°, and a gain margin of 10 dB. Use sampling time $T = 0.10$ sec. Show the effect of the phase-lag-lead controller on the closed-loop system stability and step response.

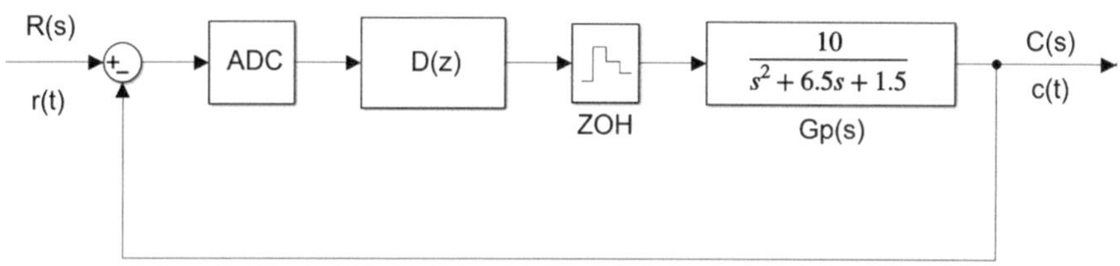

Fig. P12.7 Model of a discrete-time closed-loop system for Problem 12.7

12.8 It is required to design a phase-lag-lead controller for a system shown in Fig. P12.8 with $G_p(s) = \frac{K}{(s+1)(s^2+10s+4)}$, $H(s) = 1$. Obtain K such that the closed-loop system has a steady-state error of 5% for a unit step input with the controller. Design a phase-lag-lead controller such that the closed-loop system has a gain cross-over frequency of 2 rad/s, a phase margin of 60°, and a gain margin of 10 dB. Use sampling time $T = 0.10$ sec. Use suitable value for the integral gain. Show the effect of the PID controller on the closed-loop system stability and step response.

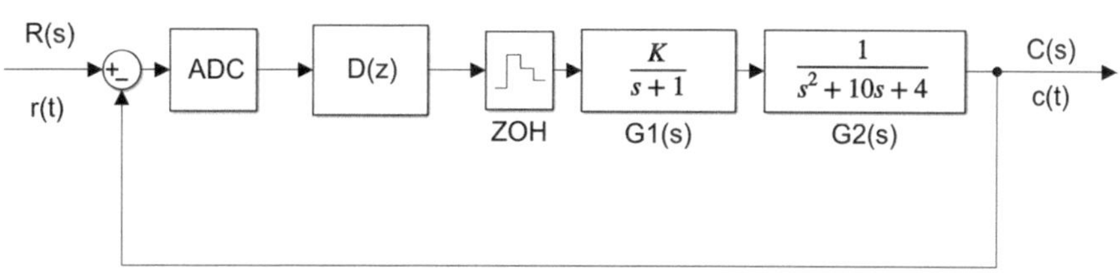

Fig. P12.8 Model of a discrete-time closed-loop system for Problem 12.8

12.9 Design a phase-lag controller for the system shown in Fig. P12.9 with $G_p(s) = \frac{10}{s^2+6.5s+1.5}$, $H(s) = 1$ such that the closed-loop system has a maximum steady-state error of 5% due to a step input, a phase margin of 60°, and a gain margin of 10 dB. Use sampling time $T = 0.10$ sec. Show the effect of the phase-lag controller on the closed-loop system stability and step response.

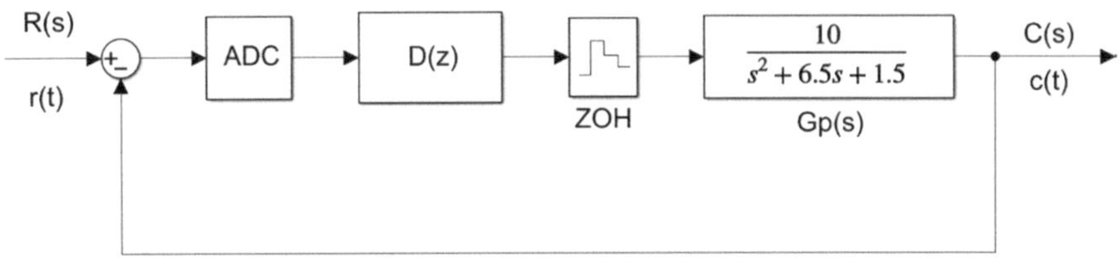

Fig. P12.9 Model of a discrete-time closed-loop system for Problem 12.9

12.10 Design a phase-lead controller for the system shown in Fig. P12.10 with $G_p(s) = \frac{10}{s^2+6.5s+1.5}$, $H(s) = 1$ such that the closed-loop system has a maximum steady-state error of 5% due to a step input, a phase margin of 60°, and a gain margin of 10 dB. Use sampling time $T = 0.10$ sec. Show the effect of the phase-lag controller on the closed-loop system stability and step response.

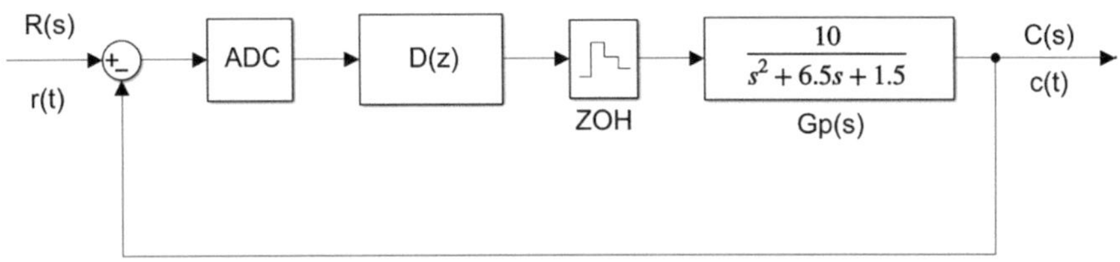

Fig. P12.10 Model of a discrete-time closed-loop system for Problem 12.10

12.11 Design a phase-lag-lead controller for the system shown in Fig. P12.11 with $G_p(s) = \frac{10}{s^2+6.5s+1.5}$, $H(s) = 1$ such that the closed-loop system has a maximum steady-state error of 5% due to a step input, a phase margin of 60°, and a gain margin of 10 dB. Use sampling time $T = 0.10$ sec. Use suitable specifications for the phase-lag part. Show the effect of the phase-lag-lead controller on the closed-loop system stability and step response.

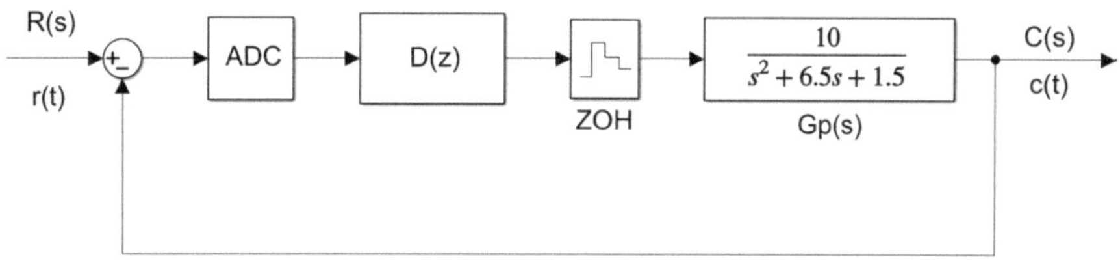

Fig. P12.11 Model of a discrete-time closed-loop system for Problem 12.11

12.12 It is required to design a phase-lag-lead controller for a system shown in Fig. P12.12 with $G_p(s) = \frac{K}{(s+1)(s^2+10s+4)}$, $H(s) = 1$. Obtain K such that the closed-loop system has a steady-state error of 5% for a unit step input with the controller. Design a phase-lag-lead controller such that the closed-loop system has a phase margin of 60° and a gain margin of 10 dB. Use sampling time $T = 0.10$ sec. Use suitable specifications for the phase-lag part. Show the effect of the phase-lag-lead controller on the closed-loop system stability and step response.

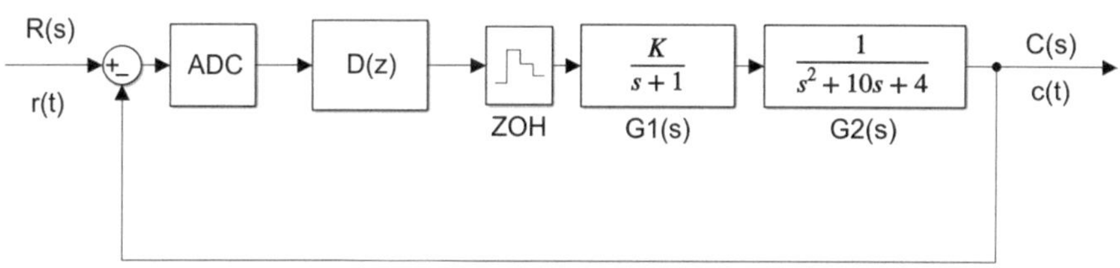

Fig. P12.12 Model of a discrete-time closed-loop system for Problem 12.12

12.13 It is required to design a phase-lag-lead controller for a position control system of an automated platform shown in Fig. P12.13 with $G_p(s) = \frac{K}{s(s+3)(s^2+9s+6)}$. The sampling time ($T$) is 0.1 sec. Obtain K such that the closed-loop system has a steady-state error of 15% for a ramp input with the controller. Design a phase-lag-lead controller such that the closed-loop system has a phase margin of 60° and a gain margin of 10 dB. Use suitable specifications for the phase-lag part. Show the effect of the phase-lag-lead controller on the closed-loop system stability and step response.

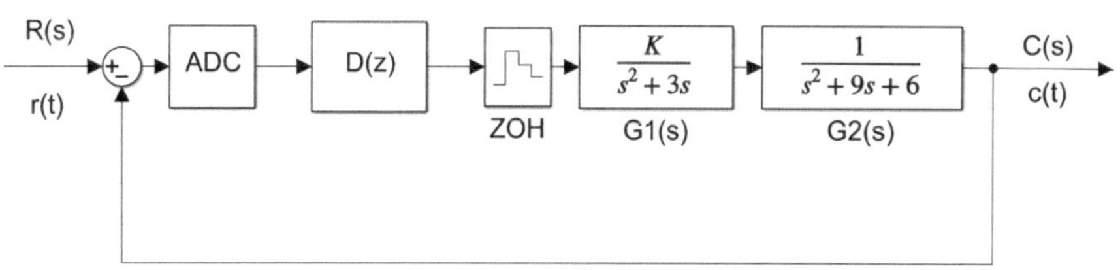

Fig. P12.13 Model of a discrete-time closed-loop system for Problem 12.13

12.14 It is required to design a phase-lag-lead controller for a position control system of an automated platform shown in Fig. P12.14 with $G_p(s) = \frac{K}{s(s+2)(s^2+7s+5)}$. The sampling time ($T$) is 0.1 sec. Obtain K such that the closed-loop system has a steady-state error of 10% for a ramp input with the controller. Design a phase-lag-lead controller such that the closed-loop system has a phase margin of 60° and a gain margin of 10 dB. Use suitable specifications for the phase-lag part. Show the effect of the phase-lag-lead controller on the closed-loop system stability and step response.

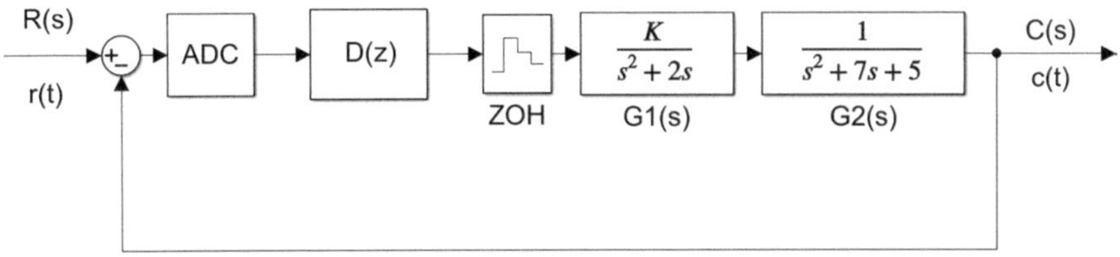

Fig. P12.14 Model of a discrete-time closed-loop system for Problem 12.14

Further Reading

1. Asadi F (2023) Signals and systems with MATLAB® and Simulink®. Springer, Cham
2. Astrom KJ, Murray RM (2021) Feedback systems: an introduction for scientists and engineers, 2nd edn. Princeton University Press, Princeton
3. Chen C-T (2006) Analog and digital control system design: transfer-function, state-space, and algebraic methods. Oxford University Press, New York
4. Dorf R, Bishop R (2016) Modern control systems, 13th edn. Pearson, Essex
5. Fadali MS, Visioli A (2020) Digital control engineering analysis and design, 3rd edn. Academic Press, London
6. Franklin GF, Powell JD, Emami-Naeini A (2019) Feedback control of dynamic systems, 8th edn. Pearson, New Jersey
7. Franklin GF, Powell JD, Workman M (1998) Digital control of dynamic systems, 3rd edn. Addison-Wesley, Boston
8. Isermann R (1981) Digital control systems. Springer-Verlag, Berlin
9. Jalili N, Candelino NW (2023) Dynamic systems and control engineering. Cambridge University Press, Cambridge
10. Keviczky L, Bars R, Hetthéssy J, Bányász (2019) Control engineering. Springer, Singapore
11. Koeing DM (2009) Practical control engineering: a guide for engineers, managers, and practitioners. McGraw-Hill, New York
12. Kuo BC (1995) Digital control systems, 2nd edn. Oxford University Press, Oxford
13. Landau ID, Gito G (2006) Digital control systems: design, identification and implementation. Springer-Verlag, London
14. Leigh JR (2006) Applied digital control: theory, design and implementation, 2nd edn. Dover, Mineola
15. Mathworks (2024a) Control system toolbox. https://www.mathworks.com/products/control.html
16. Mathworks (2024b) Matlab. https://www.mathworks.com/products/matlab.html
17. National Instruments (2024a) LabVIEW. https://www.ni.com/en/support/downloads/software-products/download.labview.html#487445

18. National Instruments (2024b) NI-ELVIS-II. https://www.ni.com/en-us/support/model.ni-elvis-ii.html
19. National Instruments (2024c) Labs for Quanser QNET DC Motor Add-On Board. https://education.ni.com/teach/resources/180/labs-for-quanser-qnet-dc-motor-add-on-board
20. Nise NS (2019) Control systems engineering, 8th edn. Wiley, Hoboken, New Jersey
21. Nof SY (ed) (2009) Springer handbook of automation. Springer-Verlag, Berlin
22. Ogata K (1995) Discrete-time control systems, 2nd edn. Pearson, New Jersey
23. Phillips CL, Nagle HT, Chakrabortty A (2015) Digital control system analysis and design, 4th edn. Pearson, Essex
24. Quanser (2024) QUBE Servo 2. https://www.quanser.com/products/qube-servo-2/
25. Rabbath CA, Lechevin N (2014) Discrete-time control system design with applications. Springer, New York
26. Samanta B (2023) Introduction to mechatronics- an integrated approach. Springer, Cham
27. Shertukde HM (2015) Digital control applications illustrated with MATLAB®. CRC Press, Boca Raton
28. Starr G (2020) Introduction to applied digital controls. Springer, Cham
29. Veloni A, Miridakis NI (2018) Digital control systems-theoretical problems and simulation tools. CRC Press, Boca Raton
30. Vilanova R, Visioli A (2012) PID control in the third millennium - lessons learned and new approaches. Springer-Verlag, London
31. Visioli A (2006) Practical PID control. Springer, London

Chapter 13
State Space Representation of Discrete-Time Systems

13.1 Introduction

In this chapter, the state-space representation of systems in continuous time domain is presented first. Next, conversion from continuous-time to discrete-time systems in state space is discussed. The conversion from transfer functions to state space is covered next. The concepts of controllability and observability of systems represented in state space are discussed. Finally, the conversion from state space to transfer function form is covered. Examples are presented along with the corresponding codes in Matlab to illustrate the procedures. End-of-chapter exercises are provided to enhance understanding of the topics covered and consolidate problem-solving skills.

13.2 State-Space Representation of Continuous-Time Systems

13.2.1 State-Space Equations in Continuous Time

The general form of state-space representation in continuous time for a dynamic system can be expressed as

$$\frac{d}{dt}\boldsymbol{x}(t) = \boldsymbol{A}\,\boldsymbol{x}(t) + \boldsymbol{B}\,\boldsymbol{u}(t) \tag{13.1}$$

$$\boldsymbol{y}(t) = \boldsymbol{C}\,\boldsymbol{x}(t) + \boldsymbol{D}\,\boldsymbol{u}(t) \tag{13.2}$$

where $\boldsymbol{x}(t)$ is state vector of size $n \times 1$, $\boldsymbol{u}(t)$ is the input vector of size $p \times 1$, $\boldsymbol{y}(t)$ is the output vector of size $r \times 1$, and $\boldsymbol{A}$, $\boldsymbol{B}$, $\boldsymbol{C}$, and $\boldsymbol{D}$ are matrices of size $n \times n$, $n \times p$, $r \times n$, and $r \times p$, respectively.

B. Samanta, *Introduction to Digital Control*,
https://doi.org/10.1007/978-3-031-66830-2_13

Fig. 13.1 Schematic diagram of a mass-spring-damper system

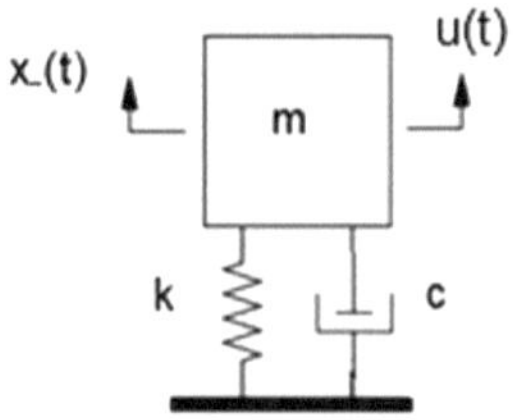

Example 13.1 Mass-Spring-Damper System A single degree-of-freedom mass-spring-damper system is shown in Fig. 13.1. Derive the equations of motion for the mass-spring-damper system. Obtain the state-space representation of the system.

Solution A mass-spring-damper system with mass (m) is supported on a spring with spring stiffness (k), and a damper with a damping coefficient (c) is placed on a fixed support, as shown in Fig. 13.1, $u(t)$ being the input force and $x(t)$ being the displacement of the mass (m).

Considering the force due to the spring, $-k(x(t))$, and the force due to the damper, $-c(\dot{x}(t))$, the equation of motion can be written as

$$(\uparrow +)\sum F_x = m\,\ddot{x}(t) \tag{13.3}$$

$$u(t) - k(x(t)) - c(\dot{x}(t)) = m\,\ddot{x}(t) \tag{13.4}$$

Equation (13.4) can be rearranged as

$$m\,\ddot{x}(t) + c\,\dot{x}(t) + k\,x(t) = u(t) \tag{13.5}$$

Equation (13.5) can be rewritten as

$$\ddot{x}(t) + \frac{c}{m}\,\dot{x}(t) + \frac{k}{m}\,x(t) = \frac{1}{m}u(t) \tag{13.6}$$

Substituting $\frac{c}{m} = 2\zeta\omega_n$, $\frac{k}{m} = \omega_n^2$, Eq. (13.6) can be written as

$$\ddot{x}(t) + 2\zeta\omega_n\,\dot{x}(t) + \omega_n^2\,x(t) = \frac{1}{m}u(t) \tag{13.7}$$

where the undamped natural frequency is given as

$$\omega_n = \sqrt{\frac{k}{m}} \tag{13.8}$$

And the damping ratio is given as

$$\zeta = \frac{c}{2\sqrt{km}} \tag{13.9}$$

The damping ratio (ζ) can be either $0 < \zeta < 1$, underdamped, $\zeta = 1$ critically damped, or $\zeta > 1$, overdamped system. In general, the damping ratio for control systems is less than 1: $0 < \zeta < 1$ (underdamped).

The second-order ordinary differential Eq. (13.7) can be converted to state-space representation in the form of two first-order ordinary differential equations defining state vectors as

$$x_1(t) = x(t) \tag{13.10}$$

$$x_2(t) = \dot{x}_1(t) = \dot{x}(t) \tag{13.11}$$

The state-space representation is as

$$\frac{d}{dt}\begin{bmatrix} x_1(t) \\ x_2(t) \end{bmatrix} = \begin{bmatrix} 0 & 1 \\ -\omega_n^2 & -2\zeta\omega_n \end{bmatrix}\begin{bmatrix} x_1(t) \\ x_2(t) \end{bmatrix} + \begin{bmatrix} 0 \\ \frac{1}{m} \end{bmatrix} u(t) \tag{13.12}$$

Measurement of position can be expressed as

$$y(t) = \begin{bmatrix} 1 & 0 \end{bmatrix}\begin{bmatrix} x_1(t) \\ x_2(t) \end{bmatrix} + [0]u(t) \tag{13.13}$$

State-space equations of (13.12) and (13.13) can be expressed as in Eqs. (13.1) and (13.2) where the state vector and the matrices are represented, with $n = 2, p = 1$, $r = 1$, , $x_1(t) = x(t)$, $x_2(t) = \dot{x}(t)$, as

$$\boldsymbol{x}(t) = \begin{bmatrix} x_1(t) \\ x_2(t) \end{bmatrix} \tag{13.14}$$

$$\boldsymbol{A} = \begin{bmatrix} 0 & 1 \\ -\omega_n^2 & -2\zeta\omega_n \end{bmatrix} \tag{13.15}$$

$$\boldsymbol{B} = \begin{bmatrix} 0 \\ \frac{1}{m} \end{bmatrix} \tag{13.16}$$

$$\boldsymbol{C} = \begin{bmatrix} 1 & 0 \end{bmatrix} \tag{13.17}$$

$$\boldsymbol{D} = 0 \tag{13.18}$$

Example 13.2 Inverted Pendulum The schematic of an inverted pendulum on a cart is shown in Fig. 13.2. Derive equations of motion for translation motion of cart $x(t)$ and small rotational motion of the pendulum around the upright position $\theta(t)$.

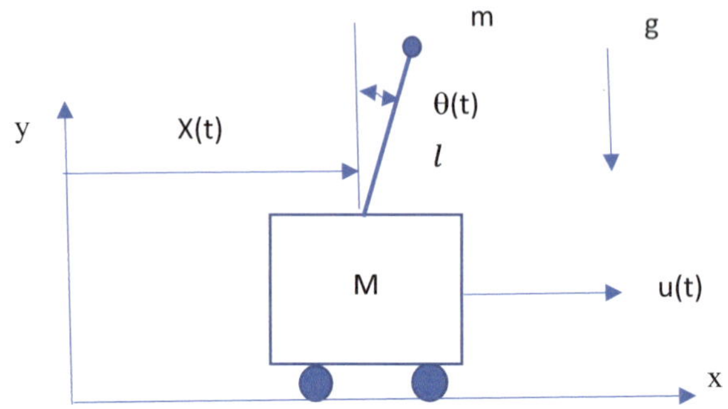

Fig. 13.2 Schematic diagram of an inverted pendulum cart system

Use M and m as the mass of the cart and the pendulum point mass, respectively, l as the length of the massless pendulum rod, and c_1 and c_2 as coefficients of damping at the cart-support interface and the pendulum-cart hinge joint, respectively. Obtain the state-space representation of the system.

Solution The equation for the translation motion of the system can be written as

$$(\rightarrow +) \quad \sum F_x \; = M\,\ddot{x}(t) + m\frac{d^2}{dt^2}(x(t) + l\;\sin(\theta(t))) \tag{13.19}$$

Assuming $\theta(t)$ to be small so that $\sin(\theta(t)) \approx \theta(t)$, Eq. (13.19) can be expressed as

$$u(t) - c_1(\dot{x}(t)) = (M + m)\,\ddot{x}(t) + m\,l\,\ddot{\theta}(t) \tag{13.20}$$

The equation for the rotational motion of the pendulum can be written as

$$(CW+) \quad \sum M_o \; = m\,l^2\,\frac{d^2}{dt^2}\left[\frac{(x(t) + l\;\sin(\theta(t)))}{l}\right] \tag{13.21}$$

Equation (13.21) can be rearranged as

$$mg\,l\;\sin(\theta(t)) - c_2\dot{\theta}(t) = m\,l\,\ddot{x}(t) + m\,l^2\ddot{\theta}(t) \tag{13.22}$$

After some rearrangement, Eqs. (13.20) and (13.22) can be written in matrix form as

$$\boldsymbol{M}\,\ddot{\boldsymbol{q}}(t) + \boldsymbol{G}\,\dot{\boldsymbol{q}}(t) + \boldsymbol{K}\,\boldsymbol{q}(t) = \boldsymbol{R}\,u(t) \tag{13.23}$$

where

$$\boldsymbol{q}(t) = \begin{bmatrix} x(t) \\ (t) \end{bmatrix} \tag{13.24}$$

$$\boldsymbol{M} = \begin{bmatrix} M+m & ml \\ ml & ml^2 \end{bmatrix} \tag{13.25}$$

$$\boldsymbol{G} = \begin{bmatrix} c_1 & 0 \\ 0 & c_2 \end{bmatrix} \tag{13.26}$$

$$\boldsymbol{K} = \begin{bmatrix} 0 & 0 \\ 0 & -mgl \end{bmatrix} \tag{13.27}$$

$$\boldsymbol{R} = \begin{bmatrix} 1 \\ 0 \end{bmatrix} \tag{13.28}$$

Equation (13.23) can be expressed, assuming $\boldsymbol{M}$ to be non-singular with non-zero inverse $\boldsymbol{M}^{-1}$, as

$$\ddot{\boldsymbol{q}}(t) = \boldsymbol{M}^{-1}[-\boldsymbol{G}\,\dot{\boldsymbol{q}}(t) - \boldsymbol{K}\,\boldsymbol{q}(t) + \boldsymbol{R}\,u(t)] \tag{13.29}$$

Equation (13.29) can be expressed in state space as in Eqs. (13.1) and (13.2) where the state vector and the matrices are represented with $n = 4$, $p = 1$, $r = 2, x_1(t) = x(t), x_2(t) = \dot{x}(t), x_3(t) = \theta(t), x_4(t) = \dot{\theta}(t),$ as

$$\boldsymbol{x}(t) = \begin{bmatrix} x_1(t) \\ x_2(t) \\ x_3(t) \\ x_4(t) \end{bmatrix} \tag{13.30}$$

$$\boldsymbol{A} = \begin{bmatrix} 0 & 1 & 0 & 0 \\ f_{11} & e_{11} & f_{12} & e_{12} \\ 0 & 0 & 0 & 1 \\ f_{21} & e_{21} & f_{22} & e_{22} \end{bmatrix} \tag{13.31}$$

$$\boldsymbol{B} = \begin{bmatrix} 0 \\ h_1 \\ 0 \\ h_2 \end{bmatrix} \tag{13.32}$$

$$\boldsymbol{y}(t) = \begin{bmatrix} x(t) \\ (t) \end{bmatrix} \tag{13.33}$$

$$\boldsymbol{C} = \begin{bmatrix} 1 & 0 & 0 & 0 \\ 0 & 0 & 1 & 0 \end{bmatrix} \tag{13.34}$$

$$\boldsymbol{D} = \begin{bmatrix} 0 \\ 0 \end{bmatrix} \tag{13.35}$$

$$\boldsymbol{E} = -\boldsymbol{M}^{-1}\boldsymbol{G} \tag{13.36}$$

$$\boldsymbol{F} = -\boldsymbol{M}^{-1}\boldsymbol{K} \tag{13.37}$$

$$\boldsymbol{H} = \boldsymbol{M}^{-1}\boldsymbol{R} \tag{13.38}$$

Example 13.3 Mass-Spring-Damper System Obtain mass-spring-damper system equations in state-space form using the following parameters: $m = 0.500\ \mathrm{kg}, c = 4.5\ \mathrm{N}.\frac{\mathrm{s}}{\mathrm{m}}, k = 50\frac{\mathrm{N}}{\mathrm{m}}$.

Solution Matlab code and results are presented.

```
% Ch 13-Example 3-Mass-Spring-Damper System
clear all; close all; clc
% System Parameters
m=0.5; c=4.5; k=50;
wn = sqrt(k/m) % undamped natural frequency (rad/s)
zeta = c/(2*sqrt(k*m))
% State-Space Model
A = [0              1;
    -wn^2   -2*zeta*wn]
B = [0;
     1/m]
C = [1  0]
D = 0

Results:
wn = 10, zeta = 0.4500

A =[  0      1
   -100     -9]

B = [0
      2]

C = [1      0]


D = [0]
```

Example 13.4 Inverted Pendulum Obtain inverted pendulum system equations in state-space form using the following parameters: $M = 8.085\ \mathrm{kg}, m = 0.825\ \mathrm{kg}, l = 0.098\ \mathrm{m}, g = 9.81\frac{\mathrm{m}}{\mathrm{s}^2}, c_1 = 0.35\ \mathrm{N}.\frac{\mathrm{s}}{\mathrm{m}},\ \ c_2 = 0.25\ \mathrm{N.m}.\frac{\mathrm{s}}{\mathrm{rad}}$.

Solution Matlab code and results are presented.

```
% Example 13.3 state-space for inverted pendulum
% Parameters
clear all
M=8.085; m=0.825; l=0.098; g=9.81; c1=0.35; c2=0.25;


MM = [M+m m*l;
    m*l  m*l^2]

G = [c1  0;
    0   c2]

K = [0  0;
    0  -m*g*l]

R =[1; 0]
Minv = inv(MM)

E = -Minv*G
F = -Minv*K
H =  Minv*R

A = [0         1          0            0;
     F(1,1)    E(1,1)     F(1,2)       E(1,2);
     0         0          0            1;
     F(2,1)    E(2,1)     F(2,2)       E(2,2)]

B = [0;
    H(1);
    0;
    H(2)]

C = [1  0  0  0;
     0  0  1  0]

D = [0; 0]

Results:

A =      0    1.0000         0         0
         0   -0.0433   -1.0010    0.3155
         0         0         0    1.0000
         0    0.4417  110.3165  -34.7722

B =      0
    0.1237
         0
   -1.2621
C =  1     0     0     0
     0     0     1     0

D =  0
     0
```

13.2.2 Conversion from Continuous-Time to Discrete-Time Systems

The solution of the state-space equation can be expressed as

$$\boldsymbol{x}(t) = e^{\boldsymbol{A}(t-t_0)}\boldsymbol{x}(t_0) + \int_{t_0}^{t} e^{\boldsymbol{A}(t-\tau)}Bu(\tau)d\tau \tag{13.39}$$

The system output can be expressed as

$$\boldsymbol{y}(t) = \boldsymbol{C}e^{\boldsymbol{A}(t-t_0)}\boldsymbol{x}(t_0) + \boldsymbol{C}\int_{t_0}^{t} e^{\boldsymbol{A}(t-\tau)}Bu(\tau)d\tau + \boldsymbol{D}\,\boldsymbol{u}(t) \tag{13.40}$$

Considering $t_0 = kT$ and $t = kT + T$, where k and T are sampling instant and sampling time, respectively, Eq. (13.39) can be expressed for the discrete-time system as

$$\boldsymbol{x}(kT+T) = e^{\boldsymbol{A}T}\boldsymbol{x}(kT) + \int_{kT}^{kT+T} e^{\boldsymbol{A}(kT+T-\tau)}Bu(\tau)d\tau \tag{13.41}$$

Assuming $u(t) = u(kT)$ for $kT \le t \le kT + T$ and $\gamma = kT + T - \tau$, Eq. (13.41) can be expressed as

$$\boldsymbol{x}(kT+T) = e^{\boldsymbol{A}T}\boldsymbol{x}(kT) + \int_{0}^{T} e^{\boldsymbol{A}\gamma}Bd\gamma\; u(kT) \tag{13.42}$$

Defining

$$\Phi = e^{\boldsymbol{A}T} \tag{13.43}$$

$$\Gamma = \int_{0}^{T} e^{\boldsymbol{A}\gamma}Bd\gamma \tag{13.44}$$

The state-space representation for the discrete-time system can be expressed as

$$\boldsymbol{x}(kT+T) = \Phi\,\boldsymbol{x}(kT) + \Gamma\,\boldsymbol{u}(kT) \tag{13.45}$$

$$\boldsymbol{y}(kT) = \boldsymbol{C}\,\boldsymbol{x}(kT) + \boldsymbol{D}\,\boldsymbol{u}(kT) \tag{13.46}$$

In Matlab, c2d can be used to convert the state-space system from continuous-time to discrete-time domain as

$$[a, b] = c2d(A, B, T)$$

Example 13.5 Obtain discrete-time state-space representation of the mass-spring-damper system of Example 13.3 with sampling time $T = 0.0.06$ s.

Solution The Matlab code added to Example 13.3 and the results are presented.

```
% Ch 13-Ex3_5-Mass-Spring_Damper System
clear all; close all; clc
% System Parameters
m=0.5; c=4.5; k=50;
wn = sqrt(k/m) % undamped natural frequaency (rad/s)
zeta = c/(2*sqrt(k*m))
% State-Space Model
A = [0             1;
    -wn^2   -2*zeta*wn]
B = [0;
     1/m]
C = [1  0]
D = 0

% Conversion from continuous-time to discrete-time
T=0.06 % sampling time
[a, b] = c2d(A,B,T) % state-space model a(k), b(k)

Results:

a = 0.8528    0.0436
   -4.3642    0.4600

b = 0.0029
    0.0873
```

Example 13.6 Obtain discrete-time state-space representation of the inverted pendulum system of Example 13.3 with sampling time $T = 0.1$ s.

Solution The Matlab code added to Example 13.4 and results are presented.

```
% Discrete-Time Systems
T=0.1
[a,b]=c2d(A,B,T)    % ss (k)
Results:
a = 1.0000    0.0998    -0.0022    0.0006
         0    0.9960    -0.0324    0.0080
         0    0.0010     1.2451    0.0324
         0    0.0143     3.5756    0.1180


b = 0.0006
    0.0116
   -0.0028
   -0.0408
```

13.2.3 From s-Domain Transfer Functions to State-Space Representation

In Matlab, the tf2ss command can be used to convert s-domain transfer functions to state-space form. The state-space representation can be converted from continuous time to discrete time using the c2d command. An example is presented to illustrate the procedure.

Example 13.7 The s-domain transfer function of a DC motor position dynamics is given as $G_p(s) = \frac{K}{s(\tau s+1)}$, $K = 30$, $\tau = 0.1$ s, $T = 0.01$ s. Obtain state-space representation from transfer function in continuous time. Next, convert from continuous-time to discrete-time domain.

Solution Matlab code and results are presented.

```
% Ch 13- Example 7 state space method DC Motor
% State space model Controllability and observability
clear all; close all;
clc
%G(s)=K/s(s+tau)
K=30; %Static gain
tau=0.1; % time constant
T=0.01 ; % sampling time
ngs=K; dgs=[tau 1 0];
printsys(ngs,dgs)
%ss model [ac,bc,,c,d]
[ac,bc,c,d ]=tf2ss(ngs,dgs) % ss (s)
%ss model in discrete-time
[a,b]=c2d(ac,bc,T)   % ss (k)

Results:
                    30
G(s) =    ---------------
             0.1s^2 + s

Continuous-time state-space
ac =   -10      0
         1     0
bc =     1
         0
c =      0    300
d =      0

Discrete-time state-space
a =     0.9048         0
          0.0095    1.0000

b =   0.0095
        0.0000
```

13.3 Controllability and Observability of Discrete-Time Systems

A linear time-invariant (LTI) discrete-time system in state-space form can be expressed, denoting Φ and Γ as **A** and **B** and omitting T in Eqs. (13.45) and (13.46) as

$$\boldsymbol{x}(k+1) = \mathbf{A}\,\boldsymbol{x}(k) + \mathbf{B}\,\boldsymbol{u}(k) \tag{13.47}$$

$$\boldsymbol{y}(k) = \boldsymbol{C}\,\boldsymbol{x}(k) + \boldsymbol{D}\,\boldsymbol{u}(k) \tag{13.48}$$

The LTI system with system matrices **A**, **B**, **C**, and **D** can be controllable for the input vector $\boldsymbol{u}(k)$ if the controllability matrix ***CM*** given in Eq. (13.49) has a rank equal to the order of the system, n:

$$\boldsymbol{CM} = \left[\boldsymbol{B}\ \ \boldsymbol{AB}\ \ \boldsymbol{A}^2\boldsymbol{B}\ldots\boldsymbol{A}^{n-2}\boldsymbol{B}\ \ \boldsymbol{A}^{n-1}\boldsymbol{B}\right] \tag{13.49}$$

where **A** and ***B*** are matrices of size $n \times n$ and $n \times p$ resulting in the controllability matrix of size $n \times np$. In other words, for the LTI system to be controllable, the following condition, Eq. (13.50), should be satisfied:

$$\text{rank}(\boldsymbol{CM}) = n \tag{13.50}$$

The LTI system can be observable if the observability matrix ***OM*** given in Eq. (13.51) has a rank equal to the order of the system, n:

$$\boldsymbol{OM} = \begin{bmatrix} \boldsymbol{C} \\ \boldsymbol{CA} \\ \boldsymbol{CA}^2 \\ \cdot \\ \cdot \\ \cdot \\ \boldsymbol{CA}^{n-2} \\ \boldsymbol{CA}^{n-1} \end{bmatrix} \tag{13.51}$$

where **A** and ***C*** are matrices of size $n \times n$ and $r \times n$ resulting in the observability matrix ***OM*** of size $nr \times n$. In other words, for the system to be observable, the following condition, Eq. (13.52), should be satisfied:

$$rank(\boldsymbol{OM}) = n \tag{13.52}$$

Example 13.8 For the discrete-time system state-space model of the DC motor of Example 13.7, check if the system is controllable and observable.

Solution Matlab code and results are presented.

```
% Ch 13- Example 8 state space method DC Motor- controllability and
observability
% State space model Controllability and observability
clear all; close all;
clc
%G(s)=K/s(s+tau)
K=30; %Static gain
tau=0.1; % time constant
T=0.01 ; % sampling time
ngs=K; dgs=[tau 1 0];
printsys(ngs,dgs)
%ss model [ac,bc,,c,d]
[ac,bc,c,d ]=tf2ss(ngs,dgs) % ss (s)
%ss model in discrete-time
[a,b]=c2d(ac,bc,T)   % ss (k)
% controllability
cont=[b a*b ]
cr=rank(cont)
%observabilty
obsr=[c;c*a]
or=rank(obsr)

Results:

CM = 0.0095    0.0086
     0.0000    0.0001

cr = 2, The system is controllable

OM =      0  300.0000
    2.8549  300.0000

or = 2, The system is observable.
```

Example 13.9 For the discrete-time system state-space model of the inverted pendulum of Example 13.6, check if the system is controllable and observable.

Solution Matlab code and results are presented.

```
% Example 13.9 state-space for inverted pendulum-controllability and
observability
% Parameters
clear all
M=8.085; m=0.825; l=0.098; g=9.81; c1=0.35; c2=0.25;

MM = [M+m m*l;
    m*l  m*l^2]
G = [c1  0;
    0   c2]
    1

K = [0  0;
    0  -m*g*l]

R =[1; 0]
Minv = inv(MM)

E = -Minv*G
F = -Minv*K
H =  Minv*R

A = [0         1         0          0;
     F(1,1)    E(1,1)    F(1,2)     E(1,2);
     0         0         0          1;
     F(2,1)    E(2,1)    F(2,2)     E(2,2)]

B = [0;
    H(1);
    0;
    H(2)]

C = [1  0  0  0;
     0  0  1  0]

D = [0; 0]

% Discrete-Time Systems
T=0.1
[a,b]=c2d(A,B,T)   % ss (k)
% controllability
ab =a*b;
a2b =a*ab;
a3b=a*a2b;
cont=[b ab a2b a3b]
cr=rank(cont)

ca  = C*a;
ca2 = ca*a;
ca3 = ca2*a;
obsr=[C;ca; ca2; ca3]
or=rank(obsr)
```

```
Results:
CM = 0.0006     0.0017     0.0028     0.0040
     0.0116     0.0113     0.0113     0.0113
    -0.0028    -0.0048    -0.0064    -0.0086
    -0.0408    -0.0147    -0.0187    -0.0251

cr = 4, the system is controllable

OM = 1.0000          0          0          0
          0          0     1.0000          0
     1.0000     0.0998    -0.0022     0.0006
          0     0.0010     1.2451     0.0324
     1.0000     0.1992    -0.0060     0.0014
          0     0.0027     1.6661     0.0442
     1.0000     0.2982    -0.0111     0.0022
          0     0.0049     2.2324     0.0593

or = 4, the system is observable.
```

13.4 Transfer Function Representation from State-Space

The linear time-invariant (LTI) system represented in state-space form, Eqs. (13.47) and (13.48), can be expressed in z-domain, using z-transform as

$$z\boldsymbol{I}\boldsymbol{X}(z) = \boldsymbol{A}\,\boldsymbol{X}(z) + \boldsymbol{B}\,\boldsymbol{U}(z) \tag{13.53}$$

$$\boldsymbol{Y}(z) = \boldsymbol{C}\,\boldsymbol{X}(z) + \boldsymbol{D}\,\boldsymbol{U}(z) \tag{13.54}$$

Equation (13.53) can be used to express $\boldsymbol{X}(z)$ in terms of $\boldsymbol{U}(z)$ as

$$\boldsymbol{X}(z) = (z\boldsymbol{I} - \boldsymbol{A})^{-1}\boldsymbol{B}\,\boldsymbol{U}(z) \tag{13.55}$$

where $\boldsymbol{I}$ represents an identity matrix of size $n \times n$

$$\boldsymbol{I} = \begin{bmatrix} 1 & 0 & 0\ldots & 0 \\ 0 & 1 & 0\ldots & 0 \\ & & \cdot & \\ & & \cdot & \\ & & \cdot & \\ 0 & 0 & 0\ldots & 1 \end{bmatrix} \tag{13.56}$$

Equation (13.54) can be rewritten to express $\boldsymbol{Y}(z)$ in terms of $\boldsymbol{U}(z)$ as

$$\boldsymbol{Y}(z) = \left(\boldsymbol{C}\,(z\boldsymbol{I} - \boldsymbol{A})^{-1}\boldsymbol{B} + \boldsymbol{D}\right)\,\boldsymbol{U}(z) \tag{13.57}$$

Equation (13.57) can be used to get the transfer function of the discrete-time system, $\boldsymbol{G}(z)$, recognizing the input and the output of the system, $\boldsymbol{U}(z)$ and $\boldsymbol{Y}(z)$, as

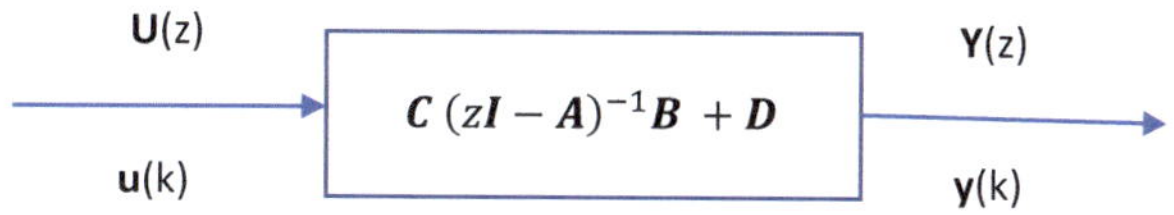

Fig. 13.3 Block diagram of a discrete-time system represented in state space

$$G(z) = \frac{Y(z)}{U(z)} = C\,(zI - A)^{-1}B + D \tag{13.58}$$

The size of the system transfer function matrix, $G(z)$, representing the multi-input multi-output (MIMO) system is $r \times p$, where r and p represent numbers of output and input, respectively. The block diagram for the closed-loop system transfer function can be represented in Fig. 13.3.

Exercises

13.1 The closed-loop vehicle distance control system for a traffic control system can be represented as shown in Fig. P13.1. The plant transfer function of the distance control system can be represented in s-domain as $G_p(s) = \frac{1.5}{s^2+12s+18}$. The sampling time is given as $T = 0.10$ s. Obtain state-space representation of the open-loop system in continuous time, and convert it to discrete-time domain. Check if the system is observable and controllable.

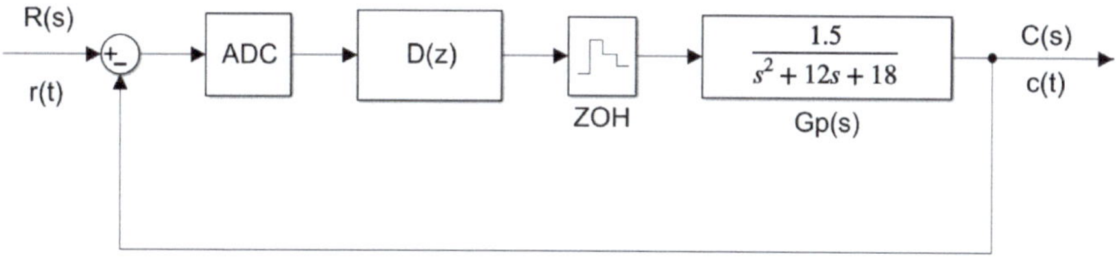

Fig. P13.1 Model of closed-loop discrete-time system for Problem 13.1

13.2 The closed-loop roll angle control system for a jet airplane can be represented as shown in Fig. P13.2. The plant transfer function of the roll control system is represented in the s-domain as $G_p(s) = \frac{10}{(s+10)(s^2+2.5s+25)}$. The sampling time is given as $T = 0.05$ s. Obtain state-space representation of the open-loop system in continuous time, and convert it to discrete-time domain. Check if the system is observable and controllable.

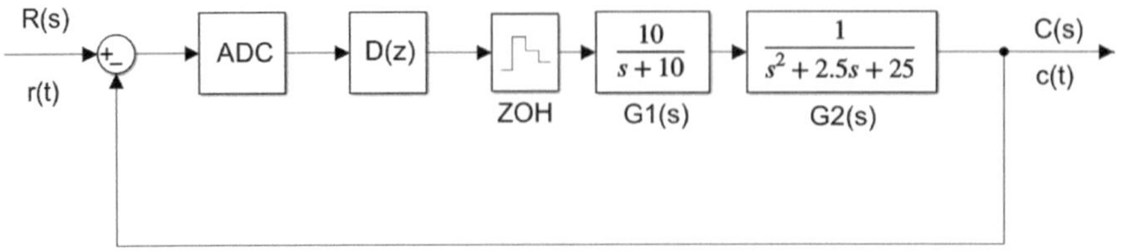

Fig. P13.2 Model of closed-loop discrete-time system for Problem 13.2

13.3 The closed-loop tilt control system in a high-speed train for passenger comfort can be represented as shown in Fig. P13.3. The plant transfer function of the train tilt control system is represented in the s-domain as $G_p(s) = \frac{5}{s(s+5)(s+35)}$. The sampling time is given as $T = 0.01$ sec. Obtain state-space representation of the open-loop system in continuous time, and convert it to discrete-time domain. Check if the system is observable and controllable.

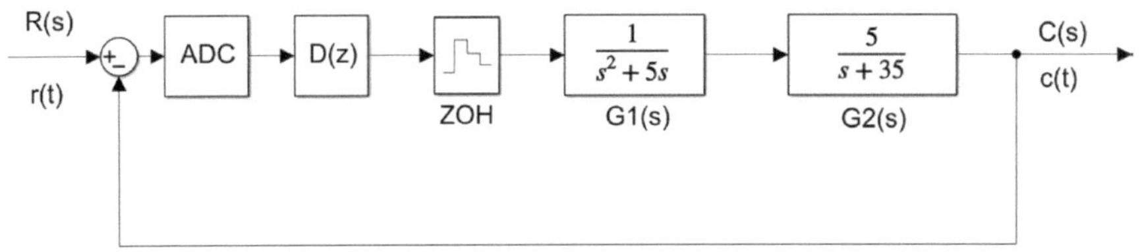

Fig. P13.3 Model of closed-loop discrete-time system for Problem 13.3

13.4 The closed-loop position control system in a machine tool can be represented as shown in Fig. P13.4. The plant transfer function of the position control system is represented in the s-domain as $G_p(s) = \frac{5K}{s(0.25s+1)\left(1.5\times 10^{-3}s+2.5\times 10^{-2}\right)}$. The sampling time is given as $T = 0.10$ s. The amplifier gain K is such that the steady-state error to a ramp input is limited to 2.5%. Obtain state-space representation of the open-loop system in continuous time, and convert it to discrete-time domain. Check if the system is observable and controllable.

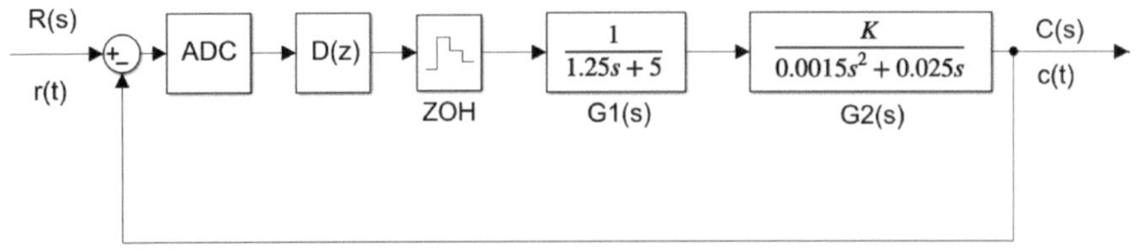

Fig. P13.4 Model of closed-loop discrete-time system for Problem 13.4

13.5 The state-space model matrices (**A**, **B**, **C**, **D**) for lateral motions of a ship moving at a constant velocity can be represented as [7]

$$\boldsymbol{A}=\begin{bmatrix}-0.0895 & -0.2860 & 0\\ -0.0439 & -0.2720 & 0\\ 0 & 1.0000 & 0\end{bmatrix}, \boldsymbol{B}=\begin{bmatrix}0.0145\\ -0.0122\\ 0\end{bmatrix}, \boldsymbol{C}=[0\ \ 0\ \ 1], \boldsymbol{D}=[0]$$

where the state variables $(x_1(t), x_2(t), x_3(t))$ are ship side slip angle (degrees), ship heading angle (degrees), and ship yaw rate, respectively. The rudder angle (degrees) is the input ($u(t)$), and the heading angle is the measurement variable ($y(t)$). The sampling time is given as $T = 0.05$ s. Obtain state-space representation of the open-loop system in discrete time. Check if the system is observable and controllable.

13.6 The state-space model matrices (**A**, **B**, **C**, **D**) for the linearized longitudinal motion of a helicopter can be represented as [7]

$$\boldsymbol{A}=\begin{bmatrix}-0.4000 & 0 & -0.0100\\ 1.0000 & 0 & 0\\ -1.4000 & 9.8 & -0.0200\end{bmatrix}, \boldsymbol{B}=\begin{bmatrix}6.3000\\ 0\\ 9.8000\end{bmatrix}, \boldsymbol{C}=[0\ \ 0\ \ 1], \boldsymbol{D}=[0]$$

where the state variables $(x_1(t), x_2(t), x_3(t))$ are the rate of pitch angle $(\dot{\theta}(t))$, pitch angle ($\theta(t)$), and horizontal velocity ($v(t)$), respectively. The rudder angle (degrees) is the input ($u(t)$), and the horizontal velocity ($v(t)$) is the measurement variable ($y(t)$). The sampling time is given as $T = 0.20$ s. Obtain state-space representation of the open-loop system in discrete time. Check if the system is observable and controllable.

13.7 The state-space model matrices (**A**, **B**, **C**, **D**) for a linearized model of liquid levels in a 3-tank system can be represented as [5]

$$\boldsymbol{A}=\begin{bmatrix}-1 & 0 & 0\\ 1 & -1 & 0\\ 0 & 1 & -1\end{bmatrix}, \boldsymbol{B}=\begin{bmatrix}1\\ 0\\ 0\end{bmatrix}, \boldsymbol{C}=[0\ \ 0\ \ 1], \boldsymbol{D}=[0]$$

where the state variables $(x_1(t), x_2(t), x_3(t))$ are liquid levels $(h_1(t), h_2(t), h_3(t))$ of the top, the middle, and the bottom tank, respectively. The inlet liquid flow rate in the top tank is the input ($u(t)$), and the liquid level of the bottom tank is the measurement variable ($y(t)$). The sampling time is given as $T = 010$ s. Obtain state-space representation of the open-loop system in discrete time. Check if the system is observable and controllable.

13.8 The closed-loop thickness control system in a metal processing plant can be represented as shown in Fig. P13.5. The plant transfer function of the thickness control system is represented in s-domain as $G_p(s) = \frac{5}{s(s+2)(s^2+s+10)}$. The sampling time is given as $T = 0.10$ s. Obtain state-space representation of the open-loop system in continuous time, and convert it to discrete-time domain. Check if the system is observable and controllable.

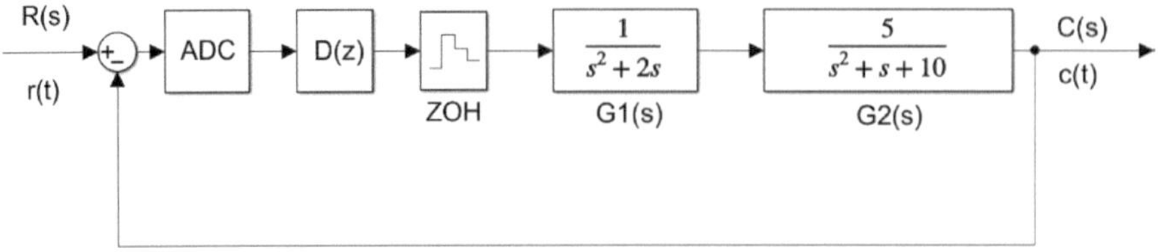

Fig. P13.5 Model of closed-loop discrete-time system for Problem 13.8

13.9 The closed-loop position control system of a typical Mars rover can be represented as shown in Fig. P13.6. The plant transfer function of the position control system is represented in the s-domain as $G_p(s) = \frac{1}{s^2(s+5)}$ [4]. The sampling time is given as $T = 0.02$ s. Obtain state-space representation of the open-loop system in continuous time, and convert it to discrete-time domain. Check if the system is observable and controllable.

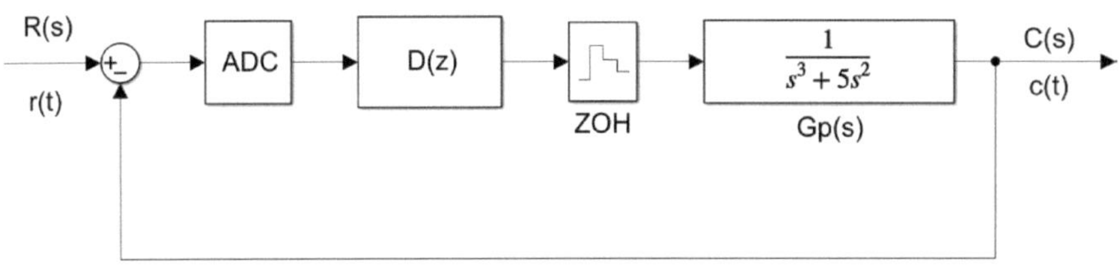

Fig. P13.6 Model of closed-loop discrete-time system for Problem 13.9

13.10 The closed-loop position control system for one of the axes of an experimental automated surface-grinding wheel can be represented as shown in Fig. P13.7. The plant transfer function of the axis position control system can be represented in the s-domain as $G_p(s) = \frac{5}{s(0.2s+1)(0.3s+1)}$ [4]. The sampling time is given as $T = 0.02$ s. Obtain state-space representation of the open-loop system in continuous time, and convert it to discrete-time domain. Check if the system is observable and controllable.

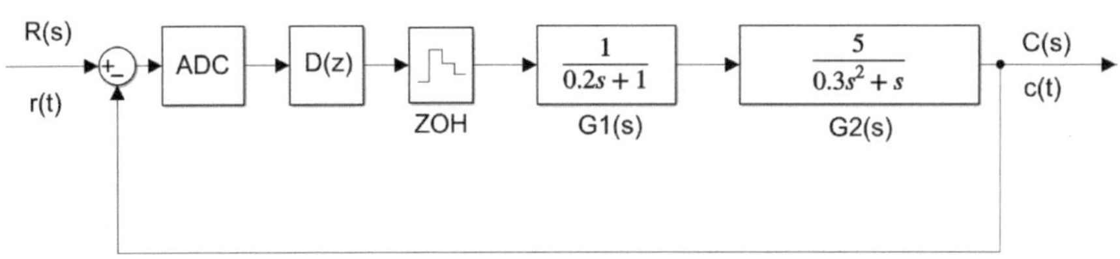

Fig. P13.7 Model of closed-loop discrete-time system for Problem 13.10

13.11 The closed-loop steering control system for a second-order system with a time delay can be represented as shown in Fig. P13.8, $G_p(s) = \frac{1}{s(s+1)} e^{-sT}$. The sampling time is given as $T = 0.2$ s. Obtain state-space representation of the open-loop system in continuous time, and convert it to discrete-time domain. Check if the system is observable and controllable.

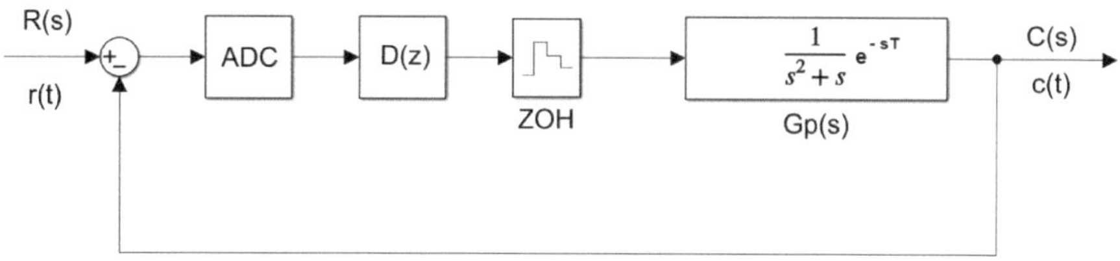

Fig. P13.8 Model of closed-loop discrete-time system for Problem 13.11

13.12 The closed-loop position control system with a time delay can be represented as shown in Fig. P13.9. The plant transfer function can be represented in the s-domain with a delay $G_p(s) = \frac{2}{s^2+5s+10} e^{-sT}$. The sampling time is given as $T = 0.2$ s. Obtain state-space representation of the open-loop system in continuous time, and convert it to discrete-time domain. Check if the system is observable and controllable.

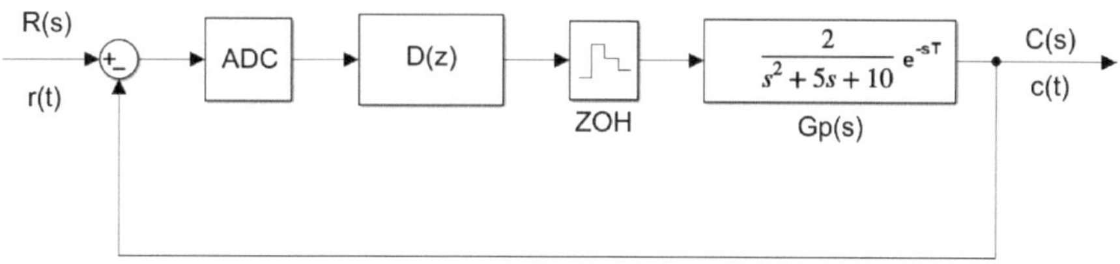

Fig. P13.9 Model of closed-loop discrete-time system for Problem 13.12

Further Reading

1. Asadi F (2023) Signals and systems with MATLAB® and Simulink®. Springer, Cham
2. Astrom KJ, Murray RM (2021) Feedback systems: an introduction for scientists and engineers, 2nd edn. Princeton University Press, Princeton
3. Chen C-T (2006) Analog and digital control system design: transfer-function, state-space, and algebraic methods. Oxford University Press, New York
4. Dorf R, Bishop R (2016) Modern control systems, 13th edn. Pearson, Harlow
5. Fadali MS, Visioli A (2020) Digital control engineering analysis and design, 3rd edn. Academic, London
6. Franklin GF, Powell JD, Workman M (1998) Digital control of dynamic systems, 3rd edn. Addison-Wesley, Boston

7. Franklin GF, Powell JD, Emami-Naeini A (2019) Feedback control of dynamic systems, 8th edn. Pearson, Upper Saddle River
8. Isermann R (1981) Digital control systems. Springer, Berlin
9. Jalili N, Candelino NW (2023) Dynamic systems and control engineering. Cambridge University Press, Cambridge
10. Keviczky L, Bars R, Hetthéssy J, Bányász (2019) Control engineering. Springer, Singapore
11. Koeing DM (2009) Practical control engineering: a guide for engineers, managers, and practitioners. McGraw-Hill, New York
12. Kuo BC (1995) Digital control systems, 2nd edn. Oxford University Press, Oxford
13. Landau ID, Gito G (2006) Digital control systems: design, identification and implementation. Springer, London
14. Leigh JR (2006) Applied digital control: theory, design and implementation, 2nd edn. Dover, Mineola
15. Mathworks (2023a) Control System Toolbox. https://www.mathworks.com/products/control.html
16. Mathworks (2023b) Matlab. https://www.mathworks.com/products/matlab.html
17. Nise NS (2019) Control systems engineering, 8th edn. Wiley, Hoboken
18. Nof SY (ed) (2009) Springer handbook of automation. Springer, Berlin
19. Ogata K (1995) Discrete-time control systems, 2nd edn. Pearson, Hoboken
20. Phillips CL, Nagle HT, Chakrabortty A (2015) Digital control system analysis and design, 4th edn. Pearson, Essex
21. Rabbath CA, Lechevin N (2014) Discrete-time control system design with applications. Springer, New York
22. Samanta B (2023) Introduction to mechatronics – an integrated approach. Springer, Cham
23. Shertukde HM (2015) Digital control applications illustrated with MATLAB®. CRC Press, Boca Raton
24. Starr G (2020) Introduction to applied digital controls. Springer, Cham
25. Veloni A, Miridakis NI (2018) Digital control systems-theoretical problems and simulation tools. CRC Press, Boca Raton
26. Vilanova R, Visioli A (2012) PID control in the third millennium – lessons learned and new approaches. Springer, London
27. Visioli A (2006) Practical PID control. Springer, London

Chapter 14
Design of Controllers for Discrete-Time Systems in State Space

14.1 Introduction

State-space methods to design controllers for discrete-time systems are presented in this chapter. Full-state feedback pole placement controllers with full-order prediction observers, reduced-order prediction observers, and full-order current observers are covered. Examples are presented along with Matlab codes and Simulink models to illustrate the procedures. Laboratory experiments for analysis, design in state space, simulation, and real-time implementation of the controllers on a DC motor module are presented. Analysis, design, and simulation are presented using Matlab, and implementation of the discrete-time controllers is presented using LabVIEW. End-of-chapter exercises are provided to enhance understanding of the topics covered and consolidate problem-solving skills.

14.2 Full State Feedback Pole Placement Control

Full state feedback (FSF) pole placement control for discrete-time systems is a method employed in feedback control system theory to place the closed-loop poles of a system in pre-determined locations in the z-plane. Placing poles is desirable because the location of the poles corresponds directly to the eigenvalues of the system, which control the characteristics of the response of the system. These closed-loop poles (z_i, $i = 1, \ldots, , n$, n being the order of the system) can be obtained from desired transient characteristics like overshoot (α) and settling time (t_{set}) discussed in Chap. 10. The system must be considered controllable, as discussed in Chap. 13, in order to implement this method.

B. Samanta, *Introduction to Digital Control*,
https://doi.org/10.1007/978-3-031-66830-2_14

Let the closed-loop discrete-time system be represented in state space form as

$$\boldsymbol{x}(k+1) = \boldsymbol{A}\,\boldsymbol{x}(k) + \boldsymbol{B}\,\boldsymbol{u}(k) \tag{14.1}$$

$$\boldsymbol{y}(k) = \boldsymbol{C}\boldsymbol{x}(k) + \boldsymbol{D}\boldsymbol{u}(k) \tag{14.2}$$

The poles of the closed-loop system are the roots of the characteristic equation given by

$$|z\boldsymbol{I} - \boldsymbol{A}| = 0 \tag{14.3}$$

Full state feedback (FSF) is utilized by commanding the input vector $\boldsymbol{u}(k)$ to achieve the desired poles. The control input $\boldsymbol{u}(k)$ can be made proportional (in the matrix sense) to the state vector $\boldsymbol{x}(k)$ with a gain matrix $\boldsymbol{K}$ of size $p \times n$ as

$$\boldsymbol{u}(k) = -\boldsymbol{K}\,\boldsymbol{x}(k) \tag{14.4}$$

Substituting $\boldsymbol{u}(k)$ from Eq. (14.4), Eqs. (14.1) and (14.2) can be expressed as

$$\boldsymbol{x}(k+1) = (\boldsymbol{A} - \boldsymbol{B}\,\boldsymbol{K})\,\boldsymbol{x}(k) \tag{14.5}$$

$$\boldsymbol{y}(k) = (\boldsymbol{C} - \boldsymbol{D}\,\boldsymbol{K})\,\boldsymbol{x}(k) \tag{14.6}$$

Equations (14.5) and (14.6) can be rewritten in z-domain, using z-transform as

$$z\boldsymbol{I}\boldsymbol{X}(z) = (\boldsymbol{A} - \boldsymbol{B}\,\boldsymbol{K})\,\boldsymbol{X}(z) \tag{14.7}$$

$$\boldsymbol{Y}(z) = (\boldsymbol{C} - \boldsymbol{D}\,\boldsymbol{K})\,\boldsymbol{X}(z) \tag{14.8}$$

It should be mentioned that most commonly the output vector $\boldsymbol{y}(k)$ in Eq. (14.2) is not directly related to the input vector $\boldsymbol{u}(k)$, i.e., the matrix $\boldsymbol{D}$ is a null matrix of size $r \times p$, $\boldsymbol{D} = \boldsymbol{0}$. For the rest of this chapter, this assumption ($\boldsymbol{D} = \boldsymbol{0}$) will be followed for simplicity.

Controller Gain Matrix

The characteristic equation for the closed-loop system with full state feedback can be expressed from Eq. (14.7) as

$$|z\boldsymbol{I} - (\boldsymbol{A} - \boldsymbol{B}\,\boldsymbol{K})| = 0 \tag{14.9}$$

Let the desired closed-loop system poles be designated as

$$z = p_1, p_2, \ldots, p_n \tag{14.10}$$

The closed-loop system characteristic polynomial can be expressed as

$$\lambda(z) = (z - p_1)(z - p_2)\ldots(z - p_n) \tag{14.11}$$

The elements of the gain matrix $\boldsymbol{K}$ can be obtained by equating the characteristic polynomial, Eqs. (14.9) and (14.11), as

$$\lambda(z) = |z\boldsymbol{I} - (\boldsymbol{A} - \boldsymbol{B}\,\boldsymbol{K})| = (z - p_1)(z - p_2)\ldots(z - p_n) \tag{14.12}$$

In Matlab, the controller gain matrix $\boldsymbol{K}$ can be obtained using the command *place*() as

$$\boldsymbol{K} = place(\boldsymbol{A}, \boldsymbol{B}, \boldsymbol{p}) \tag{14.13}$$

where desired poles of the closed-loop discrete-time system are represented in the vector $\boldsymbol{p}$ as

$$\boldsymbol{p} = [p_1, p_2, \ldots, p_n] \tag{14.14}$$

Representation of a System with Full State Feedback Pole Placement Control
A system with full state feedback pole placement control, given by Eqs. (14.1), (14.2), and (14.4), can be represented schematically in Fig. 14.1.

Equation (14.1) can be written in z-domain using z-transform as

$$z\boldsymbol{I}\,\boldsymbol{X}(z) = \boldsymbol{A}\,\boldsymbol{X}(z) + \boldsymbol{B}\,\boldsymbol{U}(z) \tag{14.15}$$

Equation (14.15) can be used to express state vector $\boldsymbol{X}(z)$ in terms of input vector $\boldsymbol{U}(z)$ in z-domain as

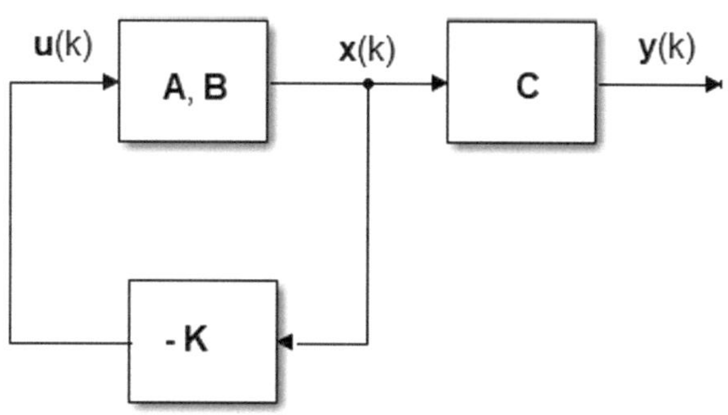

Fig. 14.1 Schematic of a system with full state feedback pole placement control

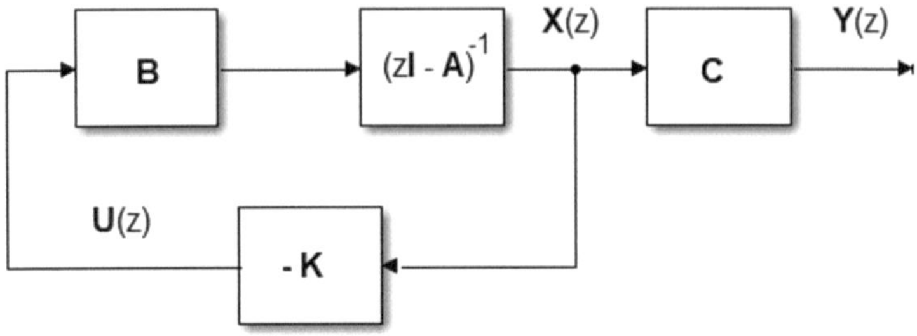

Fig. 14.2 Block diagram of a system with full state feedback pole placement control

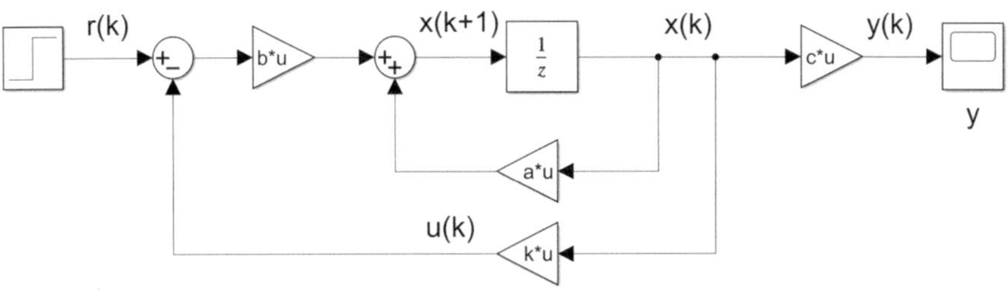

Fig. 14.3 Simulink model of a system with pole placement control and a reference input vector

$$\boldsymbol{X}(z) = (z\boldsymbol{I} - \boldsymbol{A})^{-1}\boldsymbol{B}\,\boldsymbol{U}(z) \tag{14.16}$$

Equation (14.2), with $\boldsymbol{D} = \boldsymbol{0}$, can be rewritten the output vector in z-domain as

$$\boldsymbol{Y}(z) = \boldsymbol{C}\boldsymbol{X}(z) \tag{14.17}$$

The control law, Eq. (14.4), can be written in z-domain as

$$\boldsymbol{U}(z) = -\boldsymbol{K}\boldsymbol{X}(z) \tag{14.18}$$

Equations (14.15)–(14.18) can be represented in a block diagram as shown in Fig. 14.2.

A Simulink model for the discrete-time system in state space with full state feedback pole placement control, given in Eqs. (14.1), (14.2), and (14.4), with a reference input vector $\boldsymbol{r}(k)$ of size $p \times 1$ is shown in Fig. 14.3. It should be noted that state vector $\boldsymbol{x}(k)$, the output vector $\boldsymbol{y}(k)$, and the input vector $\boldsymbol{u}(k)$ are of size $n \times 1$, $r \times 1$, and $p \times 1$, respectively. The matrices $\boldsymbol{a}, \boldsymbol{b}, \boldsymbol{c}$, and $\boldsymbol{k}$ are the same as $\boldsymbol{A}, \boldsymbol{B}, \boldsymbol{C}$, and $\boldsymbol{K}$, respectively, as defined in Eqs. (14.1), (14.2), and (14.4).

Equivalent State-Space Representation

The closed-loop discrete-time system represented in Eqs. (14.1) and (14.2) and the full state feedback control Eq. (14.4) can be expressed as a transfer function in z-domain, using z-transform, as

$$\boldsymbol{G}_{pp}(z) = \frac{\boldsymbol{Y}(z)}{\boldsymbol{U}(z)} = \boldsymbol{C}\left(z\boldsymbol{I} - (\boldsymbol{A} - \boldsymbol{B}\boldsymbol{K})\right)^{-1}\boldsymbol{B} + \boldsymbol{D} \tag{14.19}$$

Comparing Eq. (14.15) with the standard state-space representation of a discrete-time system in state space, Eq. (13.58) in Chap. 13, the equivalent system matrices of the new system can be expressed as

$$\boldsymbol{A}_{pp} = \boldsymbol{A} - \boldsymbol{B}\boldsymbol{K} \tag{14.20}$$

$$\boldsymbol{B}_{pp} = \boldsymbol{B} \tag{14.21}$$

$$\boldsymbol{C}_{pp} = \boldsymbol{C} \tag{14.22}$$

$$\boldsymbol{D}_{pp} = \boldsymbol{D} \tag{14.23}$$

The full state feedback pole placement control algorithm can be implemented provided the full state vector is available. However, In the following section, it will be shown that an estimate of the full state vector can be used to obtain the feedback control.

Example 14.1 The s-domain transfer function of a DC motor position dynamics is given as $G_p(s) = \frac{K}{s(\tau s+1)}$, $K = 30$, $\tau = 0.1$ s, $T = 0.01$ s. Obtain state-space representation from transfer function in continuous time, and convert from continuous-time to discrete-time domain. Design a full state feedback pole placement controller such that the closed-loop system with the controller will have an overshoot of 15% and a settling time of 0.25 s when subjected to a step input. Verify from closed-loop system step response if the specifications are met.

Solution Matlab code and results are presented.

```
% Ch 14- Example 1 DC Motor Pole Placement Controller
% Pole placement controller
clear all; close all;clc
%G(s)=K/s(s+tau)
K=30; %Static gain
tau=0.1; % time constant
T=0.01 ; % sampling time
ngs=K; dgs=[tau 1 0];
printsys(ngs,dgs)
%ss model [ac,bc,,c,d]
[ac,bc,c,d ]=tf2ss(ngs,dgs) % ss (s)
%ss model in discrete-time
[a,b]=c2d(ac,bc,T)   % ss (k)
% controllability
cont=[b a*b ]
cr=rank(cont)
%observabilty
obsr=[c;c*a]
or=rank(obsr)
% pole placement controller design
% desired pole locations
alpha=0.15; tset=0.25; beta=0.02;
zeta=abs(log(alpha))/sqrt(pi^2+log(alpha)^2)
wn=-log(beta)/(zeta*tset)
s1=-zeta*wn+j*wn*sqrt(1-zeta^2);
z1=exp(s1*T)
s2=-zeta*wn-j*wn*sqrt(1-zeta^2);
z2=exp(s2*T)
p=[z1, z2];
k=place(a,b,p)
%pause
% closed loop response (with the state feedback only)
at=a-b*k;bt=b;ct=c;dt=d;
[zt,pt,kt]=ss2zp(at,bt,ct,dt)
n=200;
dcg=ddcgain(at,bt,ct,dt);
figure
dstep(at,bt,ct/dcg,dt,1,n);
yc=dstep(at,bt,ct,dt,1,n);
t=linspace(0,n*T,n);
yc=yc/dcg;
figure
plot(t,yc)
title('Closed-loop System Step Response')
xlabel('Time (seconds)')
ylabel('y(t)')

Results:
System matrices
a = 0.9048         0
    0.0095    1.0000

b = 0.0095
    0.0000
Desired poles in z-plane
zeta = 0.5169, wn = 30.2712, z1 = 0.8266 + 0.2191i, z2 = 0.8266 - 0.2191i

Controller gain matrix
k = [22.2725  820.5221]
```

The closed-loop step response of the system is presented in Fig. 14.4. The percentage overshoot and settling are close to the specifications.

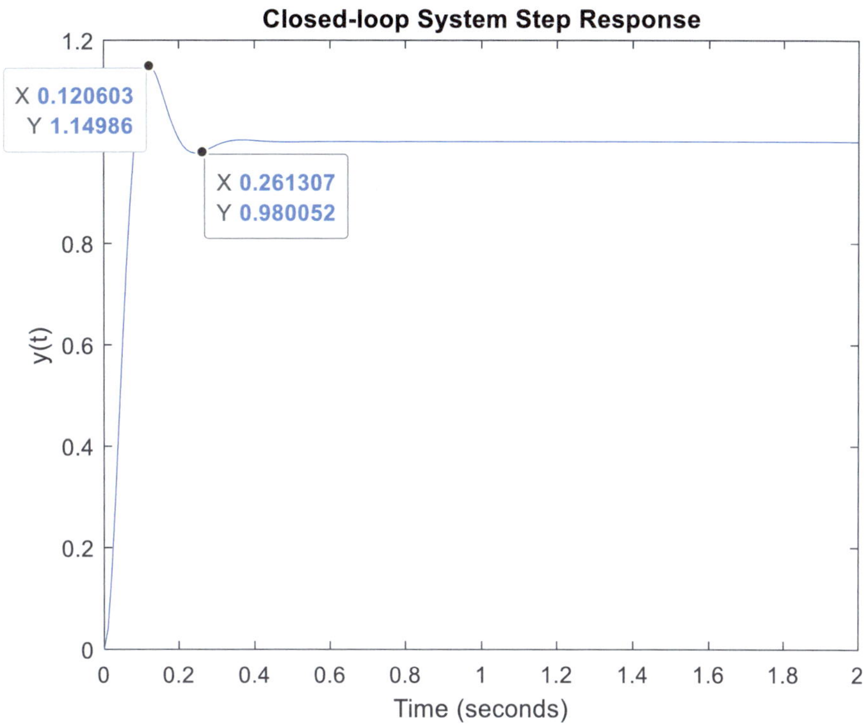

Fig. 14.4 Closed-loop system step response with the pole placement controller

14.3 Design of Full-Order Prediction Observers

It is necessary to have the full state vector available for realizing the full state feedback pole placement control algorithm presented in Sect. 14.2. However, very often in real-life situations, it may not be feasible or practical to measure the entire state vector. Instead, if the system is observable, as discussed in Sect. 13.3, it is possible to measure a few of the states and estimate the entire state vector using these measured values through an estimator or observer which is a mathematical model of the discrete-time system. The design of the full-order state estimator or observer is presented in this section.

The discrete-time system of order n with an input vector $\boldsymbol{u}(k)$ of size $p \times 1$, an output (measurement) vector $\boldsymbol{y}(k)$ of size $r \times 1$, can be represented in state-space form as in Eqs. (14.1) and (14.2) where the matrix $\boldsymbol{D}$ of size $r \times p$ is a null matrix, i.e., $\boldsymbol{D} = \boldsymbol{0}$.

System model:

$$\boldsymbol{x}(k+1) = \boldsymbol{A}\,\boldsymbol{x}(k) + \boldsymbol{B}\,\boldsymbol{u}(k) \tag{14.24}$$

$$\boldsymbol{y}(k) = \boldsymbol{C}\boldsymbol{x}(k) \tag{14.25}$$

It is assumed that system matrices , $\boldsymbol{B}$, $\boldsymbol{C}$ and measurement vector $\boldsymbol{y}(k)$ are known. The state vector $\boldsymbol{x}(k)$ can be estimated as $\boldsymbol{q}(k)$ using an observer or estimator model that is similar to the system model with system matrices , $\boldsymbol{B}$, $\boldsymbol{C}$ but with an additional gain matrix $\boldsymbol{L}$ of size $n \times r$ to minimize the error between the measurement vector $\boldsymbol{y}(k)$ and its estimate $\widehat{\boldsymbol{y}}(k)$.

Observer model:

$$\boldsymbol{q}(k+1) = \boldsymbol{A}\,\boldsymbol{q}(k) + \boldsymbol{B}\,\boldsymbol{u}(k) + \boldsymbol{L}(\boldsymbol{y}(k) - \widehat{\boldsymbol{y}}(k)) \tag{14.26}$$

$$\widehat{\boldsymbol{y}}(k) = \boldsymbol{C}\,\boldsymbol{q}(k) \tag{14.27}$$

The controller output is obtained, similar to Eq. (14.4), where the state vector $\boldsymbol{x}(k)$ is replaced with its estimate $\boldsymbol{q}(k)$ as

Control law:

$$\boldsymbol{u}(k) = -\boldsymbol{K}\,\boldsymbol{q}(k) \tag{14.28}$$

Equation (14.26) can be rewritten, in association with Eqs. (14.25) and (14.27), as

$$\boldsymbol{q}(k+1) = \boldsymbol{A}\,\boldsymbol{q}(k) + \boldsymbol{B}\,\boldsymbol{u}(k) + \boldsymbol{L}\boldsymbol{C}(\boldsymbol{x}(k) - \boldsymbol{q}(k)) \tag{14.29}$$

Error Dynamics

The difference between the actual state vector $\boldsymbol{x}(k)$ and its estimate $\boldsymbol{q}(k)$ can be written as

$$\boldsymbol{e}(k) = \boldsymbol{x}(k) - \boldsymbol{q}(k) \tag{14.30}$$

Equations (14.24) and (14.29) can be used to represent the error dynamics as

$$\begin{aligned}\boldsymbol{e}(k+1) &= \boldsymbol{x}(k+1) - \boldsymbol{q}(k+1) = \boldsymbol{A}\,\boldsymbol{x}(k) + \boldsymbol{B}\,\boldsymbol{u}(k) \\ &- [\boldsymbol{A}\,\boldsymbol{q}(k) + \boldsymbol{B}\,\boldsymbol{u}(k) + \boldsymbol{L}\boldsymbol{C}(\boldsymbol{x}(k) - \boldsymbol{q}(k))] = (\boldsymbol{A} - \boldsymbol{L}\,\boldsymbol{C})\,\boldsymbol{e}(k)\end{aligned} \tag{14.31}$$

Using z-transform, Eq. (14.31) can be written as

$$z\boldsymbol{E}(z) = (\boldsymbol{A} - \boldsymbol{L}\,\boldsymbol{C})\,\boldsymbol{E}(z) \tag{14.32}$$

Or

$$(z\boldsymbol{I} - (\boldsymbol{A} - \boldsymbol{L}\,\boldsymbol{C}))\,\boldsymbol{E}(z) = \boldsymbol{0} \tag{14.33}$$

The characteristic equation for error dynamics is obtained, for non-trivial solutions, as

$$|z\boldsymbol{I} - (\boldsymbol{A} - \boldsymbol{L}\,\boldsymbol{C})| = 0 \tag{14.34}$$

Observer Gain Matrix

Equation (14.29) can also be rewritten, with $\boldsymbol{u}(k)$ and $\boldsymbol{y}(k)$ as inputs, to describe observer dynamics as

$$\boldsymbol{q}(k+1) = (\boldsymbol{A} - \boldsymbol{LC}\,)\boldsymbol{q}(k) + \boldsymbol{B}\,\boldsymbol{u}(k) + \boldsymbol{L}\,\boldsymbol{y}(k) \tag{14.35}$$

The observer characteristic equation can be expressed, similar to the system characteristic Eq. (14.9), as

$$|z\boldsymbol{I} - (\boldsymbol{A} - \boldsymbol{L}\,\boldsymbol{C})| = 0 \tag{14.36}$$

which is the same as the characteristic equation of error dynamics Eq. (14.34). The characteristic equation of the observer dynamics can be expressed in terms of the closed-loop poles of the observer, p_{ei}, $i = 1, \ldots, n$ as

$$\lambda_e(z) = (z - p_{e1})(z - p_{e2}) \ldots (z - p_{en}) \tag{14.37}$$

The elements of the gain matrix $\boldsymbol{L}$ can be obtained by equating the characteristic polynomial, Eqs. (14.34) and (14.36), as

$$\lambda_e(z) = |z\boldsymbol{I} - (\boldsymbol{A} - \boldsymbol{L}\,\boldsymbol{C})| = (z - p_{e1})(z - p_{e2}) \ldots (z - p_{en}) \tag{14.38}$$

In Matlab, the observer gain matrix $\boldsymbol{L}$ can be obtained using the command *place*(), rewriting Eq. (14.34) or (14.36), after taking transpose, as

$$\left|z\boldsymbol{I} - \left(\boldsymbol{A}^T - \boldsymbol{C}^T\,\boldsymbol{L}^T\right)\right| = 0 \tag{14.39}$$

$$\boldsymbol{L}^T = place\left(\boldsymbol{A}^T, \boldsymbol{C}^T, \boldsymbol{p_e}\right) \tag{14.40}$$

where desired poles of the closed-loop discrete-time observer are represented in the vector $\boldsymbol{p_e}$ as

$$\boldsymbol{p_e} = [p_{e1}, p_{e2}, \ldots, p_{en}] \tag{14.41}$$

Combined Controller-Estimator Dynamics

The system-controller dynamics can be expressed by rewriting Eq. (14.24), in association with Eqs. (14.28) and (14.30), as

$$\boldsymbol{x}(k+1) = \boldsymbol{A}\,\boldsymbol{x}(k) + \boldsymbol{B}\,(-\,\boldsymbol{K}\,(\boldsymbol{x}(k) - \boldsymbol{e}(k))) \tag{14.42}$$

After some rearrangement, Eq. (14.42) can be written in terms of state vector $\boldsymbol{x}(k)$ and error vector $\boldsymbol{e}(k)$, as

$$\boldsymbol{x}(k+1) = (\boldsymbol{A} - \boldsymbol{B}\,\boldsymbol{K})\,\boldsymbol{x}(k) + \boldsymbol{B}\,\boldsymbol{K}\,\boldsymbol{e}(k) \tag{14.43}$$

Equation (14.43) can be combined with the error dynamics Eq. (14.31) to get combined controller-estimator dynamics as

$$\begin{bmatrix} \boldsymbol{x}(k+1) \\ \boldsymbol{e}(k+1) \end{bmatrix} = \begin{bmatrix} (\boldsymbol{A} - \boldsymbol{B}\,\boldsymbol{K}) & \boldsymbol{B}\,\boldsymbol{K} \\ \boldsymbol{0}_n & (\boldsymbol{A} - \boldsymbol{L}\,\boldsymbol{C}) \end{bmatrix} \begin{bmatrix} \boldsymbol{x}(k) \\ \boldsymbol{e}(k) \end{bmatrix} \tag{14.44}$$

where each submatrix is of size $n \times n$, with $\boldsymbol{0}_n$ being the null matrix of size $n \times n$.

Using z-transform, Eq. (14.44) can be expressed as

$$\begin{bmatrix} z\boldsymbol{I} - (\boldsymbol{A} - \boldsymbol{B}\,\boldsymbol{K}) & -\boldsymbol{B}\,\boldsymbol{K} \\ \boldsymbol{0}_n & z\boldsymbol{I} - (\boldsymbol{A} - \boldsymbol{L}\,\boldsymbol{C}) \end{bmatrix} \begin{bmatrix} \boldsymbol{X}(z) \\ \boldsymbol{E}(z) \end{bmatrix} = \begin{bmatrix} \boldsymbol{0}_n \\ \boldsymbol{0}_n \end{bmatrix} \tag{14.45}$$

The characteristic equation of the combined controller-estimator dynamics can be expressed as

$$\begin{vmatrix} z\boldsymbol{I} - (\boldsymbol{A} - \boldsymbol{B}\,\boldsymbol{K}) & -\boldsymbol{B}\,\boldsymbol{K} \\ \boldsymbol{0}_n & z\boldsymbol{I} - (\boldsymbol{A} - \boldsymbol{L}\,\boldsymbol{C}) \end{vmatrix} = 0 \tag{14.46}$$

Equation (14.46) can be rewritten as

$$|z\boldsymbol{I} - (\boldsymbol{A} - \boldsymbol{B}\,\boldsymbol{K})|\;\;|z\boldsymbol{I} - (\boldsymbol{A} - \boldsymbol{L}\,\boldsymbol{C})| = 0 \tag{14.47}$$

Equation (14.47) can be satisfied by each determinant as

$$|z\boldsymbol{I} - (\boldsymbol{A} - \boldsymbol{B}\,\boldsymbol{K})| = 0 \tag{14.48}$$

$$|z\boldsymbol{I} - (\boldsymbol{A} - \boldsymbol{L}\,\boldsymbol{C})| = 0 \tag{14.49}$$

Equation (14.48) is the same as that obtained in Eq. (14.9) as the characteristic equation of the system with the full state feedback control, and Eq. (14.49) is the same as that obtained in (14.36) as the characteristic equation of the observer. The implication of Eqs. (14.48) and (14.49) is that the controller gain matrix $\boldsymbol{K}$ and the observer gain matrix $\boldsymbol{L}$ can be calculated independently.

Representation of a System with Full-Order Prediction Observer

A system with a full-order prediction observer and pole placement controller, given by Eqs. (14.24)–(14.28), can be represented schematically in Fig. 14.5.

Following the procedure of Sect. 14.2, Eqs. (14.24), (14.25), (14.28), and (14.29) can be expressed in z-domain, using z-transform, respectively, as

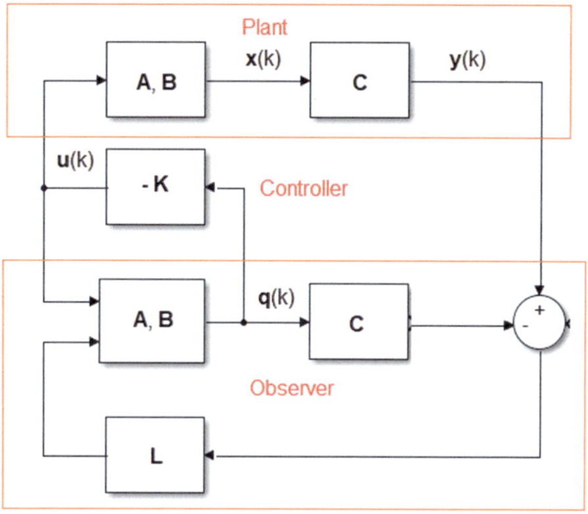

Fig. 14.5 Schematic diagram of a full-order prediction controller-observer system

$$X(z) = (zI - A)^{-1} B\, U(z) \tag{14.50}$$

$$Y(z) = CX(z) \tag{14.51}$$

$$U(z) = -KQ(z) \tag{14.52}$$

$$Q(z) = (zI - A)^{-1} [L(Y(z) - CQ(z)) + B\, U(z)] \tag{14.53}$$

Equations (14.50)–(14.53) can be used to represent the system with the full-order prediction observer and the feedback pole placement controller in the block diagram as shown in Fig. 14.6.

A Simulink model for the discrete-time system in state-space with full-order prediction observer and feedback pole placement controller, given in Eqs. (14.24), (14.25), (14.28), and (14.29), with a reference input vector $\boldsymbol{r}(k)$ of size $p \times 1$ is shown in Fig. 14.7. It should be noted that state vector $\boldsymbol{x}(k)$, the output vector $\boldsymbol{y}(k)$, and the input vector $\boldsymbol{u}(k)$ are of size $n \times 1$, $r \times 1$, and $p \times 1$, respectively. The matrices $\boldsymbol{a}$, $\boldsymbol{b}$, $\boldsymbol{c}$, $\boldsymbol{k}$, and $\boldsymbol{l}$ are the same as $\boldsymbol{A}$, $\boldsymbol{B}$, $\boldsymbol{C}$, $\boldsymbol{K}$, and $\boldsymbol{L}$, respectively, as defined in Eqs. (14.24), (14.25), (14.28), and (14.29).

Controller-Estimator Transfer Function

For the pole placement using the full-order prediction estimator, the input to the estimator is the measurement vector $\boldsymbol{y}(k)$ as shown in Eq. (14.35), and the output is the control vector $\boldsymbol{u}(k)$ as shown in Eq. (14.28). Equations (14.35) and (14.28) can be combined to get the estimator equation as

$$\boldsymbol{q}(k+1) = (\boldsymbol{A} - \boldsymbol{L}\,\boldsymbol{C})\boldsymbol{q}(k) + \boldsymbol{B}\,(-\boldsymbol{K}\,\boldsymbol{q}(k)) + \boldsymbol{L}\,\boldsymbol{y}(k) \tag{14.54}$$

Using z-transform, Eq. (14.54) can be expressed in z-domain, as

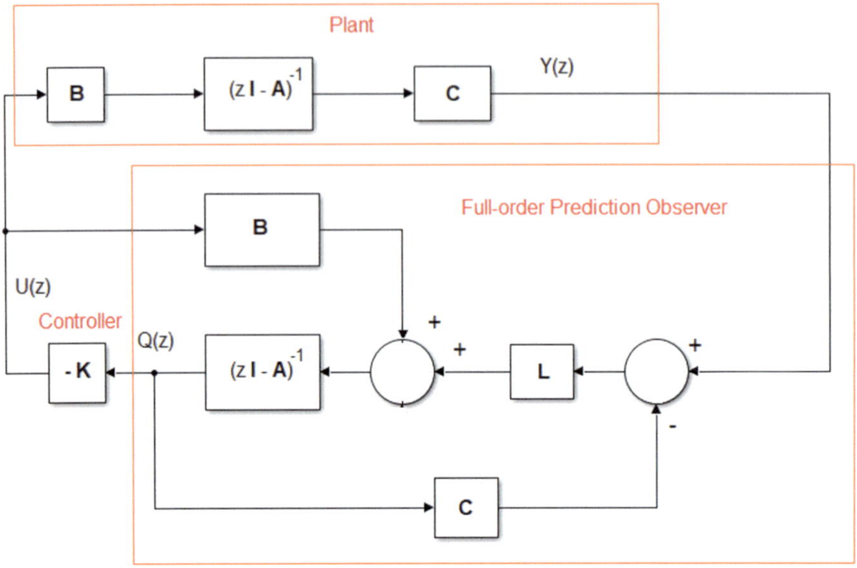

Fig. 14.6 Block diagram of a full-order prediction controller-observer system

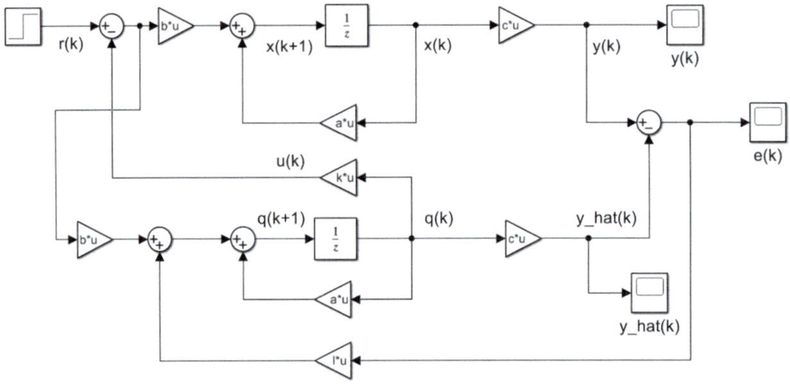

Fig. 14.7 Simulink model of a system with a full-order prediction controller-observer

$$z\boldsymbol{I}\,\boldsymbol{Q}(z) = (\boldsymbol{A} - \boldsymbol{B}\,\boldsymbol{K} - \boldsymbol{L}\,\boldsymbol{C})\boldsymbol{Q}(z) + \boldsymbol{L}\,\boldsymbol{Y}(z) \tag{14.55}$$

Equation (14.55) can be used to express $\boldsymbol{Q}(z)$ in terms of $\boldsymbol{Y}(z)$ as

$$\boldsymbol{Q}(z) = (z\boldsymbol{I} - (\boldsymbol{A} - \boldsymbol{B}\,\boldsymbol{K} - \boldsymbol{L}\,\boldsymbol{C}))^{-1}\boldsymbol{L}\,\boldsymbol{Y}(z) \tag{14.56}$$

Equation (14.28) can be expressed in the z-domain as

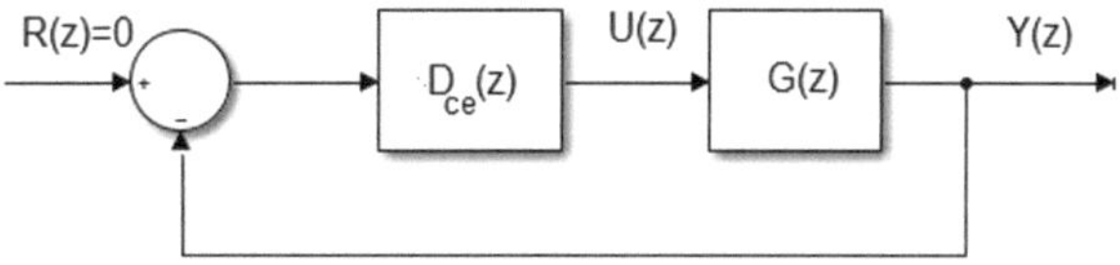

Fig. 14.8 Combined controller-estimator system

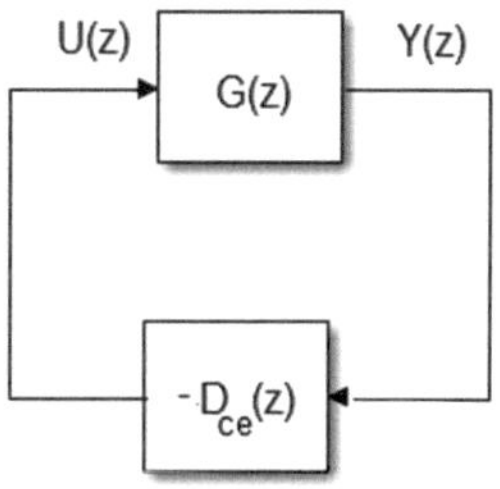

Fig. 14.9 Simplified controller-estimator system

$$\boldsymbol{U}(z) = -\boldsymbol{K}\,\boldsymbol{Q}(z) \tag{14.57}$$

Equations (14.56) and (14.57) can be combined to express $\boldsymbol{U}(z)$ in terms of $\boldsymbol{Y}(z)$ as

$$\boldsymbol{U}(z) = -\boldsymbol{K}(z\boldsymbol{I} - (\boldsymbol{A} - \boldsymbol{B}\,\boldsymbol{K} - \boldsymbol{L}\,\boldsymbol{C}\,))^{-1}\boldsymbol{L}\,\boldsymbol{Y}(z) \tag{14.58}$$

The controller-estimator transfer function can be written as

$$D_{ce}(z) = -\frac{\boldsymbol{U}(z)}{\boldsymbol{Y}(z)} = \boldsymbol{K}(z\boldsymbol{I} - (\boldsymbol{A} - \boldsymbol{B}\,\boldsymbol{K} - \boldsymbol{L}\,\boldsymbol{C}\,))^{-1}\boldsymbol{L} \tag{14.59}$$

Comparing Eq. (14.59) with the standard state-space representation of a discrete-time system, Eq. (13.58), the equivalent system matrices of the controller-estimator can be expressed as

$$\boldsymbol{A}_{ce} = \boldsymbol{A} - \boldsymbol{B}\,\boldsymbol{K} - \boldsymbol{L}\,\boldsymbol{C} \tag{14.60}$$

$$\boldsymbol{B}_{ce} = \boldsymbol{L} \tag{14.61}$$

$$\boldsymbol{C}_{ce} = \boldsymbol{K} \tag{14.62}$$

$$\boldsymbol{D}_{ce} = \boldsymbol{D} \tag{14.63}$$

The combined controller-estimator system can be represented in Fig. 14.8. The simplified version with reference input $R(z) = 0$ is shown in Fig. 14.9.

Example 14.2 The s-domain transfer function of a DC motor position dynamics is given as $G_p(s) = \frac{K}{s(\tau s+1)}$, $K = 30$, $\tau = 0.1$ s, $T = 0.01$ s. Obtain state-space representation from transfer function in continuous time, and convert from continuous-time to discrete-time domain. Design a full state feedback pole placement controller and full-order prediction observer such that the closed-loop system with the controller will have an overshoot of 15% and a settling time of 0.25 s when subjected to a step input. Verify from closed-loop system step response if the specifications are met.

Solution Matlab code and results are presented.

```
% Ch14-Ex2_Dc Motor pole placement controller with Full-order Observer
% State space model Controllability and observability
clear all; close all; clc
clear all; close all;clc
%G(s)=K/s(s+tau)
n=200;
K=30; %Static gain
tau=0.1; % time constant
T=0.01; % sampling time
ngs=K; dgs=[tau 1 0];
printsys(ngs,dgs)
%ss model [ac,bc,,c,d]
[ac,bc,c,d ]=tf2ss(ngs,dgs) % ss (s)
%ss model in discrete-time
[a,b]=c2d(ac,bc,T)    % ss (k)
% controllability
cont=[b a*b ]
cr=rank(cont)
%observabilty
obsr=[c;c*a]
or=rank(obsr)
% pole placement controller design
% desired pole locations in s-plane
alpha=0.15; tset=0.25; beta=0.02;
zeta=abs(log(alpha))/sqrt(pi^2+log(alpha)^2)
wn=-log(beta)/(zeta*tset)
s1=-zeta*wn+j*wn*sqrt(1-zeta^2);
s2=-zeta*wn-j*wn*sqrt(1-zeta^2);
%desired pole locations in z-plane
z1=exp(s1*T)
z2=exp(s2*T)
clp=[z1, z2];
k=place(a,b,clp) % controller gain matrix [K]
% closed loop response (with the state feedback only)
at=a-b*k;bt=b;ct=c;dt=d;
[zt,pt,kt]=ss2zp(at,bt,ct,dt)
dcg=ddcgain(at,bt,ct,dt);
yc=dstep(at,bt,ct,dt,1,n);
t=linspace(0,n*T,n);
yc1=yc/dcg;
figure
plot(t,yc1,'b-'),grid,hold on
title('Closed-loop System Step Response')
xlabel('Time (seconds)')
ylabel('y(t)')
% observer/estimator design
se1=2*s1;se2=2*s2;
ze1=exp(se1*T)
ze2=exp(se2*T)
cle=[ze1,ze2];

l=place(a',c',cle);l=l'
ak=a-b*k-l*c-l*d*k;bk=l;ck=k;
[tmp1,tmp2]=size(d);dk=zeros(tmp1,tmp2);
[zk,pk,kk]=ss2zp(ak,bk,ck,dk);
[nkz,dkz]=ss2tf(ak,bk,ck,dk) %Dce(z)
printsys(nkz,dkz,'z')
```

```
[agk,bgk,cgk,dgk]=series(a,b,c,d,ak,bk,ck,dk);
[at,bt,ct,dt]=feedback(a,b,c,d,ak,bk,ck,dk)

[zt,pt,kt]=ss2zp(at,bt,ct,dt)
dcg=ddcgain(at,bt,ct,dt); yc=dstep(at,bt,ct,dt,1,n);
yc=yc/dcg;
plot(t,yc,'r-'),grid
legend('without observer','with full-order observer')
% Augmented system dynamics
%X=[y x1 x2 xe1 xe2]
t=linspace(0,n*T,n)';
[ata,bta,cta,dta]=augstate(at,bt,ct,dt)
figure
x_e=dinitial(ata,bta,cta,dta,[1 1 -1 -1],n); % initial condition response
x=x_e(:,2:3);xe=x_e(:,4:5);
subplot(211)
plot(t,[x(:,1) xe(:,1)]); hold on
title('Closed-loop System Initial condition Response')
xlabel('Time (seconds)')
ylabel('x1(t)')
legend('x1','estimated x1')
subplot(212)
plot(t,[x(:,2) xe(:,2)]); hold on
title('Closed-loop System Initial condition Response')
xlabel('Time (seconds)')
ylabel('x2(t)')
legend('x2', 'estimated x2')

Results:
System matrices
a = 0.9048         0
    0.0095    1.0000

b = 0.0095
    0.0000
Desired poles in z-plane
zeta = 0.5169, wn = 30.2712, z1 = 0.8266 + 0.2191i, z2 = 0.8266 - 0.2191i

Controller gain matrix
k = [22.2725  820.5221]

Full-order observer poles
ze1 = 0.6352 + 0.3623i, ze2 = 0.6352 - 0.3623i

Observer gain matrix
l = 0.0714
    0.0021

Controller-estimator transfer function
         2.8767 z^2 - 2.1539 z
Dce(z)=  ------------------------
         z^2 - 1.0606 z + 0.43219
```

Closed-loop system step response and mismatched initial condition response are shown in Figs. 14.10 and 14.11 respectively.

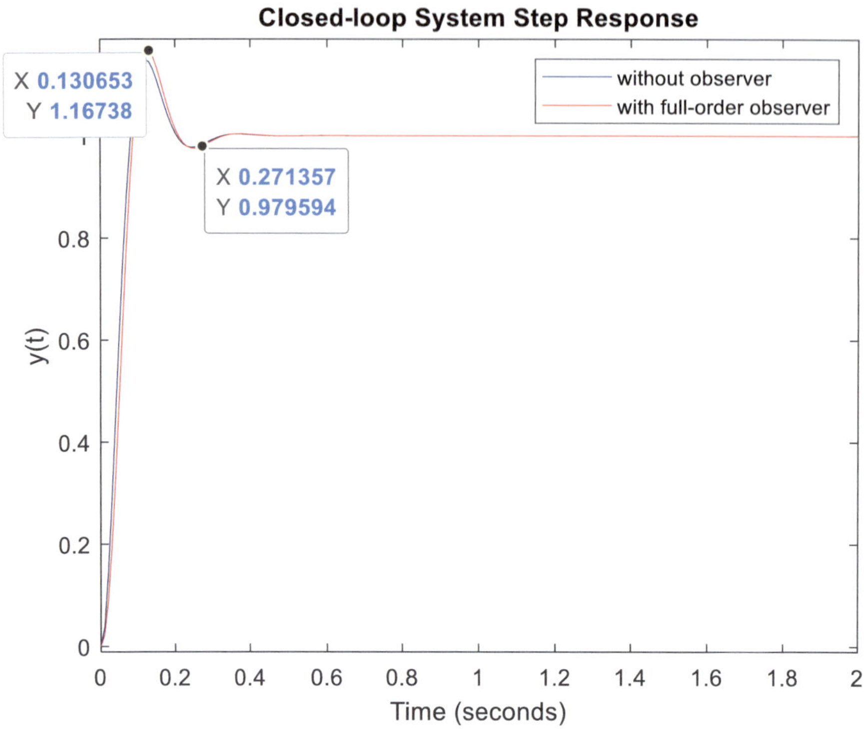

Fig. 14.10 Closed-loop system step response with the pole placement controller

14.4 Reduced-Order Observers

In reduced-order observers, some of the states $\boldsymbol{x}(k)$ are measured $\boldsymbol{x}_m(k)$, and the rest $\boldsymbol{x}_e(k)$ are estimated using the measured states. The measured and the estimated states are used in full state feedback control as in Sect. 14.2. The $n \times 1$ vector of states $\boldsymbol{x}(k)$ is partitioned into $m \times 1$ vector of measured states $\boldsymbol{x}_m(k)$ and $s \times 1$ vector of estimated states $\boldsymbol{x}_e(k)$ as

$$\boldsymbol{x}(k) = \begin{bmatrix} \boldsymbol{x}_m(k) \\ \boldsymbol{x}_e(k) \end{bmatrix} \tag{14.64}$$

The system model in state-space form, Eqs. (14.24) and (14.25), can be partitioned into measured and estimated states as

$$\begin{bmatrix} \boldsymbol{x}_m(k+1) \\ \boldsymbol{x}_e(k+1) \end{bmatrix} = \begin{bmatrix} \boldsymbol{A}_{mm} & \boldsymbol{A}_{me} \\ \boldsymbol{A}_{em} & \boldsymbol{A}_{ee} \end{bmatrix} \begin{bmatrix} \boldsymbol{x}_m(k) \\ \boldsymbol{x}_e(k) \end{bmatrix} + \begin{bmatrix} \boldsymbol{B}_m \\ \boldsymbol{B}_e \end{bmatrix} \boldsymbol{u}(k) \tag{14.65}$$

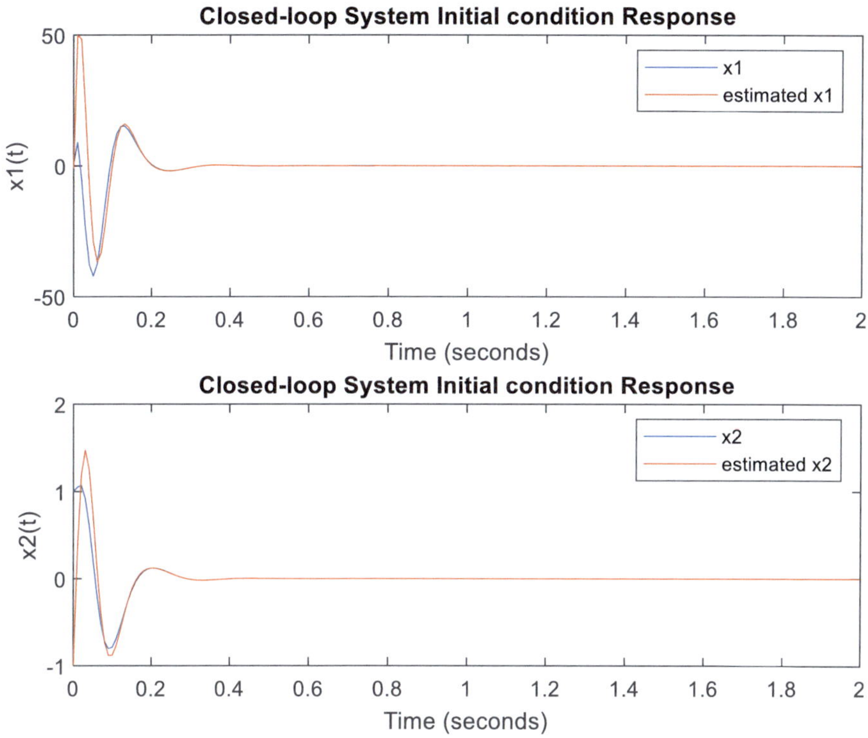

Fig. 14.11 Closed-loop system initial condition response with the pole placement controller

$$\boldsymbol{y}(k) = [\boldsymbol{I}_m \;\; \boldsymbol{0}_s] \begin{bmatrix} \boldsymbol{x}_m(k) \\ \boldsymbol{x}_e(k) \end{bmatrix} \tag{14.66}$$

where $\boldsymbol{I}_m$ and $\boldsymbol{0}_e$ are identity and null matrices of size $m \times m$ and $s \times s$, respectively.

From Eq. (14.65), the model for the measured states can be expressed in terms of known measured states $\boldsymbol{x}_m(k)$ and the input vector $\boldsymbol{u}(k)$ as

$$\boldsymbol{x}_m(k+1) - \boldsymbol{A}_{mm}\boldsymbol{x}_m(k) - \boldsymbol{B}_m\boldsymbol{u}(k) = \boldsymbol{A}_{me}\boldsymbol{x}_e(k) \tag{14.67}$$

The model for the estimated states can be expressed from Eq. (14.65) as

$$\boldsymbol{x}_e(k+1) = \boldsymbol{A}_{em}\boldsymbol{x}_m(k) + \boldsymbol{A}_{ee}\boldsymbol{x}_e(k) + \boldsymbol{B}_e\boldsymbol{u}(k) \tag{14.68}$$

Comparing the state equations of full-order observers and the reduced-order observers, Eqs. (14.24) and (14.68), (14.25), and (14.67), the following substitutions can be made to obtain reduced-order observers from full-order observers:

$$\boldsymbol{x}(k) \leftarrow \boldsymbol{x}_e(k)$$

$$\boldsymbol{A} \leftarrow \boldsymbol{A}_{ee}$$

$$\boldsymbol{B}\boldsymbol{u}(k) \leftarrow \boldsymbol{A}_{em}\boldsymbol{x}_m(k) + \boldsymbol{B}_e\boldsymbol{u}(k)$$

$$\boldsymbol{y}(k) \leftarrow \boldsymbol{x}_m(k+1) - \boldsymbol{A}_{mm}\boldsymbol{x}_m(k) - \boldsymbol{B}_m\boldsymbol{u}(k)$$

$$\boldsymbol{C} \leftarrow \boldsymbol{A}_{me}$$

Making these substitutions in the full-order observer dynamics, Eq. (14.35), the reduced-order observer dynamics can be obtained as

$$\begin{aligned}\boldsymbol{q}(k+1) = (\boldsymbol{A}_{ee} - \boldsymbol{L}\boldsymbol{A}_{me})\boldsymbol{q}(k) + \boldsymbol{A}_{em}\boldsymbol{x}_m(k) + \boldsymbol{B}_e\boldsymbol{u}(k) \\ + \boldsymbol{L}\left(\boldsymbol{x}_m(k+1) - \boldsymbol{A}_{mm}\boldsymbol{x}_m(k) - \boldsymbol{B}_m\boldsymbol{u}(k)\right)\end{aligned} \quad (14.69)$$

From Eq. (14.66)

$$\boldsymbol{y}(k) = \boldsymbol{x}_m(k) \quad (14.70)$$

Equation (14.69) can be expressed, in association with Eq. (14.70), as

$$\begin{aligned}\boldsymbol{q}(k+1) = (\boldsymbol{A}_{ee} - \boldsymbol{L}\boldsymbol{A}_{me})\boldsymbol{q}(k) + \boldsymbol{L}\,\boldsymbol{y}(k+1) + (\boldsymbol{A}_{em} - \boldsymbol{L}\,\boldsymbol{A}_{mm})\,\boldsymbol{y}(k) \\ + (\boldsymbol{B}_e - \boldsymbol{L}\boldsymbol{B}_m)\boldsymbol{u}(k)\end{aligned} \quad (14.71)$$

For the pole placement using the reduced-order estimator, the controller output vector $\boldsymbol{u}(k)$ is given as

$$\boldsymbol{u}(k) = -[\boldsymbol{K}_m \quad \boldsymbol{K}_e]\begin{bmatrix} \boldsymbol{y}(k) \\ \boldsymbol{q}(k) \end{bmatrix} \quad (14.72)$$

Observer Gain Matrix

From Eq. (14.71), the characteristic equation for the reduced-order observer can be expressed as Eq. (14.72) which is similar to that for the full-observer characteristic equation, Eq. (14.36):

$$|z\boldsymbol{I}_m - (\boldsymbol{A}_{ee} - \boldsymbol{L}\boldsymbol{A}_{me})| = 0 \quad (14.73)$$

Equation (14.72) can be used to get the gain matrix of the reduced-order observer.

In Matlab, the observer gain matrix $\boldsymbol{L}$ can be obtained using the command place, rewriting Eq. (14.73) after taking transpose, as

$$\left|z\boldsymbol{I} - \left(\boldsymbol{A}_{ee}{}^{T} - \boldsymbol{A}_{me}{}^{T}\,\boldsymbol{L}^{T}\right)\right| = 0 \quad (14.74)$$

$$\boldsymbol{L}^T = place\left(\boldsymbol{A_{ee}}^T, \boldsymbol{A_{me}}^T, \boldsymbol{p_{er}}\right) \tag{14.75}$$

where desired poles of the closed-loop discrete-time observer are represented in the vector $\boldsymbol{p_{er}}$ as

$$\boldsymbol{p_e} = [p_{er1}, p_{er2}, \ldots, p_{ers}] \tag{14.76}$$

Representation of a System with Reduced-Order Observer

Following the procedure of Sect. 14.3, Eqs. (14.65), (14.66), (14.71), and (14.72) can be expressed in z-domain, using z-transform, respectively, as

$$\boldsymbol{X}(z) = (z\boldsymbol{I} - \boldsymbol{A})^{-1}\boldsymbol{B}\,\boldsymbol{U}(z) \tag{14.77}$$

$$\boldsymbol{Y}(z) = [\boldsymbol{I}_m \quad \boldsymbol{0}_s]\boldsymbol{X}(z) \tag{14.78}$$

$$\boldsymbol{U}(z) = -[\boldsymbol{K}_m \quad \boldsymbol{K}_e]\begin{bmatrix} \boldsymbol{Y}(z) \\ \boldsymbol{Q}(z) \end{bmatrix} \tag{14.79}$$

$$\begin{aligned} \boldsymbol{Q}(z) = (z\boldsymbol{I} - \boldsymbol{A_{ee}})^{-1}[&\boldsymbol{L}(z\boldsymbol{Y}(z) - \boldsymbol{A_{me}}\,\boldsymbol{Q}(z)) + (\boldsymbol{A}_{em} - \boldsymbol{L}\,\boldsymbol{A}_{mm})\boldsymbol{Y}(z) \\ &+ (\boldsymbol{B}_e - \boldsymbol{L}\boldsymbol{B}_m)\,\boldsymbol{U}(z)] \end{aligned} \tag{14.80}$$

Equations (14.77)–(14.80) can be used to represent the system with the reduced-order observer and the feedback pole placement controller in the block diagram as shown in Fig. 14.12.

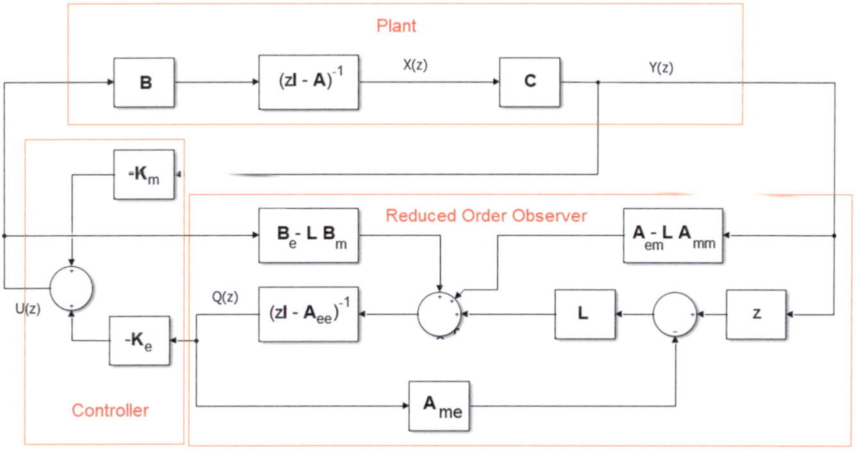

Fig. 14.12 Block diagram of a system with a reduced-order observer and a feedback pole placement controller

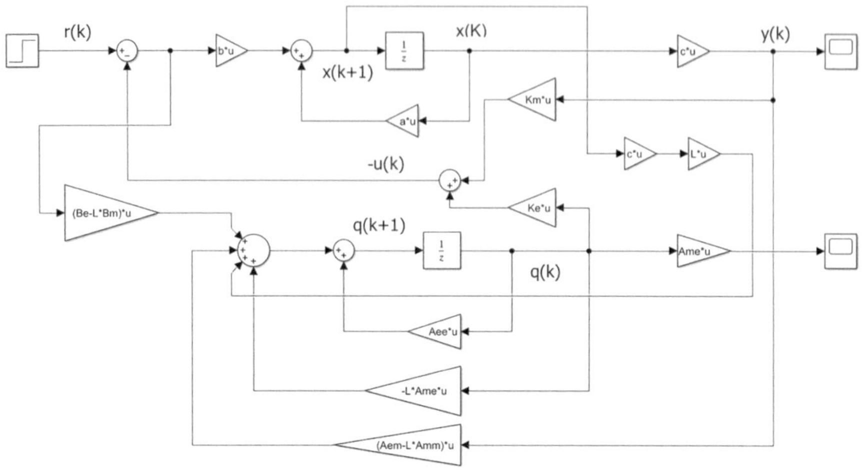

Fig. 14.13 Simulink model of a system with a reduced-order observer and a feedback pole placement controller

A Simulink model for the discrete-time system in state space with reduced-order observer and feedback pole placement controller, given in Eqs. (14.65), (14.66), (14.71), and (14.72), with a reference input vector $\boldsymbol{r}(k)$ of size $p \times 1$ is shown in Fig. 14.13. It should be noted that state vector $\boldsymbol{x}(k)$, the output vector $\boldsymbol{y}(k)$, and the input vector $\boldsymbol{u}(k)$ are of size $n \times 1$, $r \times 1$, and $p \times 1$, respectively. The matrices $\boldsymbol{a}$, $\boldsymbol{b}$, $\boldsymbol{c}$, $\boldsymbol{k}$, and $\boldsymbol{l}$ are the same as $\boldsymbol{A}$, $\boldsymbol{B}$, $\boldsymbol{C}$, $\boldsymbol{K}$, and $\boldsymbol{L}$, respectively, as defined in Eqs. (14.65), (14.66), (14.71), and (14.72).

Controller-Estimator Transfer Function For the pole placement using the reduced-order, estimator the input to the estimator is the measurement vector $\boldsymbol{y}(k)$ as shown in Eq. (14.71), and the output is the control vector $\boldsymbol{u}(k)$ as shown in Eq. (14.73):

$$\boldsymbol{u}(k) = -[\boldsymbol{K}_m \quad \boldsymbol{K}_e] \begin{bmatrix} \boldsymbol{y}(k) \\ \boldsymbol{q}(k) \end{bmatrix} \tag{14.81}$$

Equations (14.71) and (14.72) can be combined to get the estimator equation as

$$\begin{aligned} \boldsymbol{q}(k+1) &= (\boldsymbol{A}_{ee} - \boldsymbol{L}\boldsymbol{A}_{me})\boldsymbol{q}(k) + \boldsymbol{L}\,\boldsymbol{y}(k+1) + (\boldsymbol{A}_{em} - \boldsymbol{L}\,\boldsymbol{A}_{mm})\,\boldsymbol{y}(k) \\ &+ (\boldsymbol{B}_e - \boldsymbol{L}\boldsymbol{B}_m)\left(-[\boldsymbol{K}_m \quad \boldsymbol{K}_e] \begin{bmatrix} \boldsymbol{y}(k) \\ \boldsymbol{q}(k) \end{bmatrix}\right) \end{aligned} \tag{14.82}$$

Using z-transform, Eq. (14.82) can be expressed in z-domain, as

$$z\boldsymbol{I_e}\,\boldsymbol{Q}(z) = (\boldsymbol{A}_{ee} - \boldsymbol{L}\boldsymbol{A}_{me} - (\boldsymbol{B}_e - \boldsymbol{L}\,\boldsymbol{B}_m)\boldsymbol{K}_e)\boldsymbol{Q}(z) \\ + (z\boldsymbol{L} + \boldsymbol{A}_{em} - \boldsymbol{L}\,\boldsymbol{A}_{mm} - (\boldsymbol{B}_e - \boldsymbol{L}\boldsymbol{B}_m)\boldsymbol{K}_m)\,\boldsymbol{Y}(z) \tag{14.83}$$

Equation (14.83) can be used to express $\boldsymbol{Q_e}(z)$ in terms of $\boldsymbol{Y}(z)$ as

$$\boldsymbol{Q}(z) = (z\boldsymbol{I_e} - (\boldsymbol{A}_{ee} - \boldsymbol{L}\boldsymbol{A}_{me} - (\boldsymbol{B}_e - \boldsymbol{L}\,\boldsymbol{B}_m)\boldsymbol{K}_e))^{-1}(z\boldsymbol{L} + \boldsymbol{A}_{em} - \boldsymbol{L}\,\boldsymbol{A}_{mm} - (\boldsymbol{B}_e - \boldsymbol{L}\boldsymbol{B}_m)\boldsymbol{K}_m)\boldsymbol{Y}(z) \tag{14.84}$$

Equation (14.72) can be expressed in z-domain as

$$\boldsymbol{U}(z) = -\boldsymbol{K}_m\boldsymbol{Y}(z) - \boldsymbol{K}_e\,\boldsymbol{Q}(z) \tag{14.85}$$

Equations (14.84) and (14.85) can be combined to express $\boldsymbol{U}(z)$ in terms of $\boldsymbol{Y}(z)$ as

$$U(z) = -\boldsymbol{K}_m\boldsymbol{Y}(z) - \boldsymbol{K}_e\,(z\boldsymbol{I_e} - (\boldsymbol{A}_{ee} - \boldsymbol{L}\boldsymbol{A}_{me} - (\boldsymbol{B}_e - \boldsymbol{L}\,\boldsymbol{B}_m)\boldsymbol{K}_e))^{-1}(z\boldsymbol{L} + \boldsymbol{A}_{em} - \boldsymbol{L}\,\boldsymbol{A}_{mm} - (\boldsymbol{B}_e - \boldsymbol{L}\boldsymbol{B}_m)\boldsymbol{K}_m)\boldsymbol{Y}(z) \tag{14.86}$$

The controller-estimator transfer function can be written as

$$\begin{aligned} D_{ce}(z) &= -\frac{\boldsymbol{U}(z)}{\boldsymbol{Y}(z)} \\ &= \boldsymbol{K}_m + \boldsymbol{K}_e\,(z\boldsymbol{I_e} - (\boldsymbol{A}_{ee} - \boldsymbol{L}\boldsymbol{A}_{me} - (\boldsymbol{B}_e - \boldsymbol{L}\,\boldsymbol{B}_m)\boldsymbol{K}_e))^{-1}(z\boldsymbol{L} + \boldsymbol{A}_{em} - \boldsymbol{L}\,\boldsymbol{A}_{mm} - (\boldsymbol{B}_e - \boldsymbol{L}\boldsymbol{B}_m)\boldsymbol{K}_m) \end{aligned} \tag{14.87}$$

Example 14.3 The s-domain transfer function of a DC motor position dynamics is given as $G_p(s) = \frac{K}{s(\tau s+1)}$, $K = 1$, $\tau = 1$ s, $T = 0.1$ s. Obtain state-space representation from transfer function in continuous time, and convert from continuous-time to discrete-time domain. Design a full state feedback pole placement controller and a reduced-order observer such that the closed-loop system with the controller will have an overshoot of 10% and a settling time of 2.0 s when subjected to a step input. Verify from closed-loop system step response if the specifications are met.

Solution Matlab code and results are presented.

```
% Ch14-Ex3_Dc Motor pole placement controller with Reduced-order Observer
clear all; close all; clc
%G(s)=K/s(tau*s+1)
K=1; %Static gain
tau=1; % time constant
T=0.1; % sampling time
ngs=K; dgs=[tau 1 0];
printsys(ngs,dgs)
Gp = tf(ngs,dgs)
%ss model [ac,bc,c,d]
[ac,bc,c,d ]=tf2ss(ngs,dgs) % ss (s)
%ss model in discrete-time
[a1,b1]=c2d(ac,bc,T)   % ss (k)
a=[a1(2,2) a1(2,1);
    a1(1,2) a1(1,1)]
b=[b1(2);
   b1(1)]
c=[c(2) c(1)]
% controllability
cont=[b a*b ]
cr=rank(cont)
%observabilty
obsr=[c;c*a]
or=rank(obsr)
% pole placement controller design
% desired pole locations in s-plane
alpha=0.10; tset=2; beta=0.02;
zeta=abs(log(alpha))/sqrt(pi^2+log(alpha)^2)
wn=-log(beta)/(zeta*tset)
s1=-zeta*wn+j*wn*sqrt(1-zeta^2);
s2=-zeta*wn-j*wn*sqrt(1-zeta^2);
%desired pole locations in z-plane
z1=exp(s1*T)
z2=exp(s2*T)
% z1=0.8893+i*0.1668;
% z2=0.8893-i*0.1668;
p=[z1, z2];
k=place(a,b,p) % controller gain matrix [K]
n = 2; p=1; r=1; m=1; e=1;
Km = k(1:m)
Ke = k(m+1:n)
% closed loop response (with the state feedback only)
at=a-b*k;bt=b;ct=c;dt=d;
[zt,pt,kt]=ss2zp(at,bt,ct,dt)
dcg=ddcgain(at,bt,ct,dt);
N=100;
yc=dstep(at,bt,ct,dt,1,N);
yc1=yc/dcg;
t=linspace(0,N*T,N);
figure
plot(t,yc1,'b-'), hold on
title('Closed-loop System Step Response')
xlabel('Time (seconds)')
ylabel('y(t)')
% reduced-order observer/estimator design
% Get Amm Ame Aem Aee
%n = 2; p=1; r=1; m=1; e=1;
```

```
Amm = a(1:m,1:m)
Ame = a(1:m,m+1:n)
Aem = a(m+1:n,1:m)
Aee = a(m+1:n,m+1:n)
Bm  = b(1:m)
Be  = b(m+1:n)

se1=1*s1;se2=1*s2;
ze1=exp(se1*T)
ze2=exp(se2*T)
pe=[ze1,ze2];
pr =(real(ze1))^2
l=place(Aee',Ame',pr);l=l';
L=l % observer gain matrix
% controller-observer (reduced-order) system
syms z
C1 = z*eye(e)-Aee+L*Ame+(Be-L*Bm)*Ke;
C2 = L*z+(Aem-L*Amm-Km*(Be-L*Bm));
Dce = (Km *C1+Ke*C2)/C1;
Dce = vpa(Dce, 4)
nzk = [17.74 -14.24]
dzk = [1  -0.3966]
Dz  = tf(nzk, dzk,T)
Gz  = c2d(Gp,T)
DGz = Dz*Gz
Tz  = feedback(DGz,1)
zpk(Tz)
figure
ycr=dstep(Tz,1);
nn=length(ycr)
tt=linspace(0,nn*T,nn);
plot(tt,yc1(1:nn),'b-',tt,ycr,'r-')
title('Closed-loop System Step Response')
xlabel('Time (seconds)')
ylabel('y(t)')
legend('without observer','with reduced-order observer')

Results:
a = 1.0000    0.0952
         0    0.9048

b = 0.0048
    0.0952

c = 1     0    d = 0
Desired pole locations
zeta = 0.5912, wn = 3.3088, z1 = 0.7932 + 0.2169i, z2 = 0.7932 - 0.2169i

controller gain matrix
k = 9.4349    2.8660
Reduced-order observer pole
pr = 0.6292

Reduced-order observer gain
L = 2.8964

Controller-estimator (reduced-order) transfer function

      17.74*z - 14.24
Dce = ----------------
       z - 0.3966
```

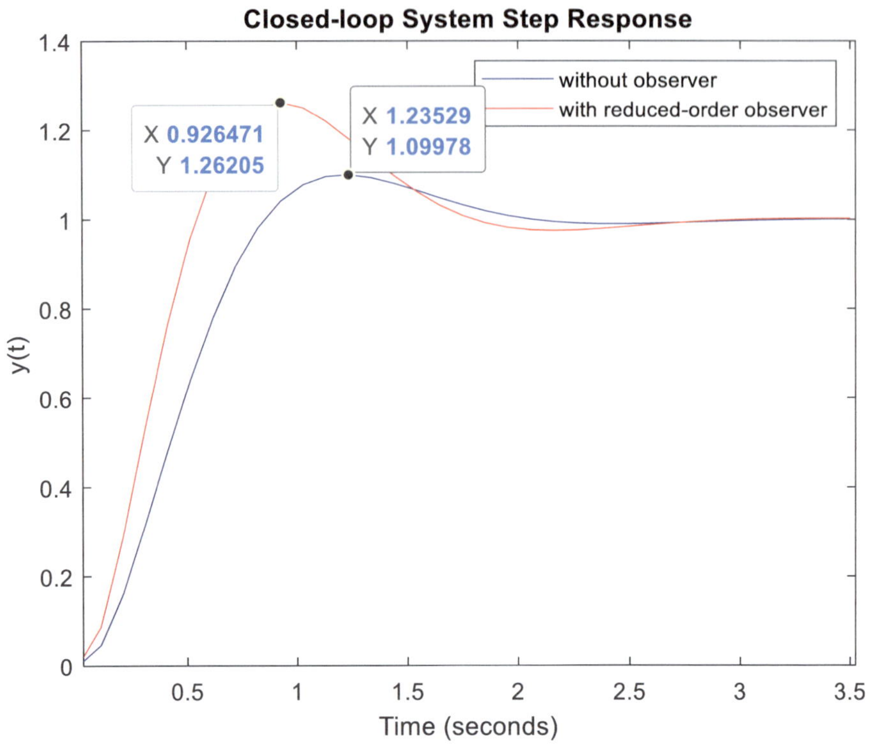

Fig. 14.14 Closed-loop system step response with the pole placement controller

The closed-loop step response with the reduced-order observer is shown in Fig. 14.14.

14.5 Design of Current Observers

The full-order observers presented in Sect. 14.3 are known as prediction observers as the state vector $\boldsymbol{x}(k + 1)$ is predicted using the measurement vector $\boldsymbol{y}(k)$. In this section, the design of full-order current observers is presented where state vector $\boldsymbol{x}(k)$ is estimated using the measurement vector $\boldsymbol{y}(k)$.

The discrete-time system of order n with an input vector $\boldsymbol{u}(k)$ of size $p \times 1$ and an output (measurement) vector $\boldsymbol{y}(k)$ of size $r \times 1$ can be represented in state-space form as in Eqs. (14.24) and (14.25) where the matrix $\boldsymbol{D}$ of size $r \times p$ is a null matrix, i.e., $\boldsymbol{D} = \boldsymbol{0}$.

System model (Eqs. (14.24) and (14.25) are repeated here for easy reference):

$$\boldsymbol{x}(k+1) = \boldsymbol{A}\,\boldsymbol{x}(k) + \boldsymbol{B}\,\boldsymbol{u}(k) \tag{14.20}$$

$$\boldsymbol{y}(k) = \boldsymbol{C}\boldsymbol{x}(k) \tag{14.21}$$

The current observer estimates the state vector in two steps: first, a first estimate of the state vector, $\overline{\boldsymbol{q}}(k+1)$, is obtained using the system matrices $\boldsymbol{A}$ and $\boldsymbol{B}$, and the estimate is updated to $\boldsymbol{q}(k + 1)$ with the current measurement $\boldsymbol{y}(k + 1)$ as

$$\overline{\boldsymbol{q}}(k+1) = \boldsymbol{A}\,\boldsymbol{q}(k) + \boldsymbol{B}\,\boldsymbol{u}(k) \tag{14.88}$$

$$\boldsymbol{q}(k+1) = \overline{\boldsymbol{q}}(k+1) + \boldsymbol{L}(\boldsymbol{y}(k+1) - \boldsymbol{C}\overline{\boldsymbol{q}}(k+1)) \tag{14.89}$$

The observer model Eq. (14.89) can be rewritten using Eq. (14.88) and system dynamics, Eqs. (14.24) and (14.25), as

$$\boldsymbol{q}(k+1) = \boldsymbol{A}\,\boldsymbol{q}(k) + \boldsymbol{B}\,\boldsymbol{u}(k) + \boldsymbol{L}(\boldsymbol{y}(k+1) - \boldsymbol{C}(\boldsymbol{A}\,\boldsymbol{q}(k) + \boldsymbol{B}\,\boldsymbol{u}(k))) \tag{14.90}$$

Equation (14.90) can be rewritten, after some rearrangements, as

$$\boldsymbol{q}(k+1) = (\boldsymbol{A} - \boldsymbol{LCA})\boldsymbol{q}(k) + (\boldsymbol{B} - \boldsymbol{LCB})\boldsymbol{u}(k) + \boldsymbol{L}\boldsymbol{y}(k+1) \tag{14.91}$$

The controller output is obtained similar to Eq. (14.4), where the state vector $\boldsymbol{x}(k)$ is replaced with its estimate $\boldsymbol{q}(k)$, as in Eq. (14.28)

$$\boldsymbol{u}(k) = -\boldsymbol{K}\,\boldsymbol{q}(k) \tag{14.92}$$

Error Dynamics

The difference between the actual state vector $\boldsymbol{x}(k)$ and its estimate $\boldsymbol{q}(k)$ can be written as the error vector, $\boldsymbol{e}(k)$, defined in Eq. (14.30). Equations (14.24) and (14.91) can be used to represent the error dynamics as

$$\begin{aligned}\boldsymbol{e}(k+1) &= \boldsymbol{x}(k+1) - \boldsymbol{q}(k+1) = \boldsymbol{A}\,\boldsymbol{x}(k) + \boldsymbol{B}\,\boldsymbol{u}(k)\\ &\quad - [(\boldsymbol{A} - \boldsymbol{LCA})\boldsymbol{q}(k) + (\boldsymbol{B} - \boldsymbol{LCB})\boldsymbol{u}(k) + \boldsymbol{LC}(\boldsymbol{A}\,\boldsymbol{x}(k) + \boldsymbol{B}\,\boldsymbol{u}(k))]\\ &= (\boldsymbol{A} - \boldsymbol{L}\,\boldsymbol{C}\,\boldsymbol{A})\,\boldsymbol{e}(k)\end{aligned} \tag{14.93}$$

Using z-transform, Eq. (14.93) can be written as

$$z\boldsymbol{E}(z) = (\boldsymbol{A} - \boldsymbol{L}\,\boldsymbol{C}\,\boldsymbol{A})\,\boldsymbol{E}(z) \tag{14.94}$$

Or

$$(z\boldsymbol{I} - (\boldsymbol{A} - \boldsymbol{L}\,\boldsymbol{C}\,\boldsymbol{A}))\,\boldsymbol{E}(z) = \boldsymbol{0} \tag{14.95}$$

The characteristic equation for error dynamics is obtained, for non-trivial solutions, as

$$|z\boldsymbol{I} - (\boldsymbol{A} - \boldsymbol{L}\,\boldsymbol{C}\,\boldsymbol{A})| = 0 \tag{14.96}$$

Current Observer Gain Matrix

The observer characteristic equation can be expressed, similar to the full-order prediction observer characteristic Eq. (14.34), as

$$|z\boldsymbol{I} - (\boldsymbol{A} - \boldsymbol{L}\,\boldsymbol{C}\,\boldsymbol{A})| = 0 \tag{14.97}$$

which is the same as the characteristic equation of error dynamics Eq. (14.96). The characteristic equation of the current observer dynamics can be expressed in terms of the closed-loop poles of the current observer, p_{coi}, $i = 1, \ldots, n$ as

$$\lambda_{co}(z) = (z - p_{co1})(z - p_{co2}) \ldots (z - p_{con}) \tag{14.98}$$

The elements of the gain matrix $\boldsymbol{L}$ can be obtained by equating the characteristic polynomial, Eqs. (14.97) and (14.98), as

$$\lambda_{co}(z) = |z\boldsymbol{I} - (\boldsymbol{A} - \boldsymbol{L}\,\boldsymbol{C}\,\boldsymbol{A})| = (z - p_{co1})(z - p_{co2}) \ldots (z - p_{con}) \tag{14.99}$$

In Matlab, the observer gain matrix $\boldsymbol{L}$ can be obtained using the command *place*(), rewriting Eq. (14.97) after taking transpose, as

$$\left|z\boldsymbol{I} - \left(\boldsymbol{A}^T - \boldsymbol{A}^T\,\boldsymbol{C}^T\,\boldsymbol{L}^T\right)\right| = 0 \tag{14.100}$$

$$\boldsymbol{L}^T = place\left(\boldsymbol{A}^T, \boldsymbol{A}^T\boldsymbol{C}^T, \boldsymbol{p_{co}}\right) \tag{14.101}$$

where desired poles of the closed-loop discrete-time current observer are represented in the vector $\boldsymbol{p_{co}}$ as

$$\boldsymbol{p_{co}} = [p_{co1}, p_{co2}, \ldots, p_{con}] \tag{14.102}$$

Combined Controller-Estimator Dynamics

The system-controller dynamics can be expressed by rewriting Eq. (14.20), in association with Eqs. (14.92) and (14.30), similar to Eq. (14.37), as

$$\boldsymbol{x}(k+1) = \boldsymbol{A}\,\boldsymbol{x}(k) + \boldsymbol{B}\,(-\boldsymbol{K}\,(\boldsymbol{x}(k) - \boldsymbol{e}(k))) \tag{14.103}$$

After some rearrangement, Eq. (14.103) can be written in terms of state vector $\boldsymbol{x}(k)$ and error vector $\boldsymbol{e}(k)$, similar to Eq. (14.38), as

$$\boldsymbol{x}(k+1) = (\boldsymbol{A} - \boldsymbol{B}\,\boldsymbol{K})\,\boldsymbol{x}(k) + \boldsymbol{B}\,\boldsymbol{K}\,\boldsymbol{e}(k) \tag{14.104}$$

Equation (14.104) can be combined with the error dynamics Eq. (14.93) to get combined current controller-estimator dynamics as

$$\begin{bmatrix} \boldsymbol{x}(k+1) \\ \boldsymbol{e}(k+1) \end{bmatrix} = \begin{bmatrix} (\boldsymbol{A}-\boldsymbol{B}\,\boldsymbol{K}) & \boldsymbol{B}\,\boldsymbol{K} \\ \boldsymbol{0}_n & (\boldsymbol{A}-\boldsymbol{L}\,\boldsymbol{C}\,\boldsymbol{A}) \end{bmatrix} \begin{bmatrix} \boldsymbol{x}(k) \\ \boldsymbol{e}(k) \end{bmatrix} \tag{14.105}$$

where each submatrix is of size $n \times n$, with $\boldsymbol{0}_n$ as the null matrix of size $n \times n$.

Using z-transform, Eq. (14.105) can be expressed as

$$\begin{bmatrix} z\boldsymbol{I}-(\boldsymbol{A}-\boldsymbol{B}\,\boldsymbol{K}) & -\boldsymbol{B}\,\boldsymbol{K} \\ \boldsymbol{0}_n & z\boldsymbol{I}-(\boldsymbol{A}-\boldsymbol{L}\,\boldsymbol{C}\,\boldsymbol{A}) \end{bmatrix} \begin{bmatrix} \boldsymbol{X}(z) \\ \boldsymbol{E}(z) \end{bmatrix} = \begin{bmatrix} \boldsymbol{0}_n \\ \boldsymbol{0}_n \end{bmatrix} \tag{14.106}$$

The characteristic equation of the combined current controller-estimator dynamics can be expressed as

$$\begin{vmatrix} z\boldsymbol{I}-(\boldsymbol{A}-\boldsymbol{B}\,\boldsymbol{K}) & -\boldsymbol{B}\,\boldsymbol{K} \\ \boldsymbol{0}_n & z\boldsymbol{I}-(\boldsymbol{A}-\boldsymbol{L}\,\boldsymbol{C}\,\boldsymbol{A}) \end{vmatrix} = 0 \tag{14.107}$$

Equation (14.107) can be rewritten as

$$|z\boldsymbol{I}-(\boldsymbol{A}-\boldsymbol{B}\,\boldsymbol{K})|\;|z\boldsymbol{I}-(\boldsymbol{A}-\boldsymbol{L}\,\boldsymbol{C}\,\boldsymbol{A})| = 0 \tag{14.108}$$

Equation (14.108) can be satisfied by each determinant as

$$|z\boldsymbol{I}-(\boldsymbol{A}-\boldsymbol{B}\,\boldsymbol{K})| = 0 \tag{14.109}$$

$$|z\boldsymbol{I}-(\boldsymbol{A}-\boldsymbol{L}\,\boldsymbol{C}\,\boldsymbol{A})| = 0 \tag{14.110}$$

Equation (14.109) is the same as that obtained in Eq. (14.9) as the characteristic equation of the system with full state feedback and Eq. (14.110) is the same as that obtained in Eq. (14.97) as the characteristic equation of the current observer. The implication of Eqs. (14.109) and (14.110) is that the controller gain matrix $\boldsymbol{K}$ and the observer gain matrix $\boldsymbol{L}$ can be calculated independently, as for full-order prediction observers.

Representation of a System with Current Observer

Following the procedure of Sect. 14.3, Eqs. (14.20), (14.21), (14.91), and (14.92) can be expressed in z-domain, using z-transform, respectively, as

$$\boldsymbol{X}(z) = (z\boldsymbol{I}-\boldsymbol{A})^{-1}\boldsymbol{B}\,\boldsymbol{U}(z) \tag{14.111}$$

$$\boldsymbol{Y}(z) = \boldsymbol{C}\,\boldsymbol{X}(z) \tag{14.112}$$

$$\boldsymbol{U}(z) = -\boldsymbol{K}\boldsymbol{Q}(z) \tag{14.113}$$

$$\boldsymbol{Q}(z) = (z\boldsymbol{I}-\boldsymbol{A})^{-1}[\boldsymbol{L}(z\boldsymbol{Y}(z)-\boldsymbol{C}\boldsymbol{A}\,\boldsymbol{Q}(z)) + (\boldsymbol{B}-\boldsymbol{L}\boldsymbol{C}\boldsymbol{B})\,\boldsymbol{U}(z)] \tag{14.114}$$

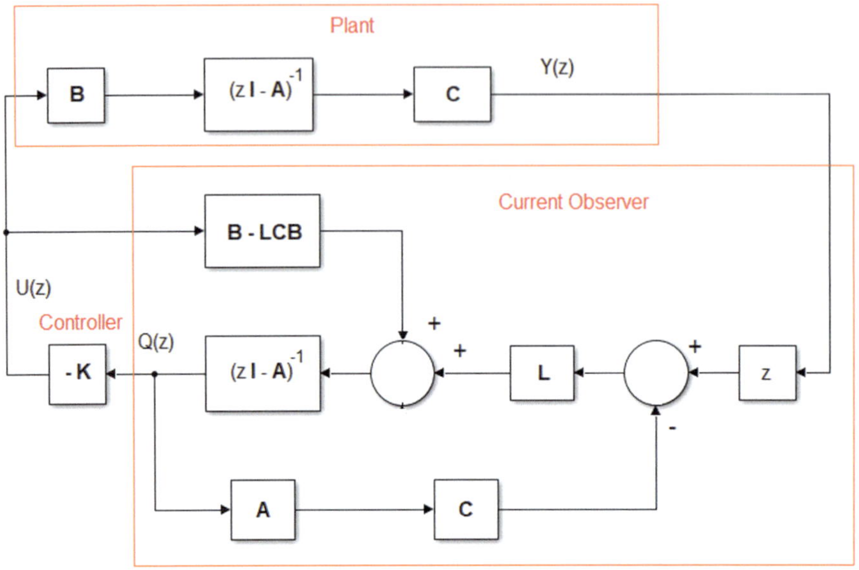

Fig. 14.15 Block diagram of a system with current observer and feedback pole placement controller

Equations (14.111)–(14.114) can be used to represent the system with the current observer and the feedback pole placement controller in the block diagram as shown in Fig. 14.15.

Equation (14.90) can be rewritten as

$$\begin{aligned} \boldsymbol{q}(k+1) = &\boldsymbol{A}\,\boldsymbol{q}(k) + \boldsymbol{B}\,\boldsymbol{u}(k) \\ &+ \boldsymbol{LC}(\boldsymbol{A}\,\boldsymbol{x}(k) + \boldsymbol{B}\,\boldsymbol{u}(k) - (\boldsymbol{A}\,\boldsymbol{q}(k) + \boldsymbol{B}\,\boldsymbol{u}(k))) \end{aligned} \tag{14.115}$$

Equation (10.115) can be expressed as

$$\boldsymbol{q}(k+1) = \boldsymbol{A}\,\boldsymbol{q}(k) + \boldsymbol{B}\,\boldsymbol{u}(k) + \boldsymbol{LCA}(\boldsymbol{e}(k)) \tag{14.116}$$

A Simulink model for the discrete-time system in state space with current observer and feedback pole placement controller, given in Eqs. (14.24), (14.25), (14.116), and (14.92), with a reference input vector $\boldsymbol{r}(k)$ of size $p \times 1$ is shown in Fig. 14.16. It should be noted that state vector $\boldsymbol{x}(k)$, the output vector $\boldsymbol{y}(k)$, and the input vector $\boldsymbol{u}(k)$ are of size $n \times 1$, $r \times 1$, and $p \times 1$, respectively. The matrices $\boldsymbol{a}$, $\boldsymbol{b}$, $\boldsymbol{c}$, $\boldsymbol{k}$, and $\boldsymbol{l}$ are the same as $\boldsymbol{A}$, $\boldsymbol{B}$, $\boldsymbol{C}$, $\boldsymbol{K}$, and $\boldsymbol{L}$, respectively, as defined in Eqs. (14.24), (14.25), (14.116), and (14.92).

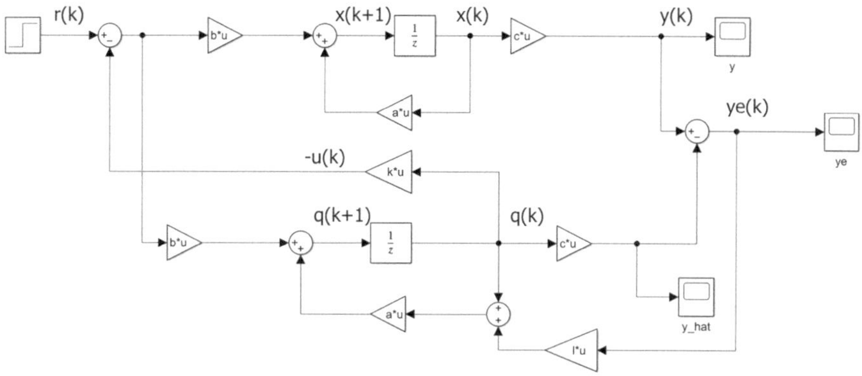

Fig. 14.16 Simulink model of a system with current observer and feedback pole placement controller

Controller-Estimator Transfer Function

For the pole placement controller using the full-order current estimator, the input to the estimator is the measurement vector $\boldsymbol{y}(k+1)$ as shown in Eq. (14.91), and the output is the control vector $\boldsymbol{u}(k)$ as shown in Eq. (14.92). Equations (14.91) and (14.92) can be combined to get the current estimator equation as

$$\begin{aligned}\boldsymbol{q}(k+1) &= (\boldsymbol{A}-\boldsymbol{L}\,\boldsymbol{C}\,\boldsymbol{A})\boldsymbol{q}(k)+(\boldsymbol{B}-\boldsymbol{LCB})\,(-\boldsymbol{K}\,\boldsymbol{q}(k))+\boldsymbol{L}\,\boldsymbol{y}(k+1)\\ &= (\boldsymbol{A}-\boldsymbol{L}\,\boldsymbol{C}\,\boldsymbol{A}-\boldsymbol{B}\,\boldsymbol{K}+\boldsymbol{L}\,\boldsymbol{C}\,\boldsymbol{B}\,\boldsymbol{K})\boldsymbol{q}(k)+\boldsymbol{L}\,\boldsymbol{y}(k+1)\end{aligned} \tag{14.117}$$

Using z-transform, Eq. (14.117) can be expressed in the z-domain, as

$$z\boldsymbol{I}\,\boldsymbol{Q}(z) = (\boldsymbol{A}-\boldsymbol{L}\,\boldsymbol{C}\,\boldsymbol{A}-\boldsymbol{B}\,\boldsymbol{K}+\boldsymbol{L}\,\boldsymbol{C}\,\boldsymbol{B}\,\boldsymbol{K})\,\boldsymbol{Q}(z)+z\boldsymbol{L}\,\boldsymbol{Y}(z) \tag{14.118}$$

Equation (14.118) can be used to express $\boldsymbol{Q}(z)$ in terms of $\boldsymbol{Y}(z)$ as

$$\boldsymbol{Q}(z) = (z\boldsymbol{I}-(\boldsymbol{A}-\boldsymbol{L}\,\boldsymbol{C}\,\boldsymbol{A}-\boldsymbol{B}\,\boldsymbol{K}+\boldsymbol{L}\,\boldsymbol{C}\,\boldsymbol{B}\,\boldsymbol{K}))^{-1}z\boldsymbol{L}\,\boldsymbol{Y}(z) \tag{14.119}$$

Equation (14.92) can be expressed in z-domain as

$$\boldsymbol{U}(z) = -\boldsymbol{K}\,\boldsymbol{Q}(z) \tag{14.120}$$

Equations (14.119) and (14.120) can be combined to express $\boldsymbol{U}(z)$ in terms of $\boldsymbol{Y}(z)$ as

$$\boldsymbol{U}(z) = -z\boldsymbol{K}(z\boldsymbol{I} - (\boldsymbol{A} - \boldsymbol{L}\,\boldsymbol{C}\,\boldsymbol{A} - \boldsymbol{B}\,\boldsymbol{K} + \boldsymbol{L}\,\boldsymbol{C}\,\boldsymbol{B}\,\boldsymbol{K}))^{-1}\boldsymbol{L}\,\boldsymbol{Y}(z) \tag{14.121}$$

The controller-current estimator transfer function can be written as

$$D_{cco}(z) = -\frac{\boldsymbol{U}(z)}{\boldsymbol{Y}(z)} = z\boldsymbol{K}(z\boldsymbol{I} - (\boldsymbol{A} - \boldsymbol{L}\,\boldsymbol{C}\,\boldsymbol{A} - \boldsymbol{B}\,\boldsymbol{K} + \boldsymbol{L}\,\boldsymbol{C}\,\boldsymbol{B}\,\boldsymbol{K}))^{-1}\boldsymbol{L} \tag{14.122}$$

Comparing Eq. (14.122) with the standard state-space representation of a discrete-time system, Eq. (13.58), the equivalent system matrices of the controller-current estimator can be expressed for

$$D_{ccop}(z) = \boldsymbol{K}(z\boldsymbol{I} - (\boldsymbol{A} - \boldsymbol{L}\,\boldsymbol{C}\,\boldsymbol{A} - \boldsymbol{B}\,\boldsymbol{K} + \boldsymbol{L}\,\boldsymbol{C}\,\boldsymbol{B}\,\boldsymbol{K}))^{-1}\boldsymbol{L} \tag{14.123}$$

$$\boldsymbol{A}_{ccop} = \boldsymbol{A} - \boldsymbol{L}\,\boldsymbol{C}\,\boldsymbol{A} - \boldsymbol{B}\,\boldsymbol{K} + \boldsymbol{L}\,\boldsymbol{C}\,\boldsymbol{B}\,\boldsymbol{K} \tag{14.124}$$

$$\boldsymbol{B}_{ccop} = \boldsymbol{L} \tag{14.125}$$

$$\boldsymbol{C}_{ccop} = \boldsymbol{K} \tag{14.126}$$

$$\boldsymbol{D}_{ccop} = \boldsymbol{D} \tag{14.127}$$

Once the transfer function for $D_{ccop}(z)$ is obtained, the numerator can be multiplied by z to get $D_{cco}(z)$ as

$$D_{cco}(z) = zD_{ccop}(z) \tag{14.128}$$

Example 14.4 The s-domain transfer function of a DC motor position dynamics is given as $G_p(s) = \frac{K}{s(\tau s+1)}$, $K = 30$, $\tau = 0.1$ s, $T = 0.01$ s. Obtain state-space representation from transfer function in continuous time, and convert from continuous-time to discrete-time domain. Design a full state feedback pole placement controller and a full-order current observer such that the closed-loop system with the controller will have an overshoot of 15% and a settling time of 0.25 s when subjected to a step input. Verify from closed-loop system step response if the specifications are met.

Solution Matlab code and results are presented.

```
% Ch14-Ex4_DC Motor pole placement controller with Current Observer
clear all; close all; clc
%G(s)=K/s(s+tau)
n=200;
K=30; %Static gain
tau=0.1; % time constant
T=0.01; % sampling time
ngs=K; dgs=[tau 1 0];
printsys(ngs,dgs)
%ss model [ac,bc,,c,d]
[ac,bc,c,d ]=tf2ss(ngs,dgs) % ss (s)
%ss model in discrete-time
[a,b]=c2d(ac,bc,T)    % ss (k)
% controllability
cont=[b a*b ]
cr=rank(cont)
%observabilty
obsr=[c;c*a]
or=rank(obsr)
% pole placement controller design
% desired pole locations in s-plane
alpha=0.15; tset=0.25; beta=0.02;
zeta=abs(log(alpha))/sqrt(pi^2+log(alpha)^2)
wn=-log(beta)/(zeta*tset)
s1=-zeta*wn+j*wn*sqrt(1-zeta^2);
s2=-zeta*wn-j*wn*sqrt(1-zeta^2);
%desired pole locations in z-plane
z1=exp(s1*T)
z2=exp(s2*T)
clp=[z1, z2];
k=place(a,b,clp) % controller gain matrix [K]
% closed loop response (with the state feedback only)
at=a-b*k;bt=b;ct=c;dt=d;
[zt,pt,kt]=ss2zp(at,bt,ct,dt)
dcg=ddcgain(at,bt,ct,dt);
yc=dstep(at,bt,ct,dt,1,n);
t=linspace(0,n*T,n);
yc1=yc/dcg;
figure
plot(t,yc1,'b-'),grid,hold on
title('Closed-loop System Step Response')
xlabel('Time (seconds)')
ylabel('y(t)')
% observer/estimator design

%Current Observer Design
se1=2*s1;se2=2*s2;
ze1=exp(se1*T);
ze2=exp(se2*T);
cle=[ze1,ze2];
ca=c*a;
l=place(a',ca',cle);l=l' % current observer gain matrix

ak=a-b*k-l*c*a+l*c*b*k;bk=l;ck=k;
[tmp1,tmp2]=size(d);dk=zeros(tmp1,tmp2);
[nkz,dkz]=ss2tf(ak,bk,ck,dk) %
nkz=[nkz(2) nkz(3) 0]
```

```
printsys(nkz,dkz,'z') % Dce(z)
[at,bt,ct,dt]=feedback(a,b,c,d,ak,bk,ck,dk)
%pause
[zt,pt,kt]=ss2zp(at,bt,ct,dt)
dcg=ddcgain(at,bt,ct,dt); ycc=dstep(at,bt,ct,dt,1,n);
ycc=ycc/dcg;
figure
plot(t,yc1,'b-'),grid,hold on
title('Closed-loop System Step Response')
xlabel('Time (seconds)')
ylabel('y(t)')
plot(t,ycc,'r-')
legend('without observer','with current observer')
%hold off
t=linspace(0,n*T,n)';
[ata,bta,cta,dta]=augstate(at,bt,ct,dt)
%pause
figure
x_e=dinitial(ata,bta,cta,dta,[1 1 -1 -1],n);
x=x_e(:,2:3);xe=x_e(:,4:5);
subplot(211)
plot(t,[x(:,1) xe(:,1)]);
title('Closed-loop System Initial condition Response')
xlabel('Time (seconds)')
ylabel('x1(t)')
legend('x1','estimated x1')
subplot(212)
plot(t,[x(:,2) xe(:,2)]);
title('Closed-loop System Initial condition Response')
xlabel('Time (seconds)')
ylabel('x2(t)')
legend('x2','estimated x2')

Results:
Current observer gain matrix
l =
    0.0789
    0.0014

Controller-current estimator transfer function

           2.8767 z^2 - 2.1539 z
 Dce(z) = -----------------------
           z^2 - 1.0606 z + 0.43219
```

The closed-loop step response and the mismatched initial condition response are shown in Figs. 14.17 and 14.18.

Example 14.5 The transfer function of the camera and laser position control system in a laser eye surgery setup is represented in the s-domain as $G_p(s) = \frac{1.5}{s(s+1.5)(s+3.5)}$. The sampling time is given as $T = 0.025$ s. Obtain state-space representation from transfer function in continuous time, and convert from continuous-time to discrete-time domain. Design a full state feedback pole placement controller and a full-order prediction observer such that the closed-loop system with the controller will have an overshoot of 5% and a settling time of 0.5 s when subjected to a step input. Design a current observer with the same transient time domain specifications. Verify from closed-loop system step response if the specifications are met, and compare time response with full-order and current observers.

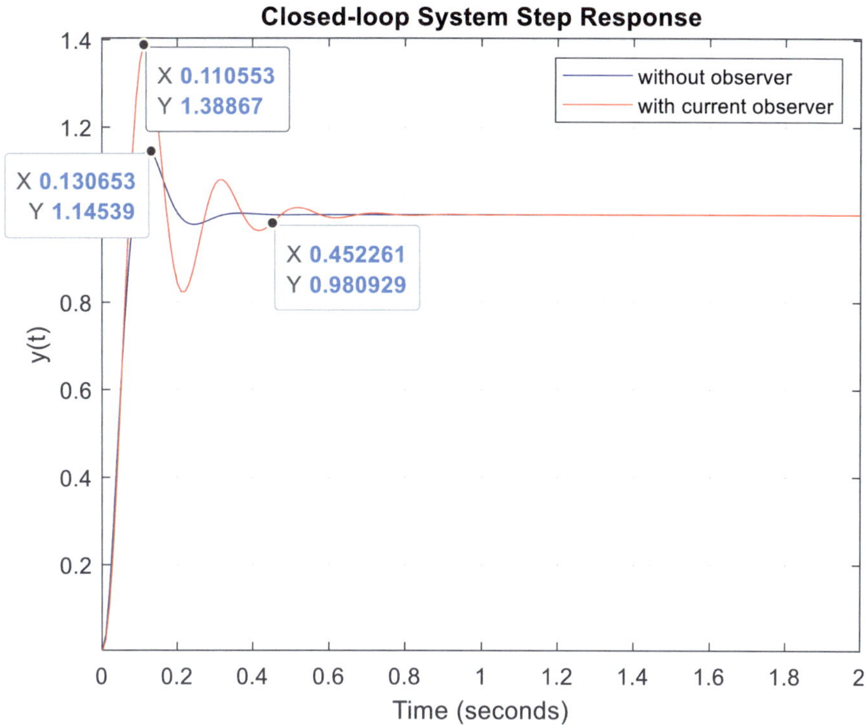

Fig. 14.17 Closed-loop system step response with the pole placement controller

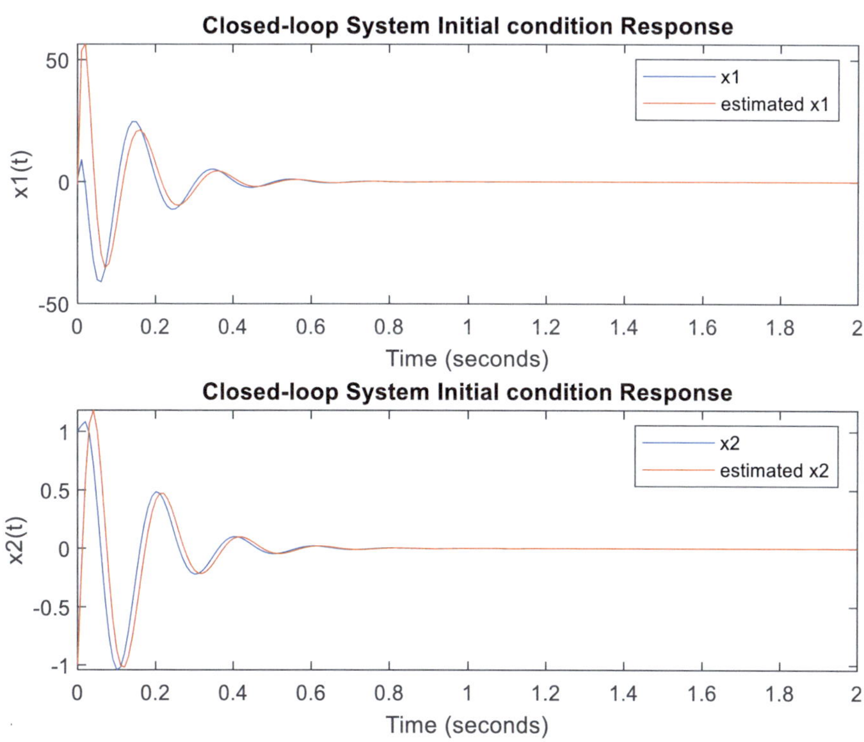

Fig. 14.18 Closed-loop system initial condition response with the pole placement controller

Solution Matlab code and results are presented.

```
% Ch 14- ex14.5
% State space model Controllability and observability
clear all; close all;clc
%G(s)=K/(s+12)(s^22s+24)
K=1.5;%Static gain
T=0.025 ; % sampling time
G1=tf(K,[1 1.5 0]);
G2=tf(1,[1 3.5])
Gs=G1*G2
Gz=c2d(Gs,T)
[ngz,dgz]=tfdata(Gz,'v')
%dgz=[dgz 0]
%ss model in discrete-time [a,b,c,d]
[a,b,c,d ]=tf2ss(ngz,dgz) % ss (k)

% controllability
ab=a*b;
a2b=a*ab;
cont=[b ab a2b]
cr=rank(cont)
%observabilty
ca=c*a;
ca2=ca*a;
obsr=[c;ca;ca2]
or=rank(obsr)
N=100;
% full-state feedback pole placement controller design
% desired pole locations
alpha=0.05; tset=0.5; beta=0.02;
zeta=abs(log(alpha))/sqrt(pi^2+log(alpha)^2)
wn=-log(beta)/(zeta*tset)
s1=-zeta*wn+j*wn*sqrt(1-zeta^2);
z1=exp(s1*T)
s2=-zeta*wn-j*wn*sqrt(1-zeta^2);
z2=exp(s2*T)
s3=2*real(s1);
z3=exp(s3*T)
clp=[z1 z2 z3];
k=place(a,b,clp)
% closed loop response (with the state feedback only)
at=a-b*k;bt=b;ct=c;dt=d;
%[zt,pt,kt]=ss2zp(at,bt,ct,dt)
dcg=ddcgain(at,bt,ct,dt); yc=dstep(at,bt,ct,dt,1,N);
yc=yc/dcg;
t=linspace(0,N*T,N)';
figure
plot(t,yc),grid,hold on
title('Closed-loop step response')
xlabel('Time (seconds)')
ylabel('y(t)')

% full-order observer/estimator design
se1=2*s1;se2=2*s2;se3=2*s3;
%n = length(dgz)
ze1=exp(se1*T)
ze2=exp(se2*T)
ze3=exp(se3*T)
```

```
cle=[ze1 ze2 ze3];
l=place(a',c',cle);l=l'
ak=a-b*k-l*c-l*d*k;bk=l;ck=k;
[tmp1,tmp2]=size(d);dk=zeros(tmp1,tmp2);
[zk,pk,kk]=ss2zp(ak,bk,ck,dk);
[nkz,dkz]=ss2tf(ak,bk,ck,dk) %Dce(z)
printsys(nkz,dkz,'z')
[agk,bgk,cgk,dgk]=series(a,b,c,d,ak,bk,ck,dk);
[at,bt,ct,dt]=feedback(a,b,c,d,ak,bk,ck,dk)
[zt,pt,kt]=ss2zp(at,bt,ct,dt)
dcg=ddcgain(at,bt,ct,dt); ycf=dstep(at,bt,ct,dt,1,N);
ycf=ycf/dcg;
plot(t,ycf,'r-'),grid
legend('without observer','with full-order observer')
% Augmented system dynamics
%X=[y x1 x2 xe1 xe2]
[ata,bta,cta,dta]=augstate(at,bt,ct,dt)
% initial state response
figure
x_e=dinitial(ata,bta,cta,dta,[0.1 0.1 0.1 -0. -0.1 -0.1],N); % initial
condition response
x=x_e(:,2:4);xe=x_e(:,5:7);
subplot(311)
plot(t,[x(:,1) xe(:,1)]); hold on
title('Closed-loop System Initial condition Response')
xlabel('Time (seconds)')
% ylabel('x1(t)')
legend('x1','estimated x1')
subplot(312)
plot(t,[x(:,2) xe(:,2)]); hold on
title('Closed-loop System Initial condition Response')
xlabel('Time (seconds)')
ylabel('x2(t)')
legend('x2', 'estimated x2')
subplot(313)
plot(t,[x(:,3) xe(:,3)]); hold on
title('Closed-loop System Initial condition Response')
xlabel('Time (seconds)')
ylabel('x3(t)')
legend('x3', 'estimated x3')

%Current Observer Design
se1=2*s1;se2=2*s2;se3=2*s3
ze1=exp(se1*T);
ze2=exp(se2*T);
ze3=exp(se3*T);
cle=[ze1,ze2 ze3];
ca=c*a;
g=place(a',ca',cle);g=g' % current observer gain matrix
ak=a-b*k-g*c*a+g*c*b*k-l*d*k;bk=g;ck=k;
[tmp1,tmp2]=size(d);dk=zeros(tmp1,tmp2);
[nkz,dkz]=ss2tf(ak,bk,ck,dk) %
nkz=[nkz(2) nkz(3) nkz(4) 0]
printsys(nkz,dkz,'z') % Dce(z)
[at,bt,ct,dt]=feedback(a,b,c,d,ak,bk,ck,dk)
%pause
[zt,pt,kt]=ss2zp(at,bt,ct,dt)
```

```
dcg=ddcgain(at,bt,ct,dt); ycc=dstep(at,bt,ct,dt,1,N);
ycc=ycc/dcg;
figure
plot(t,yc,'b-'),grid,hold on
title('Closed-loop System Step Response')
xlabel('Time (seconds)')
ylabel('y(t)')
plot(t,ycc,'r-')
legend('without observer','with current observer')
t=linspace(0,N*T,N)';
[ata,bta,cta,dta]=augstate(at,bt,ct,dt)
% initial state response
figure
x_e=dinitial(ata,bta,cta,dta,[0.1 0.1 0.1 -0. -0.1 -0.1],N); % initial
condition response
x=x_e(:,2:4);xe=x_e(:,5:7);
subplot(311)
plot(t,[x(:,1) xe(:,1)]); hold on
title('Closed-loop System Initial condition Response')
xlabel('Time (seconds)')
ylabel('x1(t)')
legend('x1','estimated x1')
subplot(312)
plot(t,[x(:,2) xe(:,2)]); hold on
title('Closed-loop System Initial condition Response')
xlabel('Time (seconds)')
ylabel('x2(t)')
legend('x2', 'estimated x2')
subplot(313)
plot(t,[x(:,3) xe(:,3)]); hold on
title('Closed-loop System Initial condition Response')
xlabel('Time (seconds)')
ylabel('x3(t)')
legend('x3', 'estimated x3')

Results:
System matrices
a = 2.8794   -2.7619    0.8825
    1.0000         0         0
         0    1.0000         0
b = 1
    0
    0
c =
   1.0e-04 *[ 0.0379    0.1468    0.0356] d = 0

Controllability
cont =
    1.0000    2.8794    5.5291
         0    1.0000    2.8794
         0         0    1.0000

cr = 3, the system is controllable

Observability
obsr =
   1.0e-04 *
```

```
0.0379    0.1468    0.0356
    0.2559   -0.0690    0.0334
    0.6677   -0.6732    0.2258

or = 3, the system is observable

Desired poles in z-plane
zeta = 0.6901,wn = 11.3374, z1 = 0.8051 + 0.1675i, z2 = 0.8051 - 0.1675i,
z3 = 0.6762
zeta =

Full-state feedback Controller gain matrix
k = [0.5930   -0.9968    0.4252]

Full-order observer poles
ze1 = 0.6201 + 0.2697i, ze2 = 0.6201 - 0.2697i, ze3 = 0.4573

Observer gain matrix

l = 1.0e+04 * [7.4956; 5.2953; 3.3894]

Controller-estimator transfer function

         6076.1574 z^2 - 10700.4494 z + 4738.6378
Dce(z)=  ----------------------------------------
         z^3 - 1.1046 z^2 + 0.7055 z - 0.089269
```

The closed-loop step response and the mismatched initial condition response with full-order prediction estimator are shown in Figs. 14.19 and 14.20.

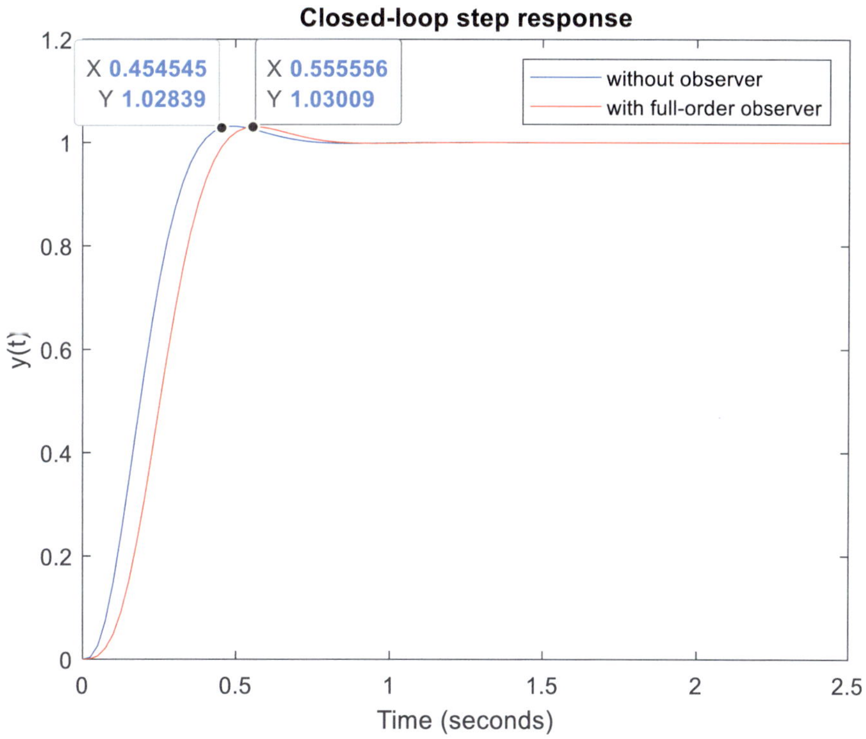

Fig. 14.19 Closed-loop system step response with the pole placement controller and full-order prediction observer

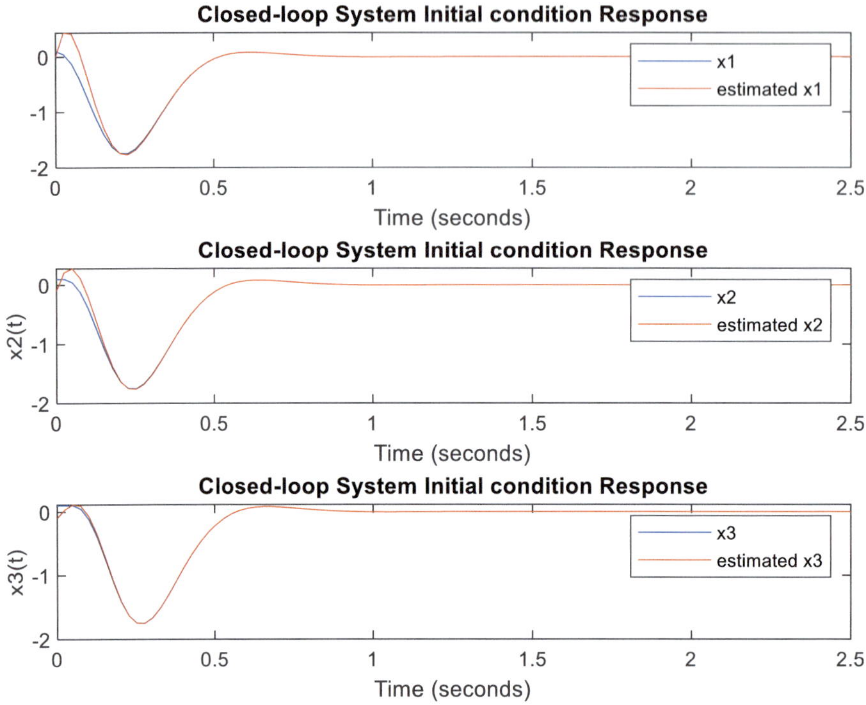

Fig. 14.20 Closed-loop system initial condition response with the pole placement controller and full-order prediction observer

```
Current Observer gain matrix
g = 1.0e+04 *[5.2953; 3.3894; 1.8240]


Controller-estimator transfer function

            5369.5801 z^3 - 9385.0828 z^2 + 4129.8485 z
Dce(z)=     -------------------------------------------
              z^3 - 1.1249 z^2 + 0.62666 z - 0.10837
```

The closed-loop step response and the mismatched initial condition response with the current estimator are shown in Figs. 14.21 and 14.22.

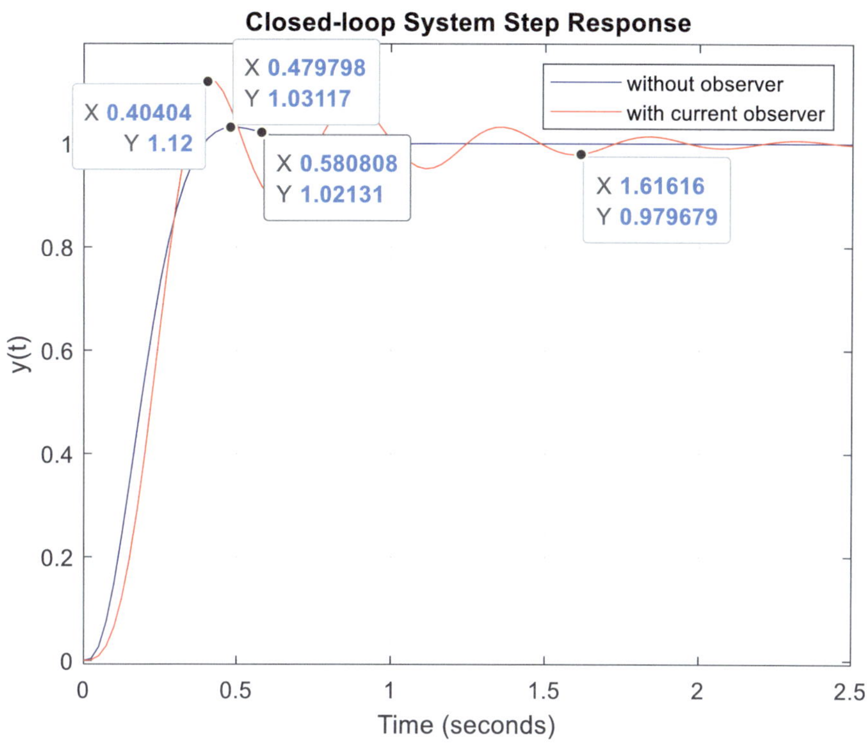

Fig. 14.21 Closed-loop system step response with the pole placement controller and current observer

14.6 Laboratory Experiment for Analysis, Design, Simulation, and Implementation of State-Space Controllers for Discrete-Time Systems

Objectives The objectives are to design and simulate controllers with full-order prediction and current observers in state space for desired transient response specifications and implement and validate these controllers for a DC motor position control system.

Introduction The full-order prediction and current observers can be designed in state space to achieve desired transient response specifications. The controller-estimator transfer function $D_{ce}(z)$ in z-domain can be implemented for a DC motor position control system. In Part I, it is required to design the controllers with full-order prediction and current observers and simulate the performance of each controller in Matlab as presented in Chap. 14. In Part II, it is required to implement each of these controllers in LabVIEW and validate its performance with reference to the design specifications.

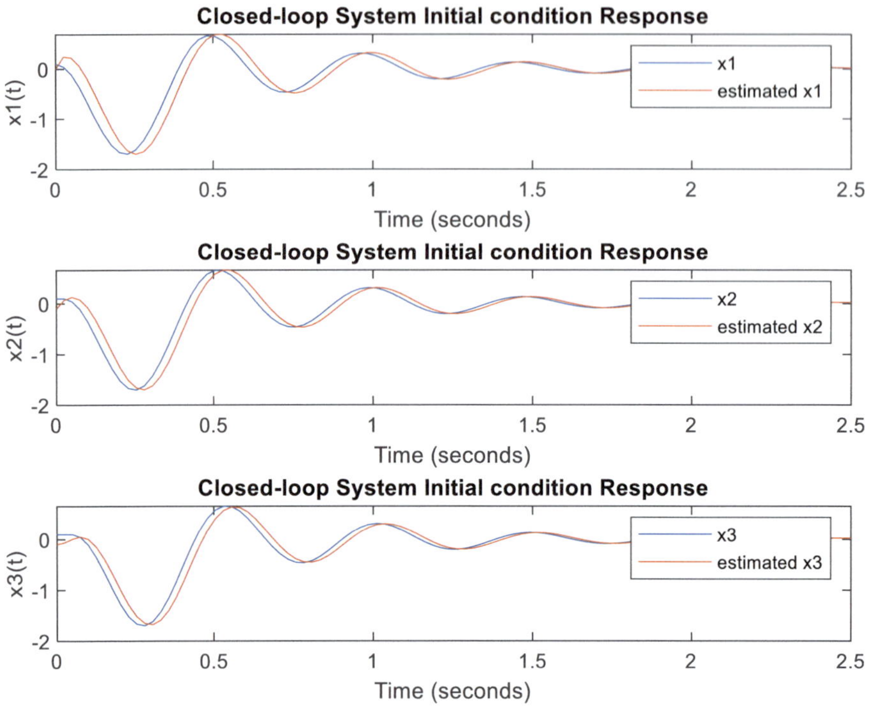

Fig. 14.22 Closed-loop system initial condition response with the pole placement controller and current observer

I. Controller Design and Simulation.

1. Use the DC motor position control model with previously estimated values of K and τ to get the discrete-time open-loop transfer function in z-domain $G(z)$ selecting a suitable sampling time T. For the DC motor control trainer (DCMCT) module, the sampling time (T) is taken as 0.01 s. Convert $G(z)$ to state-space system matrices (a, b, c, d). Refer to Chap. 13 for details.
2. Determine the desired pole location in both s- and z-planes such that closed-loop DC motor position response has an overshoot (α) within 15% and a settling time (t_{set}) of 0.25 s for an error limit (β) of $\pm 2\%$ for a step input. Refer to Sect. 10.3 for details.
3. Design a full state feedback pole placement controller for the position control system in MATLAB for the desired transient specifications. Refer to Sect. 14.2 for details. Note down the controller gain matrix and the controller transfer function $D(z)$ in Table 14.1.
4. Assign full-order prediction observer poles in the s-domain at least two times the corresponding s-domain poles with the full state feedback pole placement controller of step (3). Convert the s-domain observer poles to the z-domain, and design a full-order prediction observer. Refer to Sect. 14.3 for details

Table 14.1 Summary of controller parameters and performance

Controller	Gain matrix	Closed-loop poles	Controller-estimator transfer function $D_{ce}(z)$	Closed-loop system step response			
				Simulation		Implementation	
				Overshoot α (%)	Settling time$t_{set}(s)$	Overshoot α (%)	Settling time$t_{set}(s)$
Pole placement							
Full-order prediction observer							
Full-order current observer							

Note down the observer gain matrix and controller-estimator transfer function $D_{ce}(z)$ in Table 14.1.

5. Design a full-order current observer with the same z-domain poles of the full-order prediction observer of step (4). Refer to Sect. 14.5 for details. Note down the observer gain matrix and controller-estimator transfer function $D_{ce}(z)$ in Table 14.1.
6. Simulate closed-loop step response with each controller. Locate response characteristics by right-clicking the figure in Matlab, selecting peak response, settling time, and steady state (right-click → response characteristics → peak response → settling time → steady state → cursor on identified point). Note down in Table 14.1 overshoot (α) and settling time (t_{set}) from the simulated response for each type of controller. Copy closed-loop step response with the controller for presentation in the lab report.
7. Obtain the closed-loop system response for each state with mismatched initial conditions for each observer. Copy these response graphs for presentation in the lab report.

II. Implementation and Validation.

1. Set up NI ELVIS II board and DC motor module. Select the device, and set the sampling rate. For the DC motor control trainer (DCMCT) module, a sampling rate of 100 Hz is used.
2. Study the block diagram for the position control experiment (*01-QNET_DCMCT_Position_Control.vi*).
3. Open the block diagram VI and identify the different components of the control system, including the input, error, controller, plant, sensor, and output in Fig. 14.23.
4. Modify the block diagram VI of Fig. 14.23 by removing the control parameters block and adding a subtraction block to calculate the error, as well as a discrete transfer function block for implementation of the designed controller, as shown in the VI of Fig. 14.24.
5. Put the discrete-time z-domain controller-estimator transfer function $D_{ce}(z)$, and follow the format of setting the controller transfer function coefficients. Save the VI.
6. Select your input parameters. For the DCMCT module, typical values for the input square wave voltage are as follows: amplitude of 1.0 V, frequency of 0.40 Hz, and offset of 3.0 V.
7. Run the VI, and observe the closed-loop system response. Copy graphs of closed-loop system response from the front panel VI for presentation in the lab report.
8. Make sure to set the number of significant figures to at least 5 for the display format of each graph on the front panel VI (right-click on the graph → display format → decimal → 6). Export the time response graphs, both motor rotation $\theta(t)$ and motor input voltage $V(t)$ to Excel by right-clicking on each graph and selecting the Export to Excel option (right-click → Export to Excel). Estimate overshoot and settling time from closed-loop system response graph.

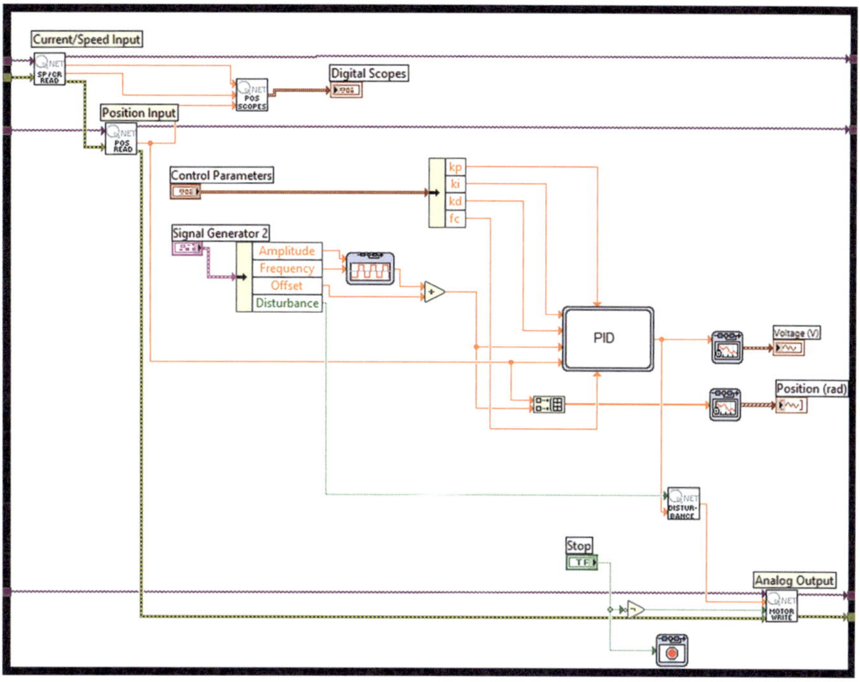

Fig. 14.23 Block diagram VI for DC motor control trainer position control VI before modification

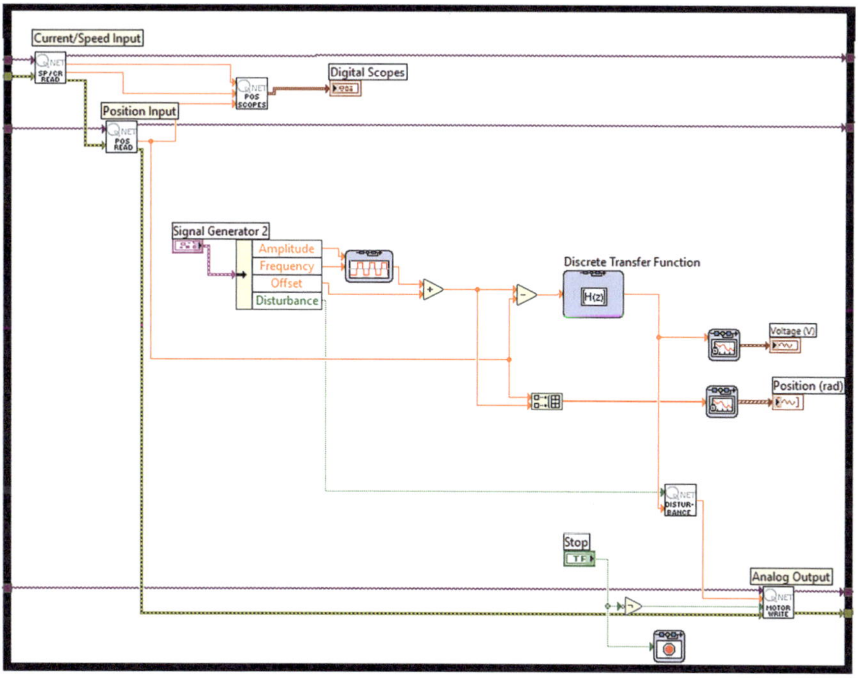

Fig. 14.24 Block diagram VI for DC motor control trainer position control VI after modification

9. Repeat steps (5)–(8) for each type of controller. Complete the last two columns of Table 14.1.

III. Deliverables.

1. Complete Table 14.1.
2. Present graphs of simulated closed-loop step response for each controller with response characteristics (peak response, settling time, and steady-state value) clearly annotated.
3. Present graphs of state responses with mismatched initial conditions for each controller-observer combination.
4. Present the closed-loop step response of the DC motor for each implemented controller-observer combination.
5. Discuss results comparing the performance of controller-observer combinations in simulation and experimental validation.

Lab Report

Present all deliverables in the lab report and discuss your results.

Exercises

14.1 The closed-loop vehicle distance control system for a traffic control system can be represented as shown in Fig. P14.1. The plant transfer function of the distance control system can be represented in the s-domain as $G_p(s) = \frac{1.5}{s^2+12s+18}$. The sampling time is given as $T = 0.10$ s. Obtain state-space representation of the open-loop system in continuous time, and convert it to discrete-time domain. Design a full state feedback pole placement controller and a full-order prediction observer such that the closed-loop system with the controller will have an overshoot of 10% and a settling time of 2.5 s when subjected to a step input. Design a current observer with the same transient time domain specifications. Verify from closed-loop system step response if the specifications are met, and compare time response with full-order prediction and current observers.

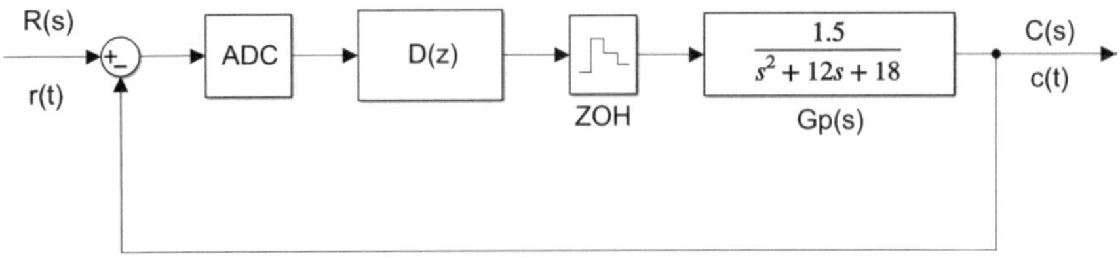

Fig. P14.1 Model of closed-loop discrete-time system for Problem 14.1

14.2 The closed-loop roll angle control system for a jet airplane can be represented as shown in Fig. P14.2. The plant transfer function of the roll control system is represented in the s-domain as $G_p(s) = \frac{10}{(s+10)(s^2+2.5s+25)}$. The sampling time is given as $T = 0.05$ s. Obtain state-space representation of the open-loop system in continuous time, and convert it to discrete-time domain. Design a full state feedback pole placement controller and a full-order prediction observer such that the closed-loop system with the controller will have an overshoot of 5% and a settling time of 1.5 s when subjected to a step input. Design a current observer with the same transient time domain specifications. Verify from closed-loop system step response if the specifications are met, and compare time response with full-order prediction and current observers.

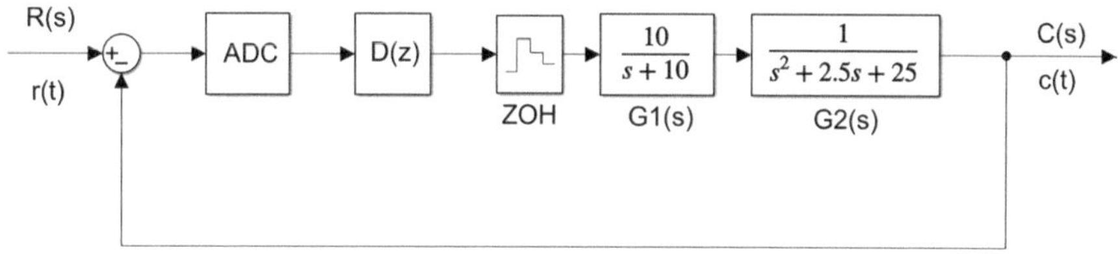

Fig. P14.2 Model of closed-loop discrete-time system for Problem 14.2

14.3 The closed-loop tilt control system in a high-speed train for passenger comfort can be represented as shown in Fig. P14.3. The plant transfer function of the train tilt control system is represented in the s-domain as $G_p(s) = \frac{5}{s(s+5)(s+35)}$. The sampling time is given as $T = 0.01$ s. Obtain state-space representation of the system in continuous time, and convert it to the discrete-time domain. Design a full state feedback pole placement controller and a full-order prediction observer such that the closed-loop system with the controller will have an overshoot of 5% and a settling time of 0.65 s when subjected to a step input. Design a current observer with the same transient time domain specifications. Verify from closed-loop system step response if the specifications are met, and compare time response with full-order prediction and current observers.

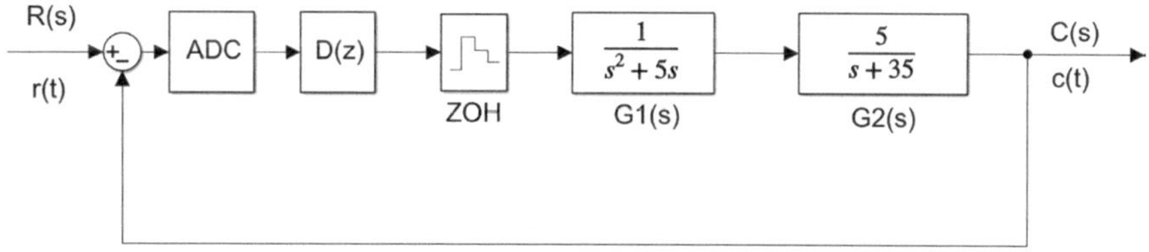

Fig. P14.3 Model of closed-loop discrete-time system for Problem 14.3

14.4 The closed-loop position control system in a machine tool can be represented as shown in Fig. P14.4. The plant transfer function of the position control system is represented in the s-domain as $G_p(s) = \frac{K}{5s(0.25s+1)\left(1.5\times10^{-3}s+2.5\times10^{-2}\right)}$. The sampling time is given as $T = 0.10$ s. The amplifier gain K is such that the steady-state error to a ramp input is limited to 2.5%. Obtain state-space representation of the open-loop system in continuous time, and convert it to discrete-time domain. Design a full state feedback pole placement controller and a full-order prediction observer such that the closed-loop system with the controller will have an overshoot of 5% and a settling time of 1.0 s when subjected to a step input. Design a current observer with the same transient time domain specifications. Verify from closed-loop system step response if the specifications are met, and compare time response with full-order prediction and current observers.

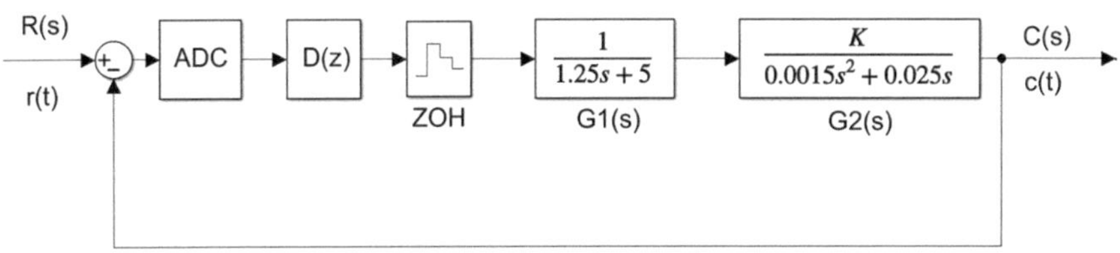

Fig. P14.4 Model of closed-loop discrete-time system for Problem 14.4

14.5 The state-space model matrices (**A**, **B**, **C**, **D**) for lateral motions of a ship moving at a constant velocity can be represented as [6]

$$\boldsymbol{A} = \begin{bmatrix} -0.0895 & -0.2860 & 0 \\ -0.0439 & -0.2720 & 0 \\ 0 & 1.0000 & 0 \end{bmatrix}, \boldsymbol{B} = \begin{bmatrix} 0.0145 \\ -0.0122 \\ 0 \end{bmatrix}, \boldsymbol{C} = [0\ \ 0\ \ 1], \boldsymbol{D} = [0]$$

where the state variables $(x_1(t),\ x_2(t), x_3(t))$ are ship side slip angle (degrees), ship heading angle (degrees), and ship yaw rate, respectively. The rudder angle (degrees) is the input $(u(t))$, and the heading angle is the measurement variable $(y(t))$. The sampling time is given as $T = 0.05$ s. Obtain state-space representation of the open-loop system in discrete time. Design a full state feedback pole placement controller and a full-order observer such that the closed-loop system with the controller will have an overshoot of 10% and a settling time of 20 s when subjected to a step input. Design a current observer with the same transient time domain specifications. Verify from closed-loop system step response if the specifications are met, and compare time response with full-order and current observers.

14.6 The state-space model matrices (**A**, **B**, **C**, **D**) for the linearized longitudinal motion of a helicopter can be represented as [6]

$$A = \begin{bmatrix} -0.4000 & 0 & -0.0100 \\ 1.0000 & 0 & 0 \\ -1.4000 & 9.8 & -0.0200 \end{bmatrix}, B = \begin{bmatrix} 6.3000 \\ 0 \\ 9.8000 \end{bmatrix}, C = [0 \ 0 \ 1], D = [0]$$

where the state variables $(x_1(t),\ x_2(t), x_3(t))$ are the rate of pitch angle $(\dot{\theta}(t))$, pitch angle $(\theta(t))$, and horizontal velocity $(v(t))$, respectively. The rudder angle (degrees) is the input $(u(t))$, and the horizontal velocity $(v(t))$ is the measurement variable $(y(t))$. The sampling time is given as $T = 0.20$ s. Obtain state-space representation of the open-loop system in discrete time. Design a full state feedback pole placement controller and a full-order observer such that the closed-loop system with the controller will have an overshoot of 10% and a settling time of 5 s when subjected to a step input. Design a current observer with the same transient time domain specifications. Verify from closed-loop system step response if the specifications are met, and compare time response with full-order prediction and current observers.

14.7 The state-space model matrices (**A**, **B**, **C**, **D**) for a linearized model of liquid levels in a 3-tank system can be represented as [5]

$$A = \begin{bmatrix} -1 & 0 & 0 \\ 1 & -1 & 0 \\ 0 & 1 & -1 \end{bmatrix}, B = \begin{bmatrix} 1 \\ 0 \\ 0 \end{bmatrix}, C = [0 \ 0 \ 1], D = [0]$$

where the state variables $(x_1(t),\ x_2(t), x_3(t))$ are liquid levels $(h_1(t),\ h_2(t), h_3(t))$ of the top, the middle, and the bottom tank, respectively. The inlet liquid flow rate in the top tank is the input $(u(t))$, and the liquid level of the bottom tank is the measurement variable $(y(t))$. The sampling time is given as $T = 010$ s. Obtain state-space representation of the open-loop system in discrete time. Design a full state feedback pole placement controller and a full-order observer such that the closed-loop system with the controller will have an overshoot of 10% and a settling time of 5 s when subjected to a step input. Design a current observer with the same transient time domain specifications. Verify from closed-loop system step response if the specifications are met, and compare time response with full-order and current observers.

14.8 The closed-loop thickness control system in a metal processing plant can be represented as shown in Fig. P14.5. The plant transfer function of the thickness control system is represented in the s-domain as $G_p(s) = \frac{5}{s(s+2)(s^2+s+10)}$. The sampling time is given as $T = 0.10$ s. Obtain state-space representation of the open-loop system in continuous time, and convert it to discrete-time domain. Design a full state feedback pole placement controller and a full-order prediction observer such that the closed-loop system with the controller will have an overshoot of 10% and a settling time of 4 s when subjected to a step input. Design a current observer with the same transient time domain specifications. Verify from closed-loop system step response if the specifications are met, and compare time response with full-order prediction and current observers.

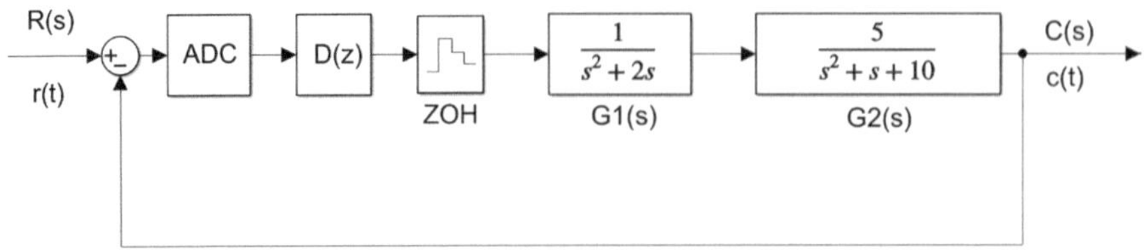

Fig. P14.5 Model of closed-loop discrete-time system for Problem 14.8

14.9 The closed-loop position control system of a typical Mars rover can be represented as shown in Fig. P14.6. The plant transfer function of the position control system is represented in the s-domain, similar to Dorf and Bishop [4], as $G_p(s) = \frac{1}{s^2(s+5)}$. The sampling time is given as $T = 0.02$ s. Obtain state-space representation of the open-loop system in continuous time, and convert it to discrete-time domain. Design a full state feedback pole placement controller and a full-order prediction observer such that the closed-loop system with the controller will have an overshoot of 5% and a settling time of 5 s when subjected to a step input. Design a current observer with the same transient time domain specifications. Verify from closed-loop system step response if the specifications are met, and compare time response with full-order prediction and current observers.

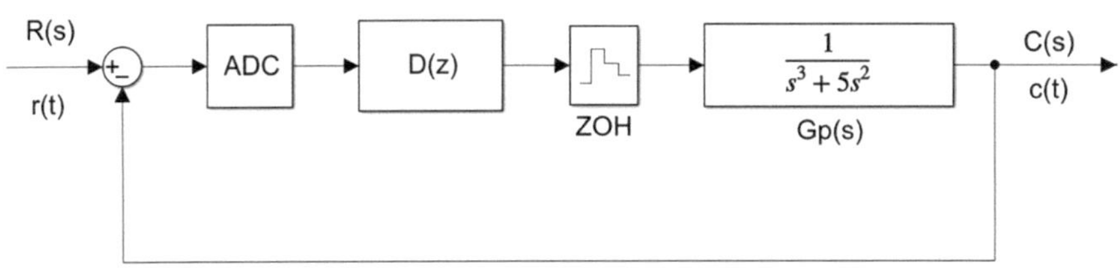

Fig. P14.6 Model of closed-loop discrete-time system for Problem 14.9

14.10 The closed-loop position control system for one of the axes of an experimental automated surface-grinding wheel can be represented as shown in Fig. P14.7. The plant transfer function of the axis position control system can be represented in the s-domain, similar to Dorf and Bishop [4], as $G_p(s) = \frac{5}{s(0.2s+1)(0.3s+1)}$. The sampling time is given as $T = 0.02$ s. Obtain state-space representation of the open-loop system in continuous time, and convert it to discrete-time domain. Design a full state feedback pole placement controller and a full-order prediction observer such that the closed-loop system with the controller will have an overshoot of 5% and a settling time of 2 s when subjected to a step input. Design a current observer with the same transient time domain specifications. Verify from closed-loop system step response if the specifications are met, and compare time response with full-order prediction and current observers.

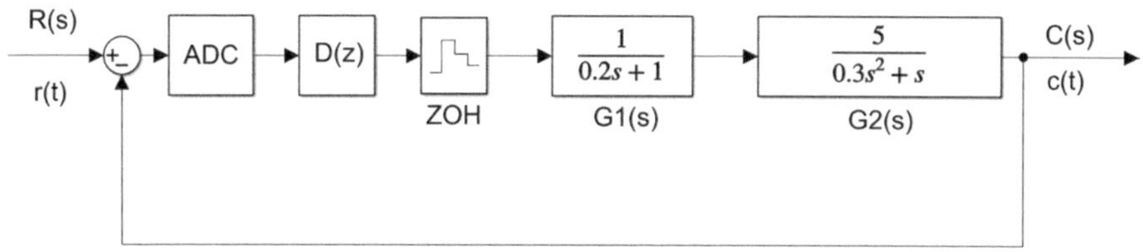

Fig. P14.7 Model of closed-loop discrete-time system for Problem 14.10

14.11 The closed-loop vehicle distance control system for a traffic control system can be represented as shown in Fig. P14.8. The plant transfer function of the distance control system can be represented in the s-domain as $G_p(s) = \frac{1.5}{s^2+12s+18}$. The sampling time is given as $T = 0.10$ s. Obtain state-space representation of the open-loop system in continuous time, and convert it to discrete-time domain. Design a full state feedback pole placement controller and a reduced-order prediction observer such that the closed-loop system with the controller will have an overshoot of 10% and a settling time of 2.5 s when subjected to a step input. Verify from closed-loop system step response if the specifications are met, and compare time response with full-order and reduced-order observers.

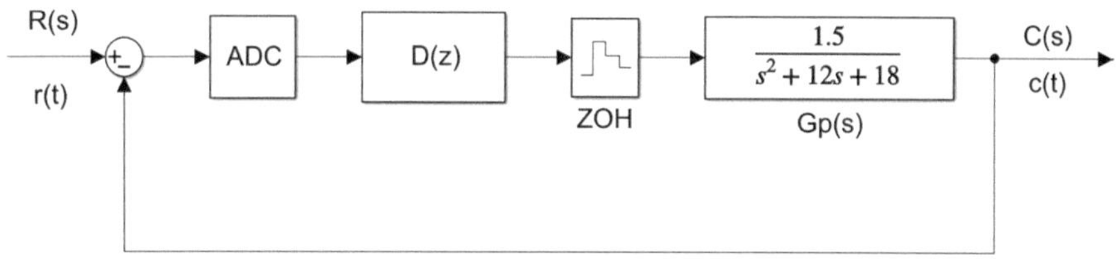

Fig. P14.8 Model of closed-loop discrete-time system for Problem 14.11

14.12 The closed-loop steering control system for a second-order system with a time delay can be represented as shown in Fig. P14.9, $G_p(s) = \frac{1}{s(s+1)}e^{-sT}$. The sampling time is given as $T = 0.2$ s. Obtain state-space representation of the open-loop system in continuous time, and convert it to discrete-time domain. Design a full state feedback pole placement controller and a full-order prediction observer such that the closed-loop system with the controller will have an overshoot of 5% and a settling time of 2 s when subjected to a step input. Verify from closed-loop system step response if the specifications are met, and compare time response with full-order prediction observer.

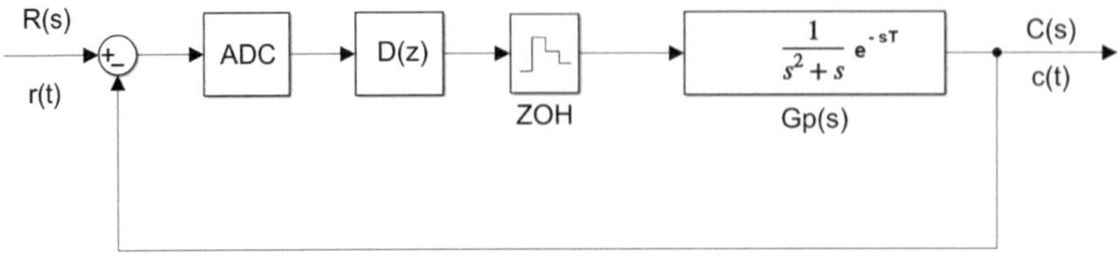

Fig. P14.9 Model of closed-loop discrete-time system for Problem 14.12

Further Reading

1. Asadi F (2023) Signals and systems with MATLAB® and Simulink®. Springer, Cham
2. Astrom KJ, Murray RM (2021) Feedback systems: an introduction for scientists and engineers, 2nd edn. Princeton University Press, Princeton
3. Chen C-T (2006) Analog and digital control system design: transfer-function, state-space, and algebraic methods. Oxford University Press, New York
4. Dorf R, Bishop R (2016) Modern control systems, 13th edn. Pearson, Essex
5. Fadali MS, Visioli A (2020) Digital control engineering analysis and design, 3rd edn. Academic, London
6. Franklin GF, Powell JD, Emami-Naeini A (2019) Feedback control of dynamic systems, 8th edn. Pearson, Upper Saddle River
7. Franklin GF, Powell JD, Workman M (1998) Digital control of dynamic systems, 3rd edn. Addison-Wesley, Boston
8. Isermann R (1981) Digital control systems. Springer, Berlin
9. Jalili N, Candelino NW (2023) Dynamic systems and control engineering. Cambridge University Press, Cambridge
10. Keviczky L, Bars R, Hetthéssy J, Bányász (2019) Control engineering. Springer, Singapore
11. Koeing DM (2009) Practical control engineering: a guide for engineers, managers, and practitioners. McGraw-Hill, New York
12. Kuo BC (1995) Digital control systems, 2nd edn. Oxford University Press, Oxford
13. Landau ID, Gito G (2006) Digital control systems: design, identification and implementation. Springer, London
14. Leigh JR (2006) Applied digital control: theory, design and implementation, 2nd edn. Dover, Mineola, New York
15. Mathworks (2024a) Control System Toolbox. https://www.mathworks.com/products/control.html
16. Mathworks (2024b) Matlab. https://www.mathworks.com/products/matlab.html
17. National Instruments (2024a) LabVIEW. https://www.ni.com/en/support/downloads/software-products/download.labview.html#487445
18. National Instruments (2024b) NI-ELVIS-II. https://www.ni.com/en-us/support/model.ni-elvis-ii.html
19. National Instruments (2024c) Labs for Quanser QNET DC Motor Add-On Board. https://education.ni.com/teach/resources/180/labs-for-quanser-qnet-dc-motor-add-on-board
20. Nise NS (2019) Control systems engineering, 8th edn. Wiley, Hoboken
21. Nof SY (ed) (2009) Springer handbook of automation. Springer, Berlin
22. Ogata K (1995) Discrete-time control systems, 2nd edn. Pearson, Hoboken
23. Phillips CL, Nagle HT, Chakrabortty A (2015) Digital control system analysis and design, 4th edn. Pearson, Essex
24. Quanser (2024) QUBE Servo 2. https://www.quanser.com/products/qube-servo-2/

25. Rabbath CA, Lechevin N (2014) Discrete-time control system design with applications. Springer, New York
26. Samanta B (2023) Introduction to mechatronics – an integrated approach. Springer, Cham
27. Shertukde HM (2015) Digital control applications illustrated with MATLAB®. CRC Press, Boca Raton
28. Starr G (2020) Introduction to applied digital controls. Springer, Cham
29. Veloni A, Miridakis NI (2018) Digital control systems-theoretical problems and simulation tools. CRC Press, Boca Raton
30. Vilanova R, Visioli A (2012) PID control in the third millennium – lessons learned and new approaches. Springer, London
31. Visioli A (2006) Practical PID control. Springer, London

Chapter 15
Discrete-Time System Identification

15.1 Introduction

Methods of model parameter identification for discrete-time systems are presented in this chapter. Methods of least squares, weighted least squares, and recursive least squares for system model parameter identification are covered. Examples are presented along with Matlab codes to illustrate the procedures. Laboratory experiments for system identification are presented for a DC motor module. End-of-chapter exercises are provided to enhance understanding of covered concepts and consolidate problem-solving skills.

15.2 Least-Squares System Identification

The discrete-time transfer function (TF) of a dynamic system of order n can be represented using z-transform as follows:

$$D(z) = \frac{Y(z)}{U(z)} = \frac{b_1 z^{n-1} + b_2 z^{n-2} + \ldots + b_n}{z^n - a_1 z^{n-1} - a_2 z^{n-2} \quad \ldots \quad a_n} \tag{15.1}$$

where the coefficients a_i and b_i, $i = 1, \ldots, n$ are not known and are to be found out through system identification. The model in Eq. (15.1) is generally known as the autoregressive moving average (ARMA) model.

The system response $y(k)$ can be represented as the following difference equation:

$$\begin{aligned} y(k) = {} & a_1 y(k-1) + a_2 y(k-2) + \ldots \quad + a_n y(k-n) + b_1 u(k-1) \\ & + b_2 u(k-2) + \ldots + b_n u(k-n) \end{aligned} \tag{15.2}$$

B. Samanta, *Introduction to Digital Control*,
https://doi.org/10.1007/978-3-031-66830-2_15

The objective is to find the $2n$ vector of coefficients $\boldsymbol{\theta}$, Eq. (15.3), from $(N + 1)$ measurement pairs, Eq. (15.4):

$$\boldsymbol{\theta} = [a_1 \ a_2 \ \dots \ a_n \ b_1 \ b_2 \ \dots \ b_n \ ]^T \tag{15.3}$$

$$M(N) = [\ \{u(0)\ y(0)\}, \{u(1)\ y(1)\}, \ \dots, \ \{u(N)\ y(N)\}] \tag{15.4}$$

where $N \geq n$.

The measured system response vector can be represented in terms of the ARMA model and the error vector $\boldsymbol{e}(N)$ as follows:

$$\boldsymbol{y}(N) = \boldsymbol{F}(N)\boldsymbol{\theta} + \boldsymbol{e}(N) \tag{15.5}$$

where

$$\boldsymbol{y}(N) = \begin{bmatrix} y(n) \\ y(n+1) \\ \cdot \\ \cdot \\ \cdot \\ y(N) \end{bmatrix} \tag{15.6}$$

$$\boldsymbol{F}(N) = \begin{bmatrix} \boldsymbol{f}^T(n) \\ \boldsymbol{f}^T(n+1) \\ \cdot \\ \cdot \\ \cdot \\ \boldsymbol{f}^T(N) \end{bmatrix} \tag{15.7}$$

$$\boldsymbol{e}(N) = \begin{bmatrix} e(n) \\ e(n+1) \\ \cdot \\ \cdot \\ \cdot \\ e(N) \end{bmatrix} \tag{15.8}$$

$$\boldsymbol{f}^T(k) = [y(k-1) \quad y(k-2) \dots \ y(k-n) \quad u(k-1) \quad u(k-2) \dots \ u(k-n)] \tag{15.9}$$

The vectors/matrices $\boldsymbol{y}(N)$, $\boldsymbol{F}(N)$, $\boldsymbol{\theta}$, and $\boldsymbol{e}(N)$ are of size $(N - n + 1) \times 1$, $(N - n + 1) \times 2n$, $2n \times 1$, and $(N - n + 1) \times 1$, respectively. The error in the estimation of system response at sampling instant k is represented as $e(k)$.

The cost function is constructed from the error vector as

$$J_N(\boldsymbol{\theta}) = \sum_{k=n}^{N} e^2(k) = \boldsymbol{e}^T(N)\,\boldsymbol{e}(N) \tag{15.10}$$

The cost function can be expressed using Eq. (15.5) as

$$J_N(\boldsymbol{\theta}) = \boldsymbol{y}^T(N)\boldsymbol{y}(N) - \boldsymbol{2\theta}^T\boldsymbol{F}^T(N)\boldsymbol{y}(N)) + (\boldsymbol{\theta}^T\boldsymbol{F}^T(N)\boldsymbol{F}(N)\boldsymbol{\theta}) \tag{15.11}$$

The least-squares estimate $\hat{\boldsymbol{\theta}}$ of the parameter vector $\boldsymbol{\theta}$ can be obtained by minimizing the cost function as

$$\frac{\partial J_N(\boldsymbol{\theta})}{\partial \boldsymbol{\theta}} = -2\boldsymbol{F}^T(N)\boldsymbol{y}(N) + 2(\boldsymbol{F}^T(N)\boldsymbol{F}(N)\boldsymbol{\theta}) = 0 \tag{15.12}$$

$$\hat{\boldsymbol{\theta}} = [\boldsymbol{F}^T(N)\boldsymbol{F}(N)]^{-\mathbf{1}}(\boldsymbol{F}^T(N)\boldsymbol{y}(N)) \tag{15.13}$$

It is assumed that the input sequence $u(k)$ is persistently exciting so that all dynamics of the system are sufficiently excited giving rise to the dynamically rich system response ensuring the matrix $[\boldsymbol{F}^T(N)\boldsymbol{F}(N)]$ is non-singular.

Example 15.1 Least-Squares System Identification A second-order system with the discrete-time transfer function

$$G(z) = \frac{Y(z)}{U(z)} = \frac{0.0861\,z}{z^2 - 1.7235z + 0.7408}$$

is to be identified using the least-squares system identification algorithm. A general input sequence, $\{u(k)\}$,

$$u(k) = [0.5\ 0.75\ 1.0\ 0.75\ 1.0\ 1.25\ 0.95\ 0.65\ 0.50\ 0.45\ 0.65]$$

was used to simulate the system output $y(k)$ that was used to form the measurement vector. Use least-squares system identification to estimate the system parameters: the coefficients (a_i, b_i, $i = 1, 2$) of the denominator and the numerator of the transfer function $G(z)$.

Solution Matlab code and results are presented.

```
% ch15_ex 1-Least-squares system identification
% G(z) =0.0861z/(z-0.9048)(z-0.8187)
% G(z)=Y(z)/U(z)=0.0861z^-1/(1-1.7235z^-1+0.7408z^-2)
% y(k)=1.7235 y(k-1)-0.7408y(k-2)+0.0861u(k)
% generate output y(k) for u(k) k>=0
clear all; clc
% initial values
uk1=0;
uk2=0;
yk1=0;
yk2=0;
% input sequence (persistently exciting the system)
uk=[0.5 0.75 1.0 0.75 1.0 1.25 0.95 0.65 0.50 0.45 0.65];
n=4; N=4;nu=length(uk)
f= zeros(nu,n);
yM=zeros(nu,2);
for i=1:nu
    yk =1.7235*yk1-0.7408*yk2+0.0861*uk1 +0*uk2;
    yk2=yk1;
    yk1=yk;
    uk2=uk1;
    uk1=uk(i);
    yM(i,:)= [yk uk(i)];
    f(i,:) = [yk1  yk2 uk1 uk2];
end
% Least-Squares Estimation  a1, a2, b1, b2
nS = 4 % starting sample nStart
F = zeros(N,n);
F(1:N,:) = f(nS:N+nS-1,:) % nStart:N+nStart-1
y =zeros(N,1);
y = yM(nS+1:N+nS,1) % nStart+1:N+nStart
FtF = F'*F;
Fty = F'*y;
th = FtF\Fty % estimated parameters (ai, bi)

Results:
F = 0.2934    0.1388    0.7500    1.0000
    0.4674    0.2934    1.0000    0.7500
    0.6744    0.4674    1.2500    1.0000
    0.9236    0.6744    0.9500    1.2500

y = 0.4674
    0.6744
    0.9236
    1.1741

th = 1.7235
    -0.7408
     0.0861
     0.0000
The estimated parameters are same as the actual parameters.
```

15.3 Weighted Least-Squares System Identification

Very often, the cost function of Eq. (15.10) is modified to give weight to the errors as

$$J_N(\boldsymbol{\theta}) = \sum_{k=n}^{N} w(k)e^2(k) = \boldsymbol{e}^T(N)\boldsymbol{W}(N)\boldsymbol{e}(N) \tag{15.14}$$

where $\boldsymbol{W}(N)$ is a $N \times N$ diagonal weighting matrix with elements $w_{ii} = w(i + n - 1)$, $i = 1, \ldots, N + 1$. Following the process presented in the previous section, the weighted least-squares estimate of the parameter vector $\widehat{\boldsymbol{\theta}}_{WLS}$ can be obtained as

$$\hat{\boldsymbol{\theta}}_{WLS} = \left[\boldsymbol{F}^T(N)\boldsymbol{W}(N)\boldsymbol{F}(N)\right]^{-1}\left(\boldsymbol{F}^T(N)\boldsymbol{W}(N)\boldsymbol{y}(N)\right) \tag{15.15}$$

A most commonly used weighting term is given as

$$w(k) = a\gamma^{N-k}, \quad \gamma \le 1 \tag{15.16}$$

For exponential weighting, the factor a is chosen as

$$a = 1 - \gamma \tag{15.17}$$

If $\gamma < 1$, the most recent data gets dominance over the distant past data, and as γ approaches 1, the distant past data gets prominence over the most recent data in the estimation process. If $a = \gamma = 1$, all errors are equally weighted resulting in the ordinary least-squares estimation process.

Example 15.2 Weighted Least-Squares System Identification A second-order system with the discrete-time transfer function

$$G(z) = \frac{Y(z)}{U(z)} = \frac{0.0861\, z}{z^2 - 1.7235z + 0.7408}$$

to be identified using a weighted least-squares system identification algorithm. A general input sequence

$$u(k) = [0.5\ 0.75\ 1.0\ 0.75\ 1.0\ 1.25\ 0.95\ 0.65\ 0.50\ 0.45\ 0.65]$$

was used to simulate the system output $y(k)$ that was used to form the measurement vector. Use weighted least-squares system identification to estimate the system parameters: the coefficients (a_i, b_i, $i = 1, 2$) of the denominator and the numerator of the transfer function $G(z)$. Use $g = 0.35$; $a = 1 - g$.

Solution Matlab code and results are presented.

```
% ch15_ex 2-Weighted Least-squares system identification
% G(z) =0.0861z/(z-0.9048)(z-0.8187)
% G(z)=Y(z)/U(z)=0.0861z^-1/(1-1.7235z^-1+0.7408z^-2)
% y(k)=1.7235 y(k-1)-0.7408y(k-2)+0.0861u(k)
% generate output y(k) for u(k) k>=0
clear all; clc
% Intial conditions
uk1=0;
uk2=0;
yk1=0;
yk2=0;
% input sequence
uk=[0.5 0.75 1.0 0.75 1.0 1.25 0.95 0.65 0.50 0.45 0.65]
n=4; N=4;nu=length(uk)
f= zeros(nu,n);
yM=zeros(nu,2);

% Weighting Matrix
W=zeros(N,N);
g=0.35; a=1-g;
for i=1:N
    W(i,i)=a*g^(N-i);
end
W
for i=1:nu
    yk =1.7235*yk1-0.7408*yk2+0.0861*uk1 +0*uk2;
    yk2=yk1;
    yk1=yk;
    uk2=uk1;
    uk1=uk(i);
    yM(i,:)= [yk uk(i)];
    f(i,:) = [yk1  yk2 uk1 uk2];

end
% Least-Squares Estimation a0, a1, a2, b0, b1, b2
nS = 3 % starting sample nStart
F = zeros(N,n);
F(1:N,:) = f(nS:N+nS-1,:) % nStart:N+nStart-1
y =zeros(N,1);
y = yM(nS+1:N+nS,1) % nStart+1:N+nStart
FtWF = F'*W*F
FtWy = F'*W*y
th = FtWF\FtWy


Results:

W =          0.0279            0         0            0
                  0       0.0796         0            0
                  0            0    0.2275            0
                  0            0              0  0.6500

F =      0.1388    0.0430    1.0000    0.7500
         0.2934    0.1388    0.7500    1.0000
         0.4674    0.2934    1.0000    0.7500
         0.6744    0.4674    1.2500    1.0000
```

```
y =        0.2934
           0.4674
           0.6744
           0.9236

FtWF =           0.3527    0.2395    0.6756    0.5443
                 0.2395    0.1632    0.4560    0.3658
                 0.6756    0.4560    1.3158    1.0637
                 0.5443    0.3658    1.0637    0.8733

FtWy =           0.4886
                 0.3311
                 0.9399
                 0.7588
th =      1.7235
         -0.7408
          0.0861
         -0.0000
```

15.4 Recursive Least-Squares System Identification

In Sects. 15.2 and 15.3, it is assumed that the entire measurement datasets are available for batch processing in the estimation of system parameters. However, there is an alternative process to perform the estimation of the system parameter vector $\boldsymbol{\theta}$ as each new data is available resulting in a recursive least-squares estimation.

Following Eqs. (15.2), (15.3), and (15.9), the measurement at sampling instant k, $y(k)$, can be expressed as

$$y(k) = \boldsymbol{f}^T(k)\boldsymbol{\theta} + \mathrm{e(k)} \tag{15.18}$$

From Eq. (15.7), the measurement matrix at $N + 1$ can be expressed as

$$\boldsymbol{F}^T(N+1) = [\boldsymbol{f}(n)\,\boldsymbol{f}(n+1)\ldots\boldsymbol{f}(N)\,\boldsymbol{f}(N+1)\,] \tag{15.19}$$

The weighting factor of Eq. (15.16) is written as

$$w(k) = a\gamma^{N+1-k}, \quad \gamma \leq 1 \tag{15.20}$$

The weighted matrix can be written as

$$\boldsymbol{F}^T(N+1)\boldsymbol{W}(N+1)\boldsymbol{F}\ (N+1)=\sum_{k=n}^{N+1}\boldsymbol{f}(k)\ a\gamma^{N+1-k}\boldsymbol{f}^T(k) \tag{15.21}$$

Equation (15.21) can be expressed as

$$\sum_{k=n}^{N+1}\boldsymbol{f}(k)\ a\gamma^{N+1-k}\boldsymbol{f}^T(k)=\sum_{k=n}^{N}\boldsymbol{f}(k)\ a\gamma\gamma^{N-k}\boldsymbol{f}^T(k)+\boldsymbol{f}(N+1)a\boldsymbol{f}^T(N+1) \tag{15.22}$$

Equation (15.22) can be expressed recursively as

$$\begin{aligned}\boldsymbol{F}^T(N+1)\boldsymbol{W}(N+1)\boldsymbol{F}\ (N+1)&=\gamma\boldsymbol{F}^T(N)\boldsymbol{W}(N)\boldsymbol{F}\ (N)\\&+\boldsymbol{f}(N+1)a\boldsymbol{f}^T(N+1)\end{aligned} \tag{15.23}$$

A $2n \times 2n$ matrix is defined as

$$\boldsymbol{P}(k)=\left[\boldsymbol{F}^T(k)\boldsymbol{W}(k)\boldsymbol{F}\ (k)\right]^{-1} \tag{15.24}$$

Equation (15.23) can be expressed, using Eq. (15.24), as

$$\boldsymbol{P}^{-1}(N+1)=\gamma\boldsymbol{P}^{-1}(N)+\boldsymbol{f}(N+1)a\boldsymbol{f}^T(N+1) \tag{15.25}$$

Next, the matrix inversion relation of Eq. (15.26) is examined, in relation to Eq. (15.25), as

$$(\boldsymbol{A}+\boldsymbol{BCD})^{-1}=\boldsymbol{A}^{-1}-\boldsymbol{A}^{-1}\boldsymbol{B}\left(\boldsymbol{C}^{-1}+\boldsymbol{DA}^{-1}\boldsymbol{B}\right)^{-1}\boldsymbol{DA}^{-1} \tag{15.26}$$

$$\boldsymbol{A}=\gamma\boldsymbol{P}^{-1}(N) \tag{15.27}$$

$$\boldsymbol{B}=\boldsymbol{f}(N+1) \tag{15.28}$$

$$\boldsymbol{C}=a \tag{15.29}$$

$$\boldsymbol{D}=\boldsymbol{f}^T(N+1) \tag{15.30}$$

Equation (15.25) can be expressed as

$$\begin{aligned}\boldsymbol{P}(N+1)&=\left(\gamma\boldsymbol{P}^{-1}(N)+\boldsymbol{f}(N+1)a\boldsymbol{f}^T(N+1)\right)^{-1}\\&=\frac{1}{\gamma}\boldsymbol{P}(N)-\frac{1}{\gamma}\boldsymbol{P}(N)\boldsymbol{f}(N+1)\left(\frac{1}{a}+\boldsymbol{f}^T(N+1)\frac{1}{\gamma}\boldsymbol{P}(N)\boldsymbol{f}(N+1)\right)^{-1}\boldsymbol{f}^T(N+1)\frac{1}{\gamma}\boldsymbol{P}(N)\end{aligned} \tag{15.31}$$

The second factor in Eq. (15.15) can be written for $k = N + 1$ as

$$\boldsymbol{F}^T(N+1)\ \boldsymbol{W}(N+1)\boldsymbol{y}(N+1) = [\boldsymbol{f}(n)\ \boldsymbol{f}(n+1)\ldots\boldsymbol{f}(N)\ \boldsymbol{f}(N+1)\ ]\times$$

$$\begin{bmatrix} a\gamma^{N+1-n} & & & & \\ & a\gamma^{N-n} & & & \\ & & \cdots & & \\ & & & a\gamma & \\ & & & & a \end{bmatrix}\begin{bmatrix} y(n) \\ y(n+1) \\ . \\ . \\ . \\ y(N+1) \end{bmatrix}$$

$$= \gamma\boldsymbol{F}^T(N)\boldsymbol{W}(N)\boldsymbol{y}(N) + \boldsymbol{f}(N+1)a\boldsymbol{f}^T(N+1) \tag{15.32}$$

The weighted least-squares estimate from Eq. (15.15) can be written, using Eqs. (15.24), (15.31), and (15.32), in recursive least-squares form as

$$\boldsymbol{P}(N) = \left[\boldsymbol{F}^T(N)\ \boldsymbol{W}(N)\boldsymbol{F}(N)\right]^{-1} \tag{15.33}$$

$$\boldsymbol{L}(N+1) = \frac{1}{\gamma}\boldsymbol{P}(N)\boldsymbol{f}(N+1)\left(\frac{1}{a} + \boldsymbol{f}^T(N+1)\frac{1}{\gamma}\boldsymbol{P}(N)\boldsymbol{f}(N+1)\right)^{-1} \tag{15.34}$$

$$\begin{aligned} &\hat{\boldsymbol{\theta}}_{WLS}(N+1) \\ &\quad = \left[\boldsymbol{F}^T(N+1)\boldsymbol{W}(N+1)\boldsymbol{F}(N+1)\right]^{-1}\left(\boldsymbol{F}^T(N+1)\boldsymbol{W}(N+1)\boldsymbol{y}(N+1)\right) \\ &\quad = \hat{\boldsymbol{\theta}}_{WLS}(N) + \boldsymbol{L}(N+1)\left(y(N+1) - \boldsymbol{f}^T(N+1)\hat{\boldsymbol{\theta}}_{WLS}(N)\right) \end{aligned} \tag{15.35}$$

$$\boldsymbol{P}(N+1) = \frac{1}{\gamma}\left(\boldsymbol{I} - \boldsymbol{L}(N+1)\boldsymbol{f}^T(N+1)\boldsymbol{P}(N)\right) \tag{15.36}$$

Equations (15.33)–(15.36) are valid for $N \geq n$; these can be updated by replacing N with k. The process starts with some initial values of $\boldsymbol{P}(k)$and $\hat{\boldsymbol{\theta}}_{WLS}(k)$. For example, the matrix $\boldsymbol{P}$ can be initialized as an identity matrix of order N, $\boldsymbol{I}_N$, with the main diagonal of all 1 and the rest of the elements being 0.

$$P = \boldsymbol{I}_N \tag{15.37}$$

Next, the vector $\boldsymbol{f}(k+1)$ is obtained from the measurements as $\boldsymbol{f}^T(k+1) = [y(k)\ y(k-1)\ldots\ y(k-n)\quad u(k)\quad u(k-1)\ldots\ u(k-n)]$. The next set of measurements form $y(k+1)$ and Eqs. (15.34)–(15.36) are used recursively. It should be noted that the initial values of $\boldsymbol{P}(k)$ and $\hat{\boldsymbol{\theta}}_{WLS}(k)$ are assumed and the recursive relations are iterated a number of times till the estimates of $\hat{\boldsymbol{\theta}}_{WLS}(k)$ converge to nearly constant values.

Example 15.3 Recursive Least-Squares System Identification A second-order system with the discrete-time transfer function

$$G(z) = \frac{Y(z)}{U(z)} = \frac{0.0861\,z}{z^2 - 1.7235z + 0.7408}$$

is to be identified using the recursive least-squares system identification algorithm. A general input sequence

$$u(k) = [0.5\ 0.75\ 1.0\ 0.75\ 1.0\ 1.25\ 0.95\ 0.65\ 0.50\ 0.45\ 0.65\ 0.35\ 0.45\ 0.25\ 0.55]$$

was used to simulate the system output $y(k)$ that was used to form the measurement vector. Use weighted least-squares system identification to estimate the system parameters: the coefficients (a_i, b_i, $i = 1, 2$) of the denominator and the numerator of the transfer function $G(z)$. Use $g = 0.35$; $a = 1 - g$.

Solution Matlab code and results are presented.

```
% ch15_ex 3-Recursive Least-squares system identification
% G(z) =0.0861z/(z-0.9048)(z-0.8187)
% G(z)=Y(z)/U(z)=0.0861z^-1/(1-1.7235z^-1+0.7408z^-2)
% y(k)=1.7235 y(k-1)-0.7408y(k-2)+0.0861u(k)
% generate output y(k) for u(k) k>=0
clear all; clc

uk1=0;
uk2=0;
yk1=0;
yk2=0;
uk=[0.5 0.75 1.0 0.75 1.0 1.25 0.95 0.65 0.50 0.45 0.65 0.35 0.45 0.25
0.55]
n=4; N=4;nu=length(uk)
f= zeros(nu,n);
yM=zeros(nu,2);
W=zeros(N,N);
% weighting matrix
g=0.35; a=1-g;
for i=1:N
    W(i,i)=a*g^(N-i);
end
W
% generation of measurement signals
for i=1:nu
    yk =1.7235*yk1-0.7408*yk2+0.0861*uk1 +0*uk2;
    yk2=yk1;
    yk1=yk;
    uk2=uk1;
    uk1=uk(i);
    yM(i,:)= [yk uk(i)];
    f(i,:) = [yk1  yk2 uk1 uk2];

end
```

```
% Least-Squares Estimation a0, a1, a2, b0, b1, b2
nS = 3 % starting sample nStart
F = zeros(N,n);
F(1:N,:) = f(nS:N+nS-1,:) % nStart:N+nStart-1
y =zeros(N,1);
y = yM(nS+1:N+nS,1) % nStart+1:N+nStart
FtWF = F'*W*F
FtWy = F'*W*y
th = FtWF\FtWy
% Recursive WLS RWLS
P = eye(N)
for i=1:3
    fN1 = (f(nS+i,:))'
    fPf =fN1'*P*fN1
    Lpi = (1/a+(1/g)*fPf)
    L = (1/g)*P*fN1*(Lpi)^-1
    th = th + L*(yM(nS+i+1,1)-fN1'*th) %RWLS
    P = (1/g)*(eye(N)-L*fN1'*P)
End
```

```
Results of one iteration are shown. As the initial estimated values of
system parameters were close to the actual ones based on first batch of
measured data, the corrections to the estimate were not significant.
F =
```

```
0.1388     0.0430     1.0000     0.7500
    0.2934     0.1388     0.7500     1.0000
    0.4674     0.2934     1.0000     0.7500
    0.6744     0.4674     1.2500     1.0000
y =
    0.2934
    0.4674
    0.6744
    0.9236

FtWF =
    0.3527     0.2395     0.6756     0.5443
    0.2395     0.1632     0.4560     0.3658
    0.6756     0.4560     1.3158     1.0637
    0.5443     0.3658     1.0637     0.8733

FtWy =
    0.4886
    0.3311
    0.9399
    0.7588


th =
    1.7235
   -0.7408
    0.0861
   -0.0000
```

```
P =
       1     0     0     0
       0     1     0     0
       0     0     1     0
       0     0     0     1

fN1 =
     0.2934
     0.1388
     0.7500
     1.0000

fPf =
     1.6678

Lpi =
     6.3037

L =
     0.1330
     0.0629
     0.3399
     0.4532

th =
     1.7235
    -0.7408
     0.0861
    -0.0000

 P =
     2.7457  -0.0527  -0.2849  -0.3799
    -0.0527   2.8322  -0.1348  -0.1797
    -0.2849  -0.1348   2.1287  -0.9712
    -0.3799  -0.1797  -0.9712   1.5621
```

Example 15.4 The measurement pairs for a third-order discrete-time system $G(z)$ with unknown coefficients a_i, b_i, $i = 1, \ldots, 3$ are given in Table 15.1.

$$G(z) = \frac{Y(z)}{U(z)} = \frac{b_1 z^2 + b_2 z + b_3}{z^3 + a_1 z^2 + a_2 z + a_3}$$

Table 15.1 Measurement pairs for a third order system

k	$u(k)$	$y(k)$
0	0.50	$0.0000e + 00$
1	0.75	$2.6649e - 04$
2	1.00	$6.6546e - 04$
3	0.75	$1.1380e - 03$
4	1.00	$1.1429e - 03$
5	1.25	$1.8055e - 03$
6	0.95	$5.8369e - 05$
7	0.65	$7.2713e - 03$
8	0.50	$-1.9122e - 02$
9	0.45	7.5862e − 02

Estimate the system parameters through system identification algorithms of (a) least squares, (b) weighted least squares, and (c) recursive least squares.

Solution Matlab code and results are presented.

```
% ch15_ex15_4_system identification using LS, WLS, RWLS
clear all; clc
% initial values
uk1=0;
uk2=0;
uk3=0;
yk1=0;
yk2=0;
yk3=0;

% Maeasurement pairs
Uk=[0.5          0.75         1.0          0.75         1.0          1.25
0.95         0.65         0.50         0.45]
Yk=[0.0000e+00  2.6649e-04  6.6546e-04  1.1380e-03  1.1429e-03  1.8055e-03
5.8369e-05  7.2713e-03  -1.9122e-02  7.5862e-02]
% system dimension
n=3; n2=2*n; N=n2;nu=length(Uk)
f= zeros(nu,n2);
yM=zeros(nu,2);

% Weighting Matrix
W=zeros(N,N);
g=0.35; a=1-g;
for i=1:N
    W(i,i)=a*g^(N-i);
end
W;
for i=1:nu
    yk=Yk(i);
    uk=Uk(i);
    yk3=yk2;
    yk2=yk1;
    yk1=yk;
    uk3=uk2;
    uk2=uk1;
    uk1=uk;
    yM(i,:)= [yk uk];
    f(i,:) = [yk1  yk2 yk3 uk1 uk2 uk3];
end

nS = 1 % starting sample nStart
F = zeros(N,n2);
F(1:N,:) = f(nS:N+nS-1,:) % nStart:N+nStart-1
y =zeros(N,1);
%%%%%%%%%%%%%%%%%%%%%%%%%%%%%  LS %%%%%%%%%%%%%%%%%%%%%%%%
y = yM(nS+1:N+nS,1) % nStart+1:N+nStart
FtF = F'*F;
Fty = F'*y;
th = FtF\Fty % estimated parameters (ai, bi)
fprintf('%8.4e\n', th)
%%%%%%%%%%%%%%%%%%%%%%%%%%%%% WLS %%%%%%%%%%%%%%%%%%
FtWF = F'*W*F
FtWy = F'*W*y
thw = FtWF\FtWy
fprintf('%8.4e\n', thw)
%%%%%%%%%%%%%%%%%%%%%%%%%%%%% RWLS %%%%%%%%%%%%%%%%%
FtWF = F'*W*F;
FtWy = F'*W*y;
th = FtWF\FtWy;
% Recursive WLS RWLS
P = eye(N);
```

```
for i=1:6
    fN1 = (f(nS+i,:))';
    fPf =fN1'*P*fN1;
    Lpi = (1/a+(1/g)*fPf);
    L = (1/g)*P*fN1*(Lpi)^-1;
    th = th + L*(yM(nS+i+1,1)-fN1'*th) %RWLS
    P = (1/g)*(eye(N)-L*fN1'*P)
end
fprintf('%8.4e\n', th)
%%%%%%%%%%%%%%%%%%%%%%%%%%%%%%%%%%%%%%%%%%%%%%%%%%%%
```

```
Results:
(a)   Least-squares system identification

      F =

               0         0         0    0.5000         0         0
          0.0003         0         0    0.7500    0.5000         0
          0.0007    0.0003         0    1.0000    0.7500    0.5000
          0.0011    0.0007    0.0003    0.7500    1.0000    0.7500
          0.0011    0.0011    0.0007    1.0000    0.7500    1.0000
          0.0018    0.0011    0.0011    1.2500    1.0000    0.7500


      y =

          0.0003
          0.0007
          0.0011
          0.0011
          0.0018
          0.0001


      th =

         -2.8407
          2.6875
         -0.8468
          0.0005
          0.0020
          0.0005

      -2.8407e+00
      2.6875e+00
      -8.4680e-01
      5.3298e-04
      2.0455e-03
      4.9015e-04

(b)   Weighted least-squares system identification

      FtWF =

          0.0000    0.0000    0.0000    0.0018    0.0015    0.0012
          0.0000    0.0000    0.0000    0.0012    0.0010    0.0009
          0.0000    0.0000    0.0000    0.0011    0.0009    0.0007
          0.0018    0.0012    0.0011    1.3221    1.0674    0.8956
          0.0015    0.0010    0.0009    1.0674    0.8757    0.7283
          0.0012    0.0009    0.0007    0.8956    0.7283    0.6449
```

```
FtWy =

    1.0e-03 *

      0.0007
      0.0006
             0.0003
             0.5635
             0.4640
             0.5233


      thw =

          -2.8407
           2.6875
          -0.8468
           0.0005
           0.0020
           0.0005

      -2.8407e+00
      2.6875e+00
      -8.4680e-01
      5.3298e-04
      2.0455e-03
      4.9015e-04

(c)   Recursive least-squares system identification

      P =

          2.8571   -0.0000    0.0000   -0.0164    0.0017   -0.0336
         -0.0000    2.8571    0.0000   -0.0273    0.0028   -0.0560
          0.0000    0.0000    2.8571    0.0317   -0.0032    0.0651
         -0.0164   -0.0273    0.0317  -48.7856    5.2406 -105.9192
          0.0017    0.0028   -0.0032    5.2406    2.3253   10.7484
         -0.0336   -0.0560    0.0651 -105.9192   10.7484 -214.3831

      -2.8407e+00
      2.6875e+00
      -8.4680e-01
      5.3274e-04
      2.0455e-03
      4.8966e-04

                 0.000533 z^2 + 0.002045 z + 0.000490
      G(z)= -------------------------------------------------
                     z^3 - 2.84 z^2 + 2.687 z - 0.8465
```

15.5 Laboratory Experiment for Identification of Discrete-Time Systems

Objectives The objectives are to collect open-loop speed response data for a general input dataset using a DC motor module and to estimate model parameters of the discrete-time transfer function in z-domain $G(z)$ for the DC motor module using least-squares system identification algorithms.

Introduction The discrete-time model of an n-order dynamic system can be represented as a z-domain transfer function $G(z)$:

$$G(z) = \frac{Y(z)}{U(z)} = \frac{b_1 z^{n-1} + b_2 z^{n-2} + \ldots + b_n}{z^n - a_1 z^{n-1} - a_2 z^{n-2} - \ \ldots\ - a_n}$$

For a lab scale DC motor module similar to the DC motor control trainer (DCMCT) from Quanser, the motor speed dynamics can be represented by a first-order model ($n = 1$) with two parameters (a_1, b_1). In Part I, it is required to set up a data collection system in LabVIEW for the DC motor module and collect motor speed response data for a general input voltage sequence. In part II, the collected dataset of speed response and the input voltage is used to estimate the model parameters (a_1, b_1) using least-squares system identification algorithms.

I. Set up Data Collection.

1. Set up NI ELVIS II board and DC motor module. Select the device, and set the sampling rate. For the DC motor control trainer (DCMCT) module, a sampling rate of 1 kHz is used for data collection.
2. Set up the DC motor module for open-loop speed response configuration so that it can take a given input signal and give the motor speed as output. Modify the VI from the DCMCT module library as necessary.
3. Set up the system to gather DC motor speed response for a 3 V_{pp}, 0.5 Hz sine wave input for a period of 10 s with a sampling period of 0.001 s. Add a "write to measurement file" to store the collected dataset.
4. Add a low-pass filter block (with a corner frequency of around 10 Hz) from the VI library of LabVIEW to filter the DC motor speed response. Store both filtered and unfiltered data using the "write to measurement file" block. Refer to the block diagram VI shown in Fig. 15.1.
5. Export both unfiltered and filtered datasets to Excel that can be exported to Matlab for further processing for system identification.
6. Collect datasets of DC motor speed response for (a) 3 V_{pp}, 1 Hz sine wave and (b) 3 V_{pp}, 1 Hz triangular wave voltage inputs.

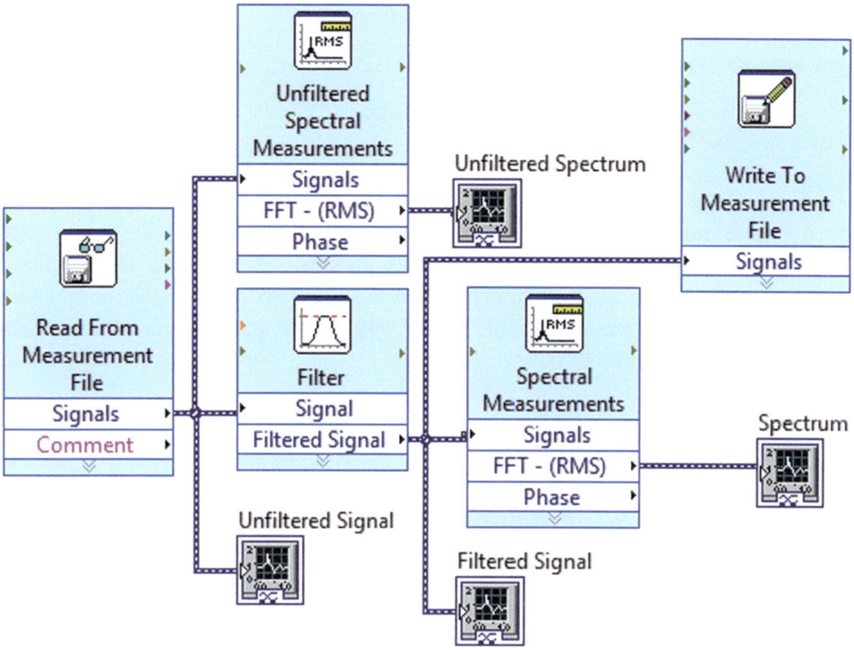

Fig. 15.1 LabVIEW VI for filtering tachometer data of DC motor module

Table 15.2 Summary of estimated model parameters

System identification algorithm	Estimated model parameters	
	a_1	b_1
Least squares		
Weighted least squares		
Recursive least squares		

II. System Identification.

1. Use the MATLAB code developed for least squares estimation for the parameters of the motor discrete-time transfer function from the measured input and speed response datasets.
2. Compare the system parameters with both unfiltered and filtered motor response data. Note down the estimated model parameters in Table 15.2.
3. Repeat steps (1) and (2) for model parameter estimation using weighted least-squares and recursive least-squares system identification algorithms.
4. Simulate the DC motor speed response for the 3 V_{pp}, 0.5 Hz sine wave voltage input, and compare it with the actual DC motor speed response dataset for the same sine wave voltage input. Estimate root-mean-square error (RMSE) between the simulated and the measured speed response and record in Table 15.3. Copy the comparison graph for presentation in the lab report.

Table 15.3 Summary of system response RMSE for different input waveforms

Input voltage waveform	RMSE
3 V_{pp}, 0.5 Hz sine	
3 V_{pp}, 1.0 Hz sine	
3 V_{pp}, 1.0 Hz triangular	

5. Simulate the open-loop DC motor speed response for (a) 3 V_{pp}, 1 Hz sine wave and (b) 3 V_{pp}, 1 Hz triangular voltage inputs. Compare these simulated responses with the collected datasets for the corresponding input voltage waveforms. Estimate RMSE between the simulated and the measured speed response, and record in Table 15.3. Copy the comparison graphs for presentation in the lab report.

III. Deliverables.

1. Complete Tables 15.2 and 15.3.
2. Present graph for comparison of simulated and actual DC motor open-loop speed responses for original 3 V_{pp}, 0.5 Hz sine wave voltage input.
3. Present graphs of comparison of simulated and actual DC motor open-loop speed responses for (a) 3 V_{pp}, 1 Hz sine wave and (b) 3 V_{pp}, 1 Hz triangular wave voltage inputs.
4. Discuss results comparing the performance of different system identification algorithms.
5. Discuss the results of comparing the simulated and measured speed responses for different input voltage waveforms.

Lab Report
Present all deliverables in the lab report and discuss your results.

Exercises

15.1 A second-order system with the discrete-time transfer function

$$G(z) = \frac{Y(z)}{U(z)} = \frac{0.07446\, z + 0.05515}{z^2 - 1.2705z + 0.4046}$$

is to be identified using the least-squares system identification algorithm. A general input sequence

$$u(k) = [0.5\ 0.75\ 1.0\ 0.75\ 1.0\ 1.25\ 0.95\ 0.65\ 0.50\ 0.45\ 0.65]$$

was used to simulate the system output $y(k)$ that was used to form the measurement vector. Use least-squares system identification to estimate the system parameters: the coefficients (a_i, b_i, $i = 1, 2$) of the denominator and the numerator of the transfer function $G(z)$.

15.2 A third-order system with the discrete-time transfer function

$$G(z) = \frac{Y(z)}{U(z)} = \frac{0.0173\, z^2 + 0.05745z + 0.01189}{z^3 - 2.337z^2 + 1.828z - 0.4724}$$

is to be identified using the least-squares system identification algorithm. A general input sequence

$$u(k) = [0.5\ 0.75\ 1.0\ 0.75\ 1.0\ 1.25\ 0.95\ 0.65\ 0.50\ 0.45\ 0.65]$$

was used to simulate the system output $y(k)$ that was used to form the measurement vector. Use least-squares system identification to estimate the system parameters: the coefficients (a_i, b_i, $i = 1, 2, 3$) of the denominator and the numerator of the transfer function $G(z)$.

15.3 A fourth-order system with the discrete-time transfer function

$$G(z) = \frac{Y(z)}{U(z)} = \frac{3.693\text{e} - 05\, z^3 + 0.0003599z^2 + 0.0003194\, z + 2.577\text{e} - 05}{z^4 - 3.409z^3 + 4.406z^2 - 2.545z + 0.5488}$$

to be identified using the least-squares system identification algorithm. A general input sequence

$$u(k) = [0.5\ 0.75\ 1.0\ 0.75\ 1.0\ 1.25\ 0.95\ 0.65\ 0.50\ 0.45\ 0.65]$$

was used to simulate the system output $y(k)$ that was used to form the measurement vector. Use least-squares system identification to estimate the system parameters: the coefficients (a_i, b_i, $i = 1, \ldots, 4$) of the denominator and the numerator of the transfer function $G(z)$.

15.4 The measurement pairs for a second-order discrete-time system $G(z)$ with unknown coefficients a_i, b_i, $i = 1, 2$ are given in Table P15.1.

Table P15.1 Measurement pairs for a second order system

k	$u(k)$	$y(k)$
0	0.50	$0.0000e + 00$
1	0.75	$0.2575e - 02$
2	1.00	$0.2507e - 02$
3	0.75	$0.5516e - 02$
4	1.00	$0.1468e - 02$
5	1.25	$0.7646\, e - 02$
6	0.95	$0.1179e - 02$
7	0.65	$1.0106e - 02$
8	0.50	$-0.5118e - 02$
9	0.45	$1.3998\text{e} - 02$

$$G(z) = \frac{Y(z)}{U(z)} = \frac{b_1 z + b_2}{z^2 + a_1 z + a_2}$$

Estimate the system parameters through system identification algorithms of (a) least squares, (b) weighted least squares, and (c) recursive least squares.

15.5 The measurement pairs for a third-order discrete-time system $G(z)$ with unknown coefficients a_i, b_i, $i = 1, 2, 3$ are given in Table P15.2.

$$G(z) = \frac{Y(z)}{U(z)} = \frac{b_1 z^2 + b_2 z + b_3}{z^3 + a_1 z^2 + a_2 z + a_3}$$

Estimate the system parameters through system identification algorithms of (a) least squares, (b) weighted least squares, and (c) recursive least squares.

Table P15.2 Measurement pairs for a third order system

k	$u(k)$	$y(k)$
0	0.50	$0.0000e + 00$
1	0.75	$0.0877e - 03$
2	1.00	$0.2168e - 03$
3	0.75	$0.3290e - 03$
4	1.00	$0.3992e - 03$
5	1.25	$0.3092e - 03$
6	0.95	$0.7647e - 03$
7	0.65	$-0.4089e - 03$
8	0.50	$3.1242e - 03$
9	0.45	$-8.0633e - 03$

15.6 The measurement pairs for a third-order discrete-time system $G(z)$ with unknown coefficients a_i, b_i, $i = 1, \ldots, 4$ are given in Table P15.3.

Table P15.3 Measurement pairs for a fourth order system

k	$u(k)$	$y(k)$
0	0.50	$0.0000e + 00$
1	0.75	1.9426e-05
2	1.00	1.5724e – 04
3	0.75	4.5332e-05
4	1.00	1.2879e-03
5	1.25	−4.2270e-0
6	0.95	2.2699e-02
7	0.65	−1.0744e-01
8	0.50	5.2268e-01
9	0.45	−2.5278e+00

$$G(z) = \frac{Y(z)}{U(z)} = \frac{b_1 z^3 + b_2 z^2 + b_3 z + b_4}{z^4 + a_1 z^3 + a_2 z^2 + a_3 z + a_4}$$

Estimate the system parameters through system identification algorithms of (a) least squares, (b) weighted least squares, and (c) recursive least squares.

Further Reading

1. Asadi F (2023) Signals and systems with MATLAB® and Simulink®. Springer, Cham
2. Astrom KJ, Murray RM (2021) Feedback systems: an introduction for scientists and engineers, 2nd edn. Princeton University Press, Princeton
3. Chen C-T (2006) Analog and digital control system design: transfer-function, state-space, and algebraic methods. Oxford University Press, New York
4. Dorf R, Bishop R (2016) Modern control systems, 13th edn. Pearson, Essex
5. Fadali MS, Visioli A (2020) Digital control engineering analysis and design, 3rd edn. Academic, London
6. Franklin GF, Powell JD, Workman M (1998) Digital control of dynamic systems, 3rd edn. Addison-Wesley, Boston
7. Franklin GF, Powell JD, Emami-Naeini A (2019) Feedback control of dynamic systems, 8th edn. Pearson, Upper Saddle River
8. Isermann R (1981) Digital control systems. Springer, Berlin
9. Jalili N, Candelino NW (2023) Dynamic systems and control engineering. Cambridge University Press, Cambridge
10. Keviczky L, Bars R, Hetthéssy J, Bányász (2019) Control engineering. Springer, Singapore
11. Koeing DM (2009) Practical control engineering: a guide for engineers, managers, and practitioners. McGraw-Hill, New York
12. Kuo BC (1995) Digital control systems, 2nd edn. Oxford University Press, Oxford
13. Landau ID, Gito G (2006) Digital control systems: design, identification and implementation. Springer, London
14. Leigh JR (2006) Applied digital control: theory, design and implementation, 2nd edn. Dover, Mineola
15. Mathworks (2024a) Control System Toolbox. https://www.mathworks.com/products/control.html
16. Mathworks (2024b) Matlab. https://www.mathworks.com/products/matlab.html
17. National Instruments (2024a) LabVIEW. https://www.ni.com/en/support/downloads/software-products/download.labview.html#487445
18. National Instruments (2024b) NI-ELVIS-II. https://www.ni.com/en-us/support/model.ni-elvis-ii.html
19. National Instruments (2024c) Labs for Quanser QNET DC Motor Add-On Board. https://education.ni.com/teach/resources/180/labs-for-quanser-qnet-dc-motor-add-on-board
20. Nise NS (2019) Control systems engineering, 8th edn. Wiley, Hoboken
21. Nof SY (ed) (2009) Springer handbook of automation. Springer, Berlin
22. Ogata K (1995) Discrete-time control systems, 2nd edn. Pearson, Hoboken
23. Phillips CL, Nagle HT, Chakrabortty A (2015) Digital control system analysis and design, 4th edn. Pearson, Essex
24. Quanser (2024) QUBE Servo 2. https://www.quanser.com/products/qube-servo-2/
25. Rabbath CA, Lechevin N (2014) Discrete-time control system design with applications. Springer, New York
26. Samanta B (2023) Introduction to mechatronics – an integrated approach. Springer, Cham

27. Shertukde HM (2015) Digital control applications illustrated with MATLAB®. CRC Press, Boca Raton
28. Starr G (2020) Introduction to applied digital controls. Springer, Cham
29. Turner J, Samanta B (2012) Least squares estimation of dynamic system parameters using LabVIEW. In: IEEE South East conference, Orlando, FL, March 15–18, 2012. https://doi.org/10.1109/SECon.2012.6196962
30. Veloni A, Miridakis NI (2018) Digital control systems-theoretical problems and simulation tools. CRC Press, Boca Raton
31. Vilanova R, Visioli A (2012) PID control in the third millennium – lessons learned and new approaches. Springer, London
32. Visioli A (2006) Practical PID control. Springer, London

Chapter 16
Design of Optimal Controllers for Discrete-Time Systems in State Space

16.1 Introduction

Methods for designing optimal controllers for discrete-time systems are presented in this chapter. Linear quadratic optimal controllers and linear quadratic optimal estimators (Kalman filters) for discrete-time systems are covered. Examples along with Matlab codes and Simulink models are presented to illustrate the procedures. Laboratory experiments for analysis, design in state space, simulation, and real-time implementation of the optimal controllers on a DC motor module are presented. Analysis, design, and simulation are presented using Matlab, and implementation of the discrete-time controllers is presented using LabVIEW. End-of-chapter exercises are provided to enhance understanding of the topics covered and consolidate problem-solving skills.

16.2 Design of Linear Quadratic Optimal Controllers

The n-order discrete-time system can be represented in state space form as

$$\boldsymbol{x}(k+1) = \boldsymbol{A}\,\boldsymbol{x}(k) + \boldsymbol{B}\,\boldsymbol{u}(k) \tag{16.1}$$

$$\boldsymbol{y}(k) = \boldsymbol{C}\boldsymbol{x}(k) + \boldsymbol{D}\,\boldsymbol{u}(k) \tag{16.2}$$

where the system matrices $\boldsymbol{A}$, $\boldsymbol{B}$, and $\boldsymbol{C}$ are of size $n \times n$, $n \times p$, and $r \times n$, respectively, and the vectors of states $\boldsymbol{x}(k)$, inputs $\boldsymbol{u}(k)$, and outputs $\boldsymbol{y}(k)$ are of size $n \times 1$, $p \times 1$, and $r \times 1$, respectively. Here, matrix $\boldsymbol{D}$ of size $r \times p$ is assumed to be a null matrix.

B. Samanta, *Introduction to Digital Control*,
https://doi.org/10.1007/978-3-031-66830-2_16

The optimal control law is given as

$$\boldsymbol{u}(k) = -\boldsymbol{K}\,\boldsymbol{x}(k) \tag{16.3}$$

where the optimal controller gain matrix $\boldsymbol{K}$ of size $p \times n$ is obtained such that the quadratic cost function J_N presented in Eq. (16.4) is minimized. The quadratic cost function J_N is constructed taking into account the contributions from the state vector $\boldsymbol{x}(k)$ and the input vector $\boldsymbol{u}(k)$ with the corresponding weighting matrices: $\boldsymbol{Q}$ of size $n \times n$ and $\boldsymbol{R}$ of size $p \times p$.

$$J_N = \sum\nolimits_{k=0}^{N} \boldsymbol{x}^T(k)\boldsymbol{Q}\,\boldsymbol{x}(k) + \boldsymbol{u}^T(k)\boldsymbol{R}\,\boldsymbol{u}(k) \tag{16.4}$$

where N is large approaching ∞, $N \rightarrow \infty$ and the algorithm is called infinite horizon linear quadratic regulator (IH-LQR).

The weighting matrices $\boldsymbol{Q}$ and $\boldsymbol{R}$ are positive semidefinite and positive definite with eigenvalues $\lambda_i \geq 0$ and $\lambda_i > 0$, respectively. If the weighting matrix $\boldsymbol{Q}$ is high compared to $\boldsymbol{R}$, then the states have a higher contribution than the inputs in the cost function. On the other hand, if the weighting matrix $\boldsymbol{R}$ is high compared to $\boldsymbol{Q}$, then the inputs have a higher contribution than the states in the cost function. These matrices are decided based on the relative importance of states and inputs for a particular system under consideration.

For a linear time-invariant (LTI) system, the steady-state value of the optimal controller gain matrix $\boldsymbol{K}$ can be obtained using the solution of the Riccati equation as

$$\boldsymbol{P} = \boldsymbol{A}^T\boldsymbol{P}\boldsymbol{A} + \boldsymbol{Q} - \boldsymbol{A}^T\boldsymbol{P}\boldsymbol{B}\left[\boldsymbol{R} + \boldsymbol{B}^T\boldsymbol{P}\boldsymbol{B}\right]^{-1}\boldsymbol{B}^T\boldsymbol{P}\boldsymbol{A} \tag{16.5}$$

$$\boldsymbol{K} = \left[\boldsymbol{R} + \boldsymbol{B}^T\boldsymbol{P}\boldsymbol{B}\right]^{-1}\boldsymbol{B}^T\boldsymbol{P}\boldsymbol{A} \tag{16.6}$$

The $n \times n$ $\boldsymbol{P}$ matrix can be obtained as a solution of the Riccati Eq. (16.5). First, an initial guess of $\boldsymbol{P_0} = \boldsymbol{I}_n$, $\boldsymbol{I}_n$ being the $n \times n$ identity matrix is used on the right-hand side of Eq. (16.5) to get an update of $\boldsymbol{P}$. Next, the process is repeated with the updated $\boldsymbol{P}$ till the matrix $\boldsymbol{P}$ converges to a steady state. The steady-state $\boldsymbol{P}$ matrix is used in Eq. (16.6) to get the steady-state optimal controller gain matrix $\boldsymbol{K}$.

In Matlab, the discrete-time steady-state optimal controller gain matrix is obtained using the command *dlqr* (*d*iscrete *l*inear *q*uadratic *r*egulator) as

$$[K, P, E] = dlqr(A, B, Q, R)$$

where A, B, Q, and R represent the matrices $\boldsymbol{A}, \boldsymbol{B}, \boldsymbol{Q}$, and $\boldsymbol{R}$, respectively, as inputs to the command. The outputs K and P are $p \times n$ matrix of optimal controller gain $\boldsymbol{K}$, and $n \times n$ matrix of discrete Riccati equation solution matrix $\boldsymbol{P}$, respectively. E represents the vector of closed-loop eigenvalues of $(\boldsymbol{A} - \boldsymbol{BK})$.

16.3 Design of Linear Quadratic Optimal Estimators (Kalman Filters)

In this section, optimal state estimators with Gaussian white noise as the system disturbance and system measurements are discussed. These estimators are also commonly known as Kalman filters as well as infinite horizon linear quadratic Gaussian estimators (IH-LQG).

The system dynamics model in state space with white-noise disturbance and measurement uncertainties is represented as

$$\boldsymbol{x}(k+1) = \boldsymbol{A}\,\boldsymbol{x}(k) + \boldsymbol{B}\,\boldsymbol{u}(k) + \boldsymbol{B}_1\boldsymbol{w}(k) \tag{16.7}$$

$$\boldsymbol{y}(k) = \boldsymbol{C}\boldsymbol{x}(k) + \boldsymbol{\nu}(k) \tag{16.8}$$

where $\boldsymbol{w}(k)$ and $\boldsymbol{\nu}$ (k) are random plant disturbances of size $(s \times 1)$ and measurement noise (uncertainties) of size $(r \times 1)$, respectively. The random inputs $\boldsymbol{w}(k)$ and $\boldsymbol{\nu}$ (k) are assumed to be uncorrelated white-noise sequences with Gaussian distributions. The matrix $\boldsymbol{B}_1$ of size $(n \times s)$ takes into account the effect of random disturbance on the system response. The system matrices $\boldsymbol{A}$, $\boldsymbol{B}$, and $\boldsymbol{C}$ and vectors $\boldsymbol{x}(k)$, $\boldsymbol{y}(k)$ are as defined earlier in Sect. 16.2.

The expectation and covariance relations of random inputs $\boldsymbol{w}(k)$ and $\boldsymbol{\nu}$ (k) can be expressed as

$$E[\boldsymbol{w}(k)] = 0, E[\boldsymbol{\nu}(k)] = 0, \tag{16.9}$$

$$cov[\boldsymbol{w}(j), \boldsymbol{w}(k)] = E\left[\boldsymbol{w}(j)\boldsymbol{w}^{\boldsymbol{T}}(k)\right] = \boldsymbol{R}_w\delta_{jk} \tag{16.10}$$

$$\text{cov}[\boldsymbol{\nu}(j), \boldsymbol{\nu}(k)] = E[\boldsymbol{\nu}(j)\boldsymbol{\nu}^{\boldsymbol{T}}(k)] = \boldsymbol{R}_\nu\delta_{jk} \tag{16.11}$$

where δ_{jk} is the Kronecker delta function defined as

$$\delta_{jk} = \begin{cases} 0, j \neq k \\ 1, j = k \end{cases} \tag{16.12}$$

The infinite horizon linear quadratic Gaussian (IH-LQG) estimator design can be studied using the concept of expectation. All matrices and vectors are assumed to represent the average values. Here, the linear quadratic Gaussian estimator, commonly known as the Kalman filter, in steady state with matrices as time-invariant, is discussed.

The estimator dynamics can be expressed, similar to the current observer discussed in Sect. 14.4, in a two-step process. A first estimate of the state vector, $\overline{\boldsymbol{q}}(k+1)$, is obtained using the system matrices $\boldsymbol{A}$ and $\boldsymbol{B}$, and in the second step, the estimate is updated to $\boldsymbol{q}(k+1)$ with the current measurement $\boldsymbol{y}(k+1)$ as

$$\overline{\boldsymbol{q}}(k+1) = \boldsymbol{A}\,\boldsymbol{q}(k) + \boldsymbol{B}\,\boldsymbol{u}(k) \tag{16.13}$$

$$\boldsymbol{q}(k+1) = \overline{\boldsymbol{q}}(k+1) + \boldsymbol{L}(\boldsymbol{y}(k+1) - \boldsymbol{C}\overline{\boldsymbol{q}}(k+1)) \tag{16.14}$$

where $\boldsymbol{L}$ is the LQG estimator gain matrix of size $n \times r$.

The error in estimation can be expressed as the difference between the actual state vector $\boldsymbol{x}(k)$ and its estimate $\boldsymbol{q}(k)$ as

$$\boldsymbol{e}(k) = \boldsymbol{x}(k) - \boldsymbol{q}(k) \tag{16.15}$$

The objective of the estimator is to minimize a cost function that is the sum of expected squared estimation error given as

$$J(k) = E\left[e^{T}(k)\boldsymbol{e}(k)\right] \tag{16.16}$$

It can be shown, following the process similar to that presented for current observers in Sect. 14.5, that the IH-LQG estimator gain matrix $\boldsymbol{L}$ and the linear quadratic regulator (optimal controller) gain matrix $\boldsymbol{K}$ can be calculated independently.

The IH-LQG estimator model Eq. (16.13) can be rewritten using Eq. (16.14) as

$$\boldsymbol{q}(k+1) = \boldsymbol{A}\,\boldsymbol{q}(k) + \boldsymbol{B}\,\boldsymbol{u}(k) + \boldsymbol{L}(\boldsymbol{y}(k+1) - \boldsymbol{C}(\boldsymbol{A}\,\boldsymbol{q}(k) + \boldsymbol{B}\,\boldsymbol{u}(k))) \tag{16.17}$$

Equation (16.17) can be rewritten, after some rearrangements, as

$$\boldsymbol{q}(k+1) = (\boldsymbol{A} - \boldsymbol{LCA})\boldsymbol{q}(k) + (\boldsymbol{B} - \boldsymbol{LCB})\boldsymbol{u}(k) + \boldsymbol{Ly}(k+1) \tag{16.18}$$

The optimal controller can be used in combination with the IH-LQG estimator where controller output is obtained, similar to Eq. (16.3), replacing the state vector $\boldsymbol{x}(k)$ with its estimate $\boldsymbol{q}(k)$, as

$$\boldsymbol{u}(k) = -\boldsymbol{K}\,\boldsymbol{q}(k) \tag{16.19}$$

The $p \times n$ matrix of steady-state optimal controller gain $\boldsymbol{K}$ can be obtained using the *dlqr()* command in Matlab as discussed in Sect. 16.2.

The characteristic equation for the steady-state optimal estimator (Kalman filter) is obtained from Eq. (16.18) as

$$|z\boldsymbol{I} - (\boldsymbol{A} - \boldsymbol{L}\,\boldsymbol{C}\,\boldsymbol{A})| = 0 \tag{16.20}$$

Steady-State Optimal Estimator (IH-LQG Filter) Gain Matrix
In Matlab, the discrete-time steady-state optimal estimator gain matrix $\boldsymbol{L}$ can be obtained using the command *dlqe* (*d*iscrete *l*inear *q*uadratic *e*stimator) as

$$[L, P, Z, E] = dlqe(A, B1, C, Rw, Rv)$$

where A, $B1$, and C represent the matrices $\boldsymbol{A}$, $\boldsymbol{B1}$, and $\boldsymbol{C}$, respectively, and Rw and Rv as standard deviations of white-noise system disturbance and measurement uncertainties as inputs to the command. The outputs L, P, and Z are matrices of optimal controller gain $\boldsymbol{L}$ ($n \times r$), Riccati equation solution matrix $\boldsymbol{P}$ ($n \times n$), and error covariance matrix ($n \times n$). E represents the vector of eigenvalues of ($\boldsymbol{A} - \boldsymbol{L}\,\boldsymbol{C}\,\boldsymbol{A}$).

Kalman Filter System Matrices in State Space
Matlab command *destim* can also be used to obtain Kalman filter matrices Ae, Be, Ce, and De from the original system matrices A, B, C, and D and Kalman filter gain matrix L

$$[Ae, Be, Ce, De] = destim(A, B, C, D, L)$$

Kalman state estimator model in state space is given as

$$\boldsymbol{q}(k+1) = \boldsymbol{A_e}\,\boldsymbol{q}(k) + \boldsymbol{B_e}\,\boldsymbol{y}(k) \tag{16.21}$$

$$\begin{bmatrix} \widehat{\boldsymbol{y}}(k) \\ \widehat{\boldsymbol{x}}(k) \end{bmatrix} = \boldsymbol{C_e}\boldsymbol{q}(k) + \boldsymbol{D_e}\boldsymbol{y}(k) \tag{16.22}$$

where the Kalman filter system matrices are defined as

$$\boldsymbol{A_e} = \boldsymbol{A} - \boldsymbol{A}\,\boldsymbol{L}\,\boldsymbol{C} \tag{16.23}$$

$$\boldsymbol{B_e} = \boldsymbol{A}\,\boldsymbol{L} \tag{16.24}$$

$$\boldsymbol{C_e} = \begin{bmatrix} \boldsymbol{C} - \boldsymbol{C}\,\boldsymbol{L}\,\boldsymbol{C} \\ \boldsymbol{I} - \boldsymbol{L}\,\boldsymbol{C} \end{bmatrix} \tag{16.25}$$

$$\boldsymbol{D_e} = \begin{bmatrix} \boldsymbol{C}\,\boldsymbol{L} \\ \boldsymbol{L} \end{bmatrix} \tag{16.26}$$

Representation of a System with Optimal Controller and IH-LQG Filter
Following the procedure of Sect. 14.5, Eqs. (16.7), (16.8), (16.18), and (16.19), with expected white-noise system disturbance and measurement uncertainty as 0, can be expressed in the z-domain, using z-transform, respectively, as

$$\boldsymbol{X}(z) = (z\boldsymbol{I} - \boldsymbol{A})^{-1}\boldsymbol{B}\,\boldsymbol{U}(z) \tag{16.27}$$

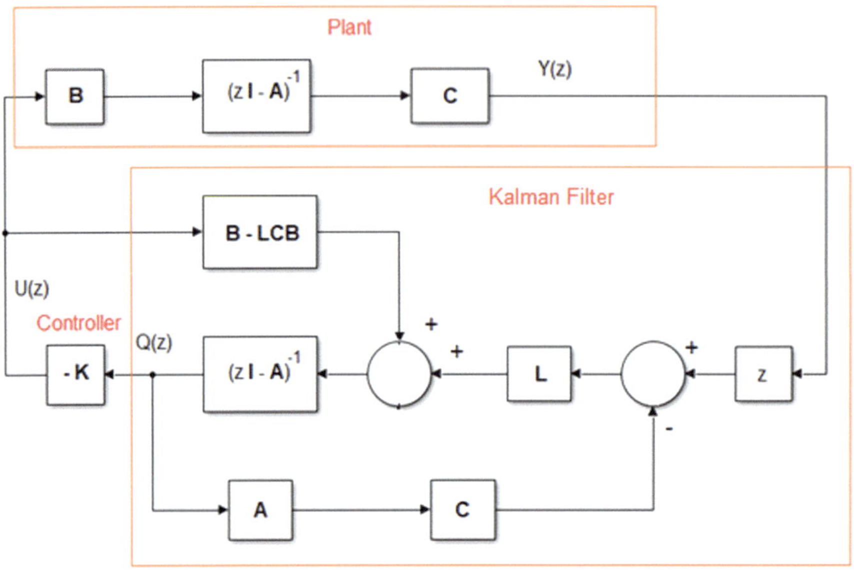

Fig. 16.1 Block diagram of a system with steady-state optimal controller and Kalman filter

$$\boldsymbol{Y}(z) = \boldsymbol{C}\,\boldsymbol{X}(z) \tag{16.28}$$

$$\boldsymbol{Q}(z) = (z\boldsymbol{I} - \boldsymbol{A})^{-1}[\boldsymbol{L}(z\boldsymbol{Y}(z) - \boldsymbol{CA}\,\boldsymbol{Q}(z)) + (\boldsymbol{B} - \boldsymbol{LCB})\,\boldsymbol{U}(z)] \tag{16.29}$$

$$\boldsymbol{U}(z) = -\boldsymbol{KQ}(z) \tag{16.30}$$

Equations (16.27)–(16.30) can be used to represent the system with a steady-state optimal estimator and optimal controller in the block diagram as shown in Fig. 16.1.

The optimal estimator model, as given in Eq. (16.17), can be expressed as

$$\boldsymbol{q}(k+1) = \boldsymbol{A}\,\boldsymbol{q}(k) + \boldsymbol{B}\,\boldsymbol{u}(k) + \boldsymbol{L}(\boldsymbol{C}(\boldsymbol{A}\,\boldsymbol{x}(k) + \boldsymbol{B}\,\boldsymbol{u}(k) - \boldsymbol{C}(\boldsymbol{A}\,\boldsymbol{q}(k) + \boldsymbol{B}\,\boldsymbol{u}(k)) \tag{16.31}$$

Equation (16.31) can be written, with some rearrangements, as

$$\boldsymbol{q}(k+1) = \boldsymbol{A}\,\boldsymbol{q}(k) + \boldsymbol{B}\,\boldsymbol{u}(k) + \boldsymbol{LCA}(\,\boldsymbol{x}(k) - \boldsymbol{q}(k)) \tag{16.32}$$

Equation (16.32) can be further expressed in compact form as

$$\boldsymbol{q}(k+1) = \boldsymbol{A}\,(\boldsymbol{q}(k) + \boldsymbol{L}\,\boldsymbol{y}_e(k)) + \boldsymbol{B}\,\boldsymbol{u}(k) \tag{16.33}$$

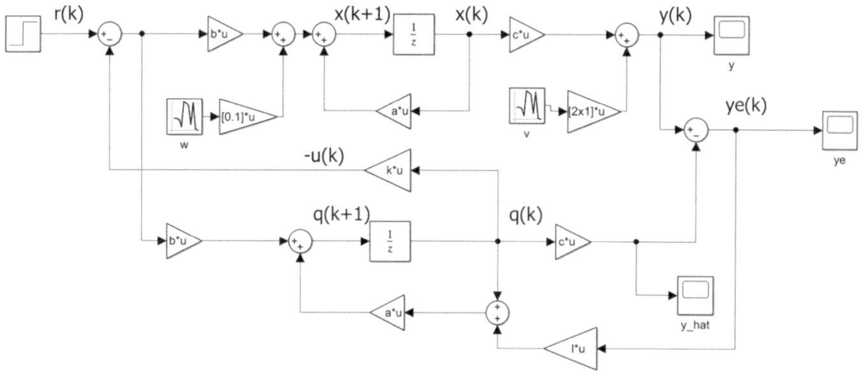

Fig. 16.2 Simulink model of a system with steady-state optimal controller and Kalman filter

where the difference between the vector of measured states and the corresponding estimated values is expressed as

$$\boldsymbol{y}_e(k) = \boldsymbol{C}(\boldsymbol{x}(k) - \boldsymbol{q}(k)) \tag{16.34}$$

A Simulink model for the discrete-time system in state space with steady-state optimal estimator and optimal controller, given in Eqs. (16.7), (16.8), (16.33), and (16.19), with a reference input vector $\boldsymbol{r}(k)$ of size $p \times 1$ is shown in Fig. 16.2. It should be noted that state vector $\boldsymbol{x}(k)$, the output vector $\boldsymbol{y}(k)$, and the input vector $\boldsymbol{u}(k)$ are of size $n \times 1$, $r \times 1$, and $p \times 1$, respectively. The matrices $\boldsymbol{a}$, $\boldsymbol{b}$, $\boldsymbol{c}$, $\boldsymbol{k}$, and $\boldsymbol{l}$ are the same as $\boldsymbol{A}$, $\boldsymbol{B}$, $\boldsymbol{C}$, $\boldsymbol{K}$, and $\boldsymbol{L}$, respectively, as defined in Eqs. (16.7), (16.8), (16.33), and (16.19).

Optimal Controller-Estimator Transfer Function

Combining Eqs. (16.18) and (16.19), IH-LQG estimator dynamics can be expressed as

$$\begin{aligned}\boldsymbol{q}(k+1) &= (\boldsymbol{A} - \boldsymbol{L}\,\boldsymbol{C}\,\boldsymbol{A}\,)\boldsymbol{q}(k) + (\boldsymbol{B} - \boldsymbol{L}\boldsymbol{C}\boldsymbol{B})\,(-\boldsymbol{K}\,\boldsymbol{q}(k)) + \boldsymbol{L}\,\boldsymbol{y}(k+1)\\ &= (\boldsymbol{A} - \boldsymbol{L}\,\boldsymbol{C}\,\boldsymbol{A} - \boldsymbol{B}\,\boldsymbol{K} + \boldsymbol{L}\,\boldsymbol{C}\,\boldsymbol{B}\,\boldsymbol{K})\boldsymbol{q}(k) + \boldsymbol{L}\,\boldsymbol{y}(k+1)\end{aligned} \tag{16.35}$$

Using z-transform, Eq. (16.35) can be expressed in z-domain, as

$$z\boldsymbol{I}\,\boldsymbol{Q}(z) = (\boldsymbol{A} - \boldsymbol{L}\,\boldsymbol{C}\,\boldsymbol{A} - \boldsymbol{B}\,\boldsymbol{K} + \boldsymbol{L}\,\boldsymbol{C}\,\boldsymbol{B}\,\boldsymbol{K})\,\boldsymbol{Q}(z) + z\boldsymbol{L}\,\boldsymbol{Y}(z) \tag{16.36}$$

Equation (16.36) can be used to express $\boldsymbol{Q}(z)$ in terms of $\boldsymbol{Y}(z)$ as

$$\boldsymbol{Q}(z) = (z\boldsymbol{I} - (\boldsymbol{A} - \boldsymbol{L}\,\boldsymbol{C}\,\boldsymbol{A} - \boldsymbol{B}\,\boldsymbol{K} + \boldsymbol{L}\,\boldsymbol{C}\,\boldsymbol{B}\,\boldsymbol{K}))^{-1} z\boldsymbol{L}\,\boldsymbol{Y}(z) \tag{16.37}$$

Equations (16.30) and (16.37) can be combined to express $\boldsymbol{U}(z)$ in terms of $\boldsymbol{Y}(z)$ as

$$\boldsymbol{U}(z) = -z\boldsymbol{K}(z\boldsymbol{I} - (\boldsymbol{A} - \boldsymbol{L}\,\boldsymbol{C}\,\boldsymbol{A} - \boldsymbol{B}\,\boldsymbol{K} + \boldsymbol{L}\,\boldsymbol{C}\,\boldsymbol{B}\,\boldsymbol{K}))^{-1}\boldsymbol{L}\,\boldsymbol{Y}(z) \tag{16.38}$$

The controller-IH-LQG estimator transfer function can be written as

$$D_{cLQG}(z) = -\frac{\boldsymbol{U}(z)}{\boldsymbol{Y}(z)} = z\boldsymbol{K}(z\boldsymbol{I} - (\boldsymbol{A} - \boldsymbol{L}\,\boldsymbol{C}\,\boldsymbol{A} - \boldsymbol{B}\,\boldsymbol{K} + \boldsymbol{L}\,\boldsymbol{C}\,\boldsymbol{B}\,\boldsymbol{K}))^{-1}\boldsymbol{L} \tag{16.39}$$

This is similar to the controller-estimator transfer function for the current observer presented in Sect. 14.5.

Comparing Eq. (16.39) with the standard state-space representation of a discrete-time system, Eq. (13.58), the equivalent system matrices of the controller-IH-LQG estimator can be expressed for

$$D_{cLQGp}(z) = \boldsymbol{K}(z\boldsymbol{I} - (\boldsymbol{A} - \boldsymbol{L}\,\boldsymbol{C}\,\boldsymbol{A} - \boldsymbol{B}\,\boldsymbol{K} + \boldsymbol{L}\,\boldsymbol{C}\,\boldsymbol{B}\,\boldsymbol{K}))^{-1}\boldsymbol{L} \tag{16.40}$$

$$\boldsymbol{A}_{cLQGp} = \boldsymbol{A} - \boldsymbol{L}\,\boldsymbol{C}\,\boldsymbol{A} - \boldsymbol{B}\,\boldsymbol{K} + \boldsymbol{L}\,\boldsymbol{C}\,\boldsymbol{B}\,\boldsymbol{K} \tag{16.41}$$

$$\boldsymbol{B}_{cLQGp} = \boldsymbol{L} \tag{16.42}$$

$$\boldsymbol{C}_{cLQGp} = \boldsymbol{K} \tag{16.43}$$

$$\boldsymbol{D}_{cLQGp} = \boldsymbol{D} \tag{16.44}$$

Once the transfer function for $D_{cLQGp}(z)$ is obtained, the numerator can be multiplied by z to get $D_{cLQG}(z)$ as

$$D_{cLQG}(z) = zD_{cLQGp}(z) \tag{16.45}$$

Example 16.1 For DC motor position control model $G_p(s) = \frac{K}{s(\tau s+1)}$ with $K = 30$, $\tau = 0.10$, design an optimal controller using $Q = [2\ 0;\ 0\ 2]$; $R = 10$. Design an optimal estimator for Gaussian white-noise system disturbance and measurement uncertainties given by standard deviations as $R_w = 0.10$, $R_v = 0.01$.

Solution Matlab code along with results are presented next.

```
% Ch16_OptimalControl
% LQG Controller (LQR+KALMAN filter)

clear all; close all; clc
%G(s)=K/s(s+tau)
ns=25000;

K=30; %Static gain
tau=0.1; % time constant
T=0.01 ; % sampling time
t=linspace(0,T*ns,ns);
ngs=K; dgs=[tau 1 0];
printsys(ngs,dgs)
%ss model [ac,bc,,c,d]
[ac,bc,c,d ]=tf2ss(ngs,dgs)% ss (s)
%ss modelin discrete-time
[a,b]=c2d(ac,bc,T) % ss (k)
% controllability
cont=[b a*b ]
cr=rank(cont)
%observabilty
obsr=[c;c*a]
or=rank(obsr)

n=size (a,1);p=size(b,2); r=size(c,1);
Q=zeros(n,n);P=zeros(n,n); R=zeros(p,p);
q=2; rho=10;
% initialization
for i=1:n
Q(i,i)=q
end
for i=1:p
R(i,i)=rho;
end

% calculate optimal controller gain k (ss) from dlqr
[k,P,E] = dlqr(a,b,Q,R)

% closed loop response (with the state feedback only)
at=a-b*k;bt=b;ct=c;dt=d;
[zt,pt,kt]=ss2zp(at,bt,ct,dt)
dcg=ddcgain(at,bt,ct,dt); yc=dstep(at,bt,ct,dt,1,ns);
yc=yc/dcg;
figure
plot(t,yc),grid
title('Closed-loop system step response with optimal controller')
xlabel('Time (s)')
ylabel('System output y(k)')'
% estimator design using dlqe (Kalman filter)
Rw=0.1; Rv=0.01;b1=b;
[l,P,Z,E]=dlqe(a,b1,c,Rw,Rv)
ak=a-b*k-l*c*a+l*c*b*k;bk=l;ck=k;
[tmp1,tmp2]=size(d);dk=zeros(tmp1,tmp2);

[nkz,dkz]=ss2tf(ak,bk,ck,dk)%Dce(z)/z
nkz=[nkz 0]%numerator of Dce(z) putting z in the numerator
dkz=[0 dkz]% to keep order of nkz, dkz same
printsys(nkz,dkz,'z') % Dce(z)
```

```
%Controller-Estimator Combination
[ak,bk,ck,dk]=tf2ss(nkz,dkz)
[at,bt,ct,dt]=feedback(a,b,c,d,ak,bk,ck,dk)
dcg=ddcgain(at,bt,ct,dt); yc=dstep(at,bt,ct,dt,1,ns);
yc=yc/dcg;
figure
plot(t, yc,'r'),grid
title('Closed-loop system step response with optimal estimator-controller')
xlabel('Time (s)')
ylabel('System output y(k)')'

t=linspace(0,ns*T,ns)';
[ata,bta,cta,dta]=augstate(at,bt,ct,dt)
%pause
x_e=dinitial(ata,bta,cta,dta,[0.1e-2 0.1e-2-0.1e-2 -0.1e-2],ns);
x=x_e(:,2:3);xe=x_e(:,4:5);
figure
subplot(211);plot(t,[x(:,1) xe(:,1)]);
title('Closed-loop system initial response with optimal estimator-controller')
xlabel('Time (s)')
ylabel('System state x1(k)')'
legend('State x1', 'Estimated state xe1')

subplot(212);plot(t,[x(:,2) xe(:,2)]); %pause
title('Closed-loop system initial response with optimal estimator-controller')
xlabel('Time (s)')
ylabel('System state x2(k)')'
legend('State x2', 'Estimated state xe2')
% Kalman Estimator System Matrices
[Ae, Be,Ce, De] = destim(a, b, c, d, l)

Results
a = 0.9048         0
    0.0095    1.0000

b = 0.0095
    0.0000

cont = 0.0095    0.0086
       0.0000    0.0001

cr =   2


obsr =  0      300.0000
       2.8549  300.0000

or =     2
Q =  2     0
     0     2
R=10

Optimal controller gain matrix
k = 0.0541   0.4471

P = 1.0e+03 *
    0.0556    0.4472
    0.4472    4.4975
```

```
E = 0.9996
    0.9047

Optimal estimator gain matrix

l = 0.0168
    0.0010

P = 1.0e-04 *
    0.3184    0.0079
    0.0079    0.0005

Z = 1.0e-04 *
    0.2783    0.0056
    0.0056    0.0003

E=  0.7821 + 0.1687i
    0.7821 - 0.1687i

             0.0013468 z^2- 0.0012336 z
Dce(z)  =  ---------------------------
              z^2 - 1.5637 z + 0.63978
Kalman estimator system matrices
Ae = 0.9048   -4.5730
     0.0095    0.659
Be =  0.0152
      0.0011

Ce =       0  212.2356
      1.0000   -5.0539
           0    0.7075

De =  0.2925
      0.0168
      0.0010
```

The closed-loop system step response and mismatched initial condition response are shown in Figs. 16.3, 16.4 and 16.5.

16.4 Laboratory Experiment for Analysis, Design, Simulation, and Implementation of Optimal Controllers for Discrete-Time Systems

Objectives The objectives are to design and simulate a linear quadratic optimal controller with a linear quadratic optimal estimator in state space and implement and validate the controller for a DC motor position control system.

Introduction The linear quadratic optimal controller with a linear quadratic estimator or Kalman filter can be designed in state space with appropriate weighting matrices for state and control input vectors. The controller-estimator transfer function $D_{ce}(z)$ in z-domain can be implemented for a DC motor position control system. In Part I, it is required to design the linear quadratic optimal controller with a linear

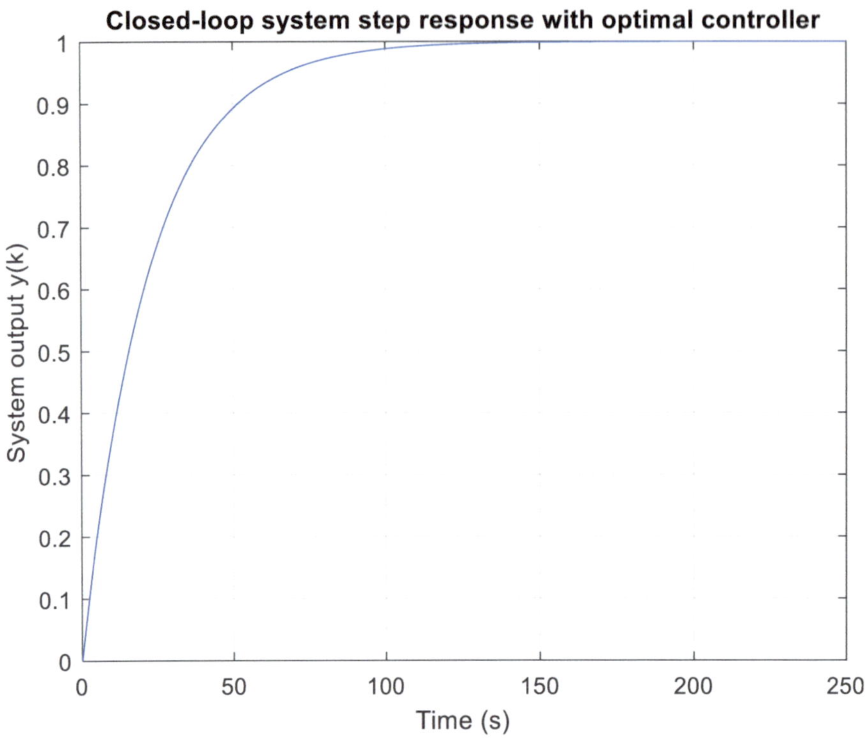

Fig. 16.3 Closed-loop system step response with optimal controller

quadratic estimator (Kalman filter) and simulate its performance in Matlab as presented in this chapter. In Part II, it is required to implement the controller in LabVIEW and validate its performance.

I. Controller Design and Simulation.

1. Use the DC motor position control model with previously estimated values of K and τ to get the discrete-time open-loop transfer function in z-domain $G(z)$ selecting a suitable sampling time T. For the DC motor control trainer (DCMCT) module, the sampling time (T) is taken as 0.01 s. Convert $G(z)$ to state-space system matrices (a, b, c, d). Refer to Chap. 13 for details.
2. Design a linear quadratic regulator (LQR) in a discrete-time domain. Use suitable weighting matrices (Q and R) for the state and the control vectors. Check the effects of the Q and R matrices on the controller performance. Refer to Sect. 16.2 for details. Note down the controller gain matrix and closed-loop poles in Table 16.1.
3. Design a linear quadratic estimator (LQE) (Kalman filter) for the system (use suitable variances R_w and R_v for the controller and the sensor noise). Refer to

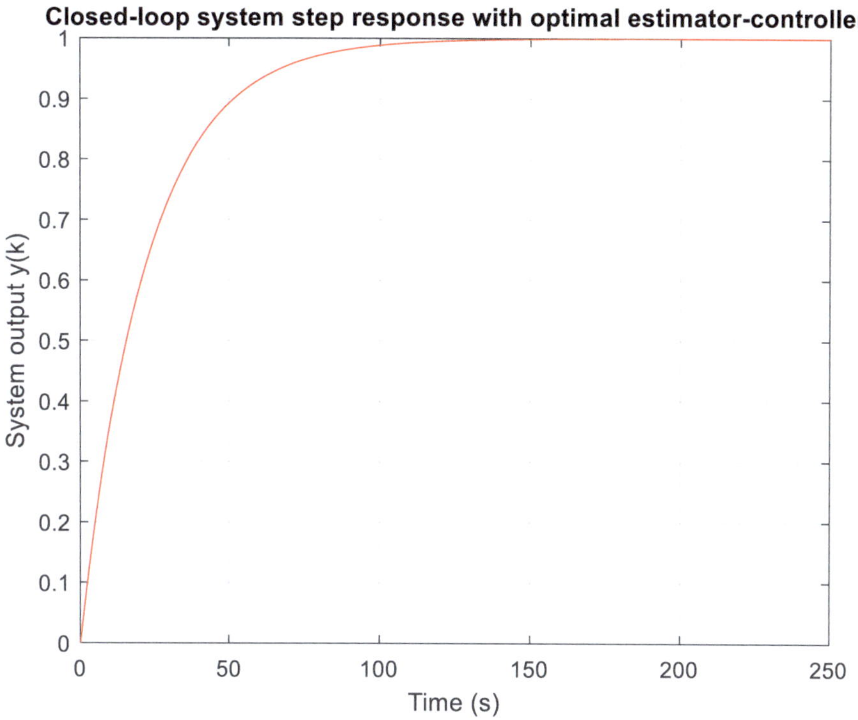

Fig. 16.4 Closed-loop system step response with optimal estimator-controller

Sect. 16.3 for details. Note down the estimator gain matrix and the controller-estimator transfer function $D_{ce}(z)$ in Table 16.1.

4. Simulate closed-loop step response with the controller-estimator combination. Locate response characteristics by right-clicking the figure in Matlab, selecting peak response, settling time, and steady state (right-click → response characteristics → peak response → settling time → steady state → cursor on identified point). Note down in Table 16.1 overshoot (α) and settling time (t_{set}) from the simulated response for the controller. Copy closed-loop step response with the controller-estimator for presentation in the lab report.
5. Obtain closed-loop system response for the controller-estimator state with mismatched initial conditions. Copy these response graphs for presentation in the lab report.

II. Implementation and Validation.

1. Set up NI ELVIS II board and DC motor module. Select the device, and set the sampling rate. For the DC motor control trainer (DCMCT) module, a sampling rate of 100 Hz is used.

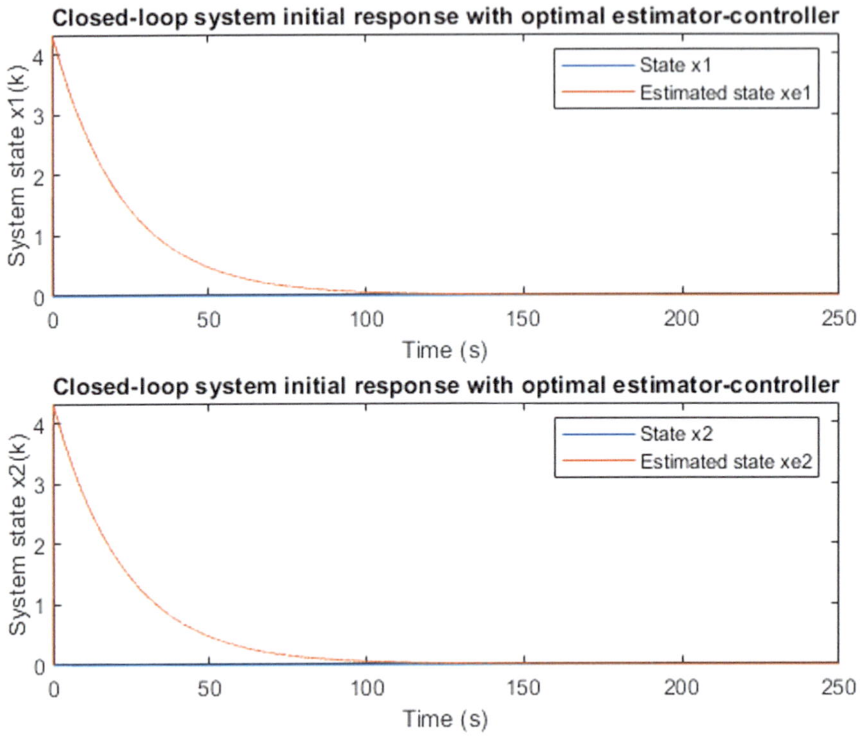

Fig. 16.5 Closed-loop system initial state response with optimal estimator-controller

2. Study the block diagram for the position control experiment (*01-QNET_DCMCT_Position_Control.vi*).
3. Open the block diagram VI and identify the different components of the control system, including the input, error, controller, plant, sensor, and output in Fig. 16.6.
4. Modify the block diagram VI of Fig. 16.6 by removing the control parameters block and adding a subtraction block to calculate the error, as well as a discrete transfer function block for implementation of the designed controller, as shown in the VI of Fig. 16.7.
5. Put the discrete-time z-domain controller-estimator transfer function $D_{ce}(z)$, and follow the format of setting the controller transfer function coefficients. Save the VI.
6. Select your input parameters. For the DCMCT module, typical values for the input square wave voltage are as follows: amplitude of 1.0 V, frequency of 0.40 Hz, and offset of 3.0 V.
7. Run the VI, and observe the closed-loop system response. Copy graphs of closed-loop system response from the front panel VI for presentation in the lab report.

Table 16.1 Summary of controller parameters and performance

Controller	Gain matrix	Zeros (z_i)	Poles (p_i)	Controller-estimator transfer function $D_{ce}(z)$	Closed-loop system step response			
					Simulation		Implementation	
					Overshoot α (%)	Settling time $t_{set}(s)$	Overshoot α (%)	Settling time $t_{set}(s)$
Optimal controller								
Optimal observer								

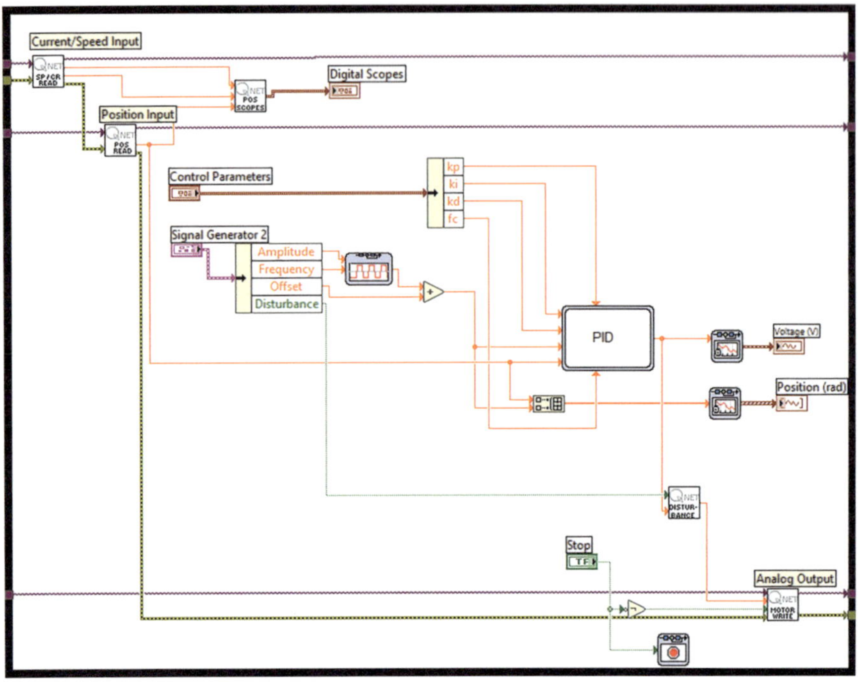

Fig. 16.6 Block diagram VI for DC motor control trainer position control VI before modification

8. Make sure to set the number of significant figures to at least 5 for the display format of each graph on the front panel VI (right-click on the graph → display format → decimal → 6). Export the time response graphs, both motor rotation $\theta(t)$ and motor input voltage $V(t)$, to Excel by right-clicking on each graph and selecting the Export to Excel option (right-click → Export to Excel). Estimate overshoot and settling time from closed-loop system response graph.
9. Complete the last two columns of Table 16.1.

III. Deliverables.

1. Complete Table 16.1.
2. Present graphs of simulated closed-loop step response for the controller-estimator with response characteristics (peak response, settling time, and steady-state value) clearly annotated.
3. Present graphs of state responses with mismatched initial conditions for each controller-observer combination.
4. Present the closed-loop step response of the DC motor module for the implemented controller-estimator combination.
5. Discuss results comparing the performance of controller-observer combinations in simulation and experimental validation.

Lab Report

Present all deliverables in the lab report and discuss your results.

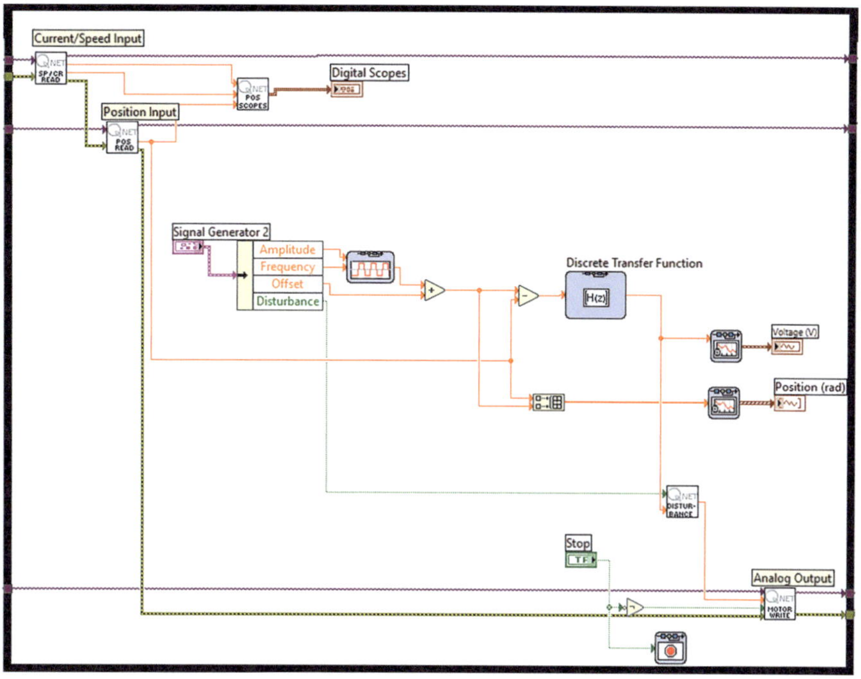

Fig. 16.7 Block diagram VI for DC motor control trainer position control VI after modification

Exercises

16.1 The closed-loop vehicle distance control system for a traffic control system can be represented as shown in Fig. P16.1. The plant transfer function of the distance control system can be represented in s-domain as $G_p(s) = \frac{1.5}{s^2+12s+18}$. The sampling time is given as $T = 0.10$ s. Design an optimal controller using $Q = [1\ 0; 0\ 1]$; $R = 10$. Design an optimal estimator for Gaussian white-noise system disturbance and measurement uncertainties given by standard deviations as $R_w = 0.10$, $R_v = 0.01$.

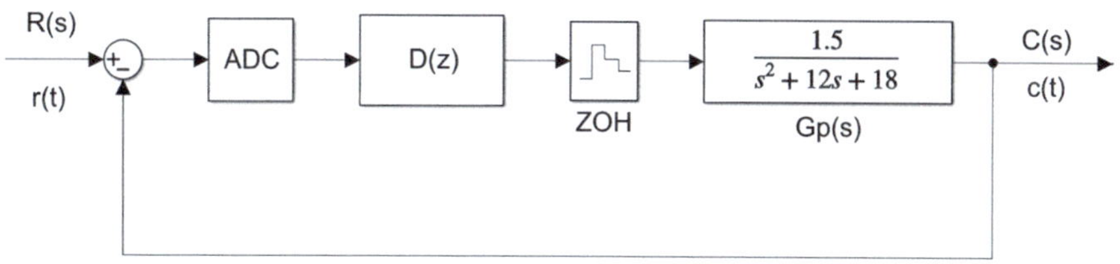

Fig. P16.1 Model of closed-loop discrete-time system for Problem 16.1

16.2 The closed-loop roll angle control system for a jet airplane can be represented as shown in Fig. P16.2. The plant transfer function of the roll control system is represented in the s-domain as $G_p(s) = \frac{10}{(s+12)(s^2+2s+24)}$. The sampling time is given as $T = 0.05$ s. Design an optimal controller using $Q = [200\ 0\ 0; 0200\ 0; 0\ 0200]$; $R = 1$. Design an optimal estimator for Gaussian white-noise system disturbance and measurement uncertainties given by standard deviations as $R_w = 0.10$, $R_v = 0.02$.

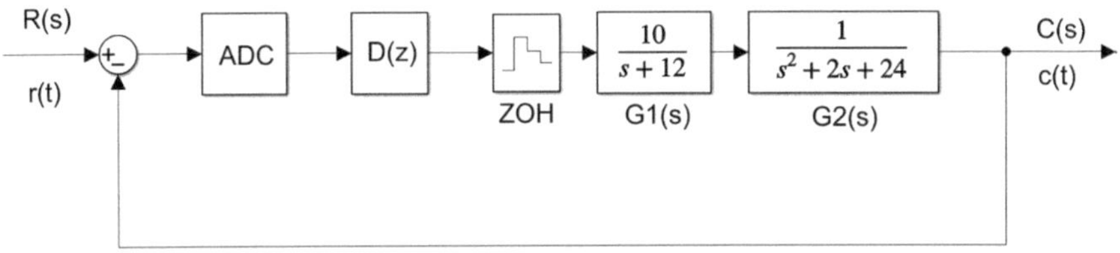

Fig. P16.2 Model of closed-loop discrete-time system for Problem 16.2

16.3 The closed-loop tilt control system in a high-speed train for passenger comfort can be represented as shown in Fig. P16.3. The plant transfer function of the train tilt control system is represented in the s-domain as $G_p(s) = \frac{5}{s(s+5)(s+35)}$. The sampling time is given as $T = 0.1$ s. Design an optimal controller using $Q = [1\ 0\ 0; 0\ 1\ 0; 0\ 0\ 1]$; $R = 5$. Design an optimal estimator for Gaussian white-noise system disturbance and measurement uncertainties given by standard deviations as $R_w = 0.50$, R_v=0.05.

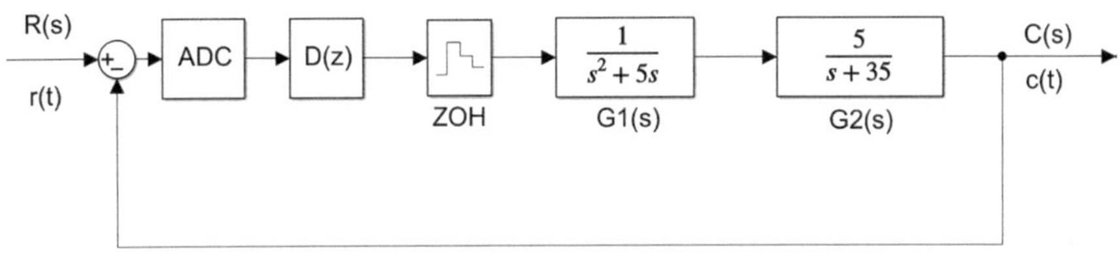

Fig. P16.3 Model of closed-loop discrete-time system for Problem 16.3

16.4 The closed-loop position control system in a machine tool can be represented as shown in Fig. P16.4. The plant transfer function of the position control system is represented in the s-domain as $G_p(s) = \frac{K}{5s(0.25s+1)\left(1.5\times10^{-3}s+2.5\times10^{-2}\right)}$. The sampling time is given as $T = 0.10$ s. The amplifier gain K is such that the steady-state error to a ramp input is limited to 2.5%. Design an optimal controller using $Q = [1\ 0\ 0; 0\ 1\ 0; 0\ 0\ 1]$; $R = 5$. Design an optimal estimator for Gaussian white-noise system disturbance and measurement uncertainties given by standard deviations as $R_w = 0.20$, R_v=0.02.

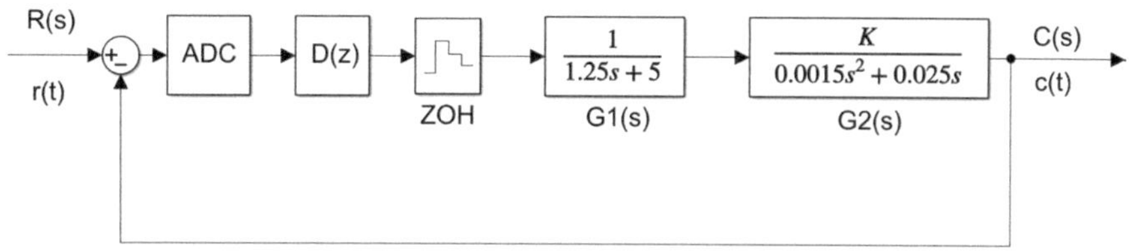

Fig. P16.4 Model of closed-loop discrete-time system for Problem 16.4

16.5 The state-space model matrices (**A**, **B**, **C**, **D**) for lateral motions of a ship moving at a constant velocity can be represented as [7]

$$\boldsymbol{A}=\begin{bmatrix}-0.0895 & -0.2860 & 0\\ -0.0439 & -0.2720 & 0\\ 0 & 1.0000 & 0\end{bmatrix},\boldsymbol{B}=\begin{bmatrix}0.0145\\ -0.0122\\ 0\end{bmatrix},\boldsymbol{C}=[0\ \ 0\ \ 1],\boldsymbol{D}=[0]$$

where the state variables $(x_1(t),\ x_2(t), x_3(t))$ are ship side slip angle (degrees), ship heading angle (degrees), and ship yaw rate, respectively. The rudder angle (degrees) is the input $(u(t))$, and the heading angle is the measurement variable $(y(t))$. The sampling time is given as $T = 0.05$ s. Design an optimal controller using $Q = [5\ 0\ 0; 0\ 5\ 0; 0\ 0\ 5]$; $R = 1$. Design an optimal estimator for Gaussian white-noise system disturbance and measurement uncertainties given by standard deviations as $R_w = 0.20$, $R_\nu = 0.01$.

16.6 The state-space model matrices (**A**, **B**, **C**, **D**) for the linearized longitudinal motion of a helicopter can be represented as [7]

$$\boldsymbol{A}=\begin{bmatrix}-0.4000 & 0 & -0.0100\\ 1.0000 & 0 & 0\\ -1.4000 & 9.8 & -0.0200\end{bmatrix},\boldsymbol{B}=\begin{bmatrix}6.3000\\ 0\\ 9.8000\end{bmatrix},\boldsymbol{C}=[0\ \ 0\ \ 1],\boldsymbol{D}=[0]$$

where the state variables $(x_1(t),\ x_2(t), x_3(t))$ are the rate of pitch angle $(\dot{\theta}(t))$, pitch angle $(\theta(t))$, and horizontal velocity $(v(t))$, respectively. The rudder angle (degrees) is the input $(u(t))$, and the horizontal velocity $(v(t))$ is the measurement variable $(y(t))$. The sampling time is given as $T = 0.20$ s. Obtain state-space representation of the open-loop system in discrete time. Design an optimal controller using $Q = [1\ 0\ 0; 0\ 1\ 0; 0\ 0\ 1]$; $R = 10$. Design an optimal estimator for Gaussian white-noise system disturbance and measurement uncertainties given by standard deviations as $R_w = 0.10$, $R_\nu = 0.05$.

16.7 The state-space model matrices (**A**, **B**, **C**, **D**) for a linearized model of liquid levels in a 3-tank system can be represented as [5]

$$A = \begin{bmatrix} -1 & 0 & 0 \\ 1 & -1 & 0 \\ 0 & 1 & -1 \end{bmatrix}, B = \begin{bmatrix} 1 \\ 0 \\ 0 \end{bmatrix}, C = [0\ 0\ 1], D = [0]$$

where the state variables $(x_1(t),\ x_2(t), x_3(t))$ are liquid levels $(h_1(t),\ h_2(t), h_3(t))$ of the top, the middle, and the bottom tank, respectively. The inlet liquid flow rate in the top tank is the input $(u(t))$, and the liquid level of the bottom tank is the measurement variable $(y(t))$. The sampling time is given as $T = 010$ s. Design an optimal controller using $Q = [1\ 0\ 0;\ 0\ 1\ 0;\ 0\ 0\ 1]$; $R = 5$. Design an optimal estimator for Gaussian white-noise system disturbance and measurement uncertainties given by standard deviations as $R_w = 0.20$, R_v=0.01.

16.8 The closed-loop thickness control system in a metal processing plant can be represented as shown in Fig. P16.5. The plant transfer function of the thickness control system is represented in the s-domain as $G_p(s) = \frac{5}{s(s+2)(s^2+s+10)}$. The sampling time is given as $T = 0.10$ s. Design an optimal controller using $Q = [1\,0\,0\,0;\,0\,1\,0\,0;\,0\,0\,1\,0;\,0\,0\,0\,1]$; $R = 10$. Design an optimal estimator for Gaussian white-noise system disturbance and measurement uncertainties given by standard deviations as $R_w = 0.50$, $R_v = 0.05$.

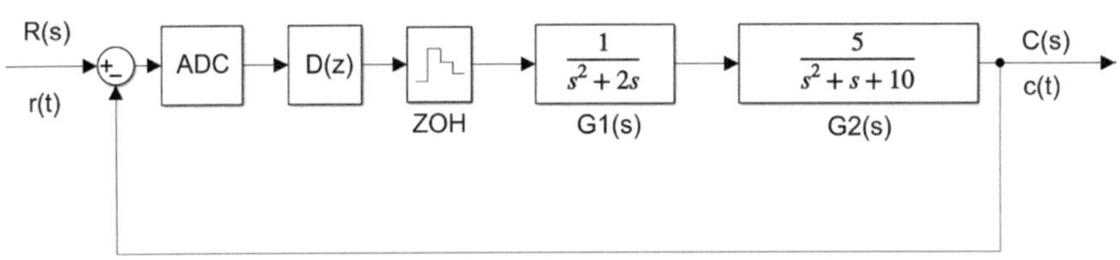

Fig. P16.5 Model of closed-loop discrete-time system for Problem 16.8

16.9 The closed-loop position control system of a typical Mars rover can be represented as shown in Fig. P16.6. The plant transfer function of the position control system is represented in the s-domain, similar to Dorf and Bishop [4], as $G_p(s) = \frac{1}{s^2(s+5)}$. The sampling time is given as $T = 0.02$ s. Design an optimal controller using $Q = [1\ 0\ 0;\ 0\ 1\ 0;\ 0\ 0\ 1]$; $R = 10$. Design an optimal estimator for Gaussian white-noise system disturbance and measurement uncertainties given by standard deviations as $R_w = 0.50$, $R_v = 0.05$.

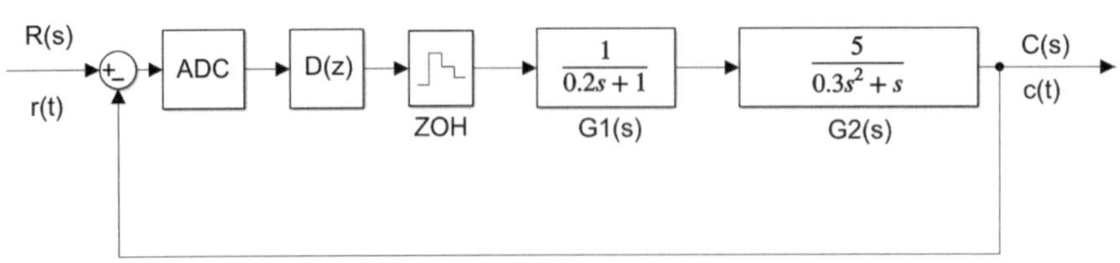

Fig. P16.6 Model of closed-loop discrete-time system for Problem 16.9

16.10 The closed-loop position control system for one of the axes of an experimental automated surface-grinding wheel can be represented as shown in Fig. P16.7. The plant transfer function of the axis position control system can be represented in the s-domain, similar to Dorf and Bishop [4], as $G_p(s) = \frac{5}{s(0.2s+1)(0.3s+1)}$. The sampling time is given as $T = 0.02$ s. Design an optimal controller using $Q = [1\ 0\ 0;\ 0\ 1\ 0;\ 0\ 0\ 1]$; $R = 5$. Design an optimal estimator for Gaussian white-noise system disturbance and measurement uncertainties given by standard deviations as $R_w = 0.10$, $R_\nu = 0.01$.

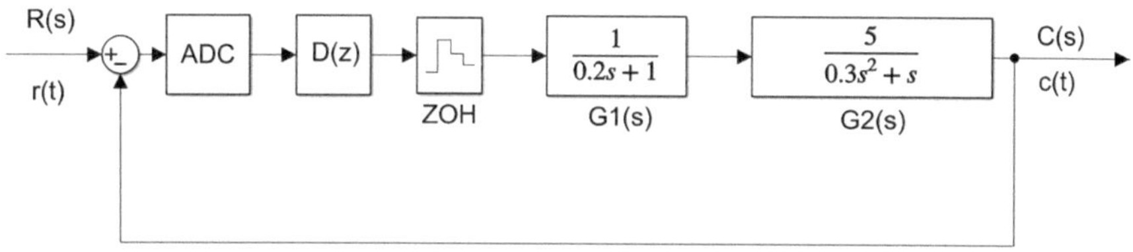

Fig. P16.7 Model of closed-loop discrete-time system for Problem 16.10

16.11 The closed-loop position control system with a time delay can be represented as shown in Fig. P16.8. The plant transfer function can be represented in s-domain with a delay $G_p(s) = \frac{1}{s(s+1)} e^{-sT}$. The sampling time is given as $T = 0.2$ s. Design an optimal controller using $Q = [1\ 0\ 0;\ 0\ 1\ 0;\ 0\ 0\ 1]$; $R = 10$. Design an optimal estimator for Gaussian white-noise system disturbance and measurement uncertainties given by standard deviations as $R_w = 0.10$, $R_\nu = 0.01$.

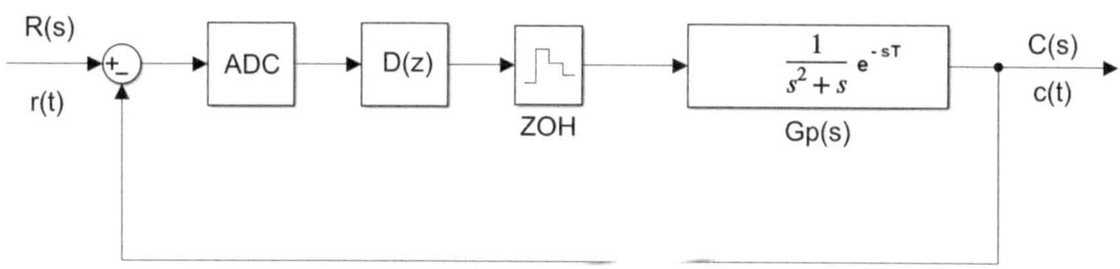

Fig. P16.8 Model of closed-loop discrete-time system for Problem 16.11

16.12 The closed-loop position control system with a time delay can be represented as shown in Fig. P16.9. The plant transfer function can be represented in s-domain with a delay $G_p(s) = \frac{2}{s^2+5s+10} e^{-sT}$. The sampling time is given as $T = 0.2$ s. Design an optimal controller using $Q = [2\ 0\ 0;\ 0\ 2\ 0;\ 0\ 0\ 2]$; $R = 10$. Design an optimal estimator for Gaussian white-noise system disturbance and measurement uncertainties given by standard deviations as $R_w = 0.10$, $R_\nu = 0.01$.

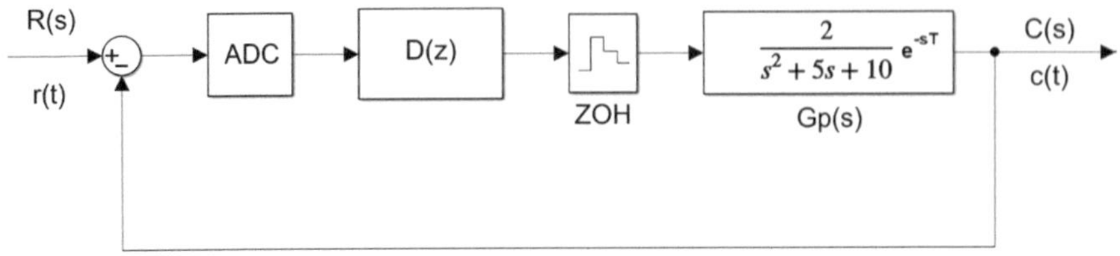

Fig. P16.9 Model of closed-loop discrete-time system for Problem 16.12

Further Reading

1. Asadi F (2023) Signals and systems with MATLAB® and Simulink®. Springer, Cham
2. Astrom KJ, Murray RM (2021) Feedback systems: an introduction for scientists and engineers, 2nd edn. Princeton University Press, Princeton
3. Chen C-T (2006) Analog and digital control system design: transfer-function, state-space, and algebraic methods. Oxford University Press, New York
4. Dorf R, Bishop R (2016) Modern control systems, 13th edn. Pearson, Essex
5. Fadali MS, Visioli A (2020) Digital control engineering analysis and design, 3rd edn. Academic, London
6. Franklin GF, Powell JD, Workman M (1998) Digital control of dynamic systems, 3rd edn. Addison-Wesley, Boston
7. Franklin GF, Powell JD, Emami-Naeini A (2019) Feedback control of dynamic systems, 8th edn. Pearson, Upper Saddle River
8. Isermann R (1981) Digital control systems. Springer, Berlin
9. Jalili N, Candelino NW (2023) Dynamic systems and control engineering. Cambridge University Press, Cambridge
10. Keviczky L, Bars R, Hetthéssy J, Bányász (2019) Control engineering. Springer, Singapore
11. Koeing DM (2009) Practical control engineering: a guide for engineers, managers, and practitioners. McGraw-Hill, New York
12. Kuo BC (1995) Digital control systems, 2nd edn. Oxford University Press, Oxford
13. Landau ID, Gito G (2006) Digital control systems: design, identification and implementation. Springer, London
14. Leigh JR (2006) Applied digital control: theory, design and implementation, 2nd edn. Dover, Mineola
15. Mathworks (2024a) Control System Toolbox. https://www.mathworks.com/products/control.html
16. Mathworks (2024b) Matlab. https://www.mathworks.com/products/matlab.html
17. National Instruments (2024a) LabVIEW. https://www.ni.com/en/support/downloads/software-products/download.labview.html#487445
18. National Instruments (2024b) NI-ELVIS-II. https://www.ni.com/en-us/support/model.ni-elvis-ii.html
19. National Instruments (2024c) Labs for Quanser QNET DC Motor Add-On Board. https://education.ni.com/teach/resources/180/labs-for-quanser-qnet-dc-motor-add-on-board
20. Nise NS (2019) Control systems engineering, 8th edn. Wiley, Hoboken
21. Nof SY (ed) (2009) Springer handbook of automation. Springer, Berlin
22. Ogata K (1995) Discrete-time control systems, 2nd edn. Pearson, Hoboken
23. Phillips CL, Nagle HT, Chakrabortty A (2015) Digital control system analysis and design, 4th edn. Pearson, Essex
24. Quanser (2024) QUBE Servo 2. https://www.quanser.com/products/qube-servo-2/

25. Rabbath CA, Lechevin N (2014) Discrete-time control system design with applications. Springer, New York
26. Samanta B (2023) Introduction to mechatronics – an integrated approach. Springer, Cham
27. Shertukde HM (2015) Digital control applications illustrated with MATLAB®. CRC Press, Boca Raton
28. Starr G (2020) Introduction to applied digital controls. Springer, Cham
29. Veloni A, Miridakis NI (2018) Digital control systems-theoretical problems and simulation tools. CRC Press, Boca Raton
30. Vilanova R, Visioli A (2012) PID control in the third millennium – lessons learned and new approaches. Springer, London
31. Visioli A (2006) Practical PID control. Springer, London

Chapter 17
Case Studies

17.1 Introduction

Case studies covering analysis, design, and simulation of discrete-time controllers are presented in this chapter. The speed control of a hybrid electric vehicle (HEV) model is considered for the analysis, design, and simulation of controllers. Both transfer function– and state space–based controllers are considered project activities integrating the concepts covered in the book. The position control of a rotary inverted pendulum model is considered for modeling, analysis, design, and simulation of controllers in state space as a second project.

17.2 A Hybrid Electric Vehicle (HEV) Model

The model for speed control of a hybrid electric vehicle (HEV) [24] with unity feedback can be represented as a transfer function $G_{\mathrm{p}}(s) = \frac{0.11(s+0.6)}{6s^2+3.613s+0.0572}$ [20]. It is proposed to create a closed-loop discrete-time system for the speed control of a HEV as shown in Fig. 17.1 with a discrete-time controller $D(z)$ to be designed using the methods covered in this book for the given specifications. Use a sampling time $T = 0.10$ s.

1. Design a PID controller such that steady-state error due to a ramp input is limited to 2%: the overshoot is limited to 4% and the settling time is within 4 s for a step input. Examine the effectiveness of the controller by comparing system stability and closed-loop step response without and with the controller. Present the results in Table 17.1(a, b).
2. Design a phase-lag-lead controller such that the steady-state error is limited to 2%: the overshoot is limited to 4% and the settling time is within 4 s for a step input. Examine the effectiveness of the controller by comparing system stability

B. Samanta, *Introduction to Digital Control*,
https://doi.org/10.1007/978-3-031-66830-2_17

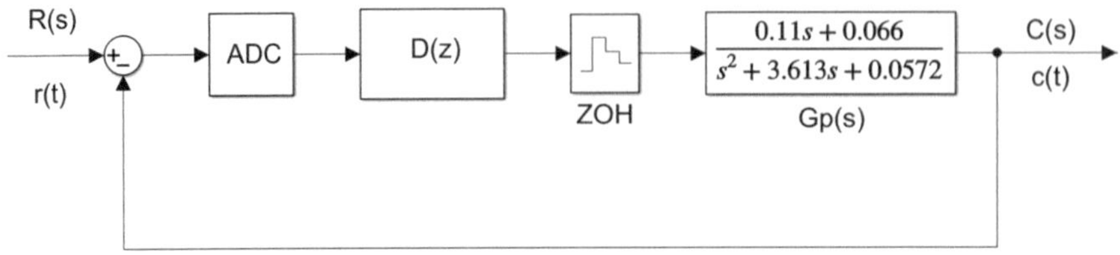

Fig. 17.1 A closed-loop discrete-time system for speed control of a HEV

and closed-loop step response without and with the controller. Present the results in Table 17.2(a, b).

3. Design a PID controller such that steady-state error due to a ramp input is limited to 2%: the phase margin is 75° at a gain cross-over frequency of 3.5 rad/s. Examine the effectiveness of the controller by comparing system stability and closed-loop step response without and with the controller. Present the results in Table 17.3(a, b).
4. Design a phase-lag-lead controller such that the steady-state error due to a step input is limited to 2%: the phase margin is 75° at a gain cross-over frequency of 3.5 rad/s. Examine the effectiveness of the controller by comparing system stability and closed-loop step response without and with the controller. Present the results in Table 17.4(a, b).
5. Design a phase-lead controller such that the steady-state error due to a step input is limited to 2%: the phase margin is 75° and the gain margin is 10 dB. Examine the effectiveness of the controller by comparing system stability and closed-loop step response without and with the controller. Present the results in Table 17.5(a, b).
6. Convert the system to state space form in discrete time. Check the observability and controllability of the system. Design a full-state pole placement controller so that the closed-loop system has an overshoot of 4% and a settling time of 4 s for a step input. Design a full-order prediction observer. Compare the system step response without and with the observer. Present the results in Table 17.6.
7. Design an optimal controller with suitable weighting matrices for the state and control vectors. Design an optimal estimator (Kalman filter) with suitable variances for the state and output measurement noise. Present the results in Table 17.7.
8. Create a Simulink model for each of the following in the discrete-time domain and simulate the system response.

   (a) Simulate the closed-loop system response with a controller-prediction observer combination.
   (b) Simulate the closed-loop system response with optimal controller-estimator combination. Assume reasonable models (variances) for process and measurement noise.

9. Compare the controllers, and present the results in Table 17.8. Prepare a project report with all components—Abstract, Introduction, Methodology, Results and Discussions, Conclusion, and References. Include Matlab/Simulink code for each controller design and simulation in the Appendix.

Table 17.1 (a) Summary of controller parameters and performance. (b) Summary of controller performance using root locus and Bode plots

(a)

Controller	K_P	K_I	K_D	Controller transfer function $D(z)$	Closed-loop system step response			
					Simulation		Implementation	
					Overshoot α (%)	Settling time t_{set} (s)	Overshoot α (%)	Settling time t_{set} (s)
PI			0					
PD		0						
PID								

(b)

Controller	Root locus closed-loop poles	Bode plots			
		GM	ϕ_M	ω_{gc} (rad/s)	ω_{pc} (rad/s)
Without controller					
PI					
PD					
PID					

Table 17.2 (a) Summary of controller parameters and performance. (b) Summary of controller performance using root locus and Bode plots

(a)

Controller	Zeros (z_i)	Poles (p_i)	Controller transfer function $D(z)$	Closed-loop system step response			
				Simulation		Implementation	
				Overshoot α (%)	Settling time t_{set} (s)	Overshoot α (%)	Settling time t_{set} (s)
Phase-lag							
Phase-lead							
Phase-lag-lead							

(b)

Controller	Root locus closed-loop poles	Bode plots			
		GM	ϕ_M	ω_{gc} (rad/s)	ω_{pc} (rad/s)
Without controller					
Phase-lag					
Phase-lead					
Phase-lag-lead					

Table 17.3 (a) Summary of controller parameters and performance. (b) Summary of controller performance using root locus and Bode plots

(a)

Controller	K_P	K_I	K_D	Controller transfer function $D(z)$	Closed-loop system step response			
					Simulation		Implementation	
					Overshoot α (%)	Settling time t_{set} (s)	Overshoot α (%)	Settling time t_{set} (s)
PI			0					
PD		0						
PID								

(b)

Controller	Root locus closed-loop poles	Bode plots			
		GM	ϕ_M	ω_{gc} (rad/s)	ω_{pc} (rad/s)
Without controller					
PI					
PD					
PID					

Table 17.4 (a) Summary of controller parameters and performance. (b) Summary of controller performance using root locus and Bode plots

(a)

Controller	Zeros (z_i)	Poles (p_i)	Controller transfer function $D(z)$	Closed-loop system step response			
				Simulation		Implementation	
				Overshoot α (%)	Settling time (s)	Overshoot α (%)	Settling time t_{set}(s)
Phase-lag							
Phase-lead							
Phase-lag-lead							

(b)

Controller	Root locus closed-loop poles	Bode plots			
		GM	ϕ_M	ω_{gc} (rad/s)	ω_{pc} (rad/s)
Without controller					
Phase-lag					
Phase-lead					
Phase-lag-lead					

Table 17.5 (a) Summary of controller parameters and performance. (b) Summary of controller performance using root locus and Bode plots

(a)

Controller	Zeros (z_i)	Poles (p_i)	Controller transfer function $D(z)$	Closed-loop system step response			
				Simulation		Implementation	
				Overshoot α (%)	Settling time t_{set} (s)	Overshoot α (%)	Settling time t_{set} (s)
Phase-lag							
Phase-lead							
Phase-lag-lead							

(b)

Controller	Root locus closed-loop poles	Bode plots			
		GM	ϕ_M	ω_{gc} (rad/s)	ω_{pc} (rad/s)
Without controller					
Phase-lag					
Phase-lead					
Phase-lag-lead					

Table 17.6 Summary of controller parameters and performance

Controller	Gain matrix	Closed-loop poles	Controller-estimator transfer function $D_{ce}(z)$	Closed-loop system step response			
				Simulation		Implementation	
				Overshoot α (%)	Settling time t_{set} (s)	Overshoot α (%)	Settling time t_{set} (s)
Pole placement							
Full-order prediction observer							
Current observer							

Table 17.7 Summary of controller parameters and performance

Controller	Gain matrix	Zeros (z_i)	Poles (p_i)	Controller-estimator transfer function $D_{ce}(z)$	Closed-loop system step response			
					Simulation		Implementation	
					Overshoot α (%)	Settling time t_{set} (s)	Overshoot α (%)	Settling time t_{set} (s)
Optimal controller								
Optimal observer								

Table 17.8 Summary of controller performance

	Bode plots				Closed-loop system step response			
					Simulation		Implementation	
	GM	ϕ_M	ω_{gc} (rad/s)	ω_{pc} (rad/s)	Overshoot α (%)	Settling time t_{set}(s)	Overshoot α (%)	Settling time t_{set}(s)
PID-z								
Phase-lag-lead-z								
PID-w								
Phase-lag-lead-w								
Phase-lag-lead-w (Bode)								
Pole placement								
Full-order prediction observer								
Optimal controller								
Optimal observer								

17.3 Rotary Inverted Pendulum Model

Modeling and Control of a Rotary Inverted Pendulum (Fig. 17.2).

1. Modeling

   (a) Develop a mathematical model for the rotary inverted pendulum using the Lagrange method using symbolic manipulation in Matlab. Consider the following model parameters and variables (θ, ϕ):

      J_r: Mass moment of inertia of the rotary arm about its axis of rotation (z)
      L_r: Length of the rotary arm
      J_p: Mass moment of inertia of the pendulum about its axis of rotation at its CM (B)
      m_p: Mass of the pendulum
      L_p: Length of the pendulum
      θ: Angle of rotation for the rotary arm
      ϕ: Angle of rotation for the pendulum
      c_r: Damping coefficient at the rotary arm axis
      c_p: Damping coefficient at pendulum axis
      τ: Motor torque driving the rotary arm

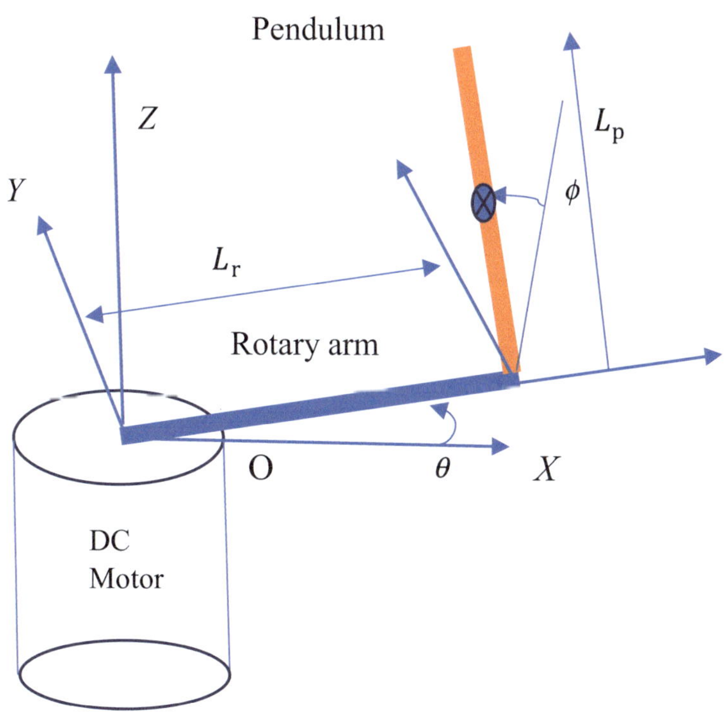

Fig. 17.2 Schematic of a rotary inverted pendulum

(i) Express the position of the end of the rotary arm (x_r, y_r, z_r) and the center of mass of the pendulum B (x_p, y_p, z_p) in terms of variables (θ, ϕ) and lengths (L_r, L_p).
(ii) Express velocity (both translational and angular) for the rotary arm and the pendulum CM.
(iii) Express kinetic energy (T) (due to both rotational and translational motion) and potential energy (V) of the system.
(iv) Obtain Lagrangian $L = T - V$
(v) Obtain partial derivatives $\frac{\partial L}{\partial \theta}, \frac{\partial L}{\partial \phi}, \frac{\partial L}{\partial \dot{\theta}}, \frac{\partial L}{\partial \dot{\phi}},$
(vi) Obtain $\frac{d}{dt}\left(\frac{\partial L}{\partial \dot{\theta}}\right), \frac{d}{dt}\left(\frac{\partial L}{\partial \dot{\phi}}\right)$
(vii) Obtain the equation for each joint: $\frac{d}{dt}\left(\frac{\partial L}{\partial \dot{\theta}}\right) - \frac{\partial L}{\partial \theta} = \tau - c_r\dot{\theta}$

$$\frac{d}{dt}\left(\frac{\partial L}{\partial \dot{\phi}}\right) - \frac{\partial L}{\partial \phi} = -c_p\dot{\phi}$$

(b) Obtain equations in the following nonlinear functions f_1 and f_2:

$$f_1\left(\theta, \varphi, \dot{\theta}, \dot{\varphi}, \ddot{\theta}, \ddot{\varphi}\right) = \tau - c_r\dot{\theta} \tag{17.1}$$

$$f_2\left(\theta, \varphi, \dot{\theta}, \dot{\varphi}, \ddot{\theta}, \ddot{\varphi}\right) = -c_p\dot{\phi} \tag{17.2}$$

(c) *Linearize* Eqs. (17.1) and (17.2) about the nominal operating points $\left(\left(\varphi, \dot{\theta}, \dot{\varphi}\right.\right.$ are small$\left.\right)$ such that $\sin(\varphi) = \varphi$, $\cos(\varphi) = 1$, and higher-order terms like $\dot{\theta}\dot{\varphi}, \dot{\theta}^2, \dot{\varphi}^2$ are negligible).
(d) Obtain the linearized equations in the form of:

$$[M]\begin{pmatrix} \ddot{\theta} \\ \ddot{\varphi} \end{pmatrix} + [K]\begin{pmatrix} \theta \\ \varphi \end{pmatrix} = \begin{Bmatrix} \tau - c_r\dot{\theta} \\ -c_p\dot{\phi} \end{Bmatrix} \tag{17.3}$$

where $[M]$ and $[K]$ are 2×2 matrices.
(e) Convert Eq. (17.3) to *state space* form (see the example of the inverted pendulum done in class earlier)

$$\{\dot{X}\} = [A]\{X\} + [B]\{u\}, \tag{17.4}$$

$$\{y\} = [C]\{X\} + [D]\{u\} \tag{17.5}$$

where $\{X\}^T = \left[\theta, \varphi, \dot{\theta}, \dot{\varphi}\right]$ and $\{y\}^T = [x_1, x_2]$, $\{u\} = \tau$.

(f) Model motor torque as follows: $\tau = \frac{k_m\left(V_{in} - k_m\,\dot{\theta}\right)}{R_m}$ where k_m, R_m, and V_{in} are motor constant, armature resistance, and voltage input to the armature, respectively.

2. Substitute numerical values

```
 %% Motor
% Resistance
Rm = 8.4 Ω;
% Current-torque (N-m/A)
kt = 0.042;
% Back-emf constant (V-s/rad)
km = 0.042;
%
%% Rotary Arm
% Mass (kg)
Mr = 0.095;
% Total length (m)
Lr = 0.085;
% Moment of inertia about pivot (kg-m^2)
Jr = Mr*Lr^2/12;
% Equivalent Viscous Damping Coefficient (N-m-s/rad)
cr = 0.0015;
%
%% Pendulum Link
% Mass (kg)
Mp = 0.024;
% Total length (m)
Lp = 0.129;
% Moment of inertia about pivot (kg-m^2)
Jp = Mp*Lp^2/12;
% Equivalent Viscous Damping Coefficient (N-m-s/rad)
cp = 0.0005;
% Gravity Constant
g = 9.81;
```

3. Obtain numerical matrices A, B, C, and D in the state space model (in continuous time domain) for the given numerical values of the system parameters.

(a) Without motor dynamics (hints: keep motor torque as an input $u(t)$ $\dot{X} = AX + Bu$, where A : 4 × 4 B : 4 × 1 C : 2 × 4, D : 2 × 1 matrices. Obtain A, B, C, and D matrices).

$$X = \left[\theta, \varphi, \dot{\theta}, \dot{\varphi}\right]^T$$

(b) Consider motor dynamics $\tau = \frac{k_m\left(V_{in} - k_m\,\dot{\theta}\right)}{R_m} = \frac{k_m(V_{in})}{R_m} - \frac{(k_m)^2}{R_m}\,\dot{\theta}$. Use this relation to update the A and B matrices.

(c) Obtain controllability and observability matrices for the continuous-time state space system with motor dynamics of step 3(ii).

4. Obtain the state space model of the inverted pendulum (Qube) system in discrete time for a sampling time of $T = 0.05$ s using the continuous-time state space system of 3c.
   (a) Obtain controllability and observability matrices for the discrete-time state space system with motor dynamics, similar to step 3c.
5. Design pole-placement controller for the state space system in the discrete-time domain.
   (a) Design a Luenberger observer (full-order prediction estimator).
   (b) Design a current observer for the system.
   (c) Design an optimal controller. Assume reasonable Q and R for optimal control. Assume reasonable models (variances) for process and measurement noise.
6. Create a Simulink model for each of the following in the discrete-time domain and simulate the system response.
   (a) Simulate the closed-loop system response with a controller-prediction observer combination.
   (b) Simulate the closed-loop system response with a controller-current observer combination.
   (c) Simulate the closed-loop system response with optimal controller-estimator combination. Assume reasonable models (variances) for process and measurement noise.
7. Prepare a project report with all components—Abstract, Introduction, Methodology, Results and Discussions, Conclusion, and References. Include Matlab/Simulink code for each controller design and simulation in the Appendix.

Further Reading

1. Asadi F (2023) Signals and systems with MATLAB® and Simulink®. Springer, Cham
2. Astrom KJ, Murray RM (2021) Feedback systems: an introduction for scientists and engineers, 2nd edn. Princeton University Press, Princeton
3. Chen C-T (2006) Analog and digital control system design: transfer-function, state-space, and algebraic methods. Oxford University Press, New York
4. Dorf R, Bishop R (2016) Modern control systems, 13th edn. Pearson, Essex
5. Fadali MS, Visioli A (2020) Digital control engineering analysis and design, 3rd edn. Academic Press, London
6. Franklin GF, Powell JD, Emami-Naeini A (2019) Feedback control of dynamic systems, 8th edn. Pearson
7. Franklin GF, Powell JD, Workman M (1998) Digital control of dynamic systems, 3rd edn. Addison-Wesley, Boston, Massachusetts
8. Isermann R (1981) Digital control systems. Springer-Verlag, Berlin
9. Jalili N, Candelino NW (2023) Dynamic systems and control engineering. Cambridge University Press, Cambridge
10. Keviczky L, Bars R, Hetthéssy J, Bányász (2019) Control engineering. Springer, Singapore

11. Koeing DM (2009) Practical control engineering: a guide for engineers, managers, and practitioners. McGraw-Hill, New York
12. Kuo BC (1995) Digital control systems, 2nd edn. Oxford University Press, Oxford
13. Landau ID, Gito G (2006) Digital control systems: design, identification and implementation. Springer-Verlag, London
14. Leigh JR (2006) Applied digital control: theory, design and implementation, 2nd edn. Dover, Mineola, New York
15. Mathworks (2024) Control system toolbox. https://www.mathworks.com/products/control.html
16. Mathworks (2024) Matlab. https://www.mathworks.com/products/matlab.html
17. National Instruments (2024) LabVIEW. https://www.ni.com/en/support/downloads/software-products/download.labview.html#487445
18. National Instruments (2024) NI-ELVIS-II. https://www.ni.com/en-us/support/model.ni-elvis-ii.html
19. National Instruments (2024) Labs for Quanser QNET DC Motor Add-On Board. https://education.ni.com/teach/resources/180/labs-for-quanser-qnet-dc-motor-add-on-board
20. Nise NS (2019) Control systems engineering, 8th edn. Wiley, Hoboken
21. Nof SY (ed) (2009) Springer handbook of automation. Springer-Verlag, Berlin
22. Ogata K (1995) Discrete-time control systems, 2nd edn. Pearson, New Jersey
23. Phillips CL, Nagle HT, Chakrabortty A (2015) Digital control system analysis and design, 4th edn. Pearson, Essex
24. Preitl Z, Bauer P, Bokor J (2007) A simple control solution for traction motor used in hybrid vehicles. In: 4th International symposium on applied computational intelligence and informatics, 17–18 May 2007, Timisoara. IEEE. https://doi.org/10.1109/SACI.2007.375502
25. Quanser (2024) QUBE Servo 2. https://www.quanser.com/products/qube-servo-2/
26. Rabbath CA, Lechevin N (2014) Discrete-time control system design with applications. Springer, New York
27. Samanta B (2023) Introduction to mechatronics- an integrated approach. Springer, Cham
28. Shertukde HM (2015) Digital control applications illustrated with MATLAB®. CRC Press, Boca Raton
29. Starr G (2020) Introduction to applied digital controls. Springer, Cham
30. Veloni A, Miridakis NI (2018) Digital control systems-theoretical problems and simulation tools. CRC Press, Boca Raton
31. Vilanova R, Visioli A (2012) PID control in the third millennium – lessons learned and new approaches. Springer-Verlag, London
32. Visioli A (2006) Practical PID control. Springer, London

Appendix A: z-Transform Table of Commonly Used Functions

B. Samanta, *Introduction to Digital Control*,
https://doi.org/10.1007/978-3-031-66830-2

Table A.1 z-Transform of commonly used functions

Time function $f(t)$	Laplace transform $F(s)$	z-Transform $F(z)$	Modified z-transform $F(z, m)$
$u(t)$	$\frac{1}{s}$	$\frac{z}{z-1}$	$\frac{1}{z-1}$
t	$\frac{1}{s^2}$	$\frac{Tz}{(z-1)^2}$	$\frac{mT}{z-1}+\frac{T}{(z-1)^2}$
$\frac{t^3}{2}$	$\frac{1}{s^3}$	$\frac{T^2(z+1)}{2(z-1)^3}$	$\frac{T^2}{2}\left[\frac{m^2}{z-1}+\frac{2m+1}{(z-1)^2}+\frac{2}{(z-1)^3}\right]$
t^{k-1}	$\frac{(k-1)!}{s^k}$	$\lim_{a\to 0}(-1)^{k-1}\frac{\partial^{k-1}}{\partial a^{k-1}}\left[\frac{z}{z-e^{-aT}}\right]$	$\lim_{a\to 0}(-1)^{k-1}\frac{\partial^{k-1}}{\partial a^{k-1}}\left[\frac{e^{-amT}}{z-e^{-aT}}\right]$
$e^{-at}, a>0$	$\frac{1}{s+a}$	$\frac{z}{z-e^{-aT}}$	$\frac{e^{-amT}}{z-e^{-aT}}$
$te^{-at}, a>0$	$\frac{1}{(s+a)^2}$	$\frac{Te^{-aT}z}{(z-e^{-aT})^2}$	$\frac{Te^{-amT}[e^{-aT}+m(z-e^{-aT})]}{(z-e^{-aT})^2}$
$t^k e^{-at}, a>0$	$\frac{(k-1)!}{(s+a)^k}$	$(-1)^k\frac{\partial^k}{\partial a^k}\left[\frac{z}{z-e^{-aT}}\right]$	$(-1)^k\frac{\partial^k}{\partial a^k}\left[\frac{e^{-amT}}{z-e^{-aT}}\right]$
$1-e^{-at}, a>0$	$\frac{a}{s(s+a)}$	$\frac{z(1-e^{-aT})}{(z-1)(z-e^{-aT})}$	$\frac{1}{z-1}-\frac{e^{-amT}}{z-e^{-aT}}$
$t-\frac{1-e^{-at}}{a}, a>0$	$\frac{a}{s^2(s+a)}$	$\frac{z[(aT-1+e^{-aT})z+(1-e^{-aT}-aTe^{-aT})]}{a(z-1)^2(z-e^{-aT})}$	$\frac{T}{(z-1)^2}+\frac{amT-1}{a(z-1)}+\frac{e^{-amT}}{a(z-e^{-aT})}$
$1-(1+at)e^{-at}, a>0$	$\frac{a^2}{s(s+a)^2}$	$\frac{z}{z-1}-\frac{z}{z-e^{-aT}}-\frac{aTe^{-aT}z}{(z-e^{-aT})^2}$	$\frac{1}{z-1}-\left[\frac{1+amT}{z-e^{-aT}}+\frac{aTe^{-aT}}{(z-e^{-aT})^2}\right]e^{-amT}$

$e^{-at} - e^{-bt}, a, b > 0$	$\frac{b-a}{(s+a)(s+b)}$	$\frac{\left(e^{-aT} - e^{-bT}\right)z}{\left(z - e^{-aT}\right)\left(z - e^{-bT}\right)}$	$\frac{e^{-amT}}{z - e^{-aT}} - \frac{e^{-bmT}}{z - e^{-bT}}$
$\sin(at)$	$\frac{a}{s^2 + a^2}$	$\frac{z\ \sin(aT)}{z^2 - 2z\ \cos(aT) + 1}$	$\frac{z\ \sin(amT) + \sin((1-m)aT)}{z^2 - 2z\ \cos(aT) + 1}$
$\cos(at)$	$\frac{s}{s^2 + a^2}$	$\frac{z\,(z - \cos(aT))}{z^2 - 2z\ \cos(aT) + 1}$	$\frac{z\ \cos(amT) - \cos((1-m)aT)}{z^2 - 2z\ \cos(aT) + 1}$
$\sinh(at)$	$\frac{a}{s^2 - a^2}$	$\frac{z\ \sinh(aT)}{z^2 - 2z\ \cosh(aT) + 1}$	$\frac{z\ \sinh(amT) + \sinh((1-m)aT)}{z^2 - 2z\ \cosh(aT) + 1}$
$\cosh(at)$	$\frac{s}{s^2 - a^2}$	$\frac{z\,(z - \cosh(aT))}{z^2 - 2z\ \cosh(aT) + 1}$	$\frac{z\ \cosh(amT) - \cosh((1-m)aT)}{z^2 - 2z\ \cosh(aT) + 1}$
$\frac{1}{b} e^{-at} \sin(bt)$	$\frac{1}{(s+a)^2 + b^2}$	$\frac{1}{b}\left[\frac{z\,e^{-aT}\sin(bT))}{z^2 - 2ze^{-aT}\cos(bT) + e^{-2aT}}\right]$	$\frac{1}{b}\left[\frac{e^{-amT}[z\ \sin(bmT)) + e^{-aT}\sin((1-m)bT)]}{z^2 - 2ze^{-aT}\cos(bT) + e^{-2aT}}\right]$
$e^{-at}\cos(bt)$	$\frac{s+a}{(s+a)^2 + b^2}$	$\frac{z^2 - z\,e^{-aT}\cos(bT))}{z^2 - 2ze^{-aT}\cos(bT) + e^{-2aT}}$	$\frac{e^{-amT}[z\ \cos(bmT)) + e^{-aT}\sin((1-m)bT)\,]}{z^2 - 2z\,e^{-aT}\cos(bT) + e^{-2aT}}$
$1 - e^{-at}\left(\cos(bt) + \frac{a}{b}\sin(bt)\right)$	$\frac{a^2 + b^2}{s\left[(s+a)^2 + b^2\right]}$	$\frac{z(Az + B)}{(z-1)(z^2 - 2ze^{-aT}\cos(bT) + e^{-2aT})}$ $A = 1 - e^{-aT}\left(\cos(bT) + \frac{a}{b}\sin(bT)\right)$ $B = e^{-2aT} + e^{-aT}\left(\frac{a}{b}\sin(bT) - \cos(bT)\right)$	$\frac{1}{z-1} - \frac{e^{-amT}[z\ \cos(bmT)) + e^{-aT}\sin((1-m)bT)]}{z^2 - 2z\,e^{-aT}\cos(bT) + e^{-2aT}}$ $+\frac{a}{b}\left[\frac{e^{-amT}[z\ \sin(bmT)) + e^{-aT}\sin((1-m)bT)]}{z^2 - 2ze^{-aT}\cos(bT) + e^{-2aT}}\right]$
$\frac{1}{ab} + \frac{e^{-at}}{a(a-b)} + \frac{e^{-bt}}{b(b-a)}$	$\frac{1}{s(s+a)(s+b)}$	$\frac{z(Az + B)}{(z-1)\left(z - e^{-aT}\right)\left(z - e^{-bT}\right)}$ $A = \frac{b\left(1 - e^{-aT}\right) - a\left(1 - e^{-bT}\right)}{ab(b-a)}$ $B = \frac{ae^{-aT}\left(1 - e^{-bT}\right) - be^{-bT}\left(1 - e^{-aT}\right)}{ab(b-a)}$	$\frac{1}{ab}\left(\frac{1}{z-1}\right)$ $+\frac{1}{a(a-b)}\left(\frac{e^{-amT}}{z - e^{-aT}}\right)$ $+\frac{1}{b(b-a)}\left(\frac{e^{-bmT}}{z - e^{-bT}}\right)$

Appendix B: Answers to Selected Exercises

Chapter 2

2.1 $F(z) = \dfrac{Tze^{Ta}}{(z\, e^{Ta})^2}$

2.3 $F(z) = \dfrac{0.08611\, z}{z^2 - 1.724\, z + 0.7408}$

2.5 $F(z) = \dfrac{z^3 - 1.896\, z^2 + 0.913\, z}{z^3 - 2.724\, z^2 + 2.464\, z - 0.7408}$

2.7 $F(z) = \dfrac{z}{4\left(z - e^{-\left(\frac{1}{5}\right)}\right)} - \dfrac{z}{4\left(z - e^{-\left(\frac{3}{3}\right)}\right)}$

2.9 $F(z) = (0.0861 \times z)/((z - 0.8187) \times (z - 0.9048))$

$f(k) = 1.0 \times 0.9048^k - 1.0 \times 0.8187^k, \geq 0$

2.11 $f(0) = \lim_{z \to \infty} F(z) = 0$

$f(\infty) = \lim_{z \to 1} (z - 1)F(z) = 1$

2.13 $f(0) = \lim_{z \to \infty} F(z) = 0$

$f(\infty) = \lim_{z \to 1} (z - 1)F(z) = 1$

2.15 $m(k) = 0.6e(k) - 1.2e(k - 1) + 0.3e(k - 2) - 0.3m(k - 1) - 0.6m(k - 2)$

2.17 $D(z) = \dfrac{M(z)}{E(z)} = \dfrac{1.5z^2 - z + 0.3}{z^2 - 1.9z + 0.9}$

B. Samanta, *Introduction to Digital Control*,
https://doi.org/10.1007/978-3-031-66830-2

2.19

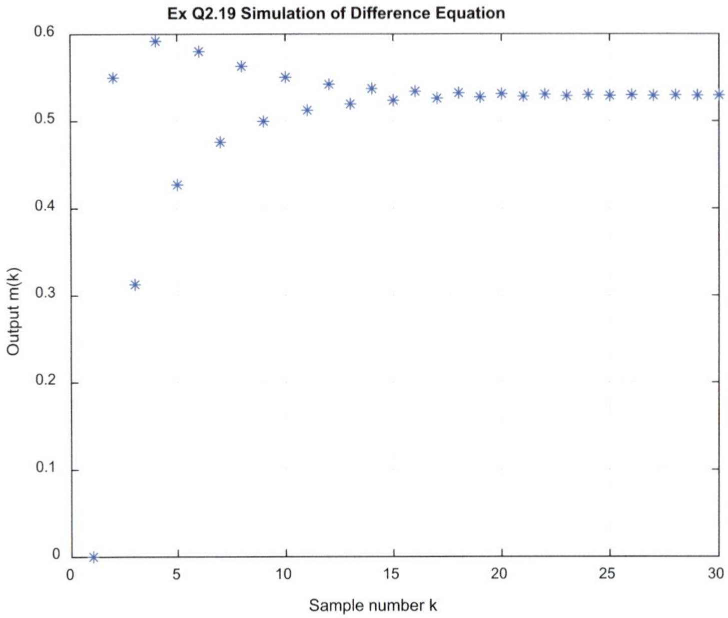

Chapter 3

3.1 $F^*(s) = \frac{1}{1 - e^{-(3+s)T}}, \left|e^{-(b+s)T}\right| < 1$

3.3 $F_2^*(s) = e^{-2Ts}F^*(s) = e^{-2Ts}\left(\frac{1}{1 - e^{-(4+s)T}}\right)$

3.5 $F^*(s) = \frac{1}{1 - e^{-(3+s)T}} - \frac{1}{1 - e^{-(4+s)T}}$

3.7 $F_n^*(s) = (1 - e^{-3Ts})F^*(s) = (1 - e^{-3Ts})\left(\frac{1}{1 - e^{-sT}} - \frac{1}{1 - e^{-(2+s)T}}\right)$

Chapter 4

4.1 $C(z) = \dfrac{0.04758\,z}{z^2 - 1.9048\,z + 0.9048}$

4.3 $C(z) = \dfrac{0.02419\,z^2 + 0.02339\,z}{z^2 - 1.905\,z + 0.9048}$

4.5 (a) $C(z) = \dfrac{0.007121\,z}{z^3 - 2.411\,z^2 + 1.908\,z - 0.4966}$

(b) $C(z) = \dfrac{0.003974\,z^2 + 0.003147\,z}{z^3 - 2.411\,z^2 + 1.908\,z - 0.4966}$

(c) $C(z) = \dfrac{0.0003738\,z + 0.0003383}{z^3 - 2.411\,z^2 + 1.908\,z - 0.4966}$

4.7 $C(z) = \dfrac{0.1857\,z^2 - 0.09286\,z}{z^3 - 2.161\,z^2 + 1.419\,z - 0.2582}$

$c(0) = \lim_{z\to\infty} C(z) = 0$

$c(\infty) = \lim_{z\to 1} (z-1)C(z) = 0.9524$

$c(k) = 0.9524 - 0.8577(0.8607)^k - 0.0946(0.3000)^k$

4.9 $C(z) = \dfrac{0.1857\,z^2 - 0.09286\,z}{z^6 - 2.161\,z^5 + 1.419\,z^4 - 0.2582\,z^3}$

4.11 $C(z) = \dfrac{0.004602\,z^3 + 0.001933\,z^2 - 0.002117\,z}{z^7 - 3.066\,z^6 + 3.374\,z^5 - 1.542\,z^4 + 0.2336\,z^3}$

4.13 $C(z) = \left(\dfrac{0.0296z + 0.1097}{z - 0.8607}\right)\left(\dfrac{0.01307z + 0.0032}{z - 0.9512}\right)\left(\dfrac{1}{z-1}\right)$

Chapter 5

5.1 (a) $GD(z) = \dfrac{0.0009934\,z^3 - 0.0005033\,z^2 - 0.0005861\,z + 0.0008881}{z^4 - 3.28\,z^3 + 4.554\,z^2 - 3.254\,z + 0.9802}$

(b) $T(z) = \dfrac{0.0009934\,z^3 - 0.0005033\,z^2 - 0.0005861\,z + 0.0008881}{z^4 - 3.279\,z^3 + 4.554\,z^2 - 3.255\,z + 0.9811}$

(c) $T(z) = \dfrac{0.0009934\,z^3 - 0.0005033\,z^2 - 0.0005861\,z + 0.0008881}{z^4 - 3.28\,z^3 + 4.554\,z^2 - 3.255z + 0.9806}$

5.2 $\mathrm{Cr}(z) = \dfrac{1.062z^2 - 0.5309z}{z^3 - 1.76z^2 + 0.8347z - 0.07438}$

$\mathrm{Cd}(z) = \dfrac{0.04424z^3 - 0.04773z^2 + 0.01034z}{z^4 - 2.539z^3 + 2.206z^2 - 0.7244z + 0.05792}$

$C(z) = \dfrac{1.106z^6 - 3.352z^5 + 3.821z^4 - 2.001z^3 + 0.4582z^2 - 0.03152z}{z^7 - 4.299z^6 + 7.51z^5 - 6.8z^4 + 3.363z^3 - 0.8706z^2 + 0.1022z - 0.004308}$

Chapter 6

6.1 $C(z) = \dfrac{0.32211z}{(z-1)(z-0.3155)}$

Inverse z-transform

$C(k) = 0.4706(1 - 0.3155^k),\ k \geq 0$

6.3 $C(z) = \dfrac{0.002341\,z^2 + 0.00219\,z}{z^3 - 2.816\,z^2 + 2.637\,z - 0.8209}$

6.5 $C(z) = \dfrac{0.04375\,z^2 + 0.03829\,z}{z^3 - 2.581\,z^2 + 2.29\,z - 0.7086}$

closed-loop zero and poles:

$z = -0.8751,\ p = 0.7907 + 0.2887i,\ 0.7907 - 0.2887i$

6.7 $Cz = \dfrac{0.02055\,z^2 + 0.01875\,z}{z^3 - 2.728\,z^2 + 2.506\,z - 0.7783}$

closed-loop zero and poles:

$z = -0.9124,\ p = 0.8641 + 0.1781i,\ 0.8641 - 0.1781i$

6.9 (a) $K_\text{p} = 1.1765$, $K_\text{v} = 0$, $K_\text{a} = 0$

essStep = 0.4595, essRamp = Inf, essPar = Inf

(b) $K_\text{p} = \text{Inf}$, $K_\text{v} = 2.3539$, $K_\text{a} = 0$

essStep = 0, essRamp = 0.4250, essPar = Inf

6.11 (a) $K_\text{p} = 9$, $K = 31.5$, $K_\text{v} = 0$, $K_\text{a} = 0$

essStep = 0.1, essRamp = Inf, essPar = Inf

(b) $K_\text{p} = \text{Inf}$, $K_\text{v} = 9.0000$, $K_\text{a} = 0$

essStep = 0, essRamp = 0.1111, essPar = Inf

6.13 $K = 13$, $e_{\text{ssStep},\ R} = 0.1875$

Chapter 7

7.1 $-0.6667 < K < 21.0868$, $p = 0.7186 - 0.6955i;\ 0.7186 + 0.6955i$, $w = 7.6906$ rad/s

7.2 $-1 < K_\text{p} < 29.566$, $p = 0.9807 + 0.0000i;\ 0.7362 - 0.6767i;\ 0.7362 + 0.6767i$, $w = 7.443$ rad/s

7.5 $G(w) = \dfrac{-0.00037313\,w^2 - 0.14191\,w + 2.9876}{w^2 + 2.9925\,w + 1.9917}$

$zw = -400.3333; 20.0000$

$pw = -1.9934;\ -0.9992,$ s-domain open-loop poles : $-1,\ -2$

7.9 $-12 < K < 202.0689$, $p = 0.0787 - 0.9969i$, $0.0787 + 0.9969i$, $w =$ 14.9205 rad/s

7.13 $K_{\text{critical}} = 1/0.0342 = 29.24$, $w = 6.56$ rad/s

Chapter 8

8.1 $0 < K < 21.1$, $p = 0.7186 + 0.6955i$, $0.7186 - 0.6955i$, $w = 7.6906$ rad/s

8.3 $0 < K < 80.7$, $p = 0.9998 + 0.0000i$, $0.3849 + 0.9254i$, $0.3849 - 0.9254i$, $w =$ 11.7661 rad/s

8.5 $0 < K < 202$, $p = 0.0787 + 0.9969i$, $0.0787 - 0.9969i$, $w = 14.9205$

$K = K_c/2 = 101$, $p = 0.3388 + 0.7320i$, $0.3388 - 0.7320i$

8.7 $0 < K < 16.2$, $p = 0.7799 + 0.6259i$, $0.7799 - 0.6259i$, $-0.1922 + 0.0000i$, $w = 3.3817$ rad/s

$K = K_c/2$, $p = 0.9677 + 0.1611i$, $0.9677 - 0.1611i$, $-0.0134 + 0.0000i$

8.9 $0 < K < 24.1730$, $p = 0.9761 + 0.2173i$, $0.9761 - 0.2173i$, $0.6038 + 0.0000i$, $w = 2.19$ rad/s

$K = K_c/3$, $p = 0.9484 + 0.1264i$, $0.9484 - 0.1264i$, $0.6615 + 0.0000i$

8.11 $0 < K < 6.91$, $p = 0.9828 + 0.1846i$, $0.9828 - 0.1846i$, $0.6575 + 0.0000i$, $w = 3.713$ rad/s

$K = K_c/3$, $p = 0.9579 + 0.1082i$, $0.9579 - 0.1082i$, $0.7088 + 0.0000i$

Chapter 9

9.1 gm = 21.0868, pm = 109.8387, wpc = 8.0935, wgc = 0.9243

wpcs = 7.6906 rad/s, wgcs = 0.9236 rad/s

9.2 gm = 29.5664, pm = 118.7153, wpc = 7.7953, wgc = 0.4193

wpcs = 7.4330 rad/s, wgcs = 0.4192 rad/s

9.5 gm = 11.4012, pm = 123.1213, wpc = 13.2657, wgc = 1.7950

wpcs = 11.7132 rad/s, wgcs = 1.7902 rad/s

9.7 gm = 202.0689, pm = Inf, wpc = 18.4840, wgc = NaN, wpcs = 14.9205, wgcs = NaN

K = gm/2
gm = 2.0000, pm = 27.4477, wpc = 18.4840, wgc = 10.9159, wpcs = 14.9205, wgcs = 9.9921

9.9 gm = 16.2528, pm = Inf, wpc = 3.5168, wgc = NaN, wpcs = 3.3817, wgcs = NaN

K = gm/2
gm = 2.0000, pm = 48.4093, wpc = 3.5168, wgc = 2.0296, wpcs = 3.3817, wgcs = 2.0024

9.11 gm = 24.1730, pm = 81.6206, wpc = 2.1988, wgc = 0.1658, wpcs = 2.1900, wgcs = 0.1658

K = gm/3
gm = 3.0000, pm = 37.7715, wpc = 2.1988, wgc = 1.1042, wpcs = 2.1900, wgcs = 1.1031

Chapter 10

10.1 $T = 0.1000$, zeta = 0.6362, $w_n = 2.4598$

$s_1 = -1.5648 + 1.8979i$, $s_2 = -1.5648 - 1.8979i$
$z_1 = 0.8398 + 0.1613i$, $z_2 = 0.8398 - 0.1613i$
$r = 0.8551$, th1 = 0.1898, th2 = −0.1898, th1d = 10.8740, th2d = −10.8740
pm = 69.443°, pm_approx = 63.6157°

10.3 $T = 0.2000$, zeta = 0.6266, $w_n = 4.1623$

$s_1 = -2.6080 + 3.2439i$, $s_2 = -2.6080 - 3.2439i$, $z_1 = 0.4730 + 0.3586i$, $z_2 = 0.4730 - 0.3586i$
$r = 0.5936$, th1 = 0.6488, th2 = −0.6488, th1d = 37.1729, th2d = −37.1729
pm = 68.7934°, pm_approx = 62.6577°

10.5 $T = 0.1000$, pm = 55, wgc = 4.5000

zeta = 0.5500, $t_{set} = 1.2448$, $w_n = 5.8424$, alpha = 0.1263
$s_1 = -3.2133 + 4.8794i$, $s_2 = -3.2133 - 4.8794i$
$z_1 = 0.6406 + 0.3400i$, $z_2 = 0.6406 - 0.3400i$
$r = 0.7252$, th1 = 0.4879, th2 = −0.4879, th1d = 27.9568, th2d = −27.9568

10.7 $T = 0.2000$, pm = 60, wgc = 3.5000, zeta = 0.6000, $t_{set} = 1.3197$, $w_n = 5.0518$

alpha = 0.0948, $s_1 = -3.0311 + 4.0415i$, $s_2 = -3.0311 - 4.0415i$
$z_1 = 0.3767 + 0.3944i$, $z_2 = 0.3767 - 0.3944i$
$r = 0.5454$, th1 = 0.8083, th2 = −0.8083, th1d = 46.3116, th2d = −46.3116

Chapter 11

11.1 kip = 0.0400, kdp = 0, kp = 0.9012, ki = 0.8000

$$D(z) = \frac{0.9412\, z - 0.8612}{z - 1}$$

11.4 kip = 0, kdp = 0.3530, kp = 0.4062, ki = 0, kd = 0.0353

$$D(z) = \frac{0.7592\, z - 0.353}{Z}$$

11.6 kip = 0.0100, kdp = 0.5578, kp = 0.4943, ki = 0.2000, kd = 0.0558

$$D(z) = \frac{1.062\, z^2 - 1.6\, z + 0.5578}{z^2 - z}$$

11.9 kip = 1.5000, kdp = 0, kp = 9.7865, ki = 30

$$Dz = \frac{11.29\, z - 8.286}{z - 1}$$

11.11 kip = 3, kdp = 15.6703, kp = 39.9503, ki = 60, kd = 1.5670

$$Dz = \frac{58.62\, z^2 - 68.29z + 15.67}{z^2 - z}$$

11.13 essStep = 0.0500, K_p = 19, a_0 = 57

$$D(w) = \frac{0.5261\, w + 1}{7.089\, w + 1}$$

$$D(z) = \frac{0.0807\, z - 0.06669}{z - 0.986}$$

11.15 essStep = 0.1200, K_p = 7.3333, a_0 = 22

$$D(w) = \frac{0.1538\, w + 1}{0.02189\, w + 1}$$

$$D(z) = \frac{2.835\, z - 1.444}{z + 0.391}$$

11.18 essStep = 0.0500 K_p = 19

Phase-lag: alpha = 0.20, t_{set} = 5

$$D(w) = \frac{1.927\, w + 1}{20.07\, w + 1}$$

$$D(z) = \frac{0.0807\, z - 0.06669}{z - 0.986}$$

Phase-lead: alpha = 0.05, $t_{set} = 1$ s

$$Dwp(w) = \frac{0.7506\ w + 1}{0.1466\ w + 1}$$

$$Dzp(z) = \frac{4.073\ z - 3.564}{z - 0.6542}$$

11.20 essStep = 0.0800 $K_p = 11.5$

Phase-lag part alpha = 0.15, $t_{set} = 10$

$$D(w) = \frac{1.125\ w + 1}{8.252\ w + 1}$$

$$Dz = \frac{0.1415\ z - 0.1295}{z - 0.988}$$

Phase-lead part alpha = 0.075, $t_{set} = 2$

$$Dwp(w) = \frac{1.354\ w + 1}{0.2969\ w + 1}$$

$$Dzp = \frac{4.046\ z - 3.758}{z - 0.8266}$$

Chapter 12

12.1 ki = 0.4000, theta = 6.2886, kd = 0, kp = 2.5290

$$D(w) = \frac{2.529\ w + 0.4}{w}$$

$$D(z) = \frac{2.549\ z - 2.509}{z-1}$$

12.2 kp = 3.29832, ki = 0.00000, kd = 0.14727

$$D(w) = \frac{0.3122\ w + 3.298}{0.05\ w + 1}$$

$$D(z) = \frac{4.771\ z - 1.473}{z}$$

12.3 kp = 3.27832, ki = 0.40000, kd = 0.16736

$$D(w) = \frac{0.3313\, w^2 + 3.298\, w + 0.4}{0.05\, s^2 + w}$$

$$D(z) = \frac{4.972\, z^2 - 6.606\, z + 1.674}{z^2 - z}$$

12.5 essStep = 0.0500, kp = 19, $c_0 = 28.5000$

$$D(w) = \frac{0.0891\,w + 1}{0.1780\,w + 1}$$

$$D(z) = \frac{0.6103\,z - 0.1717}{z - 0.5614}$$

12.6 essStep = 0.0800, kp = 11.5000, $c_0 = 1.7250$

$$Dw = \frac{1.437\,w + 1}{0.9583\,w + 1}$$

$$Dz = \frac{1.475\,z - 1.376}{z - 0.9008}$$

12.7 wgc = 4.5000, dpm = 60, essStep = 0.0200, kp = 49, $c_0 = 7.3500$

Phase-lag part
dpm = 30, wgc = 2

$$D(w) = \frac{0.3963\,w + 1}{3.494\,w + 1}$$

$$D(z) = \frac{0.1259\,z - 0.09773}{z - 0.9718}$$

Phase-lead part
dpm = 60, wgc = 4.5000,

$$Dwp(w) = \frac{0.9796\,w + 1}{0.1623\,w + 1}$$

$$Dzp(z) = \frac{4.85\,z - 4.379}{z - 0.529}$$

12.9 essStep = 0.0500, kp = 19, $c_0 = 28.5000$

$$D(w) = \frac{3.556\,w + 1}{5.254\,w + 1}$$

$$D(z) = \frac{0.6798\,z - 0.661}{z - 0.9811}$$

12.10 essStep = 0.0500, kp = 19, $c_0 = 28.5000$

$$D(w) = \frac{0.27747\,w + 1}{0.14109\,w + 1}$$

$$D(z) = \frac{1.714\,z - 1.19}{z - 0.4767}$$

12.11 essStep = 0.0200, $K_p = 49.0000$, $a_0 = 7.3500$

Phase-lag
dpm = 40

$$D_{lag}(w) = \frac{3.5557\,w + 1}{13.5497\,w + 1}$$

$$D_{lag}(z) = \frac{0.2651\,z - 0.2578}{z - 0.9926}$$

Phase-lead

$$D_{lead}(w) = \frac{0.36538\,w + 1}{0.24413\,w + 1}$$

$$D_{lead}(z) = \frac{1.412\,z - 1.072}{z - 0.66}$$

12.13 essRamp = 0.1500, $K_v = 6.6667$, $c_0 = 120$

Phase-lag
dpm = 30

$$Dw = \frac{19.4891\,w + 1}{203.3873\,w + 1}$$

$$Dz = \frac{0.09604\,z - 0.09555}{z - 0.9995}$$

Phase-lead
dpm = 60

$$Dw = \frac{2.6501\,w + 1}{0.44817\,w + 1}$$

$$Dz = \frac{5.42\,z - 5.219}{z - 0.7993}$$

Chapter 13

13.1 System matrices

$$a = \begin{matrix} 0.2597 & -1.0178 \\ 0.0565 & 0.9382 \end{matrix}$$

$$b = \begin{matrix} 0.0565 \\ 0.0034 \end{matrix}$$

$c = [0 \quad 1.5000]$, $d = 0$
Controllable, observable

13.3 System matrices

$$a = \begin{matrix} 2.6559 & -2.3262 & 0.6703 \\ 1.0000 & 0 & 0 \\ 0 & 1.0000 & 0 \end{matrix}$$

$$b = \begin{matrix} 1 \\ 0 \\ 0 \end{matrix}$$

$c = 1.0\text{e}{-05} \times [0.0756 \quad 0.2741 \quad 0.0619]$, $d = 0$
Controllable, observable

13.5 System matrices

$$a = \begin{matrix} 0.9956 & -0.0142 & 0 \\ -0.0022 & 0.9865 & 0 \\ -0.0001 & 0.0497 & 1.0000 \end{matrix}$$

$b = 1.0\text{e}{-03} \times [0.7277; \quad -0.6067; \quad -0.0152]$
$c = [0 \quad 0 \quad 1]$, $d = 0$
Controllable, observable

13.7 System matrices

$$a = \begin{matrix} 0.9048 & 0 & 0 \\ 0.0905 & 0.9048 & 0 \\ 0.0045 & 0.0905 & 0.9048 \end{matrix}$$

$b = [0.0952; \quad 0.0047; \quad 0.0002]$
$c = [0 \quad 0 \quad 1]$, $d = 0$
Controllable, observable

13.9 System matrices

$$a - \begin{matrix} 2.9048 & -2.8097 & 0.9048 \\ 1.0000 & 0 & 0 \\ 0 & 1.0000 & 0 \end{matrix}$$

$b = [1; \quad 0; \quad 0]$
$c = 1.0\text{e}{-05} \times [0.1301 \quad 0.5075 \quad 0.1237]$, $d = 0$
Controllable, observable

13.11 System matrices

$$a = \begin{matrix} 1.8187 & -0.8187 & 0 \\ 1.0000 & 0 & 0 \\ 0 & 1.0000 & 0 \end{matrix}$$

$b = [1; \quad 0; \quad 0]$
$c = [0 \quad 0.0187 \quad 0.0175]$, $d = 0$
Controllable, observable

Chapter 14

14.1 Desired poles in z-plane

zeta = 0.5912, $w_n = 2.6470$, $z_1 = 0.8357 + 0.1812i$, $z_2 = 0.8357 - 0.1812i$
Full-state feedback Controller gain matrix
$k = [-7.9153 \quad -7.5769]$
Observer gain matrix

$$l = \begin{matrix} 2.3457 \\ -0.0889 \end{matrix}$$

Controller-estimator transfer function

$$D_{\text{ce}}(z) = \frac{-17.8933\,z + 15.5235}{z^2 - 1.8048\,z + 1.1201}$$

Current Observer gain matrix

$$g = \begin{matrix} 7.0063 \\ -0.5170 \end{matrix}$$

Controller-estimator transfer function

$$D_{\text{ce}}(z) = \frac{-51.54\,z^2 + 43.8465\,z}{z^2 - 1.5394\,z + 1.2984}$$

14.3 Desired poles in z-plane

zeta = 0.6901, $w_n = 8.7211$, $z_1 = 0.9397 + 0.0594i$, $z_2 = 0.9397 - 0.0594i$, $z_3 = 0.8866$
Full-state feedback Controller gain matrix
$k = [-0.1101 \quad 0.2266 \quad -0.1157]$
Observer gain matrix
$l = 1.0\text{e+}04 \times [3.3243; \quad 2.6953; \quad 1.9083]$
Controller-estimator transfer function

$$D_{\text{ce}}(z) = \frac{240.1588\,z^2 - 448.8605\,z + 209.8406}{z^3 - 2.6552\,z^2 + 2.383\,z - 0.72435}$$

Current Observer gain matrix
$g = 1.0\text{e+}04 \times [2.6953; \quad 1.9083; \quad 0.9023]$
Controller-estimator transfer function

$$D_{\text{ce}}(z) = \frac{313.0454\,z^3 - 591.264\,z^2 + 279.3574\,z}{z^3 - 2.6555\,z^2 + 2.3822\,z - 0.72454}$$

14.5 Desired poles in z-plane

zeta = 0.5912, $w_n = 0.3309$, $z_1 = 0.9902 + 0.0132i$, $z_2 = 0.9902 - 0.0132i$, $z_3 = 0.9806$
Full-state feedback Controller gain matrix
$k = [-20.3475 \quad -58.5234 \quad -24.5196]$
Observer gain matrix
$l = [-0.3024; \quad 0.0295; \quad 0.0599]$
Controller-estimator transfer function

$$D_{ce}(z) = \frac{2.9566\,z^2 - 5.9305\,z + 2.9729}{z^3 - 2.9011\,z^2 + 2.8068\,z - 0.90544}$$

Current Observer gain matrix
$g = [-0.3033; \quad 0.0293; \quad 0.0584]$
Controller-estimator transfer function

$$D_{ce}(z) = \frac{3.0271\,z^3 - 6.0705\,z^2 + 3.0424z}{z^3 - 2.9011\,z^2 + 2.8068\,z - 0.90548}$$

14.7 Desired poles in z-plane

zeta = 0.5912, $w_n = 1.3235$, $z_1 = 0.9195 + 0.0985i$, $z_2 = 0.9195 - 0.0985i$, $z_3 = 0.8551$
Full-state feedback Controller gain matrix
$k = [0.1682 \quad 0.9206 \quad 0.6328]$
Observergain matrix
$l = 1.0e{+}04 \times [3.3243; \quad 2.6953; \quad 1.9083]$
Controller-estimator transfer function

$$D_{ce}(z) = \frac{240.1588\,z^2 - 448.8605\,z + 209.8406}{z^3 - 2.6552\,z^2 + 2.383\,z - 0.72435}$$

Current Observer gain matrix
$g = [0.8810; \quad 0.6201; \quad 0.2781]$
Controller-estimator transfer function

$$D_{ce}(z) = \frac{0.89506\,z^3 - 1.5083\,z^2 + 0.63656z}{z^3 - 2.3825\,z^2 + 1.9308\,z - 0.52788}$$

14.9 Desired poles in z-plane

zeta = 0.6901, $w_n = 1.1337$, $z_{1,2} = 0.9843 \pm 0.0162i$, $z_3 = 0.9692$
Full-state feedback Controller gain matrix
$k = [-0.0330 \quad 0.0675 \quad -0.0345]$
Observer gain matrix
$l = [411.09; \quad 370.20; \quad 326.78]$
Controller-estimator transfer function

$$D_{ce}(z) = \frac{1.5302\,z^2 - 3.0254\,z + 1.4955}{z^3 - 2.9097\,z^2 + 2.8257\,z - 0.91596}$$

Current Observer gain matrix
$g = [370.20; \quad 326.78; \quad 280.54]$
Controller-estimator transfer function

$$D_{ce}(z) = \frac{1.6528\,z^3 - 3.2708\,z^2 + 1.6183\,z}{z^3 - 2.9097\,z^2 + 2.8257\,z - 0.91597}$$

14.11 Desired pole locations

zeta $= 0.5169$, $w_n = 3.7839$, $z_1 = 0.7796 + 0.2617i$, $z_2 = 0.7796 - 0.2617i$
Controller gain matrix
$k = 2.4041 \quad -6.5350$
Reduced-order observer pole
$p_r = 0.6077$
Reduced-order observer gain
$l = -6.1555$
Controller-estimator (reduced-order) transfer function

$$D_{ce}(z) = \frac{42.63 \times z – 32.55}{z - 1.1153}$$

Chapter 15

15.1 a_i, $i = 1, 2$ b_i, $i = 1, 2$

(a) −1.2705e+00, 4.0657e−01, 7.4464e−02, 5.5148e−02
(b) −1.2705e+00, 4.0657e−01, 7.4464e−02, 5.5148e−02
(c) −1.2705e+00, 4.0657e−01, 7.4464e−02, 5.5148e−02

15.2 a_i, $i = 1, 2, 3$ b_i, $i = 1, 2, 3$

(a) −2.3369e+00, 1.8283e+00, −4.7237e−01, 1.7296e−02, 5.7451e−02, 1.1890e−02
(b) −2.3369e+00, 1.8283e+00, −4.7237e−01, 1.7296e−02, 5.7451e−02, 1.1890e−02
(c) −2.3369e+00, 1.8283e+00, −4.7237e−01, 1.7296e−02, 5.7451e−02, 1.1890e−02

15.3 a_i, $i = 1, 2, 3, 4$ b_i, $i = 1, 2, 3, 4$

(a) −3.4091e+00, 4.4057e+00, −2.5446e+00, 5.4881e−01, 3.6931e−05, 3.5995e−04, 3.1938e−04, 2.5772e−05
(b) −3.4092e+00, 4.4056e+00, −2.5445e+00, 5.4876e−01, 3.6931e−05, 3.5995e−04, 3.1940e−04, 2.5782e−05
(c) −3.4091e+00, 4.4057e+00, −2.5446e+00, 5.4880e−01, 3.6931e−05, 3.5995e−04, 3.1939e−04, 2.5775e−05

15.4 a_i, $i = 1, 2$ b_i, $i = 1, 2$

(a) −1.1980e+00, 3.0114e−01, 5.1500e−03, 3.4584e−03
(b) −1.1980e+00, 3.0114e−01, 5.1500e−03, 3.4584e−03
(c) −1.1980e+00, 3.0114e−01, 5.1504e−03, 3.4586e−03

15.5 a_i, $i = 1, 2, 3$ b_i, $i = 1, 2, 3$

(a) −2.3956e+00, 1.9234e+00, −4.9921e−01, 1.7540e−04, 5.9068e−04, 1.2253e−04
(b) −2.3956e+00, 1.9234e+00, −4.9921e−01, 1.7540e−04, 5.9068e−04, 1.2253e−04
(c) −2.3956e+00, 1.9234e+00, −4.9921e−01, 1.7542e−04, 5.9070e−04, 1.2254e−04

15.6 a_i, $i = 1, 2, 3, 4$ b_i, $i = 1, 2, 3, 4$

(a) −3.8191e+00, 4.4633e+00, −2.2824e+00, 2.7547e−01, 3.8852e−05, 4.0458e−04, 4.3371e−04, 8.9114e−05
(b) −3.8191e+00, 4.4634e+00, −2.2826e+00, 2.7556e−01, 3.8852e−05, 4.0458e−04, 4.3369e−04, 8.9095e−05
(c) −3.8191e+00, 4.4633e+00, −2.2825e+00, 2.7548e−01, 3.8852e−05, 4.0458e−04, 4.3371e−04, 8.9109e−05

Chapter 16

16.1 Optimal controller gain matrix

$k = 0.0020 \quad -0.0007;$
Optimal estimator gain matrix

$$l = \begin{matrix} 0.0006 \\ 0.0034 \end{matrix}$$

$$D_{\mathrm{ce}}(z) = \frac{-1.1412\,e-06\,z^2 - 7.5673\,e-06\,z}{z^2 - 1.193\,z + 0.29956}$$

16.3 Optimal controller gain matrix

$k = 0.7143 \quad -0.3614 \quad 0.0103$
Optimal estimator gain matrix

$$l = \begin{matrix} 8.1987 \\ 8.1655 \\ 8.1174 \end{matrix}$$

$$D_{\mathrm{ce}}(z) = \frac{2.988\,z^3 - 1.8996\,z^2 + 0.054638\,z}{z^3 - 0.91462\,z^2 + 0.29122\,z - 0.007977}$$

16.5 Optimal controller gain matrix

$k = 2.2669 \quad -8.7170 \quad -2.2283$

Optimal estimator gain matrix

$$l = \begin{matrix} -0.0037 \\ 0.0011 \\ 0.0104 \end{matrix}$$

$$D_{ce}(z) = \frac{-0.041155z^3 + 0.081511z^2 - 0.040358z}{z^3 - 2.9646z^2 + 2.9297z - 0.96512}$$

16.7 Optimal controller gain matrix

$k = 0.1586 \quad 0.0746 \quad 0.0214$

Optimal estimator gain matrix

$$l = \begin{matrix} 0.1210 \\ 0.1313 \\ 0.0895 \end{matrix}$$

$$D_{ce}(z) = \frac{0.030901z^3 - 0.054839z^2 + 0.024342z}{z^3 - 2.6057z^2 + 2.2736z - 0.66356}$$

16.9 Optimal controller gain matrix

$k = 1.4912 \quad -1.9656 \quad 0.7212$

Optimal estimator gain matrix

$$l = \begin{matrix} 1.0e + 03^{*} \\ 2.9392 \\ 2.9067 \\ 2.8742 \end{matrix}$$

$$D_{ce}(z) = \frac{742.2967z^3 - 1405.9036z^2 + 664.3785z}{z^3 - 1.3922z^2 + 0.83028z - 0.17957}$$

16.11 Optimal controller gain matrix

$k = 0.8416 \quad -0.4997 \quad 0.0000$

Optimal estimator gain matrix

$$l = \begin{matrix} 9.7135 \\ 8.6056 \\ 7.2524 \end{matrix}$$

$$D_{ce}(z) = \frac{3.8746z^3 - 2.9625z^2 - 9.7967e - 17z}{z^3 - 0.64438z^2 + 0.29045z + 1.7239e - 19}$$

Index

B. Samanta, *Introduction to Digital Control*, https://doi.org/10.1007/978-3-031-66830-2

MIX
Papier aus verantwortungsvollen Quellen
Paper from responsible sources
FSC® C105338

If you have any concerns about our products,
you can contact us on
ProductSafety@springernature.com

In case Publisher is established outside the EU,
the EU authorized representative is:
Springer Nature Customer Service Center GmbH
Europaplatz 3, 69115 Heidelberg, Germany

Printed by Libri Plureos GmbH
in Hamburg, Germany